Business Mathematics

Second Edition

Walter H. Lange
Robert D. Mason
Temoleon G. Rousos

Contributing Author
Jean B. Rogers

GLENCOE
Macmillan/McGraw-Hill

Lake Forest, Illinois Columbus, Ohio
Mission Hills, California Peoria, Illinois

Authors

Walter H. Lange Professor of Technical Science and Mathematics, Community and Technical College, The University of Toledo, Ohio

Robert D. Mason Former Professor of Statistics, College of Business Administration, The University of Toledo, Ohio

Temoleon G. Rousos Professor of Technical Science and Mathematics, Community and Technical College, The University of Toledo, Ohio

Contributing Author

Jean B. Rogers Specialist in Computer Education, Oregon Division of Continuing Education

Imprint 1991

Send all inquiries to: Glencoe Division, Macmillan/McGraw-Hill, 936 Eastwind Drive, Westerville, Ohio 43081.

Student's Edition ISBN: 0-395-37643-2

Teacher's Edition ISBN: 0-395-37644-0

7 8 9 10 11 12 13 14 15 RRD-C 00 99 98 97 96 95 94 93 92 91

Table of Contents

Memo

To: The Business Mathematics Student
From: The Authors

This is *your* textbook. It is specially designed to help you become a knowledgeable consumer and business person. This text contains information and resources to help you achieve these goals.

Each lesson begins with a brief description of the topic and highlights the key formula for you.

The Example provides a solution to a representative problem. It gives you a step-by-step method to solve for the answer. If you have any trouble or need to review, there is helpful information in the Reference Files at the back of the book. The references at the top of the example tell you just which items to study.

The Self-Check lets you use the formula on your own before you begin the lesson problems.

After you work through the example and the Self-Check, complete the problems assigned by your instructor. These problems are arranged systematically to apply gradually what you have just studied. Answers to selected problems are found at the back of your book. Use the answers to check your work.

1-1 Hourly Pay

Objective Compute the straight-time pay.

For some jobs, you are paid a fixed amount of money for each hour you work. Your **hourly rate** is the amount of money you earn per hour. **Straight-time pay** is the total amount of money you earn for a pay period at the hourly rate.

▸ **Straight-time Pay = Hourly Rate × Hours Worked**

Example

Skills 14, 8, 2 **Application** A **Term** Straight-time pay

Gladys Marzetti is an inventory control clerk. She earns \$4.30 per hour. Last week she worked her regular $37\frac{1}{2}$ hours. What is her straight-time pay for last week?

Solution

Find the **straight-time pay**.
Hourly Rate × Hours Worked
\$4.30 × $37\frac{1}{2}$
\$4.30 × 37.5 = \$161.25 straight-time pay

A. \$347.10
B. \$361.90

▸ **Self-Check** Cover the answers and complete the problems.

Find the straight-time pay.

A. \$8.90 per hour × 39 hours B. \$9.40 per hour × $38\frac{1}{2}$ hours

Problems

	Employee	Regular Hourly Rate	×	Hours Worked	=	Straight-time Pay
1.	Batson, S.	\$ 7.00	×	40	=	?
2.	Favreau, E.	\$ 4.00	×	18	=	?
3.	Levitt, R.	\$ 6.20	×	30	=	?
4.	Maguire, M.	\$ 4.50	×	$24\frac{1}{2}$	=	?
5.	Scanlon, A.	\$12.50	×	$31\frac{1}{4}$	=	?
6.	Vadola, G.	\$11.00	×	$25\frac{3}{4}$	=	?

9. Sylvia Barrett.
Financial consultant.
Earns \$18.50 per hour.
Worked $22\frac{1}{2}$ hours last week.
What was her straight-time pay?

10. Gloria Todd.
Office clerk.
Earns \$3.60 per hour.
Worked $35\frac{3}{4}$ hours last week.
What was her straight-time pay?

11. Chris Howard is a veterinary aide. She earns \$3.80 per hour. Last week she worked 34 hours. What was her straight-time pay for last week?

12. Jan Edwards has a part-time job at a delicatessen. She earns \$3.05 per hour. She works 21 hours each week. What is her straight-time pay for a week?

13. Del Martin, a construction worker, earns \$8.00 per hour. He worked his regular $36\frac{1}{4}$ hours last week. What was his straight-time pay?

14. Arthur Anderson is a mail clerk at the Donovan Marketing Corporation. He works 27 hours per week and earns \$3.72 per hour. What is his straight-time pay each week?

15. Eve Wright, a guitar player for a local band, earns \$5.25 per hour. What is her straight-time pay for a week in which she worked 32 hours?

16. George Billings earns \$6.64 per hour as an automotive claims adjustor. What is his straight-time pay for this week?

Mon.	Tues.	Wed.	Thurs.	Fri.
8 hr	5 hr	7 hr	9 hr	8 hr

17. Ben Reyes was hired for the position of billing clerk. During his first week on the job, he worked $7\frac{1}{2}$ hours each day, Monday through Friday. What was his total pay for the week?

PRINTING CO. has opening for bright dependable billing clerk. Must have aptitude for figures and pay attention to detail. Prepare work for production. Start \$5.50 per hr. Call Steven, at 700-9256 Ext. 11

Maintaining Your Skills Look up the skills in parentheses if you need help or more practice.

Write as decimals. (Skill 14)

18. $\frac{1}{4}$ 19. $\frac{1}{2}$ 20. $\frac{3}{4}$ 21. $\frac{1}{10}$ 22. $\frac{3}{8}$

Multiply. Round answers to the nearest cent. (Skill 8)

23. \$4.50 × 30.25 24. \$8.30 × 3.25 25. \$14.50 × 32.5

Round to the place value indicated. (Skill 2)

Nearest ten 26. 22 27. 138 28. 8

Skill 3 Add Whole Numbers

Skills

Without Carrying

Add.

723 154 +212	723 154 +212 1089

1. 65 +41
2. 76 +32
3. 97 +40
4. 32 +25
5. 352 +837
6. 361 +834
7. 448 +351
8. 125 +604
9. 864 + 33
10. 721 + 77
11. 423 + 65
12. 108 + 91
13. 9037 +1841
14. 9520 +1379
15. 3924 +5063
16. 2840 +1152

With Carrying

Add.

8679 +9748	1 11 8 679 +9 748 18,427

17. 32 +39
18. 54 +48
19. 187 + 23
20. 49 +86
21. 728 +169
22. 527 +284
23. 845 +697
24. 697 +546
25. 3046 +1592
26. 7801 +3564
27. 5246 +6978
28. 8347 +1528

A wise consumer and business person keeps important information in orderly files for easy reference. The tabbed Reference Files at the back of your book contain all the Skills, Applications, and Terms you need for this course.

The Skills File contains 32 basic computational skills arranged in order of difficulty. This file helps you review and practice arithmetic, if needed. Study the model problems, then complete the exercises. For more practice, do the set of mixed problems at the bottom of the page.

Application A Formulas

Applications

Substituting In a Formula

To use a formula to solve a problem, substitute known values in the formula. Then solve. Remember to perform any computations within parentheses first.

▶ An airplane travels at a rate of 960 km/h. How far can the plane travel in 7 hours?

Formula: Distance = Rate × Time

Substitute: Distance = 960 × 7

Solve: Distance = 6720

The plane travels 6720 km in 7 hours.

Find the distance.

	Distance = Rate × Time		
	Rate	Time	Distance
1.	100 mi/h	6 h	?
2.	55 km/h	4 h	?

Find the percent of sales made.

	% of Sales = $\frac{\text{Sales Made}}{\text{Possible Sales}} \times 100$		
	Possible Sales	Sales Made	Percent
10.	$4600	$1900	?
11.	$19,100	$6000	?
12.	$23,000	$13,000	?
13.	$64,000	$15,000	?
14.	$8200	$3000	?

Find the price of the stereo system.

	Price = Amp + Turntable + Speakers			
	Amp	Turntable	Speakers	Price
15.	$199	$99.95	$139	?
16.	$395	$119.99	$259	?

The Applications File will help you recall other math-related topics that are applied in consumer and business situations. The applications are coded from A to Z.

Terms accelerated cost recovery system/base price

Terms

accelerated cost recovery system (ACRS) (p. 434) A method of computing depreciation that allows depreciation according to fixed percents over the shortest period of time.
accumulated depreciation (p. 428) The total depreciation to date.
age group (p. 198) A category to which an automobile is assigned by an insurance company depending on its age. Age group affects the base premium.
air freight (p. 394) A way of shipping goods to customers.
amount (p. 119) The balance in a savings account at the end of an interest period, computed by adding the principal and interest.
amount financed (p. 178) The portion of the purchase price of an item that you still owe after making the down payment.
amount of change (p. 450) The difference between the base figure and the corresponding figure on the current income statement.
annual expenses (p. 264) Amounts for real estate taxes, insurance payments, contributions, and so on, which occur once a year.
annual interest rate (p. 117) The percent of the principal that you earn as interest based on one year.
annual percentage rate (APR) (p. 180) An index used to compare the relative cost of borrowing money.
annual premium (p. 198) The amount you pay each year for insurance coverage.
annual yield (p. 248) The total interest or return for one year expressed as a percent of the principal or amount invested.
apportion (p. 424) To distribute among several departments.

The Terms file contains clear, precise definitions of ideas covered in your textbook. The page references show where the topic may be found in your book.

Preface

A knowledge of basic mathematics is absolutely essential for survival in today's world. Mathematics as a survival tool is implemented in almost every phase of personal and business life. How effectively one is able to use that important tool may determine how well one is able to prepare for the future, manage one's personal and business resources, and benefit from one's efforts as a consumer, worker, or business person.

Business Mathematics, Second Edition, has as its major purpose to assist students in learning to use mathematics effectively as a tool in their personal and business lives. After students have completed their study of the textbook and related ancillaries, they will be able to

1. Understand terminology relating to personal and business mathematics applications.
2. Apply basic math skills to the solution of both personal and business applications.
3. Use common mathematic formulas to solve a variety of personal and business mathematics problems.

Organization of the Textbook

While retaining the sound organization and approach of the first edition, *Business Mathematics,* Second Edition, focuses even more strongly on the review, remediation, and reinforcement of basic math skills. A new problem-solving feature enables students to approach the solving of problems in a more logical manner. Computer activities, accompanied by an optional computer diskette, provide the opportunity for students to use the computer in solving selected problems. Additional mini-simulations in the workbook add a touch of realism and heighten student interest.

The textbook is organized into four sections: Part One, Basic Math Skills Workshop: A Review of Fundamentals; Part Two, Personal Business Mathematics; Part Three, Business Mathematics; and Reference Files.

Basic Math Skills Workshop: A Review of Fundamentals

Basic math skills are an essential element of functional literacy. As an aid to students who need to review basic math skills or who need reinforcement of those skills, we have included a new feature in this second edition of our textbook: a basic skills workshop.

This new section of the textbook is a series of "workshops" covering fourteen basic math skills. Each workshop includes a description of the covered skill, a variety of examples, and numerous applications and problems that allow students to apply each specific skill. The workshops provide students with a foundation of basic skills that will be needed during their study of Parts Two and Three of the textbook.

Personal Business Mathematics

In Part Two of the textbook, Personal Business Mathematics, students will examine the personal applications of basic math skills used by workers and consumers. These applications include the calculation of gross income and net income, the use of both checking and savings accounts, cash purchases, charge purchases, loans, the cost of owning and operating an automobile, housing costs, insurance and investments, and budgets.

Business Mathematics

Part Three, Business Mathematics, uses as the application environment eleven different departments of a large business in which students apply mathematics fundamentals to realistic business situations. Representative business departments covered include personnel, production, purchasing, sales, marketing, warehousing and distribution, services, accounting, accounting records, financial, and information management.

Parts Two and Three of the textbook are organized into units covering a specific type of application or a business department. Each unit is composed of numbered lessons. Each lesson begins with a clear, concise statement of the objective to be met followed by a brief explanation of the personal or business application to be examined in the lesson. The basic mathematical formula to be used in the lesson is followed by a worked-out, sample problem and a self-check practice problem that show students in detail how to solve the application. Table, phrase, and word problems—all graded in order of difficulty—give the students a variety of opportunities to apply the lesson formula. Each lesson concludes with a set of basic math review problems that help students maintain or improve their basic math skills.

Each unit is followed by review activities; a one-page feature on such topics as careers, problem solving, or microcomputer applications; and a one-page unit test. Five mini-simulations are spaced throughout the textbook to provide interesting and realistic opportunities for students to apply their math skills.

Reference Files

Extensive reference files are included at the end of the student textbook to help students who may need to review basic skills or may want to use the files as study aids in completing individual lessons in the textbook. The reference files contain a Skills File of thirty-two numbered computational math skills with practice problems; an Applications File of twenty-six common practice applications; and a Terms File of key terms used in business and consumer situations. These skills, applications, and terms are keyed to the examples within the textbook lessons. Students who need further explanation or practice can turn to the reference files and complete the problems found there to reinforce lesson content or basic math skills.

In addition to the preceding files, the textbook also contains several tables that can be used in solving related problems, as well as the answers to odd-number problems in the text lessons.

Supporting Materials

In addition to the student textbook, the program includes the following: a teacher's edition of the textbook with teaching suggestions, annotations, and problem solutions; a workbook with additional practice, review, reinforcement, simulations, and microcomputer applications; a teacher's edition of the workbook; a solution key with worked-out solutions to all problems in the student textbook; a teacher's resource book; and a computer diskette for solving the computer activities in the textbook and workbook.

Acknowledgments

We wish to acknowledge the valuable contributions made to this textbook by teachers, students, colleagues, friends, and collaborators throughout its various stages of revision. In particular, we would like to express our appreciation to Professors James Arbaugh, Clayton L. Ziegler, and Robert A. Siddens of the University of Toledo for their expert advice in accounting, engineering, and transportation. We would also like to thank Carla Christy and Danuta L. Lange, who checked solutions and helped prepare the manuscript.

Walter H. Lange
Robert D. Mason
Temoleon G. Rousos

Part 1 Basic Math Skills Workshop: A Review of Fundamentals

▸ As you use this book, you will need to know certain mathematical fundamentals. To be sure you are prepared, here are fourteen lessons for review and practice. Each workshop uses realistic business forms to show how math is used in different job settings.

Workshop 1

Writing Numbers as Words and Rounding Numbers

▶ Look up Skill 1 on pages 504 and 505 for more practice.

millions	hundred thousands	ten thousands	thousands	hundreds	tens	ones		tenths	hundredths	thousandths
4	3	5	7	8	0	2	•	6	9	1

The place-value chart shows the value of each digit in the number 4,357,802.691. For example, the digit 7 is in the thousands place and has a value of 7000. The digit 9 is in the hundredths place and has a value of $\frac{9}{100}$, or 0.09. The place-value chart can help you write numbers.

Example	**Solution** (read as or written as)
647	six hundred forty-seven
89,065	eighty-nine thousand sixty-five
4.316	four and three hundred sixteen thousandths
$25.72	twenty-five and seventy-two one hundredths dollars
	or twenty-five and $\frac{72}{100}$ dollars
	or twenty-five dollars and seventy-two cents

▶ **Self-Check** Cover the answers and complete the problems.

Write in words.	Answers
A. 152	**A.** one hundred fifty-two
B. 4.218	**B.** four and two hundred eighteen thousandths
C. $892.15	**C.** eight hundred ninety-two and fifteen one hundredths dollars
	or eight hundred ninety-two and $\frac{15}{100}$ dollars
	or eight hundred ninety-two dollars and fifteen cents

Place value is also used in rounding numbers. If the digit to the right of the place value you want to round is 5 or more, round up by adding 1 to the number in the place value and then change all the digits to the right of the place value to zeros. If the number is 4 or less, round down by changing all the numbers to the right of the place value to zeros.

Example Round 4873 to the nearest hundred.

Solution

4**8**73	**1.** Find the digit in the hundreds place. It is 8.
48**7**3	**2.** Is the digit to the right 5 or more? Yes.
4**9**00	**3.** Add 1 to the hundreds place. Change the digits to the right to zeros.

Example Round 0.948 to the nearest tenth.

Solution

0.**9**48	**1.** Find the digit in the tenths place. It is 9.
0.9**4**8	**2.** Is the digit to the right 5 or more? No.
0.**9**	**3.** Do not change the tenths digit. Drop the digits to the right.

Dollar and cents amounts are often rounded to the nearest cent, or the hundredths place. Begin with the digit to the right of the hundredths place, or the thousandths place.

Example	**Solution**
$12.9433	$12.94
$114.5483	$114.55
$4220.215	$4220.22

▶ **Self-Check** Cover the answers and complete the problems.

Round to the nearest cent.

A. $15.186 **B.** $153.831

C. $124.1327 **D.** $737.666

Answers

A. $15.19 **B.** $153.83

C. $124.13 **D.** $737.67

Problems

Write as numbers.

1. three thousand four hundred fifty-one

2. forty-two and $\frac{36}{100}$ dollars

3. eight hundred six and twenty-five thousandths

4. thirty-seven and $\frac{9}{100}$ dollars

Write in word form.

5. 196,403 **6.** $89.12 **7.** 617.035 **8.** 500.4 **9.** $10.99

Round to the nearest place value shown.

Ten thousand	10. 346,876	11. 625,571	12. 670,409.765	13. 3,768,591
Thousand	14. 416,419	15. 50,135	16. 976,812	17. 8,618,054
Hundred	18. 57,454	19. 4239	20. 36,657	21. 37,944
Ten	22. 8681	23. 785	24. 28,834	25. 40,607
One	26. 2.201	27. 487.08	28. 41.221	29. 134.77
Tenth	30. 31.106	31. 87.4188	32. 5222.614	33. 760.95641
Hundredth	34. 4.1113	35. 5.478	36. 97.975783	37. 74.2259

Round 74,183,526 to the place value given.

38. millions
39. ten millions
40. thousands
41. hundreds
42. ten thousands
43. hundred thousands

Round to the nearest place value shown.

Cent	44. $15.8992	45. $34.441	46. $41.909	47. $73.8522
Ten cents	48. $120.774	49. $7.8578	50. $8.0820	51. $99.9971
One dollar	52. $160.432	53. $82.040	54. $99.357	55. $48.8510
Ten dollars	56. $1580	57. $2349.60	58. $673.40	59. $695.00
Hundred dollars	60. $11,262	61. $19,033	62. $3476	63. $50,547

Applications

64. In her job as an accountant for the Wilson Manufacturing Co., Wilma Montgomery needs to write numbers as words. Write each check amount in words.

a. $9.15 **b.** $8.84 **c.** $36.76 **d.** $45.61

e. $154.23 **f.** $367.45 **g.** $5722.32 **h.** $7604.14

65. Ben Cooper records the attendance data for the sporting events at Centennial Hall. Round each number to the amount indicated.

	Basketball Game	Number	Nearest Thousand	Nearest Hundred	Nearest Ten
a.	Eastern Michigan	9132	?	?	?
b.	Bowling Green	10,456	?	?	?
c.	Ohio University	8391	?	?	?
d.	Western Michigan	7868	?	?	?
e.	International	6527	?	?	?
f.	Ball State	9942	?	?	?
g.	Kent State	11,003	?	?	?
h.	Northern Ill.	8888	?	?	?

66. To get an overall sales picture of first-quarter sales, the marketing executive wants the sales figures in round numbers. Round the sales of each company to the nearest thousand dollars.

FIRST-QUARTER SALES, 19--

	Company	Sales
a.	Altex Computers	$571,114.64
b.	Crawford Electronics	434,849.17
c.	Silica Research, Inc.	350,351.58
d.	Laser Systems, Inc.	333,562.41
e.	Meltex Graphics	297,840.29
f.	Amtrex Software	168,273.71
g.	Cable Communications	146,743.12
h.	Dominion Limited	130,634.43
i.	Market Surveys, Inc.	99,753.33

67. Jane Elson, an inventory clerk, often rounds inventory figures for easier handling. Round the numbers from the inventory list to the nearest ten.

CORN INVENTORY

	Description	Quantity	Rounded
a.	Golden Cross	292 lbs	?
b.	Golden Beauty	174 lbs	?
c.	Early Glo	122 lbs	?
d.	X-tra Sweet	355 lbs	?
e.	White Sunglo	433 lbs	?
f.	Silver Queen	899 lbs	?
g.	Honey-Cream	954 lbs	?
h.	Country Gtlm	776 lbs	?

Workshop 2 Comparing Numbers

▶ Look up Skill 1 on pages 504 and 505 for more practice.

To compare numbers to see which one is greater, first find out if one of the numbers has more digits to the left of the decimal point. For example, 100 is greater than 99.62 because 100 has three digits to the left of the decimal point and 99.62 has only two. If both numbers have the same number of digits to the left of the decimal point, compare each digit, beginning at the left. One way is to write the value of each digit.

Example Compare 4932 and 4943. Which number is greater?

Solution

4932	=	4000	+	900	+	30	+	2
4943	=	4000	+	900	+	40	+	3
		⬆ same		⬆ same		⬆ 40 is greater than 30, so 4943 is greater than 4932.		

You may not have to write the value of each digit. Just compare each digit starting at the left.

Example Compare 347.51 and 347.29. Which number is greater?

Solution

3	4	7	•	5	1
3	4	7	•	2	9
⬆ same	⬆ same	⬆ same		⬆ 5 is greater than 2, so 347.51 is greater than 347.29.	

A. 3579
B. 319.62
C. 18,431
D. 29.794

▶ **Self-Check** Cover the answers and complete the problems.

Which number is greater?

A. 3579 and 3564 **B.** 319.59 and 319.62
C. 18,431 and 18,411 **D.** 29.791 and 29.794

When you are comparing decimals, you may have to write zeros to the right of the decimal point so that the numbers being compared have the same number of decimal places.

Example Compare 13.4 and 13.439 (without writing the value of each digit). Which number is greater?

Solution 13.400
13.439

3 is greater than 0, so 13.439 is greater than 13.400.

A. 710.594
B. 0.05
C. 19.131

▶ **Self-Check** Cover the answers and complete the problems.

Which number is greater?

A. 710.59 and 710.594 **B.** 0.05 and 0.005 **C.** 19.1 and 19.131

Problems

Which number is greater?

1. 43 or 34
2. 876 or 868
3. 3501 or 3510
4. 5927 or 5926
5. 3.87 or 3.78
6. 41.6 or 40.9
7. 0.764 or 0.763
8. 0.2 or 0.09
9. 8.35 or 9
10. 0.03 or 0.3
11. 0.040 or 0.004
12. 2531.7 or 2541.7
13. $7.89 or $8.50
14. $329.42 or $328.42
15. $51.87 or $200
16. "39 miles to the gallon in highway driving. 30.5 miles to the gallon in city driving." Which mileage is greater?
17. "Still available for the play-offs: 5398 box seats, 5400 grandstand seats." Which number is greater?

Write the numbers in order from lowest to highest.

18. 1.47, 1.36, 1.5
19. 5.13, 5.09, 5.10
20. 6.08, 6.38, 6.81
21. 4.86, 4.95, 4.81
22. 15.321, 15.231, 15.270
23. 40.004, 40.106, 40.037
24. 131.012, 132.009, 131.021
25. 965.51, 965.40, 965.510
26. 0.1234, 0.1342, 0.1423
27. 0.2539, 0.4139, 0.3239
28. 11.01, 10.100, 11.11
29. 67.67, 67.571, 67.6
30. 3.1, 3.062, 3.09
31. 8.001, 8.02, 8.0135

Applications

32. Denise Murphy works part-time shelving books in the school library. Arrange the following library call numbers from lowest to highest.
    - a. Science: 511.12, 509.03, 512.362, 592.411, 578.62
    - b. Literature: 879.42, 851.612, 801.3, 892.674, 829.9
    - c. Religion: 257.49, 281.6, 204.43, 221.51, 264.113
    - d. Language: 427.6, 487.11, 439.527, 429.8, 439.094
33. Fourteen books were returned in the last hour. To help Denise shelve them faster, put them in numerical order from lowest to highest.

    990.71, 826.4, 329.213, 419.52, 813.6, 122.51, 318.6, 749.02, 519.21, 927.263, 318.6676, 826.401, 615.02, 122.05
34. Denise also files the cards from checked-out books, according to the card numbers. Between which two cards would she file the checked-out books?
    - a. Card number of the checked-out book is 874.192.
      Filed cards are: 872.41, 873.15, 877.142, 879.190
    - b. Card number of the checked-out book is 332.75.
      Filed cards are: 309.8, 311.75, 332.075, 332.749, 333.54

35. Use the tax table at the right to complete the following problem.

	Amount	At least	But less than
	$174.40	$170	$180
a.	$192.51	?	?
b.	$200.40	?	?
c.	$180.24	?	?
d.	$218.00	?	?
e.	$210.00	?	?

FEDERAL WITHHOLDING TAX
WEEKLY Payroll Period

And the wages are:

At least	But less than
$170	$180
180	190
190	200
200	210
210	220
220	230

36. Shipping charges often depend on the dollar amount of the order, as shown in the table. Find the cost of shipping each order.

a. $7.90
b. $23.79
c. $18.54
d. $19.37
e. $36.24
f. $41.22
g. $45.00
h. $6.54
i. $11.70
j. $16.15

Packing & Guaranteed Delivery:
To EACH shipping address, please add:

Orders up to $10.00	$2.50
Orders from $10.01 to $15.00	$3.00
Orders from $15.01 to $25.00	$3.50
Orders from $25.01 to $35.00	$3.95
Orders from $35.01 to $45.00	$4.45

37. The minimum payment required on a charge account can be determined from a minimum payment schedule. Find the minimum payment for each balance.

a. $15.85
b. $122.82
c. $320.00
d. $162.29
e. $4.20
f. $417.20
g. $160.01
h. $426.13
i. $230.84
j. $34.43

MINIMUM PAYMENT SCHEDULE

New Balance	Minimum Monthly Payment
$ .01 to $ 5.00	Balance
5.01 to 100.00	$ 5.00
100.01 to 120.00	6.00
120.01 to 140.00	7.00
140.01 to 160.00	8.00
160.01 to 180.00	9.00
180.01 to 200.00	10.00
200.01 to 240.00	11.00
240.01 to 280.00	12.00
280.01 to 320.00	13.00
320.01 to 360.00	14.00
360.01 to 400.00	15.00
400.01 to 440.00	16.00

38. Olympic scores are often given in decimals. In racing events, the lowest time wins. List the order of finish for each event.

a. Luge

Danlin	3:04.962
Gardner	3:04.258
Doudin	3:05.012

b. 2-Man Bobsled

Hoppe	3:25.56
Manis	3:26.16
Lehman	3:26.04

c. Skiing/X-Country

Jaren	1:03:13.6
Hamel	1:01:45.0
Metani	1:02:26.7

In some events, the highest number of points wins. Determine the winner for each of the following events.

d. Ski Jumping 70 Meter

Konen	212.8
Weiss	215.2
Nyken	214.0

e. Ski Jumping 90 Meter

Weiss	213.7
Ploc	202.9
Nyken	231.2

f. Nordic Combined

Sandberg	422.595
Pulli	410.900
Karjan	416.900

Workshop 3 Adding Decimals

▶ Look up Skill 5 on page 509 for more practice.

When adding decimals, write the addition problem in vertical form. Be sure to line up the decimal points. Write a decimal point in the answer directly below the decimal points in the problem. Then add as you would whole numbers.

Example 17.93 + 18.26 + 54.03

Solution Line up decimals.

```
17.93
18.26
54.03
```

Add.

```
  17.93
  18.26
+ 54.03
-------
  90.22
```

A. 39.76
B. 65.85
C. 178.68
D. 117.14

▶ **Self-Check** Cover the answers and complete the problems.

A. 15.41 + 13.23 + 11.12 = ?
B. 21.62 + 18.91 + 25.32 = ?
C. 48.61 + 92.19 + 37.88 = ?
D. 37.08 + 15.12 + 64.94 = ?

When adding amounts with different numbers of decimal places, you may want to write zeros in the empty decimal places. Put a decimal point in any whole number included in the problem.

Example 36.4 + 581.279 + 2 + 73.08

Solution Line up decimals.

```
 36.4
581.279
  2.
 73.08
```

Write zeros.

```
 36.400
581.279
  2.000
 73.080
```

Add.

```
   36.400
  581.279
    2.000
+  73.080
---------
  692.759
```

Adding amounts of money is just like adding decimals. The decimal point separates the dollars and cents. Remember to put a dollar sign in the total.

Example \$21.85 + \$9.88

Solution Line up decimals.

```
$21.85
  9.88
```

Add.

```
  $21.85
+   9.88
--------
  $31.73
```

A. 684.802
B. 28.174
C. \$39.14
D. \$99.56

▶ **Self-Check** Cover the answers and complete the problems.

A. 130.2 + 527.592 + 27.01 = ?
B. 1.05 + 12.013 + 9.211 + 5.9 = ?
C. \$27.40 + \$2.53 + \$9.21 = ?
D. \$19.14 + \$68 + \$12.42 = ?

Problems

1. 25.53 + 60.43

2. 32.89 + 17.59

3. 663.02 + 455.17

4. 488.51 + 384.80

5. 67.098 + 53.821

6. 54.980 + 92.864

7. 2.987 + 8.082

8. 6.045 + 7.055

9. 43.25, 87.91, + 62.04

10. 33.62, 52.06, + 38.44

11. 4.5857, 7.2811, + 6.1194

12. 88.487, 20.357, + 49.047

13. 334.6, 1.882, + 37.88

14. 153.98, 89.144, + 8.4

15. 1.089, 25.4, + 654.

16. 78.721, 156.43, + 6.055

17. 9.556, 474.2, + 117.6

18. 7620.8, 55.312, + 513.8

19. 72.96, 1.4990, + 9562.9

20. 0.51, 8243., + 59.614

21. \$14.92, 28.33, + 31.08

22. \$173.84, 27.36, + 11.87

23. \$63.47, 19, + 25.80

24. \$485, 13.61, + 300.14

25. 87.35 + 19.46 + 25.01 + 33.82

26. 0.93 + 25.479 + 18 + 176.2

27. 64.7 + 92.301 + 1.53 + 464.0251

28. 3.926 + 9 + 3.12 + 87.941

29. \$1.89 + \$23.80 + \$0.46 + \$62

30. \$9.63 + \$0.87 + \$900.72 + \$5

31. \$85.09 + \$1.43 + \$121.58 + \$43

32. \$6.98 + \$6.15 + \$52.03 + \$0.78

Applications

Find the net deposit for each bank deposit slip by adding the deposits.

33.

		DOLLARS	CENTS
CASH	CURRENCY	12	00
	COINS		
CHECKS	LIST SEPARATELY 56-2	112	48
	54-10	15	02
	SUBTOTAL		
	LESS CASH RECEIVED		
	TOTAL DEPOSIT	?	

34.

		DOLLARS	CENTS
CASH	CURRENCY		
	COINS	3	50
CHECKS	LIST SEPARATELY 75-12	15	80
	75-10	45	15
	75-11	182	33
	SUBTOTAL		
	LESS CASH RECEIVED		
	TOTAL DEPOSIT	?	

35.

		DOLLARS	CENTS
CASH	CURRENCY	3	00
	COINS		
CHECKS	LIST SEPARATELY 85-76	248	80
	88-11	37	14
	75-13	120	18
	SUBTOTAL		
	LESS CASH RECEIVED		
	TOTAL DEPOSIT	?	

Complete the sales receipts by finding the subtotal and the total.

36.

DATE	AUTH. NO.	IDENTIFICATION	CLERK	REG/DEPT	☑TAKE ☐SEND
3/14/–	42		JR		

QTY	CLASS	DESCRIPTION	PRICE	AMOUNT
1		Dress shirt		19 95
1		Sweater		34 95
a.			SUB TOTAL	?
	CUSTOMER SIGNATURE X	Bernice Adams	TAX	4 22
b.		SALES SLIP	TOTAL	?

37.

DATE	AUTH. NO.	IDENTIFICATION	CLERK	REG/DEPT	☑TAKE ☐SEND
6/1/–	86430		DL		

QTY	CLASS	DESCRIPTION	PRICE	AMOUNT
1		Knit Shirt		17 98
2		Pairs Socks	2.59ea.	5 18
1		Can Tennis Balls		4 15
a.			SUB TOTAL	?
	CUSTOMER SIGNATURE X	Steve Rizzo	TAX	1 08
b.		SALES SLIP	TOTAL	?

38. The office manager for Dallas Products, Inc., received this partially completed travel expense report for Norman Minch. Find the missing totals. First add the total expenses for each category horizontally. Then add all expenses for each day vertically. Add the Daily Totals horizontally. The sum of the vertical totals should equal the sum of the horizontal totals.

Dallas Products, Inc.
10149 North Main Street
Dallas, Texas 75227

Travel Expense Report

Name: Norman Minch Position: Marketing Representative
Destination: Houston, Texas
Purpose of Trip: Marketing meeting
Date(s) of Trip: November 21-24

	Expense Item	Sun	Mon 9-3	Tues 9-4	Wed 9-5	Thurs 9-6	Fri 9-7	Sat	TOTALS
a.	Hotel		65.12	65.12	65.12	65.12			?
b.	Breakfast			7.59	6.21	5.93	8.44		?
c.	Lunch			10.21	8.50	9.38	6.19		?
d.	Dinner		17.50	14.92	54.80	22.11			?
e.	Tips		5.00	4.50	15.00	6.50	2.00		?
	Entertainment*								
f.	Taxi		11.00				12.00		?
	Public Trans.								
	Parking								
	Tolls								
	Car Rental								
g.	Air Fare		279.00						?
h.	Telephone		7.00	8.25		12.50	2.25		?
	DAILY TOTALS		?	?	?	?	?		? Total Before Mileage
			i.	j.	k.	l.	m.		n.

39. These five contestants at the Olympics received the scores listed below. Find the total for each contestant.

	Judge:	A	B	C	D	E	F	G	H	Total
	Contestant									
a.	No. 1	5.6	7.8	6.5	5.7	6.3	5.9	6.0	6.2	?
b.	No. 2	7.7	7.9	8.5	8.3	8.0	8.8	9.0	8.7	?
c.	No. 3	9.0	10.0	9.9	9.6	9.8	9.8	9.7	9.9	?
d.	No. 4	9.3	9.4	9.6	8.9	9.1	9.2	9.3	9.8	?
e.	No. 5	4.5	6.5	6.0	5.9	5.5	5.0	4.9	5.2	?

Workshop 4 Subtracting Decimals

▸ Look up Skill 6 on page 510 for more practice.

When subtracting decimals, write the subtraction problem in vertical form. Be sure to line up the decimal points. Write a decimal point in the answer directly below the decimal points in the problem. Then subtract as you would with whole numbers.

Example

97.38 − 25.54 ➡

Solution

Line up decimals.

```
97.38
25.54
```

➡ Subtract.

```
  97.38
− 25.54
  71.84
```

When subtracting amounts with different numbers of decimal places, you may want to write zeros in the empty decimal places. Put a decimal point in any whole number included in the problem.

Example

```
  584.3
−  29.167
```

➡ **Solution**

Put in zeros and subtract.

```
  584.300
−  29.167
  555.133
```

Example

```
  38.94
−  9
```

➡ **Solution**

Put in zeros and subtract.

```
  38.94
−  9.00
  29.94
```

A. 53.61
B. 38.3
C. 610.08
D. 68.76

▸ **Self-Check** Cover the answers and complete the problems.

A. 87.86 − 34.25 = ? **B.** 125.9 − 87.6 = ?
C. 675.4 − 65.32 = ? **D.** 76.76 − 8 = ?

Subtracting amounts of money is just like subtracting decimals. The decimal point separates the dollars and cents. Remember to put a dollar sign in the answer.

Example

```
  $89.59
−  56.37
```

➡ **Solution**

```
  $89.59
−  56.37
  $33.22
```

Example

```
  $572.
−   89.64
```

➡ **Solution**

```
  $572.00
−   89.64
```

➡

```
  $572.00
−   89.64
  $482.36
```

A. $35.71
B. $710.08
C. $352.14
D. $71.65

▸ **Self-Check** Cover the answers and complete the problems.

A. $56.78 − 21.07 **B.** $786.43 − 76.35 **C.** $789.12 − 436.98 **D.** $126. − 54.35

Problems

1. 45.6 − 21.2 **2.** 88.6 − 78.5 **3.** 9.87 − 7.54 **4.** 8.63 − 1.03

5. 89.54 − 70.74 **6.** 73.23 − 8.47 **7.** 88.075 − 33.526 **8.** 28.111 − 6.348

9. 18.37 − 3.067
10. 20.36 − 4.239
11. 8.1538 − .1985
12. 6.9909 − 3.9999
13. 64.392 − 31.875
14. 33.386 − 10.420
15. 654.7 − 44.332
16. 995.1 − 93.46
17. 730. − 12.78
18. 104. − 9.764
19. 4.1 − .459
20. 26. − .357
21. $99.76 − 22.36
22. $65.87 − 18.40
23. $962.21 − 186.32
24. $307.57 − 233.60
25. $147. − 72.97
26. $276. − 35.13
27. $45,552.67 − 37,872.88
28. $67,095.55 − 6,257.72
29. 235.8 − 127.9
30. 248.1 − 152.6
31. 4.21 − 0.7
32. 258.1 − 76.29
33. 13.57 − 8.6423
34. 237 − 46.11
35. $77.92 − $43.01
36. $84.88 − $21.49
37. $4 − $2.63
38. $835.66 − $81.43
39. $2439.21 − $1616.56
40. $4000 − $3353.85

Applications

41. You work as a cashier in a restaurant. Compute the correct change for each of the following orders.

	Customer's Order	Customer Gives You	Change
a.	$ 1.91	$ 2.00	?
b.	$ 2.56	$ 5.00	?
c.	$ 3.87	$ 4.00	?
d.	$ 4.32	$ 5.50	?
e.	$ 7.05	$ 7.25	?
f.	$ 13.87	$ 15.00	?
g.	$ 15.73	$ 20.00	?
h.	$ 21.06	$ 22.00	?
i.	$ 19.78	$ 20.08	?
j.	$128.78	$130.00	?

Complete the computational part of the following bank deposit tickets.

42.

		DOLLARS	CENTS
CASH	CURRENCY	2	00
	COINS		
CHECKS	LIST SEPARATELY 1-43	12	76
	1-44	17	80
a.	SUBTOTAL	?	
	LESS CASH RECEIVED	10	00
b.	TOTAL DEPOSIT	?	

43.

		DOLLARS	CENTS
CASH	CURRENCY		
	COINS	3	50
CHECKS	LIST SEPARATELY 57-12	21	50
	57-10	36	45
a.	SUBTOTAL	?	
	LESS CASH RECEIVED	20	00
b.	TOTAL DEPOSIT	?	

44.

		DOLLARS	CENTS
CASH	CURRENCY	3	00
	COINS		
CHECKS	LIST SEPARATELY 85-76	55	50
	88-11	45	45
	75-13	25	75
a.	SUBTOTAL	?	
	LESS CASH RECEIVED	35	00
b.	TOTAL DEPOSIT	?	

45.

		DOLLARS	CENTS
CASH	CURRENCY		
	COINS	7	00
CHECKS	LIST SEPARATELY 50-50	26	50
	50-75	75	33
	45-30	17	77
a.	SUBTOTAL	?	
	LESS CASH RECEIVED	40	50
b.	TOTAL DEPOSIT	?	

46.

		DOLLARS	CENTS
CASH	CURRENCY	7	00
	COINS		35
CHECKS	LIST SEPARATELY 66-40	25	67
	60-50	89	97
	65-70	111	11
a.	SUBTOTAL	?	
	LESS CASH RECEIVED	75	00
b.	TOTAL DEPOSIT	?	

47.

		DOLLARS	CENTS
CASH	CURRENCY	1	00
	COINS		75
CHECKS	LIST SEPARATELY 40-10	33	45
	45-50	27	10
	42-13	127	75
a.	SUBTOTAL	?	
	LESS CASH RECEIVED	75	00
b.	TOTAL DEPOSIT	?	

48. A partially completed budget comparison statement for the Roadway Bicycle Club is shown below.

ROADWAY BICYCLE CLUB BUDGET COMPARISON STATEMENT

For the Year 19--

ITEM	ESTIMATED	ACTUAL	DIFFERENCE	OVER/UNDER?
Receipts				
Dues	450 00	510 00	?	?
Bike-a-Thon	500 00	424 00	?	?
Raffle	950 00	1142 00	?	?
T-shirt sales	360 00	341 50	?	?
Total Receipts	?	?		

a. Find the club's total estimated and total actual receipts.

b. Complete the difference column by subtracting the actual column from the estimated column. Label the difference as being "over" or "under" the estimate.

49. The marketing department for Mountain Spring Sporting Goods Company prepared a comparison sheet that shows sales projections and actual sales for ten products. Find the difference between the projected sales and the actual sales by subtracting actual sales from projected sales.

Mountain Spring Sporting Goods Company

	Product	Sales Projections	Actual Sales	Difference
a.	Electric fishing motors	$ 2,999.64	$ 2,749.67	?
b.	Fishing tackle	12,345.50	11,897.40	?
c.	Wilderness boots	3,797.00	3,607.15	?
d.	Boat moccasin	839.10	671.28	?
e.	Enamel campware sets	359.64	329.67	?
f.	Geodesic dome tent	2,549.55	2,379.58	?
g.	Backpacking tent	2,399.28	1,999.40	?
h.	Sleeping bags	11,334.60	11,019.75	?
i.	Sleeping bag pads	4,189.20	4,189.20	?
j.	Folding cot	845.46	751.52	?

Workshop 5 Multiplying Decimals

▶ Look up Skill 8 on page 512 for more practice.

When multiplying decimals, multiply as if the decimal numbers were whole numbers. Then count the total number of decimal places to the right of the decimals in the factors. This number will be the number of decimal places in the product.

Example

$$\begin{array}{rl} 15.3 & \leftarrow \text{factor} \\ \times\ 0.24 & \leftarrow \text{factor} \\ \hline 612 & \\ 306 & \\ \hline 3672 & \leftarrow \text{product} \end{array}$$

➡

Solution

$$\begin{array}{rlr} 15.3 & \leftarrow & 1 \text{ decimal place} \\ \times\ 0.24 & \leftarrow & +\ 2 \text{ decimal places} \\ \hline 612 & & \\ 306 & & \\ \hline 3.672 & \leftarrow & 3 \text{ decimal places} \end{array}$$

If the product does not have enough digits to place the decimal in the correct position, you will need to write zeros. Start at the right of the product in counting the decimal places and write the zeros at the left.

Example

$$\begin{array}{r} 0.62 \\ \times\ 0.07 \\ \hline 434 \end{array}$$

➡

Solution

$$\begin{array}{rlr} 0.62 & \leftarrow & 2 \text{ decimal places} \\ \times\ 0.07 & \leftarrow & +\ 2 \text{ decimal places} \\ \hline 0.0434 & \leftarrow & 4 \text{ decimal places} \end{array}$$

▶ **Self-Check** Cover the answers and complete the problems.

A. $\begin{array}{r} 15.5 \\ \times\ 0.25 \\ \hline \end{array}$ **B.** $\begin{array}{r} 21.6 \\ \times\ 2.22 \\ \hline \end{array}$ **C.** $\begin{array}{r} 0.64 \\ \times\ 0.05 \\ \hline \end{array}$ **D.** $\begin{array}{r} 0.88 \\ \times\ 0.08 \\ \hline \end{array}$

A. 3.875
B. 47.952
C. 0.032
D. 0.0704

When multiplying amounts of money, round the answer off to the nearest cent. Remember to put a dollar sign in the answer.

Example

$$\begin{array}{r} \$1.47 \\ \times\ \ 2.5 \\ \hline 3675 \end{array}$$

Solution

$$\begin{array}{rlr} \$\ 1.47 & \leftarrow & 2 \text{ places} \\ \times\ \ 2.5 & \leftarrow & +\ 1 \text{ place} \\ \hline \$3.675 & \leftarrow & 3 \text{ places} \end{array}$$

$\$1.47 \times 2.5 = \$3.675$
$= \$3.68$
Rounded to nearest cent

When multiplying by 10, 100, or 1000, count the number of zeros and then move the decimal point to the right the same number of spaces.

Example 1
5.3×10

Solution
$5.3 \times 10 = 5.3 = 53.$

10 has one zero; move decimal 1 place

Example 2
2.17×100

Solution
$2.17 \times 100 = 2.17 = 217.$

100 has two zeros; move decimal 2 places

Example 3
12.1364×1000

Solution
$12.1364 \times 1000 = 12.1364$
$= 12{,}136.4$

1000 has three zeros; move decimal 3 places

A. $12.075 = $12.08
B. 432.
C. 3498.4
D. 1984.4

▸ **Self-Check** Cover the answers and complete the problems.

A. $3.45 × 3.5 B. 43.2 × 10
C. 34.984 × 100 D. 1.9844 × 1000

Problems

1. 21.2 × 0.25
2. 56.6 × 0.94
3. 79.5 × 0.22
4. 96.3 × 0.68
5. 5.80 × 0.83
6. 66.2 × 0.41
7. 95.1 × 0.27
8. 36.3 × 0.69
9. 0.62 × 0.08
10. 0.75 × 0.07
11. 0.91 × 0.04
12. 0.54 × 0.09
13. 0.46 × 0.05
14. 0.98 × 0.06
15. 0.484 × 0.003
16. 0.643 × 0.004
17. 2.864 × 0.03
18. 6.579 × 0.09
19. 1.324 × 0.54
20. 4.616 × 0.24
21. 0.024 × 0.007
22. 0.106 × 0.023
23. 0.851 × 1.897
24. 0.710 × 6.005
25. $1.78 × 25
26. $4.85 × 93
27. $4.37 × 2.5
28. $2.96 × 6.7
29. $16.28 × 1.2
30. $95.14 × 9.5

Multiply by 10 31. 24.6 32. 4.49 33. 7.435 34. 0.8971

Multiply by 100 35. 28.42 36. 33.712 37. 2.896 38. 390.31

Multiply by 1000 39. 53.663 40. 9.3581 41. 0.255 42. 897.485

Multiply by 10,000 43. 9.8891 44. 5.68145 45. 91.5321 46. 517.133

Applications

Find the total price for each item and the total order amount.

47.

	Quantity	DESCRIPTION	Price Each	TOTAL PRICE
a.	3	Leather work gloves	2.95	?
b.	2	Eyelet tool set	15.50	?
c.	2	Eyelet assortment	2.25	?
d.	4	Rechargeable battery	7.95	?
e.	1	Framing clamp	24.95	?
f.	1	Framer's brad pliers	19.95	?
g.	5	Brass belt buckles	7.97	?
h.			AMOUNT FOR GOODS	?

48. Cindy O'Keeffe is ordering tickets for the Olympic games. Complete the form and find her total amount due.

	II. Order Information		FIRST CHOICE			
	DATE (MO/DAY)	SPORT & EVENTS	EVENT CODE	NO. OF TICKETS	COST PER TICKET	TOTAL COST PER EVENT
a.	8/3	Gymnastics - Women	1103-332	2	$95.00	?
b.	8/11	Gymnastics - women	1111-327	2	60.00	?
c.	8/10	Equestrian - Individual	0810-226	1	35.00	?
d.	8/12	Diving - Platform Finals - men	1912-146	2	20.00	?
e.	8/10	Basketball - Finals - men	0410-332	1	40.00	?
f.	8/5	Basketball - Round Robin - Women	0405-324	2	15.00	?
g.	8/4	Track and Field	0304-212	2	50.00	?
			–			
			–			
h.					Total Ticket Cost	?
i.					Total × 0.06 Tax	?
j.		Non-refundable Ticket Handling Charge $1.00 × Number of Tickets (First Choices Only)			12	?
k.					Total Amount Due	?

49. The Purchasing Clerk prepares an invoice to order items from a supplier. Complete the following invoice. Find the total price for each item and record the amount in the Total Price Column. Then add the amounts in the Total Price column and record your answer in the Total for Goods box. Find the total tax. Find the total cost.

	NAME OF ITEM (1 or 2 words)	COMPLETE CATALOG NUMBER	HOW MANY (Pkgs., etc.)	COLOR NUMBER	SIZE	PRICE FOR ONE	TOTAL PRICE
a.	Ladies' Vest	2940-16A	2	N	M	27.95	?
b.	Ladies' Turtleneck	4303-14B	4	W	L	34.95	?
c.	Hooded Jacket	334-114	3	B	S	41.45	?
d.	Cossack Hat	3121-14	6	—	—	15.50	?
e.	Boot Socks	2631-17	12	N	—	5.95	?
f.	Boot Socks	2631-18	12	M	—	5.95	?
g.	Hooded Parka	5021-14	10	L	—	74.95	?
h.						TOTAL FOR GOODS	?
i.						TOTAL × 0.06 Tax	?
						Transportation and Handling	9.40
j.						TOTAL COST	?

Workshop 6 Dividing Decimals

▸ Look up Skills 9, 10, and 11 on pages 513, 514, and 515 for more practice.

```
              231  ← quotient
divisor → 3 ) 693  ← dividend
```

When dividing decimals, first check if there is a decimal point in the divisor. If there is, move the decimal point to the right to make the divisor a whole number. Then move the decimal point in the dividend to the right the same number of places you moved the decimal point in the divisor. Write the decimal point in the quotient directly above the decimal point in the dividend. Then divide as with whole numbers.

Example

21.08 ÷ 6.2

or

6.2) 21.08

Solution

```
                              3.4
6.2 ) 21.08   →        62 ) 210.8
                          - 186
                          ------
                             24 8
                           - 24 8
                           ------
```

Add zeros to the right of the decimal point in the dividend if needed.

Example

0.324 ÷ 0.45

or

0.45) 0.324

Solution

```
                           0.72
0.45 ) 0.324   →   45 ) 32.40  ← Zero added
                      - 31 5
                      ------
                        - 90
                          90
```

A. 6.55
B. 7.29
C. 5.88
D. 540

▸ **Self-Check** Cover the answers and complete the problems.

A. 34.06 ÷ 5.2

B. 12.3) 89.667

C. 1.47 ÷ 0.25

D. 0.042) 22.68

When the dividend is an amount of money, remember to place the dollar sign in the quotient and round the answer to the nearest cent.

Example

32) \$47.56

Solution

```
                  $ 1.486
32 ) $47.56  →  32 ) $47.560    $47.56 ÷ 32 = $1.49 ← Rounded to nearest cent
```

When dividing by 10, 100, or 1000, count the number of zeros in 10, 100, or 1000 and move the decimal point to the left the same number of places.

Example 1	Solution	
4.8 ÷ 10	4.8 ÷ 10 = 04.8 = 0.48	10 has 1 zero; move decimal 1 place

Example 2	Solution	
151.23 ÷ 100	151.23 ÷ 100 = 151.23 = 1.5123	100 has 2 zeros; move decimal 2 places

Example 3	Solution	
13,436.1 ÷ 1000	13,436.1 ÷ 1000 = 13436.1 = 13.4361	1000 has 3 zeros; move decimal 3 places

▶ **Self-Check** Cover the answers and complete the problems.

A. $24.12 ÷ 18 B. 4.9 ÷ 10
C. 128.7 ÷ 100 D. 2765.9 ÷ 1000

A. $1.34
B. 0.49
C. 1.287
D. 2.7659

Problems

1. 4.2) 7.56 2. 3.5) 16.1 3. 9.3) 25.11 4. 3.2) 30.72
5. 6.8) 63.24 6. 2.2) 92.84 7. 56) 408.8 8. 8.9) 596.3
9. 4.02) 26.13 10. 5.12) 39.936 11. 3.87) 10.836 12. 9.65) 29.336
13. 0.006) 0.2928 14. 0.008) .3548 15. 0.132) 4.158 16. 0.698) 0.99814

Divide. Round to the nearest hundredth or to the nearest cent.

17. 5.3) 8.891 18. 2.9) 5.343 19. 2.46) 88.34 20. 6.95) 177.9
21. 0.98) 46.98 22. 0.37) 602.7 23. 15.92) 8.027 24. 31.34) 9.109
25. 21) $82.42 26. 18) $23.98 27. 98) $678.40 28. 53) $177.79
29. 82) $267.32 30. 37) $793.28 31. 561) $9803.88 32. 217) $4635.79

Divide by 10	33. 7.3	34. 12.42	35. 626.81	36. 0.094
Divide by 100	37. 129.8	38. 433.39	39. 3462.65	40. 0.746
Divide by 1000	41. 1897.9	42. 221.29	43. 23.098	44. 9.432
Divide by 10,000	45. 43,452.18	46. 3829.17	47. 515.49	48. 34.858

49. Dwight Johnson drove 390 miles on 15.5 gallons of gas. How many miles per gallon did he get?

50. An 18-ounce box of oatmeal costs $1.28. What is the cost per ounce?

51. A 3.25-kilogram roast costs $15.99. What is the cost per kilogram?

Applications

52. As a clerk and stockperson at the Quick Stop Carryout, you are required to compute the cost per box (to nearest cent) for the following items.

	Item	Cost per Case	Number per Case	Cost per Item
a.	Cream cheese	$ 6.84	12	?
b.	Crescent rolls	5.64	12	?
c.	Margarine	19.80	30	?
d.	Orange juice, 64 oz	7.65	6	?
e.	Grape juice, 6 oz	18.96	48	?
f.	Hash browns, 2 lb	7.55	12	?
g.	Dry roasted peanuts	11.10	12	?
h.	Mayonnaise, 48 oz	16.21	8	?
i.	Greek olives	14.78	24	?

53. Compute the miles per gallon (to nearest tenth) for the types of vehicles leased by the Acme Transportation Company the first week in June.

	Type of Vehicle	Miles	Gallons of Fuel	Miles per Gallon
a.	Standard	367.4	22.0	?
b.	Intermediate	280.5	15.0	?
c.	Compact	271.4	11.5	?
d.	Subcompact	234.9	8.7	?
e.	Passenger van	283.8	21.5	?
f.	Motor home	282.9	34.5	?

54. The unit price of an item can be found by dividing the selling price by the weight. Determine the unit price of each item to the nearest thousandth.

	Item	Selling Price	Weight	Unit Price (3 decimal places)
a.	Macaroni	$1.56	3 lb	?
b.	Salad dressing	0.64	8 oz	?
c.	Catsup	0.68	14 oz	?
d.	Cranberry juice	1.10	32 oz	?
e.	Pork and beans	0.57	21 oz	?
f.	Onion soup	0.30	1.25 oz	?
g.	Chicken broth	0.55	13.75 oz	?
h.	Salt substitute	0.89	3.125 oz	?

Workshop 7 Estimation

▸ Look up Skill 2 on page 506 for more practice.

Estimation is a very valuable tool in mathematics. Many times a quick estimate is more useful than an exact computation. It can also be used as a quick method of checking a calculation. To estimate what an answer should be, round the numbers to simplify the computation.

Estimate			Calculation
87.651 − 49.023	90 − 50 = 40	The answer should be about 40.	87.651 − 49.023 = 38.628
4.78 × 6.35	5 × 6 = 30	The answer should be about 30.	4.78 × 6.35: 2390, 1434, 2868 = 30.3530

Estimate and Actual
A. 40; 36.78
B. 300; 296.93
C. 24; 21.7992
D. 500; 556.4403

▸ **Self-Check** Cover the answers and complete the problems. Estimate, then perform the computation.

A. 58.34 − 21.56 B. 684.89 − 387.96 C. 5.86 × 3.72 D. 22.41 × 24.83

If all the numbers do not have the same number of digits, round the numbers to the highest place value of the smaller number.

Estimate			Calculation
6.25) 23.75	6) 24 = 4	The answer should be about 4.	625) 2375.0 = 3.8; − 1875; 500 0; − 500 0
\$256.78 + 43.17	\$260 + 40 = \$300	The answer should be about \$300.	\$256.78 + 43.17 = \$299.95

Estimate and Actual
A. 6; 7.2
B. 40; 45.1
C. 340; 333.22
D. \$70; \$66.11

▸ **Self-Check** Cover the answers and complete the problems. Estimate, then perform the computation.

A. 4.6) 33.12 B. 2.8) 126.28

C. 387.98 − 54.76 D. \$87.65 − 21.54

Problems

Estimate first, then perform the computations.

1. $98.43 - 43.59$
2. $80.94 - 19.47$
3. $85.31 - 67.01$
4. $48.91 - 19.75$
5. $706.83 - 64.18$
6. $103.91 - 34.72$
7. $\$893.51 - 39.25$
8. $\$590.23 - 42.87$
9. 8.94×7.34
10. 6.88×4.32
11. 71.21×3.68
12. 91.52×4.81
13. 39.57×22.16
14. 51.27×26.26
15. 935.6×7.48
16. 241.9×1.84
17. $3.2\overline{)99.2}$
18. $7.1\overline{)55.38}$
19. $4.2\overline{)162.54}$
20. $7.3\overline{)74.46}$
21. $7.25\overline{)18.85}$
22. $4.17\overline{)39.615}$
23. $7.97\overline{)16.737}$
24. $7.32\overline{)65.148}$
25. $23.987 + 59.806$
26. $54.471 + 64.123$
27. $902.14 + 375.61$
28. $647.72 + 159.93$
29. $\$188.15 + 81.70$
30. $\$496.49 + 550.18$
31. $\$9333.57 + 892.94$
32. $\$2162.65 + 1122.75$
33. You purchase items for \$29.33, \$18.75, and \$42.88. Estimate the total, then calculate.
34. 22 people charter a bus for \$238. Estimate the amount each person pays, then calculate.

Applications

Estimate first, then find the total for these grocery store receipts.

35.

SUPER VALUES MARKET	
FACE TISSUE	1.05
COFFEE	2.95
FACE TISSUE	1.05
FACE TISSUE	1.05
CORN CHIPS	2.19
BACON	1.58
RYE CRACKERS	1.29
SMALL LINKS	1.98
SMALL LINKS	1.98
SMALL LINKS	1.98
CAT FOOD	2.60
DETERGENT	1.88
DETERGENT	1.08
TOTAL	?

36.

SUPER VALUES MARKET	
CORN CHIPS	2.19
VANILLA	3.09
BNL SHK HAM	25.33
ORANGE JUICE	5.48
CASING LINKS	3.65
KIELBASA	3.18
KIELBASA	2.43
PICKLES	1.69
CEREAL	1.53
MAYO	1.69
ALUMINUM FOIL	1.09
CHEDDAR	2.99
MACARONI	.65
STEAK	4.23
RIPE OLIVES	1.19
SPAN OLIVES	.99
SPAN OLIVES	.99
SPAN OLIVES	.99
VINEGAR	1.19
TOTAL	?

37.

SUPER VALUES MARKET	
BABY CLAMS	1.39
RICE	1.03
CHESTNUT	.97
CHEESE	2.65
SPROUTS	.95
CRANAPLE	1.13
SALAD DRSG	.59
HERB DRSG	.59
HERB DRSG	.59
SALAD DRSG	.59
HALF HALF PT	.85
WAX PAPER	.89
ITAL SAUS	2.17
ITAL SAUS	2.08
MINI CHP	1.95
SAUSAGE 2LB	4.35
MINTS	.35
GUM	.45
CRANPLE	1.13
TOTAL	?

Use the menu from the Columbian House to estimate the bill for the following orders.

COLUMBIAN HOUSE	
Appetizers	
Soup Romaine	$ .95
Fruit Cup	1.25
Shrimp Cocktail	3.95
Entrees	
Vegetarian Plate	5.95
Baked Ham	6.50
Red Snapper	7.95
Roast Prime Rib	12.95
Filet Mignon	11.75
Desserts	
Chocolate Torte	3.25
Apple Strudel	2.75
Walnut Dream Cake	2.50
Beverages	
Coffee	.55
Milk	.75
Herbal Tea	.55
Soft Drinks	1.00

38. Two persons

1 Soup
1 Fruit Cup
1 Baked Ham
1 Red Snapper
1 Chocolate Torte
1 Apple Strudel
1 Milk
1 Herbal Tea

39. Party of six

4 Soup
2 Fruit Cup
1 Baked Ham
3 Prime Rib
1 Red Snapper
1 Filet Mignon
3 Chocolate Torte
3 Walnut Dream Cake
2 Milk
2 Soft Drinks
2 Coffee

40. Rene Rutherford used a calculator to solve the following problems. If she entered each number properly into the calculator, her answers should be correct. Estimate each answer and decide if Rene's answers are correct.

	Problem	Her Answer on a Calculator	Your Estimate	Is She Right? Yes/No
a.	174.62 − 89.3	85.32	?	?
b.	34.6 × 15.2	5259.2	?	?
c.	2346 ÷ 10.2	230.	?	?
d.	12.3 + 14.2 + 7.6 + 0.9	43.1	?	?
e.	598 × 64	38.272	?	?
f.	632.5 − 185.03	44.747	?	?
g.	3.2 × 47.5	152.	?	?
h.	921.2 ÷ 1225	0.752	?	?
i.	3.75 + 2.98 + 12.25	18.98	?	?
j.	89.25 − 31.52	77.73	?	?

Workshop 8 Fraction to Decimal, Decimal to Fraction

▸ Look up Skill 14 on page 518 for more practice.

Any fraction can be renamed as a decimal and any decimal can be renamed as a fraction.

To rename a fraction as a decimal, use division. Think of the fraction bar in the fraction as meaning "divide by." For example, $\frac{3}{4}$ means "3 divided by 4." After the 3 write a decimal point and as many zeros as are needed. Then divide by 4.

Example Change $\frac{3}{4}$ to a decimal.

Solution

$$\frac{3}{4} \rightarrow 4\overline{)3.00} = 0.75$$

```
       0.75
4 ) 3.00
  - 2 8
      20
    - 20
```

Example Change $\frac{4}{25}$ to a decimal.

Solution

$$\frac{4}{25} \rightarrow 25\overline{)4.00} = 0.16$$

```
        0.16
25 ) 4.00
   - 2 5
     1 50
   - 1 50
```

If a fraction does not divide out evenly, divide to one more decimal place than you are rounding to.

Example Change $\frac{4}{9}$ to a decimal rounded to the nearest hundredth. (Divide to the thousandths place.)

Solution

```
          0.444  = 0.44
4/9 = 9 ) 4.000
          3 6
            40
            36
             40
             36
              4
```

$\frac{4}{9} = 0.44$ (rounded)

Example Change $\frac{5}{7}$ to a decimal rounded to the nearest thousandth. (Divide to the ten thousandths place.)

Solution

```
          0.7142  = 0.714
5/7 = 7 ) 5.0000
          4 9
            10
             7
             30
             28
              20
              14
               6
```

$\frac{5}{7} = 0.714$ (rounded)

▸ **Self-Check** Cover the answers and complete the problems. Change the fractions to decimals. Round to the nearest thousandth. (Divide to the nearest ten thousandths place.)

A. $\frac{3}{4}$ **B.** $\frac{5}{8}$ **C.** $\frac{2}{3}$ **D.** $\frac{3}{7}$

A. 0.75
B. 0.625
C. 0.667
D. 0.429

To rename a decimal as a fraction, name the place value of the digit at the far right. This is the denominator of the fraction.

$0.83 = \frac{83}{100}$ — 3 is in the hundredths place so the denominator is 100

$0.009 = \frac{9}{1000}$ — 9 is in the thousandths place so the denominator is 1000

Note that the number of zeros in the denominator is the same as the number of places to the right of the decimal point. The fraction should always be written in lowest terms.

$0.04 = \frac{4}{100} = \frac{1}{25}$ $\qquad 6.875 = 6\frac{875}{1000} = 6\frac{7}{8}$

A. $\frac{5}{10} = \frac{1}{2}$
B. $\frac{8}{100} = \frac{2}{25}$
C. $1\frac{375}{1000} = 1\frac{3}{8}$
D. $5\frac{32}{100} = 5\frac{8}{25}$

▶ **Self-Check** Cover the answers and complete the problems.

Change the decimals to fractions reduced to lowest terms.

A. 0.5 **B.** 0.08 **C.** 1.375 **D.** 5.32

Problems

Change the fractions to decimals. Round to the nearest thousandth.

1. $\frac{1}{4}$ **2.** $\frac{2}{5}$ **3.** $\frac{3}{8}$ **4.** $\frac{9}{10}$
5. $\frac{7}{20}$ **6.** $\frac{1}{2}$ **7.** $\frac{3}{20}$ **8.** $\frac{3}{5}$
9. $\frac{16}{25}$ **10.** $\frac{1}{50}$ **11.** $\frac{3}{25}$ **12.** $\frac{1}{100}$
13. $\frac{23}{40}$ **14.** $\frac{5}{16}$ **15.** $\frac{6}{125}$ **16.** $\frac{29}{500}$

Change the fractions to decimals. Round to the nearest hundredth.

17. $\frac{1}{3}$ **18.** $\frac{5}{6}$ **19.** $\frac{3}{7}$ **20.** $\frac{7}{9}$
21. $\frac{1}{6}$ **22.** $\frac{2}{3}$ **23.** $\frac{5}{12}$ **24.** $\frac{9}{16}$
25. $\frac{7}{15}$ **26.** $\frac{4}{75}$ **27.** $\frac{5}{9}$ **28.** $\frac{5}{16}$

Change the fractions to decimals. Round to the nearest thousandth.

29. $\frac{1}{9}$ **30.** $\frac{5}{7}$ **31.** $\frac{2}{3}$ **32.** $\frac{4}{7}$
33. $\frac{5}{12}$ **34.** $\frac{3}{16}$ **35.** $\frac{1}{16}$ **36.** $\frac{1}{12}$
37. $\frac{5}{9}$ **38.** $\frac{11}{15}$ **39.** $\frac{23}{365}$ **40.** $\frac{72}{365}$

Change the decimals to fractions reduced to lowest terms.

41. 0.5 **42.** 0.2 **43.** 0.25 **44.** 0.75
45. 0.8 **46.** 0.9 **47.** 0.12 **48.** 0.28
49. 0.65 **50.** 0.47 **51.** 0.35 **52.** 0.83
53. 0.875 **54.** 0.125 **55.** 0.275 **56.** 0.025
57. 0.225 **58.** 0.732 **59.** 0.3125 **60.** 0.1875
61. 1.75 **62.** 3.15 **63.** 2.12 **64.** 5.35
65. 8.25 **66.** 3.55 **67.** 8.05 **68.** 11.08
69. 3.625 **70.** 3.575 **71.** 9.4262 **72.** 7.591

73. "Serving size: 28.4 ounces." Write 28.4 as a fraction in lowest terms.

74. A baseball player had 163 hits in 500 times at bat. Write $\frac{163}{500}$ as a decimal. What was the player's batting average?

75. Three months is $\frac{1}{4}$ of a year. Write $\frac{1}{4}$ as a decimal.

76. The price of a stock is $\$8\frac{3}{4}$. Write $\$8\frac{3}{4}$ as a decimal.

77. "Area 0.675 acre." Write 0.675 as a fraction in lowest terms.

Applications

78. The seven baseball players had these batting averages. Write the batting averages as fractions reduced to lowest terms.

Player Number	Batting Average
15	0.325
43	0.300
21	0.225
46	0.125
56	0.317
16	0.235
10	0.240
22	0.170

79. Stock prices are quoted as dollars and fractions of a dollar. Change the stock prices to dollars and cents. Round to the nearest cent.

Stock	Price
Ford	$\$37\frac{3}{4}$
General Motors	$\$68\frac{3}{8}$
Chrysler	$\$29\frac{7}{8}$
K Mart	$\$30\frac{1}{2}$
Ransburg Corp.	$\$17\frac{5}{8}$
Apple Computers	$\$25\frac{1}{8}$
Champion Int.	$\$23\frac{1}{4}$

80. As an apprentice in a tool and die machine shop, Sally Hefner needs to convert fractions with denominators of 2, 4, 8, 16, 32, and 64 to decimals. She also has to convert decimals to fractions. Sometimes she works with decimals having six decimal places. Complete the chart for her.

Fraction	Decimal
$\frac{1}{32}$	?
$\frac{3}{64}$	?
$\frac{11}{32}$	?
$\frac{9}{16}$	?
$\frac{63}{64}$	?
?	0.125
?	0.3125
?	0.5625
?	0.15625
?	0.921875

81. In the metric system of measurement, fractions are expressed with denominators of 10, 100, 1000, and 10,000. Complete the chart. Convert the fractions to decimals. Express the fractions with denominators of 10, 100, 1000, or 10,000.

Fraction	Decimal
$\frac{1}{100}$	?
$\frac{3}{1000}$	?
$\frac{9}{10}$	?
$\frac{15}{1000}$	?
$\frac{32}{10{,}000}$	?
?	0.385
?	0.7
?	0.97
?	0.021
?	0.0367

Workshop 9 Percent to Decimal, Decimal to Percent

▸ Look up Skills 26 and 28 on pages 530 and 532 for more practice.

Percent is an abbreviation of the Latin words *per centum,* meaning "by the hundred." So percent means "divide by 100." A percent can be written as a decimal. To change a percent to a decimal, first write the percent as a fraction with a denominator of 100, then divide by 100.

Example Change 21% to a decimal.

Solution

$$21\% = \frac{21}{100} = 0.21$$

Example Change 15.5% to a decimal.

Solution

$$15.5\% = \frac{15.5}{100} = 0.155$$

When dividing by 100, you can just move the decimal point two places to the left. So when you write a percent as a decimal, you are moving the decimal point two places to the left and dropping the percent sign (%). If necessary, use zero as a placeholder.

	Example	Solution
1.	21%	21% = 21. = 0.21 (Drop % sign; Move decimal two places)
2.	48.2%	48.2% = 48.2 = 0.482
3.	135%	135% = 135. = 1.35
4.	5%	5% = 05. = 0.05 (Insert a zero as a placeholder)

A. 0.25
B. 0.125
C. 1.21
D. 0.07

▸ **Self-Check** Cover the answers and complete the problems.

Change the percents to decimals.

A. 25% **B.** 12.5% **C.** 121% **D.** 7%

To write a decimal as a percent, move the decimal point two places to the right and add a percent sign (%).

	Example	Solution
1.	0.94	0.94 = 0.94 = 94% (Add % sign; Move decimal two places)
2.	0.05	0.05 = 0.05 = 5%
3.	0.5	0.5 = 0.50 = 50%
4.	2.6	2.6 = 2.60 = 260%
5.	0.003	0.003 = 0.003 = 0.3%

A. 65%
B. 8%
C. 90%
D. 125%
E. 0.2%

▶ **Self-Check** Cover the answers and complete the problems.

Change the decimals to percents.

A. 0.65 B. 0.08 C. 0.9 D. 1.25 E. 0.002

Problems

Write as decimals.

1. 15%	2. 73%	3. 44%	4. 81%
5. 31%	6. 79%	7. 12%	8. 24%
9. 32.1%	10. 58.3%	11. 16.8%	12. 99.9%
13. 11.2%	14. 37.5%	15. 75.5%	16. 87.5%
17. 125%	18. 186%	19. 355%	20. 212%
21. 750%	22. 897%	23. 1244%	24. 2225%
25. 7%	26. 9%	27. 1%	28. 3%
29. 2.6%	30. 5.8%	31. 1.4%	32. 9.5%
33. 1.05%	34. 5.07%	35. 3.0125%	36. 9.0976%

Write as percents.

37. 0.58	38. 0.17	39. 0.89	40. 0.33
41. 0.25	42. 0.18	43. 0.75	44. 0.99
45. 0.04	46. 0.01	47. 0.05	48. 0.08
49. 0.015	50. 0.036	51. 0.0125	52. 0.0978
53. 0.005	54. 0.004	55. 0.0012	56. 0.0092
57. 2.125	58. 8.876	59. 3.5	60. 8.9
61. 9.81	62. 2.11	63. 1.008	64. 5.000
65. 0.002	66. 0.016	67. 12	68. 66
69. 0.3	70. 0.9	71. 340.1	72. 1,235

73. "Speakers on sale, 25% off." Write 25% as a decimal.

74. "Cost of living up 124.7%." Write 124.7% as a decimal.

75. "62 out of 100 students." Write $\frac{62}{100}$ as a percent.

76. "All blankets 20%—30% off." Write 20% and 30% as decimals.

77. "Radial tires on sale, 40% off." Write 40% as a decimal.

Applications

78. The percent changes in retail sales were reported as a decimal in the October issue of *Modern Merchandise* magazine. Change the decimals to percents.

RETAIL SALES

	Month	Change
a.	February	0.016
b.	March	0.002
c.	April	0.015
d.	May	0.045
e.	June	0.042
f.	July	0.013
g.	August	0.019

79. Jon Friedman is an investment counselor for Foxboro Investment Company. When working with clients he uses this chart. Change the percents to decimals.

	Taxable Income	Rate
a.	$20,201-$24,600	13.16%
b.	24,601- 29,900	13.90%
c.	29,901- 35,200	14.73%
d.	35,201- 45,800	16.18%
e.	45,801- 60,000	17.62%
f.	60,001- 85,600	19.35%
g.	85,601 and Up	19.74%

80. The commission rate schedule for a stockbroker is shown. Change the percents to decimals.

Commission Rate Schedule

	Dollar Amount	% of Dollar Amount
	Stocks:	
a.	$0 + $2,499	1.7%, minimum $30
b.	$2,500—$4,999	1.3%, minimum $42
c.	$5,000—$9,999	1.1%, minimum $65
d.	$10,000—$14,999	.9%, minimum $110
e.	$15,000—$24,999	.7%, minimum $135
f.	$25,000—$49,999	.5%, minimum $175
g.	$50,000 and above	negotiated

81. During the National Basketball Association season, the teams had these won—lost records. The **Pct** column shows the percent of games won, expressed as a decimal. Change the decimals to percents.

EASTERN CONFERENCE

Atlantic Division

	W	L	Pct.	GB
Boston	42	12	.778	—
Philadelphia	34	19	.642	$7\frac{1}{2}$
New York	33	20	.623	$8\frac{1}{2}$
New Jersey	26	29	.473	$16\frac{1}{2}$
Washington	25	29	.463	17

Central Division

	W	L	Pct.	GB
Milwaukee	32	23	.582	—
Detroit	30	22	.577	$\frac{1}{2}$
Atlanta	28	27	.509	4
Chicago	21	30	.412	9
Cleveland	20	32	.385	$10\frac{1}{2}$
Indiana	16	37	.302	15

WESTERN CONFERENCE

Midwest Division

	W	L	Pct.	GB
Utah	32	22	.593	—
Dallas	29	26	.527	$3\frac{1}{2}$
San Antonio	25	31	.446	8
Kansas City	23	30	.434	$8\frac{1}{2}$
Denver	22	33	.400	$10\frac{1}{2}$
Houston	21	33	.389	11

Pacific Division

	W	L	Pct.	GB
Los Angeles	34	18	.654	—
Portland	32	23	.582	$3\frac{1}{2}$
Seattle	28	24	.538	6
Phoenix	24	30	.444	11
Golden St.	24	31	.436	$11\frac{1}{2}$
San Diego	17	37	.315	18

Workshop 10 Finding a Percentage

▸ Look up Skill 30 on page 534 for more practice.

Finding a percentage means finding a percent of a number. To find a percent of a number, you change the percent to a decimal, then multiply it by the number.

Example 1 25% of 84 is what number?

Solution $25\% \times 84 = n$ In mathematics, <u>of</u> means "times" and <u>is</u> means "equals." Let n stand for the unknown number.

$0.25 \times 84 = n$ Change the percent to a decimal.
$21 = n$ Multiply.

$25\% \text{ of } 84 = 21$ Write the answer.

Example 2 The delivery charge is 5% of the selling price of $120.00. Find the delivery charge.

Solution
$5\% \times \$120.00 = n$
$0.05 \times \$120.00 = n$
$\$6.00 = n$
$5\% \times \$120.00 = \$6.00$ delivery charge

Example 3 The student had 95% correct out of 80 questions. How many answers were correct?

Solution
$95\% \times 80 = n$
$0.95 \times 80 = n$
$76 = n$
$95\% \times 80 = 76$ correct

▸ **Self-Check** Cover the answers and complete the problems.

A. Find 30% of 80.
B. Find 35% of 140.
C. Find 6% of 35.
D. Find 125% of $300.

A. 24
B. 49
C. 2.1
D. $375

Problems

Find the percentage.

1. 15% of 90
2. 75% of 80
3. 30% of 78
4. 25% of 124
5. 37% of 270
6. 84% of 896
7. 41% of 380
8. 59% of 336
9. 71% of 532
10. 9% of 70
11. 3% of 80
12. 5% of 72
13. 2% of 126
14. 1% of 321
15. 8% of 985
16. 4% of 24.5
17. 6% of 55.5
18. 7% of 70.0
19. 125% of 400
20. 150% of 640
21. 180% of 950
22. 330% of 350
23. 422% of 782
24. 670% of 322
25. 235% of 1890
26. 478% of 2220
27. 872% of 4583

28. 8.5% of 70
29. 9.1% of 84
30. 1.5% of 65
31. 4.5% of 124
32. 7.8% of 432
33. 8.5% of 606
34. 16.8% of 376
35. 25.5% of 480
36. 51.2% of 575
37. 75.4% of 867
38. 99.9% of 380
39. 32.1% of 742

Round to the nearest cent.

40. 1.2% of $40
41. 1.7% of $80
42. 1.4% of $20
43. 3.5% of $360
44. 5.7% of $240
45. 8.1% of $520
46. 1.75% of $94
47. 1.65% of $24
48. 1.83% of $60
49. 1.35% of $54.80
50. 1.66% of $98.80
51. 1.33% of $32.54
52. 0.9% of $12,000
53. 0.7% of $14,000
54. 0.5% of $38,000

Applications

55. Jennifer Barolo works for Bartz Department store as a sales clerk. When items are on sale, she computes the amount saved. She also subtracts the amount saved from the regular price to find the sale price. Complete the computations for her.

E-Z Care Acrylic Thermal Blanket

On sale now at 20% off

a.	Twin, Reg. 15.00	Save ?	Sale Price ?
b.	Full, Reg. 18.00	Save ?	Sale Price ?
c.	Queen, Reg. 22.00	Save ?	Sale Price ?
d.	King, Reg. 26.00	Save ?	Sale Price ?

56. Use the chart at the right to find the postage and handling charges for the amounts listed.

a. $4.00 **b.** $6.00
c. $8.50 **d.** $7.95
e. $12.00 **f.** $24.00
g. $15.85 **h.** $23.40
i. $28.00 **j.** $35.00
k. $75.89 **l.** $110.50

TO FIGURE POSTAGE AND HANDLING ...add to order:

Sale items, total order:
$3.00 or less, add 75¢
$3.01-$10, add 15% of total
$10.01-$25, add 10% of total
Over $25, add 5% of total

57. Student Lisa Mario received these test scores. How many answers were correct on each test?

	Subject	Test Score	Number of Items
a.	Math	85%	60
b.	English	88%	50
c.	Science	75%	44
d.	Spanish	95%	40
e.	Government	80%	120

58. Use the tax table at the right to find the joint taxes paid for each of the incomes listed.

a. $24,000 **b.** $28,000
c. $32,000 **d.** $44,000
e. $47,500 **f.** $68,900
g. $88,000 **h.** $95,000

Taxable Income	Joint Tax Bracket
$20,201-$24,600	25%
24,601- 29,900	29%
29,901- 35,200	33%
35,201- 45,800	39%
45,801- 60,000	44%
60,001- 85,600	49%
85,601 and Up	50%

59. Wilma Epson is a stockbroker. She charges a percent of the dollar amount of stocks sold. Use the rate schedule and find how much she charges on these sales amounts. Round to the nearest cent.

a. $1800 **b.** $4000
c. $7,800 **d.** $13,800
e. $42,340 **f.** $45,980
g. $3,980.40 **h.** $8,942.24
i. $2,338.43 **j.** $17,876.40

Commission Rate Schedule

Dollar Amount	% of Dollar Amount
Stocks:	
$0—$2,499	1.7%, minimum $30
$2,500—$4,999	1.3%, minimum $42
$5,000—$9,999	1.1%, minimum $65
$10,000—$14,999	.9%, minimum $110
$15,000—$24,999	.7%, minimum $135
$25,000—$49,999	.5%, minimum $175
$50,000 and above	negotiated

60. Sales taxes are found by multiplying the tax rate times the selling price of the item. Find the sales tax for each selling price. Round to the nearest cent.

	Selling Price	Tax Rate
a.	$ 28.00	5.0%
b.	$ 45.00	6.0%
c.	$ 64.00	5.5%
d.	$ 75.00	6.5%
e.	$126.00	4.75%
f.	$540.00	4.125%
g.	$ 87.87	3.0%
h.	$322.65	7.425%

Workshop 11 Average (Mean)

▶ Look up Application Q on page 547 for more practice.

The average, or mean, is a single number used to represent a group of numbers. The average, or mean, of two or more numbers is the sum of the numbers divided by the number of items added.

Example 1 Find the average of 9, 8, 3, 6, and 4.

Solution $\frac{9 + 8 + 3 + 6 + 4}{5} = \frac{30}{5} = 6$

Example 2 Find the average of 573, 487, 628, and 526.

Solution $\frac{573 + 487 + 628 + 526}{4} = \frac{2214}{4} = 553.5$

Example 3 Find the average of 4.8, 5.7, 6.2, 3.9, and 4.1. Round to the nearest tenth.

Solution $\frac{4.8 + 5.7 + 6.2 + 3.9 + 4.1}{5} = \frac{24.7}{5} = 4.94 = 4.9$

Example 4 Find the average of \$17, \$24, \$38, \$23, \$19, and \$26. Round to the nearest dollar.

Solution $\frac{\$17 + \$24 + \$38 + \$23 + \$19 + \$26}{6} = \frac{\$147}{6} = \$24.50 = \$25$

A. 5
B. 5
C. 141
D. \$31.33

▶ **Self-Check** Cover the answers and complete the problems.

Find the average for each set of numbers. Round to the nearest cent.

A. 2, 7, 3, 5, 9, 4
B. 4.5, 3.2, 7.9, 6.1, 3.3
C. 124, 156, 143
D. \$24, \$38, \$29, \$33, \$32, \$32

Problems

Find the average for each group.

1. 6, 8, 9, 11, 16
2. 7, 5, 12, 10, 11
3. 70, 75, 80, 83
4. 34, 36, 32, 35
5. 397, 408, 416
6. 125, 132, 121
7. 876, 809, 854, 833
8. 326, 487, 231, 288
9. 2.1, 3.4, 1.7, 1.9, 2.4
10. 7.8, 7.6, 7.4, 7.5, 7.7
11. 2.4, 3.5, 1.8, 2.7, 2.0, 2.9
12. 6.1, 7.3, 5.8, 4.9, 7.6, 5.2
13. \$12, \$14, \$15, \$11, \$13
14. \$54, \$48, \$32, \$41, \$55
15. \$76, \$65, \$87, \$77, \$54, \$40
16. \$33, \$33, \$36, \$39, \$35, \$31

Find the average for each group. Round to the amount indicated.

17. 4.8, 4.3, 5.3, 3.9, 6.7, 5.9, 3.1 (nearest tenth)

18. 7.2, 5.6, 3.2, 4.9, 6.5, 5.8 (nearest tenth)

19. 43.1, 44.6, 45.9, 41.0, 47.8 (nearest tenth)

20. 21.7, 43.9, 56.1, 23.8 (nearest tenth)

21. 347.5, 398.2, 356.8, 297.1, 409.9 (nearest tenth)

22. 1.37, 2.64, 3.90, 2.74 (nearest hundredth)

23. 44.87, 32.90, 56.21, 56.34, 49.89 (nearest hundredth)

24. 0.05, 0.17, 0.23, 0.09, 0.31, 0.29 (nearest hundredth)

25. $34, $79, $65, $37 (nearest dollar)

26. $27.50, $34.54, $29.65 (nearest cent)

27. $341, $479, $562, $380, $427, $398 (nearest ten cents)

28. $350,000; $286,000; $325,000; $472,000; $249,000 (nearest thousand dollars)

29. $134.76, $123.86, $167.89, $156.54 (nearest cent)

30. $0.65, $0.85, $0.71, $0.77, $0.98 (nearest cent)

31. David Weir had bowling scores of 168, 192, and 156. What was his average?

32. Tina Brooks recorded her pulse rate on four occasions as follows: 78, 87, 92, 85. What was her average pulse rate, rounded to the nearest ten?

33. Rhonda Bricker's tips from waiting on tables were $1.25, $2.25, $0.75, $1.50, and $2.25. What was her average tip?

34. The number of checks Celinda Vasquez wrote during the last six months was 25, 32, 47, 29, 35, and 36. On the average, how many checks did she write each month?

35. Shelton Glashow's telephone bills for the last three months were $38.42, $42.76, and $29.85. What was the average bill?

36. Dianne Wiltrip recorded her math test scores this quarter. What is her average?

Test Number	1	2	3	4	5	6	7	8
Score	87	75	98	95	82	77	78	88

Applications

37. Blanche Keller sells clothes at the Vivinao Boutique. Her sales for the first six months of the year are given in the chart. Determine her average monthly sales to the nearest hundred dollars.

VIVINAO BOUTIQUE

Month	Sales
January	$12,400
February	11,320
March	14,760
April	10,270
May	9,320
June	6,720

38. Roger Humbert earns extra money each spring by roto-tilling gardens. His brother, Bill, went along to record the time spent doing each part of the job. Complete the chart in order to find the average time for each part of the job and the average time for each job. Round to the nearest minute.

Task	Job 1	Job 2	Job 3	Job 4	Job 5	Average
Unload tiller	2	3	2	2	3	?
Till garden	45	55	60	40	35	?
Clean-up tiller	10	8	12	7	5	?
Load tiller	4	3	4	5	5	?
Travel time	20	25	15	10	10	?
Average	?	?	?	?	?	

39. As captain of his bowling team, Burton McKaig has to complete this form after each match. Help him by computing the total and the average for each bowler. He also computes the total and the team average for each game. Round to the nearest whole number.

Bowler	Game 1	Game 2	Game 3	Total	Average
a. McKaig	150	170	164	?	?
b. Pegorsch	210	195	201	?	?
c. Hoskinson	165	170	174	?	?
d. Goldberg	234	190	175	?	?
e. Hallauer	146	201	156	?	?
Total	?	?	?		
Team Average	?	?	?		

40. Val Winslow sells cars at Schonfeld Auto Sales. He sold five cars last week for the amounts shown on the chart. Determine the average price of the cars he sold (round to the nearest ten dollars).

SALES CHART

	Model	Value
a.	JD-11	$12,465
b.	SR-200	9,231
c.	GT-40	8,764
d.	GT-40	13,327
e.	JD/SL	11,938

Workshop 12 Elapsed Time

▶ Look up Application F on page 541 for more practice.

To find elapsed time, subtract the earlier time from the later time.

Example Find the elapsed time for Barbara Gatley who worked from:

A. 3:15 p.m. to 7:30 p.m.

Solution

```
  7:30
– 3:15
  4:15  = 4 hours 15 minutes
          written as 4 h:15 min
```

B. 9:35 a.m. to 12:58 p.m.

Solution

```
  12:58
–  9:35
   3:23  = 3 h:23 min
```

You cannot subtract 55 minutes from 20 minutes unless you borrow an hour and add it to the 20 minutes. Remember that 1 hour = 60 minutes.

Example Find the elapsed time from 2:55 p.m. to 6:20 p.m.

Solution

```
  6:20  =    5:20 + :60  =    5:80   ← Borrowed 1 hour
– 2:55  =  – 2:55        =  – 2:55
                              3:25  = 3 h:25 min
```

A. 2 h:15 min
B. 3 h:10 min
C. 2 h:40 min
D. 2 h:43 min

▶ **Self-Check** Cover the answers and complete the problems.

Find the elapsed time for a person who worked from:

A. 3:40 p.m. to 5:55 p.m. **B.** 7:15 a.m. to 10:25 a.m.
C. 2:30 p.m. to 5:10 p.m. **D.** 8:32 a.m. to 11:15 a.m.

To find elapsed time when the time period spans 1 o'clock, add 12 hours to the later time before subtracting.

Example 1 Find the elapsed time from 10:45 a.m. to 2:58 p.m.

Solution

```
   2:58  =    2:58 + 12:00  =    14:58
– 10:45  =  – 10:45         =  – 10:45
                                  4:13  = 4 h:13 min
```

Example 2 Find the elapsed time from 10:50 p.m. to 6:25 a.m.

Solution

```
   6:25  =    18:25  =    17:25 + :60  =    17:85
– 10:50  =  – 10:50  =  – 10:50        =  – 10:50
                                             7:35  =
                                                     7h:35 min
```

A. 3 h:20 min
B. 6 h:5 min
C. 4 h:50 min
D. 4 h:45 min

▶ **Self-Check** Cover the answers and complete the problems.

Find the elapsed time for a person who worked from:

A. 10:15 a.m. to 1:35 p.m. **B.** 9:35 p.m. to 3:40 a.m.
C. 9:20 a.m. to 2:10 p.m. **D.** 11:35 a.m. to 4:20 p.m.

Problems

Find the elapsed time.

1. From 1:20 p.m. to 4:30 p.m.
2. From 3:30 p.m. to 7:40 p.m.
3. From 9:32 a.m. to 11:45 a.m.
4. From 8:15 a.m. to 11:25 a.m.
5. From 1:10 a.m. to 5:15 a.m.
6. From 10:20 p.m. to 11:30 p.m.
7. From 3:30 p.m. to 8:50 p.m.
8. From 4:46 a.m. to 10:50 a.m.
9. From 11:25 a.m. to 12:45 p.m.
10. From 10:05 a.m. to 12:55 p.m.
11. From 8:35 a.m. to 11:25 a.m.
12. From 9:20 a.m. to 11:15 a.m.
13. From 9:40 p.m. to 11:30 p.m.
14. From 7:25 a.m. to 11:15 a.m.
15. From 3:35 p.m. to 6:10 p.m.
16. From 2:45 p.m. to 7:10 p.m.
17. From 10:12 a.m. to 12:05 p.m.
18. From 8:12 p.m. to 11:01 p.m.
19. From 4:44 a.m. to 9:11 a.m.
20. From 9:50 a.m. to 12:10 p.m.
21. From 10:00 a.m. to 3:00 p.m.
22. From 9:30 a.m. to 1:45 p.m.
23. From 11:20 a.m. to 4:40 p.m.
24. From 8:09 a.m. to 4:50 p.m.
25. From 7:05 a.m. to 1:25 p.m.
26. From 7:25 a.m. to 3:40 p.m.
27. From 8:12 p.m. to 1:30 a.m.
28. From 11:45 a.m. to 2:15 p.m.
29. From 10:12 p.m. to 4:04 a.m.
30. From 9:56 p.m. to 3:27 a.m.
31. Maria Andella worked from 10:52 a.m. to 4:28 p.m. How long did she work?
32. Peter Martin took a bus that left Boston at 10:35 a.m. and arrived in New York City at 2:16 p.m. How long was the trip?
33. Determine the elapsed time for each flight.

AIRLINE SCHEDULE

	From:	To:	Departure Time	Arrival Time	Elapsed Time
a.	Toledo	Boston	7:00 a.m.	10:05 a.m.	?
b.	Chicago	Houston	7:00 a.m.	11:54 a.m.	?
c.	Los Angeles	San Francisco	8:12 a.m.	10:05 a.m.	?
d.	Jacksonville	Nashville	7:15 a.m.	9:18 a.m.	?
e.	Jacksonville	Newark	12:41 p.m.	4:35 p.m.	?
f.	Syracuse	Atlanta	5:02 p.m.	7:06 p.m.	?
g.	Providence	Orlando	3:29 p.m.	8:13 p.m.	?

Workshop 13 Tables and Charts

▸ Look up Application C on page 539 for more practice.

To read a table or chart, find the *column* containing one of the pieces of information you have. Look across the *row* containing the other piece of information. Read down the column and across the row. Read the information you need where the column and row cross.

ANY FRACTION OF A POUND OVER THE WEIGHT SHOWN TAKES THE NEXT HIGHER RATE							
WEIGHT NOT TO EXCEED	RATE CHART TO GROUND ZONES						
	2	3	4	5	6	7	8
1 lb.	$1.25	$1.28	$1.32	$1.36	$1.42	$1.48	$1.55
2 lb.	1.34	1.40	1.47	1.55	1.67	1.79	1.93
3 lb.	1.43	1.52	1.63	1.75	1.92	2.11	2.32
4 lb.	1.52	1.64	1.78	1.94	2.18	2.42	2.70
5 lb.	1.61	1.75	1.93	2.14	2.43	2.74	3.09
6 lb.	1.70	1.87	2.09	2.33	2.68	3.05	3.47
7 lb.	1.79	1.99	2.24	2.53	2.94	3.37	3.86
8 lb.	1.88	2.11	2.40	2.72	3.19	3.68	4.24
9 lb.	1.97	2.23	2.55	2.92	3.44	4.00	4.63
10 lb.	2.05	2.34	2.70	3.11	3.69	4.31	5.01
11 lb.	2.14	2.46	2.86	3.31	3.95	4.63	5.40
12 lb.	2.23	2.58	3.01	3.50	4.20	4.94	5.78
13 lb.	2.32	2.70	3.17	3.70	4.45	5.26	6.17
14 lb.	2.41	2.82	3.32	3.89	4.71	5.57	6.55
15 lb.	2.50	2.93	3.47	4.09	4.96	5.89	6.94

Example What is the cost to ship a 5-lb package to Zone 7?

Solution
1. Find the Zone 7 column.
2. Find the 5-lb row.
3. Read across the 5-lb row to the Zone 7 column. The cost is $2.74.

A. $1.25
B. $1.55
C. $2.70
D. $2.70

▸ **Self-Check** Cover the answers and complete the problems.

Find the cost to ship each package to the indicated zone.

A. 1 lb, Zone 2
B. 1 lb, Zone 8
C. 10 lb, Zone 4
D. 13 lb, Zone 3

To classify an item, find the row that contains the known data. Then read the classification from the head of the column.

Size Chart—Men's Sizes

Suits and Sportcoats Sizes 36 to 46. Order by chest size. Be sure waist will fit comfortably.	order size	36	37	38	39	40	42	44	46
	if chest is (inches)	35–36	36–37	37–38	38–39	39–40	41–42	43–44	45–46
	and waist is (inches)	28–31	29–32	30–33	31–34	32–35	34–37	36–39	38–41

Jackets Sizes 36–46 Order by chest size.	order size	36	38	40	42	44	46
	if chest is (inches)	34½–36	36½–38	38½–40	40½–42	42½–44	44½–46

Example What size suit should a man with a 41-inch chest order?

Solution
1. Find the "suits and sportcoats" section.
2. Find the row "if chest is (inches)."
3. Read across the row to "41–42."
4. Read the number at the head of the column (42). A man with a 41-inch chest should order a size 42 suit.

A. Size 36
B. Size 40

▸ **Self-Check** Cover the answers and complete the problems.

Determine what size garment should be ordered.

A. Suit, chest size 35 inches. B. Jacket, chest size 39 inches.

Problems

Use the chart on page 38 to find the cost to ship each package to the indicated zone.

1. 3 lb, Zone 3
2. 2 lb, Zone 6
3. 11 lb, Zone 5
4. 9 lb, Zone 8
5. 15 lb, Zone 4
6. 7 lb, Zone 2
7. 6.5 lb, Zone 4
8. 9.75 lb, Zone 7
9. 11.2 lb, Zone 8
10. 2.6 lb, Zone 6
11. 1.35 lb, Zone 2
12. 0.75 lb, Zone 5

Applications

13. Using the same shipping chart, determine the maximum weight a package can weigh.

	a.	b.	c.	d.	e.	f.
Shipping to Zone	2	4	5	8	7	3
Shipping Cost	\$1.34	\$2.86	\$3.70	\$5.01	\$5.57	\$2.11
Maximum Weight	?	?	?	?	?	?

14. Use the size chart on page 38 to determine what size garment should be ordered.

	a.	b.	c.	d.	e.	f.
Chest Size (inches)	37	42	45	36	40	38
Waist Size (inches)	29	35	39	32	35	34
Suit Order Size	?	?	?	?	?	?
Jacket Order Size	?	?	?	?	?	?

15. What size suit should a man with a 36-inch chest and a 31-inch waist order?

16. What size sportcoat should a man with a 39-inch chest and a 33-inch waist order?

Use the New England Weather chart to answer the following.

NEW ENGLAND WEATHER

Station	Today	High	Low	Tomorrow	High
Bangor	Cloudy	36	24	Snow	32
Bedford	Snow	31	27	"	31
Boston	"	33	29	"	33
Brockton	"	33	28	"	32
Framingham	"	31	26	"	29
Gloucester	"	33	30	"	33

17. What was the high temperature in Boston today?

18. What was the low temperature in Bedford today?

19. What will be the high temperature in Framingham tomorrow?

20. In which city was a low temperature of 28 degrees recorded?

21. In which city was the highest temperature recorded for today?

Use the federal income tax table to find the following.

FEDERAL WITHHOLDING TAX—WEEKLY Payroll Period, Employee MARRIED

And the wages are—		And the number of withholding allowances claimed is—							
At least	But less than	0	1	2	3	4	5	6	7
		The amount of income tax to be withheld shall be—							
$130	$135	$ 10.40	$ 8.10	$ 5.70	$ 3.40	$ 1.10	$ 0	$ 0	$ 0
135	140	11.00	8.70	6.30	4.00	1.70	0	0	0
140	145	11.60	9.30	6.90	4.60	2.30	0	0	0
145	150	12.20	9.90	7.50	5.20	2.90	.60	0	0
150	160	13.10	10.80	8.40	6.10	3.80	1.50	0	0
160	170	14.30	12.00	9.60	7.30	5.00	2.70	.40	0
170	180	15.50	13.20	10.80	8.50	6.20	3.90	1.60	0
180	190	16.70	14.40	12.00	9.70	7.40	5.10	2.80	.50
190	200	18.40	15.60	13.20	10.90	8.60	6.30	4.00	1.70
200	210	20.10	16.80	14.40	12.10	9.80	7.50	5.20	2.90
210	220	21.80	18.50	15.60	13.30	11.00	8.70	6.40	4.10
220	230	23.50	20.20	16.90	14.50	12.20	9.90	7.60	5.30
230	240	25.20	21.90	18.60	15.70	13.40	11.10	8.80	6.50
240	250	26.90	23.60	20.30	17.10	14.60	12.30	10.00	7.70
250	260	28.60	25.30	22.00	18.80	15.80	13.50	11.20	8.90

	22.	23.	24.	25.	26.	27.
Income	$251.50	$258.75	$162.31	$155.03	$222.00	$194.00
Allowances	1	3	0	2	4	2
Amount Withheld	?	?	?	?	?	?

	Number of Allowances	Tax Withheld	Income At least	Income But less than
28.	2	$ 8.40	?	?
29.	3	$15.70	?	?
30.	0	$11.60	?	?
31.	1	$21.90	?	?
32.	6	$ 7.60	?	?
33.	4	$ 8.60	?	?

Use the above table to find the amount of tax withheld.

34. Ralph Adams earns $173.00 and claims 2 allowances.
35. Barbara Knighten earns $162.29 and claims 1 allowance.
36. Luther Whittier earns $188.60 and claims 2 allowances.
37. Stacey Hagley earns $251.30 and claims 1 allowance.
38. Verna Denbridge earns $202.19 and claims 3 allowances.
39. Shirley Sagamore earns $196.78 and claims 0 allowances.

Workshop 14 Units of Measure

Here are the abbreviations and conversions for units of measure in the U.S. Customary System.

Length	Volume	Weight
12 inches (in) = 1 foot (ft)	2 cups (c) = 1 pint (pt)	16 ounces (oz) = 1 pound (lb)
3 ft = 1 yard (yd)	2 pt = 1 quart (qt)	2000 lb = 1 ton (t)
5280 ft = 1 mile (mi)	4 qt = 1 gallon (gal)	

Here are the symbols and conversions for units of measure in the metric system.

Length	Volume
1000 millimeters (mm) = 1 meter (m)	1000 milliliters (mL) = 1 liter (L)
100 centimeters (cm) = 1 m	**Mass**
1000 m = 1 kilometer (km)	1000 grams (g) = 1 kilogram (kg)

To convert from one unit of measure to another, use the conversion lists above.

When converting to a smaller unit, multiply.

Example Convert 5 ft to inches.

Solution Use 12 in = 1 ft
5 ft: 5 × 12 = 60
5 ft = 60 in

Example Convert 3 m to centimeters.

Solution Use 100 cm = 1 m
3 m: 3 × 100 = 300
3 m = 300 cm

When converting to a larger unit, divide.

Example Convert 10 pt to quarts.

Solution Use 2 pt = 1 qt
10 pt: 10 ÷ 2 = 5
10 pt = 5 qt

Example Convert 5000 g to kg.

Solution Use 1000 g = 1 kg
5000 g: 5000 ÷ 1000 = 5
5000 g = 5 kg

A. 84 in
B. 250 ml
C. 5 yd
D. 4 m

▸ **Self-Check** Cover the answers and complete the problems.

Convert:

A. 7 ft to inches
B. 0.25 L to milliliters
C. 15 ft to yards
D. 400 cm to meters

Problems

Do the following conversions.

1. 3 yds to feet
2. 10 gal to quarts
3. 5 lb to ounces
4. 5 ft to inches
5. 15 lbs to ounces
6. 3 L to milliliters
7. 2.5 km to meters
8. 15 pts to cups
9. 6.4 kg to grams
10. 72 in to yards
11. 12 qts to gallons
12. 32 oz to pounds

13. 10 qts to gallons
14. 23 oz to pounds
15. 4000 g to kilograms
16. 80 cm to meters
17. 2300 mL to liters
18. 540 cm to meters
19. 84.6 kg to grams
20. 8.2 L to milliliters
21. 843 g to kilograms
22. 15.432 mL to liters
23. 14 cm to millimeters
24. 983.2 cm to meters
25. 2 yd 4 in to inches
26. 3 ft 5 in to inches
27. 6 qt 1 pt to pints
28. 5 lb 7 oz to ounces
29. 3 gal 2 qt to quarts
30. 1 yd 2 ft 2 in to inches
31. 4 gal 2 qt 1 pt to pints
32. 3m 142 cm 45mm to millimeters

Applications

33. How many quarts will a 2-gallon plastic bag hold?
34. How many milliliters will a one-liter bottle hold?
35. How many cups of coffee does a 4-quart coffeepot hold?
36. How many cups of hot chocolate will a 5-gallon thermos jug hold?
37. How many inches long is a $6\frac{2}{3}$-yard roll of aluminum foil?
38. Cottage cheese is sold in 1-pint containers. How many pints must be bought to have enough for a recipe that calls for 5 cups?
39. Arthur McClelland knows that his jogging stride is about 1 meter long. The jogging track he uses is 3.9 kilometers long. How many strides does it take him to go around the track once?
40. The cafeteria receives 42 cases of milk each day. Each case contains 24 half-pint cartons. How many gallons of milk are received each day?
41. A soft drink is sold in 355 mL cans. How many liters are in a six-pack?
42. George Patroulus, a pastry chef, baked a walnut cake weighing 2.4 kilograms. How many 75-gram servings can be cut from the cake?
43. Louis Koenig ordered baseboard molding for the rooms of a new house. Louis needs to complete this chart to determine the total number of feet of molding needed. How much molding is needed?

	Length	Width	2 Lengths	+	2 Widths	=	Perimeter
	12 ft	9 ft	24 ft	+	18 ft	=	42 ft
a.	10 ft	8 ft	20 ft	+	16 ft	=	?
b.	18 ft	24 ft	?	+	?	=	?
c.	10 ft 6 in	9 ft 4 in	?	+	?	=	?
d.	11 ft 3 in	7 ft 8 in	?	+	?	=	?
e.	11 ft 4 in	13 ft 2 in	?	+	?	=	?
f.	17 ft 8 in	12 ft 9 in	?	+	?	=	?
g.	8 ft 5 in	7 ft 9 in	?	+	?	=	?
					Total	=	?

Part 2 Personal Business Mathematics

▸ Part 2, Personal Business Mathematics, covers the personal applications of basic math skills that you will use in your roles as worker and knowledgeable consumer. By learning how to handle your money wisely, you can use your income to meet day-to-day living expenses and still save and invest part of your income for use in the future. Eleven consumer topics are covered in Part 2.

▸ **The following definitions provide an overview of the important areas that affect how you earn, spend, and save your money.**

Gross Income The total amount of money you earn for work done during a pay period, before any deductions are taken.

Net Income Your take-home pay after all tax withholdings and personal deductions have been taken from your gross income.

Checking Accounts Funds deposited in a bank from which you can write checks to pay for goods and services instead of paying by cash.

Savings Accounts Funds deposited in a bank for which you earn interest.

Cash Purchases Items or services that you buy at regular or sale prices.

Charge Accounts and Credit Cards Systems of buying in which you receive goods and services now and spread the payments, along with finance charges, if any, over a period of time.

Loans Money that is borrowed and must be repaid along with interest owed.

Automobile Transportation 1. The costs of buying, insuring, and maintaining your own automobile. 2. The costs of leasing an automobile from a rental agency.

Housing Costs The costs of purchasing, owning, and maintaining your own home.

Insurance and Investments 1. Financial protection purchased for your dependents. 2. Sums of money paid out to gain future income for yourself.

Recordkeeping Maintaining accurate accounts of your expenses as tools for managing your money.

Unit 1 Gross Income

▸ Your *income,* the money that you earn on a job, may be determined in one of several ways. Income may be computed on hours worked, or *hourly rate,* the number of items produced, or *piecework,* or *commission.* Commission may be a specified amount of money you are paid for selling a product, or it may be a percent of the selling price. You may also be paid on a *graduated commission,* which increases as your sales increase. Many workers are paid a fixed annual *salary,* which they receive *weekly, biweekly, semimonthly,* or *monthly.*

The piecework pay system offers an incentive to the employee. The more the employee produces, the more he or she earns.

1-1 Hourly Pay

Objective Compute the straight-time pay.

For some jobs, you are paid a fixed amount of money for each hour you work. Your **hourly rate** is the amount of money you earn per hour. **Straight-time pay** is the total amount of money you earn for a pay period at the hourly rate.

▶ **Straight-time Pay = Hourly Rate × Hours Worked**

Example

Skills 14, 8, 2 **Application** A **Term** Straight-time pay

Gladys Marzetti is an inventory control clerk. She earns $4.30 per hour. Last week she worked her regular $37\frac{1}{2}$ hours. What is her straight-time pay for last week?

Solution

Find the **straight-time pay**.

Hourly Rate × Hours Worked

$4.30 × $37\frac{1}{2}$

$4.30 × 37.5 = $161.25 straight-time pay

A. $347.10
B. $361.90

▶ **Self-Check** Cover the answers and complete the problems.

Find the straight-time pay.

A. $8.90 per hour × 39 hours

B. $9.40 per hour × $38\frac{1}{2}$ hours

Problems

	Employee	Regular Hourly Rate	×	Hours Worked	=	Straight-time Pay
1.	Batson, S.	$ 7.00	×	40	=	?
2.	Favreau, E.	$ 4.00	×	18	=	?
3.	Levitt, R.	$ 6.20	×	30	=	?
4.	Maguire, M.	$ 4.50	×	$24\frac{1}{2}$	=	?
5.	Scanlon, A.	$12.50	×	$31\frac{1}{4}$	=	?
6.	Vadola, G.	$11.00	×	$25\frac{3}{4}$	=	?

7. You earn $6.10 per hour.
Worked 30 hours last week.
What was your straight-time pay?

8. You earn $9.50 per hour.
Worked 30 hours last week.
What was your straight-time pay?

9. Sylvia Barrett.
Financial consultant.
Earns \$18.50 per hour.
Worked $22\frac{1}{2}$ hours last week.
What was her straight-time pay?

10. Gloria Todd.
Office clerk.
Earns \$3.60 per hour.
Worked $35\frac{3}{4}$ hours last week.
What was her straight-time pay?

11. Chris Howard is a veterinary aide. She earns \$3.80 per hour. Last week she worked 34 hours. What was her straight-time pay for last week?

12. Jan Edwards has a part-time job at a delicatessen. She earns \$3.05 per hour. She works 21 hours each week. What is her straight-time pay for a week?

13. Del Martin, a construction worker, earns \$8.00 per hour. He worked his regular $36\frac{1}{4}$ hours last week. What was his straight-time pay?

14. Arthur Anderson is a mail clerk at the Donovan Marketing Corporation. He works 27 hours per week and earns \$3.72 per hour. What is his straight-time pay each week?

15. Eve Wright, a guitar player for a local band, earns \$5.25 per hour. What is her straight-time pay for a week in which she worked 32 hours?

16. George Billings earns \$6.64 per hour as an automotive claims adjustor. What is his straight-time pay for this week?

Mon.	Tues.	Wed.	Thurs.	Fri.
8 hr	5 hr	7 hr	9 hr	8 hr

17. Ben Reyes was hired for the position of billing clerk. During his first week on the job, he worked $7\frac{1}{2}$ hours each day, Monday through Friday. What was his total pay for the week?

PRINTING CO. has opening for bright dependable billing clerk. Must have aptitude for figures and pay attention to detail. Prepare work for production. Start \$5.50 per hr. Call Steven, at 700-9256 Ext. 11

Maintaining Your Skills

Look up the skills in parentheses if you need help or more practice.

Write as decimals. (Skill 14)

18. $\frac{1}{4}$ 19. $\frac{1}{2}$ 20. $\frac{3}{4}$ 21. $\frac{1}{10}$ 22. $\frac{3}{8}$

Multiply. Round answers to the nearest cent. (Skill 8)

23. \$4.50 × 30.25 24. \$8.30 × 3.25 25. \$14.50 × 32.5

Round to the place value indicated. (Skill 2)

Nearest ten	26. 22	27. 138	28. 8
Nearest hundred	29. 91	30. 3092	31. 14,431
Nearest thousand	32. 15,374	33. 722	34. 158,481

1-2 Overtime Pay

Objective Compute the straight-time, overtime, and total pay.

When you work more than your regular hours, you may receive **overtime pay**. The overtime rate may be $1\frac{1}{2}$ times your regular hourly rate. This is called **time and a half**. You may receive **double time**, 2 times your regular hourly rate, for overtime on Sundays or holidays.

- **Overtime Pay = Overtime Rate × Overtime Hours Worked**
- **Total Pay = Straight-time Pay + Overtime Pay**

Example

Skill 5 Application A Term Overtime pay

Ramon Hernandez is paid $6.20 an hour for a regular 40-hour week. His overtime rate is $1\frac{1}{2}$ times his regular hourly rate. This week Ramon worked his regular 40 hours plus 10 hours of overtime. What is his total pay?

Solution

1. Find the **straight-time pay.**
   Hourly Rate × Regular Hours Worked
   $6.20 × 40 = $248.00
2. Find the **overtime pay.**
   Overtime Rate × Overtime Hours Worked
   ($1\frac{1}{2}$ × $6.20) × 10 = $93.00
3. Find the **total pay.**
   Straight-time Pay + Overtime Pay
   $248.00 + $93.00 = $341.00 total pay

A. $332.50
B. $603.20

Self-Check Cover the answers and complete the problems.

Find the total pay.

A. $7.00 per hour for 40 hours. Time and a half for 5 hours.

B. $11.60 per hour for 40 hours. Time and a half for 8 hours.

Problems

	1.	2.	3.	4.	5.	6.
Hourly Pay (40 hr)	**$8.00**	**$10.20**	**$3.60**	**$4.50**	**$15.00**	**$18.50**
Straight-time Pay	?	?	?	?	?	?
Overtime Rate	$1\frac{1}{2}$	$1\frac{1}{2}$	$1\frac{1}{2}$	2	$1\frac{1}{2}$	2
Overtime Hours	10	5	2	8	13	4
Overtime Pay	?	?	?	?	?	?
Total Pay	?	?	?	?	?	?

7. George Keller, machine operator.
Regular hourly rate \$7.60.
Time and half for overtime.
Worked 37 regular hours.
8 hours overtime.
What is his overtime pay?
What is his total pay?

8. Louise Amato, programmer.
Regular hourly rate \$9.15.
Double time for overtime.
Worked 40 regular hours.
12 hours overtime.
What is her overtime pay?
What is her total pay?

9. Pat Connaly is paid \$6.00 an hour for a regular 35-hour week. Her overtime rate is $1\frac{1}{2}$ times her regular hourly rate. This week Pat worked her regular 35 hours plus 6 hours of overtime. What is her total pay?

10. As a maintenance engineer, Alice Power earns \$6.80 an hour plus double time for weekend work. Last week she worked her regular 40 hours plus 12 hours of overtime on the weekend. What was her total pay for the week?

11. Judy Sweeney is a gas station attendant. She earns \$3.25 an hour for a regular $36\frac{1}{2}$-hour week. She earns double time for work on Sundays. Last week Judy worked her regular hours plus 7 hours on Sunday. What was her total pay for the week?

GAS STATION ATTENDANT
\$3.25/hr
Double time weekends

12. Jeb Cantlin installs swimming pools. He is paid \$5.27 an hour for a 40-hour week and time and a half for overtime. What is Jeb's total pay for a week in which he worked 52 hours?

13. Louis Kellogg earns a regular hourly rate of \$7.56. He earns time and a half for overtime work on Saturdays and double time for overtime on Sundays. This week Louis worked 40 hours from Monday through Friday, 8 hours on Saturday, and 6 hours on Sunday. What is his total pay for the week?

Maintaining Your Skills

Look up the skills in parentheses if you need help or more practice.

Add. (Skill 5)

14. $\begin{array}{r} \$42.50 \\ +\ \ 24.76 \\ \hline \end{array}$

15. $\begin{array}{r} \$160.45 \\ +\ \ 86.90 \\ \hline \end{array}$

16. $\begin{array}{r} \$10.62 \\ 9.34 \\ +\ \ 0.45 \\ \hline \end{array}$

17. $\begin{array}{r} 56.7 \\ 0.007 \\ 121.33 \\ +\ 1568.125 \\ \hline \end{array}$

Multiply. (Skills 8, 14)

18. $(1\frac{1}{2} \times \$8.40) \times 7$
19. $(2 \times \$18.40) \times 12$
20. $(1\frac{1}{4} \times \$6.00) \times 9$
21. $(1\frac{1}{2} \times \$12.50) \times 8$
22. $(2 \times \$4.50) \times 6$
23. $(1\frac{1}{4} \times \$10.00) \times 10$

1-3 Weekly Timecard

Objective

Compute the total hours on a weekly timecard.

When you work for a business that pays you on an hourly basis, you are usually required to keep a timecard. A **weekly timecard** shows the time you reported for work and the time you departed each day of the week. The hours worked are computed for each day. The daily hours are added to arrive at the total hours for the week. The hours worked each day must be rounded to the nearest quarter hour.

▸ **Total Hours = Sum of Daily Hours**

Example

Skills 2, 16 **Application** E, F **Term** Weekly timecard

Dorothy Kaatz is required to keep a weekly timecard. What are her daily hours for 9/26? What are her total hours for the week?

TEMPORARY EMPLOYEE TIMECARD	DATE	IN	OUT	IN	OUT	HOURS
	9/23	8:30	12:00	12:37	5:25	8 1/4
NAME: DOROTHY KAATZ	9/24	8:25	12:15	12:45	4:33	7 1/2
DEPT: CREDIT	9/25	8:28	12:05	12:20	5:10	8 1/4
	9/26	8:15	1:30	2:15	4:57	
Dorothy Kaatz	9/27	8:30	12:10	12:40	5:02	8
EMPLOYEE SIGNATURE					TOTAL HOURS	

Solution

1. Find the hours worked on 9/26.
   a. Time between 8:15 A.M. and 1:30 P.M.
      (1:30 + 12:00) − 8:15
      13:30 − 8:15 . . . . 5h: 15 min
   b. Time between 2:15 P.M. and 4:57 P.M.
      4:57 − 2:15 . . . . 2h: 42 min
      5h: 15 min + 2h: 42 min . . . . 7h: 57 min
      Rounded to nearest quarter hour = 8 hours

2. Find the **total hours.**
   Sum of daily hours
   $8\frac{1}{4} + 7\frac{1}{2} + 8\frac{1}{4} + 8 + 8$
   $8\frac{1}{4} + 7\frac{2}{4} + 8\frac{1}{4} + 8 + 8 = 39\frac{4}{4} = 40$ total hours

A. $8\frac{1}{2}$

B. $7\frac{1}{2}$

▸ **Self-Check** Cover the answers and complete the problems.

A. Rod Kern: Worked from 8:00 to 11:45 and from 12:30 to 5:15. Find the total hours.

B. Jon Derr: Worked from 8:30 to 11:55 and from 1:00 to 5:02. Find the total hours.

Problems

Kim Fong works as a data-entry clerk in the data processing department at Partten Corp. in Houston, Texas. He is required to keep a weekly timecard. What are his daily hours for each day? What are his total hours for the week?

	TEMPORARY EMPLOYEE TIMECARD			NAME: Kim Fong DEPARTMENT: Data Processing		
	Date	**In**	**Out**	**In**	**Out**	**Hours**
1.	11/3	8:30	12:00	12:30	4:30	?
2.	11/4	8:25	12:30	1:05	5:10	?
3.	11/5	8:35	12:40	1:15	5:05	?
4.	11/6	8:15	12:10	12:45	4:15	?
5.	11/7	8:20	1:15	1:50	5:30	?
6.					Total Hours	?

Harry Fellauer works for Whitman Auto Sales. Use his timecard below to find his total hours worked each day and his total hours for the week.

	EMPLOYEE TIMECARD WHITMAN AUTO SALES				NAME: Harley Fellauer DEPARTMENT: Automotive Parts		
	Day	**Date**	**In**	**Out**	**In**	**Out**	**Total**
7.	Mon	5/7	7:00	11:00	11:30	4:20	?
8.	Tue	5/8	7:10	11:20	12:05	5:15	?
9.	Wed	5/9	6:50	11:15	11:55	4:10	?
10.	Thur	5/10	6:58	10:45	11:10	3:50	?
11.	Fri	5/11	7:30	11:49	12:30	5:12	?
12.	Sat	5/12	7:05	11:45	—	—	?
13.						Total Hours	?

14. Dorothy Kaatz earns $6.90 per hour. Use her timecard on page 50 to find her total pay for the week.

15. Kim Fong is paid $6.80 for a regular 37-hour week. His overtime rate is $1\frac{1}{2}$ times his regular hourly rate. Use his timecard above to find his total pay for the week.

16. Harley Fellauer is paid $8.20 for a regular 40-hour week. His overtime rate is $1\frac{1}{2}$ his regular hourly rate. Use his timecard above to find his total pay for the week.

(Continue on next page.)

17. You work from 7:00 a.m. to 11:00 a.m. and from 12:00 noon to 4:00 p.m. You are not allowed to work overtime. Your hourly rate is $6.50. Find the hours worked and the total pay for the week.

Day	In	Out	In	Out	In	Out	Hrs.
Sun.							?
Mon.	6:59	11:01	12:02	4:01			?
Tues.	7:06	11:00	11:59	3:59			?
Wed.	7:01	10:58	12:04	4:02			?
Thur.	7:00	11:01	12:09	3:30			?
Fri.	6:59	10:58	11:56	3:59			?
Sat.							

18. A portion of Greg Wilkowski's earnings statement is shown below. Find the number of hours he worked each day.

Day	In	Out	In	Out	In	Out	Hrs.
Sun.	12:30	6:15					?
Mon.	1:00	5:30	6:30	9:10			?
Tues.							?
Wed.	9:45	11:59	1:20	6:10	7:30	9:12	?
Thur.	10:02	12:15	1:28	5:43			?
Fri.	1:20	6:03	7:15	9:30			?
Sat.	9:58	12:15	1:40	6:00			?

19. In Problem 18, how many hours did Greg work on Saturday and Sunday? How many hours did he work Monday through Friday? If he earns $6.20 per hour plus time-and-a-half for work on Saturday and Sunday, what is his total pay for the week?

Maintaining Your Skills

Look up the skills in parentheses if you need help or more practice.

Add. (Skill 16)

20. $\frac{1}{4} + \frac{1}{4} + \frac{1}{2}$ 21. $\frac{3}{4} + \frac{1}{8}$ 22. $\frac{3}{8} + \frac{1}{4} + \frac{1}{4}$ 23. $\frac{5}{8} + \frac{1}{3}$

Round to the nearest quarter hour. (Application E)

24. 8:09 25. 7:47 26. 11:55 27. 3:39 28. 5:23

Round to the place value indicated. (Skill 2)

Nearest thousand 29. 39,972 30. 21,944 31. 68,498

Nearest hundred 32. 842 33. 257 34. 3580

1-4 Piecework

Objective Compute the total pay on a piecework basis.

Some jobs pay on a **piecework** basis. You receive a specified amount of money for each item of work that you complete.

▶ **Total Pay = Rate per Item × Number Produced**

Example

Skill 8 **Application** A **Term** Piecework

Bill Learner works for National Chrome Plating. He is paid 7¢ for each trailer hitch he plates. What is his pay for a week in which he plates 4300 hitches?

Solution

Find the **total pay.**

Rate per Item × Number Produced

$0.07 × 4300 = $301.00 total pay

A. $483.00
B. $427.20

▶ **Self-Check** Cover the answers and complete the problems.

Find the total pay.

A. $2.30 per item × 210 items produced.

B. 12¢ per item × 3560 items produced.

Problems

	Employee	Rate per Item	×	Number Produced	=	Total Pay
1.	Aymes, A.	$0.15	×	780	=	?
2.	Campbell, J.	$0.60	×	190	=	?
3.	Farrell, K.	$0.45	×	250	=	?
4.	Gidley, L.	$0.90	×	160	=	?
5.	Jackson, L.	$0.26	×	508	=	?
6.	Medaglia, A.	$0.18	×	713	=	?
7.	Preston, S.	$1.24	×	97	=	?
8.	Stang, D.	$0.79	×	188	=	?

9. Jules Forrest.
Punch press operator.
Rate per item is $0.24.
Produced 560 items.
Find his total pay.

10. Ellen Wo.
Industrial spray painter.
Rate per piece is $2.35.
Sprayed 76 pieces.
Find her total pay.

11. Albert DelNote.
Plater of silverware.
Rate per item is $0.87.
Plated 173 items.
What is his total pay?

12. Joseph Molloy.
Shirt stitcher.
Rate per shirt is $0.45.
Stitched 366 shirts.
What is his total pay?

13. Inga Halston delivers promotional packets to town residents after school. She is paid 6.5¢ for each packet she delivers. What is Inga's total pay for a week in which she delivers 1300 packets?

14. Robert Marsh makes thermocouple probes for Central Electronics. He is paid $0.54 for every probe he makes. What is Robert's pay for a week in which he makes 300 probes?

15. Helen Angel works as a punch press operator at a tool and die company. She earns $1.25 for every three-dimensional molding she presses. What is her total pay for a week in which she presses 173 moldings?

16. Fred Sanchez makes machine gears at Westside Machine Shop. He is paid 76¢ for each gear he makes. Last week Fred made a total of 392 gears. What was his total pay for the week?

17. Marcia Resnick works as a riveter. She is paid 3¢ for each rivet she puts into place. What is her total pay for a week in which she placed the following number of rivets?

Mon.	Tues.	Wed.	Thur.	Fri.
1500	1972	1898	2243	1189

18. Carol Ying assembles wires for flex leads. She is paid $0.33 for each unit assembled during a regular work week, $0.50 for each unit assembled on Saturdays, and $0.66 for each unit assembled on Sundays. What is Carol's total pay for a week in which she assembles the following number of units?

Mon.	Tues.	Wed.	Thur.	Fri.	Sat.	Sun.
148	150	0	151	148	80	75

Maintaining Your Skills

Look up the skills in parentheses if you need help or more practice.

Multiply. Round answers to the nearest cent. (Skill 8)

19. $0.23 × 89
20. $1.10 × 240
21. $0.06 × 4192
22. $5.20 × 23
23. $0.04 × 3200
24. $0.66 × 350

Multiply. (Skill 8)

25. 1.17 × 100
26. 15.876 × 100
27. 242 × 0.04
28. 456 × 1000

1-5 Salary

Objective Compute the salary per pay period.

A **salary** is a fixed amount of money that you earn on a regular basis. Your salary may be paid weekly, biweekly, semimonthly, or monthly. Your annual salary is the total salary you earn during a year. There are 52 weekly, 26 biweekly, 24 semimonthly, and 12 monthly pay periods per year.

$$\text{Salary per Pay Period} = \frac{\text{Annual Salary}}{\text{Number of Pay Periods per Year}}$$

Example

Skill 11 **Application** K **Term** Salary

Rose Swanton is an accountant. Her annual salary is $16,320. What is her semimonthly salary?

Solution

Find the **salary per pay period.**

Annual Salary ÷ Number of Pay Periods per Year

$16,320.00 ÷ 24 = $680.00 semimonthly salary

A. $1300
B. $350

Self-Check Cover the answers and complete the problems.

A. Find the biweekly salary.
$33,800 ÷ 26 = ?

B. Find the weekly salary.
$18,200 ÷ 52 = ?

Problems

	Employee	Pay Period	Annual Salary	÷	Pay Periods per Year	=	Salary per Pay Period
1.	B. G. Arn	semimonthly	$14,400	÷	24	=	?
2.	R. J. Blue	weekly	$15,600	÷	52	=	?
3.	C. L. Crane	monthly	$15,900	÷	12	=	?
4.	W. H. Dunn	semimonthly	$31,650	÷	?	=	?
5.	J. R. Elgin	weekly	$26,598	÷	?	=	?
6.	T. C. Mendel	biweekly	$42,293	÷	?	=	?

7. You are an accountant.
Annual salary is $22,100.
What is your weekly salary?

8. You are a bus driver.
Annual salary is $18,360.
What is your semimonthly salary?

9. Sue Fung is an insurance adjuster. Her annual salary is $14,620. What is her semimonthly salary?

10. Paul McArdle is a statistical analyst. His annual salary is $21,580. What is his biweekly salary?

11. Allison Keller was just hired as an electrical engineer for the Howwel Company. Her starting salary is $23,400 per year. What is her monthly salary?

12. Cleo Harlston works as a key-tape operator. Her annual salary is $12,350. What is her weekly salary?

13. Lou Wilder is a personnel recruiter. His annual salary is $16,855. Lou is paid on a monthly basis. What is his monthly salary?

14. Jim Marks qualifies for the technical services position described in the advertisement. If he is hired, what will his monthly salary be?

TECHNICAL SERVICES CLERK
$7600 annual salary. Basic Math background helpful. Call Creative Research, 700-1212 or come to 3540 Lincoln Rd. Suite 3100.

15. Roland James is currently earning an annual salary of $14,040 at Mudgett Electronics. He has been offered a job at Delta Tech at an annual salary of $15,236. How much more would Roland earn per week at Delta Tech than at Mudgett Electronics?

16. When Gary Snell first started working at Mid-Atlantic Warehousing, he was paid a biweekly salary of $576. Mid-Atlantic is now converting to a computerized payroll system and will be paying its employees on a semimonthly basis. What will Gary's semimonthly salary be?

17. Geraldine Piela earns a weekly salary of $345 at Howard's Department Store. Next month she will be promoted from assistant buyer to buyer. In her new position, she will be paid $795 semimonthly. How much more per year will Geraldine earn as a buyer than as assistant buyer?

Maintaining Your Skills

Look up the skills in parentheses if you need help or more practice.

Divide. Round answers to the nearest hundredth. (Skill 11)

18. $50\overline{)\$14,290}$ 19. $8.6\overline{)41.62}$ 20. $14.7\overline{)191.3}$

21. $0.032\overline{)17.9}$ 22. $0.89\overline{)2.86}$ 23. $0.252\overline{)6.9146}$

Find the number of pay periods. (Application K)

24. Weekly for 2 years 25. Semimonthly for 3 years 26. Biweekly for 4 years

1-6 Commission

Objective Compute the straight commission and determine the gross pay.

A **commission** is an amount of money that you are paid for selling a product or service. Your **commission rate** may be a specified amount of money for each sale or it may be a percent of the total value of your sales. If the commission is the only pay you receive, you work on **straight commission.**

▸ **Commission = Commission Rate × Total Sales**

Example

Skills 30, 2 **Application** A **Term** Commission

Chris Jackson sells shoes at an $8\frac{1}{2}\%$ straight commission. Last month his sales totaled \$9000. What was his commission?

Solution

Find the **commission.**

Commission Rate × Total Sales

$8\frac{1}{2}\%$ × \$9000

0.085 × \$9000 = \$765.00 straight commission

A. \$325.00
B. \$733.50

▸ **Self-Check** Cover the answers and complete the problems.

Find the commission.

A. 5% commission rate × \$6500.

B. $4\frac{1}{2}\%$ commission rate × \$16,300.

Instead of working only on commission, you may be guaranteed a minimum weekly or monthly salary. The commission you earn during a week or month is compared with your minimum salary. Your gross pay is the greater of the two amounts.

▸ **Gross Pay = Salary or Commission**

Example

Skill 30 **Application** A **Term** Commission

John Navarro is guaranteed a minimum salary of \$200 a week, or 7% of his total sales, whichever is greater. What is his gross pay for a week in which his total sales were \$3542?

Solution

1. Find the **commission.**
   Commission Rate × Total Sales
   7% × \$3542.00 = \$247.94

2. Find the **gross pay.**
   Salary or Commission
   \$200.00 or \$247.94 . . . \$247.94 gross pay (greater amount)

A. $180.00
B. $1937.50

▸ **Self-Check** Cover the answers and complete the problems.

A. Minimum salary: $180.
Commission: $5\frac{1}{2}$% on $3200.

B. Minimum salary: $1200.
Commission: $6\frac{1}{4}$% on $31,000.

Problems

	Position	Commission Rate	×	Total Sales	=	Commission
1.	Fabric Sales	10%	×	$ 830	=	?
2.	Business Machine Sales	4%	×	$21,300	=	?
3.	Major Appliance Sales	13%	×	$ 7130	=	?
4.	Book Sales	7%	×	$ 420	=	?
5.	Office Supplies Sales	15%	×	$ 645	=	?
6.	Roofing Contract Sales	$11\frac{1}{2}$%	×	$ 900	=	?
7.	Carpet Sales	$5\frac{1}{2}$%	×	$16,210	=	?

	Salesperson	Minimum Monthly Salary	Commission Rate	×	Total Monthly Sales	=	Monthly Commission	Gross Pay
8.	Janet North	$2200	7.5%	×	$35,000	=	?	?
9.	Clarence Ming	$1500	6.5%	×	$27,000	=	?	?
10.	Peter Jackle	$3000	5.75%	×	$22,000	=	?	?
11.	Sue Foster	$900	8.5%	×	$12,000	=	?	?
12.	Willis Johns	$2000	5.5%	×	$15,000	=	?	?
13.	Claire Wang	$3500	7.5%	×	$30,000	=	?	?

14. Sherman Allery.
3% commission on sales.
$6500 in sales for a week.
What is his commission?

15. Mike Vincent.
$6\frac{1}{2}$% commission on sales.
$3650 in sales for a week.
What is his commission?

16. Sally Jablonski.
Minimum weekly salary is $340.
Rate of commission is 6.5%.
Weekly sales are $4320.
What is the commission?
What is the gross pay?

17. Irvin Teal.
Minimum weekly salary is $160.
Rate of commission is 6.5%.
Weekly sales are $3320.
What is the commission?
What is the gross pay?

18. Doreen Steele sells used cars for Bonded Auto. She receives a straight commission of 5% of the selling price of each car. What commission will she receive for selling a $2495 compact?

BONDED AUTO SALES
We are looking for someone to sell used cars. Earn 5% commission on every car you sell. Call 700-8081

19. Fred Lee sells real estate. His commission is $5\frac{3}{4}$% of the price of every house he sells. What is his commission when he sells a $65,750 house?

20. Auto mechanics at Brent's Garage receive a straight commission of $33\frac{1}{3}$% of the service income they bring in each week. What is the commission for a mechanic who brings in $1965 service income in a week?

21. Herman Schmidt is a salesperson in the toy department at Morris. He is guaranteed a minimum salary of $185 per week, or 6.5% of his total sales, whichever is greater. What is his gross pay for a week in which his total sales were $3015?

22. Eileen Damauer sells cosmetics for Golden Tone, Inc. She is guaranteed a salary of $860 a month, or $7\frac{1}{4}$% of her total sales, whichever is greater. What is her gross pay for a month in which her total sales were $10,987?

23. Wanda Gomez is paid either a commission or $3.50 per hour plus time-and-a-half overtime for all hours over 8 per day, whichever is greater. Her commission consists of $4\frac{1}{2}$% of sales. Find her gross pay for a week in which she worked 6 hours on Monday, 8 hours on Tuesday, 10 hours on Wednesday, 6 hours on Thursday, 10 hours on Friday, and 9 hours on Saturday. Her total sales for the week were $3800.

*24. **A Brief Case** Some jobs pay a commission plus a bonus at the end of the year. The bonus may be a percent of the salesperson's total commission for the year.

a. Madelyn Carr is a sales representative. She receives a 6% commission on all sales. At the end of the year she receives a bonus of 5% of her commission. What is her total pay for a year in which she had sales totaling $244,550?

b. What would Madelyn's total pay be if her sales were $318,700?

Maintaining Your Skills

Look up the skills in parentheses if you need help or more practice.

Write as decimals. Round to the nearest hundredth. (Skill 14)

25. $2\frac{1}{8}$ 26. $5\frac{1}{2}$ 27. $6\frac{3}{4}$ 28. $4\frac{1}{3}$ 29. $9\frac{1}{4}$

Find the percentage. (Skill 30)

30. 4% of $1250 31. $8\frac{1}{2}$% of $4300 32. $7\frac{1}{4}$% of $8200

33. 7.3% of 120 34. 8.92% of 1380 35. 39% of 281

Round to the nearest tenth. (Skill 2)

36. 0.081 37. 0.608 38. 0.92 39. 0.821 40. 0.22

1-7 Graduated Commission

Objective Compute the total graduated commission.

Your commission rate may increase as your sales increase. A **graduated commission** offers a different rate of commission for each of several levels of sales. It provides an extra incentive to sell more.

▶ **Total Graduated Commission = Sum of Commissions for All Levels of Sales**

Example

Skill 5, 28, 30 **Application** A **Term** Graduated commission

Jean Beaudry sells appliances at Four City Sales. She receives a graduated commission of 14% on her first $1000 of sales, 16% on the next $1500, and 18% on sales over $2500. Jean's sales for the past month totaled $4578. What is her commission for the month?

Solution

Find the **sum of commissions for all levels of sales.**

Commission Rate × Sales

1. First $1000: 14% of $1000 = $140.00
2. Next $1500: 16% of $1500 = $240.00
3. Over $2500: 18% of ($4578 − $2500)
   18% of $2078 = $374.04

Total Graduated Commission $754.04

A. $4200.00
B. $910.00

▶ **Self-Check** Cover the answers and complete the problems.

A. Commission: 10% on first $6000; 15% over $6000. Find the total graduated commission on $30,000.

B. Commission: 5% first $2000; 12% over $2000. Find the total graduated commission on $8750.

Problems

	1.	2.
Amount of Sales	$2800	$3200
Commission		
First $2000: 10%	?	?
Over $2000: 15%	?	?
Total Commission	?	?

	3.	4.
Amount of Sales	$4000	$4750
Commission		
First $1500: 7%	?	?
Next $2000: 12%	?	?
Over $3500: 15%	?	?
Total Commission	?	?

5. Ralph Overton, part-time sales.
30% commission on first \$500.
$35\frac{1}{2}$% on next \$900.
40% on sales over \$1400.
\$1895 total sales.
What is his total commission?

6. Shelly Gray, sporting goods sales.
5% commission on first \$1000.
$7\frac{1}{2}$% on next \$2000.
10% on amount over \$3000.
\$7325 total sales.
What is her total commission?

7. Irene Tomas sells office machines. She receives a graduated commission of 7% on her first \$2000 of sales and $8\frac{1}{2}$% on all sales over \$2000. Irene's sales for the past week totaled \$2925. What is her commission for the week?

8. Charles Westerly sells computer hardware for a computer firm. He is paid a 4% commission on the first \$6000 of sales, 6% on the next \$4000, and 8% on sales over \$10,000. What is his commission on \$13,550 in sales?

9. William Miller demonstrates cookware at the National Home Show. He is paid \$5.00 each for the first 10 demonstrations in one day and \$7.50 for each demonstration over 10. What is William's commission for a day in which he makes 22 demonstrations?

10. Mary Dees demonstrates home fire escape systems. She is paid \$8.00 per demonstration for the first 10 demonstrations given during a week and \$12.50 for each demonstration over 10. What is her commission for a week in which she gives 18 demonstrations?

* 11. **A Brief Case** Some sales positions pay a commission only if sales exceed a **sales quota.** The salesperson is rewarded only for having sales beyond an expected amount.

Barbara Evans is a manager trainee. She is paid \$750 per month plus a commission of 3.5% on all sales over a quota of \$5000 per month. What is Barbara's total pay for a month in which she has sales totaling \$6578?

Maintaining Your Skills

Look up the skills in parentheses if you need help or more practice.

Add. (Skill 5)

12. \$123.72 + 112.69 + 45.23

13. \$567.89 + 34.69 + 431.73

14. \$1350.23 + 946.00 + 18.11

15. 0.007 + 1.384 + 0.569

Write as decimals. (Skill 28)

16. $5\frac{1}{2}$% 17. $10\frac{1}{4}$% 18. 15% 19. 97% 20. 342%

Find the percentage. (Skill 30)

21. $5\frac{1}{4}$% of \$2000 22. 7% of \$4560 23. $15\frac{1}{2}$% of \$3500

Reviewing the Basics

Skills

Round to the nearest cent.

(Skill 2) **1.** \$45.672 **2.** \$469.3591 **3.** \$0.045 **4.** \$53.746

Solve.

(Skill 5) **5.** \$672.01 + \$45.32 **6.** 0.004 + 1.5 + 19.03

(Skill 8) **7.** \$45.00 × 0.055 **8.** 11.563 × 100 **9.** 0.007 × 0.03

(Skill 11) **10.** \$23,660 ÷ 26 **11.** \$55,920 ÷ 12 **12.** 27.3 ÷ 0.05

(Skill 16) **13.** $5\frac{1}{4} + 7\frac{1}{2} + 6\frac{3}{4}$ **14.** $5\frac{1}{2} + 2\frac{2}{3} + 7\frac{5}{6}$

(Skill 30) **15.** $15\frac{1}{2}\%$ of \$650 **16.** 110% of \$310 **17.** $\frac{1}{4}\%$ of \$24

Write as a decimal.

(Skill 14) **18.** $\frac{1}{2}$ **19.** $\frac{1}{4}$ **20.** $\frac{3}{4}$ **21.** $\frac{8}{10}$ **22.** $\frac{1}{8}$

(Skill 28) **23.** 11.5% **24.** 1.25% **25.** 0.125% **26.** 385%

Applications

Round to the nearest quarter hour.

(Application E) **27.** 8 hours 35 minutes **28.** 7 hours 50 minutes

Find the number of occurrences.

(Application K) **29.** Weekly for 2 years **30.** Monthly for 3 years

31. Quarterly for 4 years **32.** Semimonthly for 2 years

Terms

Write your own definition for each term.

33. Straight-time pay **34.** Piecework pay

35. Salary **36.** Commission

37. Graduated commission **38.** Overtime pay

Refer to your reference files at the back of the book if you need help.

Problem Solving Identifying Wanted, Given, and Needed Information

Before you begin to solve a word problem, first read the problem through carefully and answer these questions:

1. What are you asked to find?
2. What facts are given?
3. Are enough facts given? Do you need more information than the problem provides?

Some word problems provide more information than is needed to solve the problem. Others cannot be solved without additional information. Identifying what is wanted, what is given, and what is needed allows you to organize the information and plan your solution.

Example 1 Joan Bishop is employed by ABC Construction Company. She earns \$7.50 per hour. She is married and claims 2 withholding allowances. Last week she worked 40 hours at the regular rate and 4 hours at the weekend rate. She is 32 years old. Find her gross pay last week.

Solution

1. Asked for: Joan Bishop's gross pay last week
2. Facts given: \$7.50 hourly rate
   40 hours worked at regular rate
   4 hours worked at weekend rate
3. Facts needed: weekend rate

This problem cannot be solved.

Example 2 Tom Derby, age 26, runs 4 miles every day. How many miles does he run in a full week?

Solution

1. Asked for: number of miles run in 1 week
2. Facts given: 4 miles per day
   1 week
3. Facts needed: none

This problem can be solved. To find the solution, multiply 4 miles per day by 7 days (1 week). The answer is 28 miles.

Problems Identify the wanted, given, and needed information. If enough information is given, solve the problem.

1. Outdoor Living, Inc., is having a sale on camping equipment. They have 2-person tents for \$49.99, cooking stoves for \$27.45, and cooking sets for \$18.45. How much do 2 sleeping bags and a tent cost?

2. Assume you are driving on a 2-mile circular race track. For the first half of the track you average 30 miles per hour. What speed must you maintain for the second half of the race track to average 60 miles per hour?

3. A bottle of cider costs 25¢. The cider costs 15¢ more than the bottle. How much refund is the bottle worth?

Unit Test

Lesson 1-1

1. Lois Giovanelli, a bookkeeper, earns $5.40 an hour during a regular work week. Last week she worked her regular $36\frac{1}{2}$ hours. What was her straight-time pay for the week?

Lesson 1-2

2. Louis Glanzman works as a computer operator. He earns $7.00 per hour for a 40-hour week. His overtime rate is $1\frac{1}{2}$ times his regular hourly rate. This week Louis worked his regular 40 hours plus $8\frac{1}{4}$ hours of overtime. What is his total pay?

Lesson 1-3

3. Bettie Silverman fills out a weekly timecard. What are her total hours for the week?

Day	In	Out	In	Out	In	Out	Hrs.
Sun.							
Mon.	8:29	11:31	1:02	3:28			?
Tues.	11:30	5:28	6:59	8:42			?
Wed.	8:15	11:25	12:51	6:00			?
Thur.	8:33	11:41	12:59	5:31	7:02	9:10	?
Fri.	8:17	11:26	12:58	5:27	6:58	8:58	?
Sat.							
						Total Hours	?

Lesson 1-4

4. Allen Cordon is paid on a piecework basis. He is paid $8.60 for each chair he assembles. This week he assembled 32 chairs. What is his pay for the week?

Lesson 1-5

5. Joseph Hines is a product planner for National Foods. He earns an annual salary of $21,588 and is paid on a semimonthly basis. What is his semimonthly pay?

Lesson 1-6

6. Stella Floyd sells hospital equipment for an 8% straight commission. Her sales totaled $24,756 in September. What is her pay for the month?

Lesson 1-7

7. Mark Broden is a furniture salesman. He earns a graduated commission on sales as shown. His sales for 1 week totaled $4100. What was his total commission?

Commission	Level of Sales
7%	First $1100
9%	Next $1400
11%	Over $2500

Unit 2 Net Income

▸ Out of your *gross pay,* the total amount of money you earn, employers withhold money for federal, state, and social security taxes. These *deductions* from your gross pay result in your *net pay,* the money you actually receive each pay period. *Federal income taxes* depend on your salary and the number of people you support, or your *dependents.* Deductions from your gross pay may include cost of medical insurance. These deductions are shown on your *earnings statement,* which is attached to your paycheck.

The amount of federal income tax you pay depends on your total income, number of dependents, marital status, and total deductions.

2-1 Federal Income Tax

Objective

Use tables to find the amount withheld for FIT.

Employers are required by law to withhold a certain amount of your pay for federal **income tax** (FIT). The Internal Revenue Service provides employers with tables that show how much money to withhold. The amount withheld depends on your income, marital status, and **withholding allowances.** You may claim one allowance for yourself and one for your spouse if you are married. You may claim additional allowances for any others you support.

WEEKLY Payroll MARRIED				
Wages		Allowances		
At least	But less than	0	1	2
260	270	30.30	27.00	23.70
270	280	32.00	28.70	25.40
280	290	33.70	30.40	27.10
290	300	35.40	32.10	28.80
300	310	37.10	33.80	30.50
310	320	38.80	35.50	32.20
320	330	40.50	37.20	33.90
330	340	42.20	38.90	35.60
340	350	43.90	40.60	37.30
350	360	45.60	42.30	39.00

Example

Skill 1 **Application** C **Term** Income tax

Mary Thoman's gross pay for this week is $314.52. She is married and claims 2 allowances, herself and her husband. What amount will be withheld from Mary's pay for federal income tax?

Solution

1. Find the **income range** from the table.

At least	But less than
$310	$320

2. Find the column for 2 **allowances.**
3. **Amount of income to be withheld** is $32.20

A. $27.00
B. $37.30

Self-Check Cover the answers and complete the problems.

Find the federal income tax withheld using the tax table above.

A. Weekly income: $261.34. 1 allowance.

B. Weekly income: $345.22. 2 allowances.

Problems

Use the MARRIED Persons—WEEKLY Payroll Period table above to find the amount withheld for federal income tax.

	1.	2.	3.	4.	5.
Employee	G. Chesson	B. Brown	J. R. Foos	J. Isch	F. Restivo
Weekly Income	$293.00	$328.00	$287.50	$344.22	$288.59
Allowances	2	1	0	2	0
Amount Withheld	?	?	?	?	?

SINGLE Persons—WEEKLY Payroll Period

And the wages are—		And the number of withholding allowances claimed is—										
At least	But less than	0	1	2	3	4	5	6	7	8	9	10
		The amount of income tax to be withheld shall be—										
$210	$220	$ 27.90	$ 24.30	$ 20.90	$ 18.00	$ 15.10	$ 12.20	$ 9.30	$ 6.50	$ 4.10	$ 1.80	$ 0
220	230	29.80	26.20	22.50	19.50	16.60	13.70	10.80	8.00	5.30	3.00	.70
230	240	31.70	28.10	24.40	21.00	18.10	15.20	12.30	9.50	6.60	4.20	1.90
240	250	33.60	30.00	26.30	22.70	19.60	16.70	13.80	11.00	8.10	5.40	3.10
250	260	35.50	31.90	28.20	24.60	21.10	18.20	15.30	12.50	9.60	6.70	4.30
260	270	37.40	33.80	30.10	26.50	22.80	19.70	16.80	14.00	11.10	8.20	5.50
270	280	39.30	35.70	32.00	28.40	24.70	21.20	18.30	15.50	12.60	9.70	6.80
280	290	41.70	37.60	33.90	30.30	26.60	23.00	19.80	17.00	14.10	11.20	8.30
290	300	44.20	39.50	35.80	32.20	28.50	24.90	21.30	18.50	15.60	12.70	9.80
300	310	46.70	41.90	37.70	34.10	30.40	26.80	23.10	20.00	17.10	14.20	11.30
460	470	88.80	83.10	77.30	72.30	67.50	62.70	57.90	53.10	48.30	43.50	38.90
470	480	91.80	86.10	80.30	74.80	70.00	65.20	60.40	55.60	50.80	46.00	41.20
480	490	94.80	89.10	83.30	77.50	72.50	67.70	62.90	58.10	53.30	48.50	43.70
490	500	97.80	92.10	86.30	80.50	75.00	70.20	65.40	60.60	55.80	51.00	46.20
500	510	100.80	95.10	89.30	83.50	77.80	72.70	67.90	63.10	58.30	53.50	48.70
810	820	210.40	203.30	196.20	189.10	181.90	174.80	167.70	160.60	153.50	146.40	139.70
820	830	214.10	207.00	199.90	192.80	185.60	178.50	171.40	164.30	157.20	150.10	143.10
830	840	217.80	210.70	203.60	196.50	189.30	182.20	175.10	168.00	160.90	153.80	146.60
		37 percent of the excess over $840 plus—										
$840 and over		219.60	212.50	205.40	198.30	191.20	184.10	177.00	169.80	162.70	155.60	148.50

Use the table above to find the amount withheld.

6. Emily Adams, single.
Earns $492.15 weekly.
Claims 1 allowance.
What is the FIT withheld?

7. John Ford, single.
Earns $265 weekly.
Claims 4 allowances.
What is the FIT withheld?

8. Ted Lucius, single.
Earns $502.50 weekly.
Claims 3 allowances.
What is the FIT withheld?

9. Tina Ganecki, single.
Earns $300 weekly.
Claims 2 allowances.
What is the FIT withheld?

10. Amy Robell earns $475 a week. She is single and claims 3 allowances. What amount is withheld weekly for federal income tax?

11. Fred Koppel earns $481 a week. He is single and claims no allowances. What amount is withheld weekly for federal income tax?

12. Matt Kingsley is single and earns $1300 a week. He claims 1 allowance. What amount is withheld weekly for federal income tax?

13. Jeannie Pacetta is single, earns $920 a week, and claims 1 allowance. What amount is withheld weekly for federal income tax?

Use the MARRIED Persons—WEEKLY Payroll Period table on page 567 to find the amount withheld for federal income tax.

14. Candy Shore is married, earns $605 a week, and claims 3 allowances. What amount is withheld weekly for federal income tax?

(Continue on next page.)

15. Sid Kaplan earns $655 a week. He is married and claims 2 allowances. What amount is withheld weekly for federal income tax?

16. Louise Wecherlin is paid a weekly wage of $198.82. She is married and claims 2 allowances. Beginning next year, she will claim another allowance for supporting her mother. How much less will be withheld for federal income tax for the year?

17. Dale Ling earns $220.50 a week. He is married and claimed 1 allowance last year. He hopes to receive a refund on his next tax return by claiming no allowances this year. How much more in withholdings will be deducted weekly if he claims no allowances?

*18. **A Brief Case** Some companies use a percentage method instead of the tax tables to compute income withheld. Use the table at the right to find the amount withheld for the SINGLE employees below. Each weekly allowance is $19.23.

a. Find the **allowance amount.**
Number of Allowances × $19.23

b. Find the **taxable wages.**
Weekly Wage − Allowance Amount

c. Find the **amount withheld** using the table.

WEEKLY PAYROLL PERIOD, SINGLE person

TAXABLE WAGES		INCOME TAX WITHHELD	
Not over $27		$0	
Over	But not over		Of excess over
$ 27	$ 79	12%	$ 27
$ 79	$183	$6.24 plus 15%	$ 79
$183	$277	$21.84 plus 19%	$183
$277	$423	$39.70 plus 25%	$277
$423	$535	$76.20 plus 30%	$423
$535	$637	$109.80 plus 34%	$535
$637		$144.48 plus 37%	$637

Employee	Weekly Wage	Number of Allowances	Allowance Amount	Taxable Wages	Amount Withheld
D. Faust	$184	1	$19.23	$164.77	$19.11
J. Wray	$290	3	$57.69	$232.31	?
E. Kohn	$690	2	?	?	?
K. Bird	$545	4	?	?	?

Maintaining Your Skills

Look up the skills in parentheses if you have difficulty with the problems.

Give the place and value of the underlined digit. (Skill 1)

19. 8<u>4</u> 20. 9<u>6</u>5 21. <u>6</u>532 22. 36.9<u>4</u> 23. <u>6</u>72 24. 14.78<u>3</u>

Find the percentage. (Skill 30)

25. 22.5% × $120 26. 17.50% × $430 27. $\frac{1}{2}$% × $298 28. 128% × $82

Subtract. (Skill 6)

29. $344.50 − $68.40 30. $432.60 − $83.70 31. $1298.65 − $423.76

2-2 State Income Tax

Objective

Compute the state tax on a straight percent basis.

Most states require employers to withhold a certain amount of your pay for state income tax. In some states, the tax withheld is a percent of your taxable wages. Your taxable wages depend on **personal exemptions** allowed for supporting yourself and others in your family.

- **Taxable Wages = Annual Gross Pay − Personal Exemptions**
- **Annual Tax Withheld = Tax Rate × Taxable Wages**

Example

Skills 6, 30 **Application** A **Term** Personal exemptions

John Schneider's gross pay is $12,400 a year. The state income tax rate is 3% of taxable wages. John takes a married exemption for himself and his wife and 2 personal exemptions for his 2 children. How much is withheld a year from his gross earnings for state income tax?

PERSONAL EXEMPTIONS
Single — $1500
Married — $3000
Each Dependent — $700

Solution

1. Find the **taxable wages.**
   Annual Gross Pay − Personal Exemptions
   $12,400 − ($3000 + $700 + $700) = $8000.00
2. Find the **annual tax withheld.**
   Tax Rate × Taxable Wages
   3% × $8000.00 = $240.00 annual tax withheld

A. $508.00
B. $1215.00

Self-Check Cover the answers and complete the problems.

Find the annual tax withheld.

A. Gross pay: $14,900.
Single, 1 dependent.
State income tax rate: 4%.

B. Gross pay: $28,000.
Married, 1 dependent.
State income tax rate: 5%.

Problems

	(Annual Gross Pay	−	Personal Exemptions	=	Taxable Wages)	×	State Tax Rate	=	Tax Withheld
1.	($ 9000	−	$3000	=	$6000)	×	2%	=	?
2.	($12,000	−	$2200	=	?)	×	5%	=	?
3.	($19,500	−	$3700	=	?)	×	4%	=	?

Use the table on page 69 for personal exemptions and find the amount withheld.

4. Ruth Lambert.
Earns $14,736 annually.
Single, no dependents.
What are her personal exemptions?

5. Joseph Ryczko.
Earns $8971 annually.
Married, no dependents.
What are his personal exemptions?

6. Clay DeBrew.
Earns $28,400 annually.
Married, 3 dependents.
State tax rate is 4.6%.
What are his personal exemptions?
What is withheld for state tax?

7. Wendy Chou.
Earns $13,400 annually.
Single, 1 dependent.
State tax rate is 2.5%.
What are her personal exemptions?
What is withheld for state tax?

8. Bob Grassle earns $19,229.08 annually as a systems analyst. He is married and supports 2 children. The state tax rate in Bob's state is 2% of taxable income. What amount is withheld yearly for state income tax?

9. May Lee earns $15,600 per year. Her personal exemptions include herself and her husband. The state tax rate in May's state is 3.5% of taxable income. What amount is withheld yearly for state income tax?

10. Grace Chu is a registered pharmacist. She earns $22,600 a year and is single. The state income tax rate is 5% of taxable income. What amount is withheld yearly for state income tax?

Use the tables on pages 566–567 for federal withholdings.

11. Patrick Bonner earns $22,100 a year as a financial analyst. He is paid on a weekly basis. He is married, has no dependents, and claims 2 withholding allowances for federal income tax purposes. The state tax rate is 2% of taxable income. How much is withheld annually from Patrick's gross pay for state and federal income taxes?

12. Suzanne Ledgerwood earns $24,180 a year as an electronics engineer. She is paid on a weekly basis. She is single, has 1 dependent, and claims 2 withholding allowances. The state tax rate is 2.5% of taxable income. How much is withheld each year from her gross pay for state and federal income taxes?

Maintaining Your Skills

Look up the skills in parentheses if you need help or more practice.

Subtract. (Skill 6)

13. 82.19 − 16.32 14. 39 − 16.2 15. 46.2 − 14.297 16. 900.12 − 612.89

Find the percentage. (Skill 30)

17. 120% of 160 18. 8% of 122 19. $2\frac{1}{2}$% of 500 20. $14\frac{3}{5}$% of 200

2-3 Graduated State Income Tax

Objective Compute the state tax on a graduated income basis.

Some states have a **graduated income tax.** Graduated income tax involves a different tax rate for each of several levels of income. The tax rate increases as income increases. The tax rate on low incomes is usually 1% to 3%. The tax rate on high incomes may be as much as 20%.

▶ **Tax Withheld per Pay Period = $\dfrac{\textbf{Annual Tax Withheld}}{\textbf{Number of Pay Periods per Year}}$**

Example

Skills 5, 11 **Application** K **Term** Graduated income tax

Lisa Lininger's annual salary is $14,600. She is paid semimonthly. Her personal exemptions total $1500. How much does her employer deduct from each of Lisa's semimonthly paychecks for state income tax?

STATE TAX	
Annual Gross Pay	Tax Rate
First $1000	1.5%
Next $2000	3.0%
Next $2000	4.5%
Over $5000	5.0%

Solution

1. Find the **taxable wages.**
   Annual Gross Pay − Personal Exemptions
   $14,600 − $1500 = $13,100

2. Find the **annual tax withheld.**
   a. First $1000: 1.5% of $1000 . . . . $ 15.00
   b. Next $2000: 3.0% of $2000 . . . . $ 60.00
   c. Next $2000: 4.5% of $2000 . . . . $ 90.00
   d. Over $5000: 5.0% of ($13,100-$5000)
   5.0% of $8100 = $405.00
   Total $570.00

3. Find the **tax withheld per pay period.**
   Annual Tax Withheld ÷ Number of Pay Periods per Year
   $570.00 ÷ 24 = $23.75 tax withheld semimonthly

A. $28.75
B. $63.75

▶ **Self-Check** Cover the answers and complete the problems.

Find the tax withheld.

A. Annual salary: $17,000
Personal exemptions: $1500.
24 pay periods.

B. Annual salary: $20,000.
Personal exemptions: $3000.
12 pay periods.

Problems

1. Helen Palino.
Annual gross pay of \$6500.
Personal exemptions of \$3000.
1.5% state tax on first \$2000.
3% tax on amount over \$2000.
What is her taxable income?

2. Morgan Chesterton.
Annual gross pay of \$7800.
Personal exemptions of \$2000.
1.5% state tax on first \$2000.
3% tax on amount over \$2000.
What is his taxable income?

3. Coretta Daily.
Annual gross pay of \$12,000.
Personal exemptions of \$4400.
1% state tax on first \$2000.
3% tax on next \$3000.
4.5% tax on amount over \$5000.
How much state tax is withheld?

4. Herb McKenna.
Annual gross pay of \$8950.
Personal exemptions of \$1000.
1% state tax on first \$2000.
3% tax on next \$3000.
4.5% tax on amount over \$5000.
How much state tax is withheld?

Use the tables below to compute the state tax for exercises 5–8.

Personal Exemptions*	
Single	\$1500
Married	\$3000
Each Dependent	\$ 700

State Tax	
Annual Gross Pay	Tax Rate
First \$3500	3%
Next \$3500	4.5%
Over \$7000	7%

5. David Tuttle's annual gross pay is \$10,000. He is single and is paid on a monthly basis. How much is withheld annually for state tax?

6. Julie Fast's annual gross pay is \$14,460. She is married with no dependents. How much is withheld from her monthly paycheck for state income tax?

7. Paul Tamari earns \$22,500 a year as a welder. He is married with 3 children. How much is withheld from his weekly paycheck for state income tax?

8. Gayle Brice earns \$16,135 a year as a legal secretary. She is single and is paid on a weekly basis. How much is withheld each week for state income tax?

Maintaining Your Skills

Look up the skills in parentheses if you need help or more practice.

Add. (Skill 5)

9. 0.005 + 0.319 + 0.223

10. 465.6 + 1.627 + 15.11

11. \$ 42.60 + 187.90 + 5.42

12. \$2041.42 + 106.90 + 22.84

Divide. Round answers to the nearest hundredth. (Skill 11)

13. 397 ÷ 24
14. 28.618 ÷ 1.3
15. 0.00891 ÷ 0.31
16. 27 ÷ 15

Find the number of occurrences. (Application K)

17. Quarterly for 3 years
18. Monthly for $1\frac{1}{2}$ years
19. Weekly for 4 years

2-4 Social Security Tax

Objective

Compute the amount of income withheld for social security tax.

The Federal Insurance Contributions Act (F.I.C.A.) requires employers to deduct 7.15% of the first $42,600 of your annual income for **social security** taxes. There are no deductions on income over $42,600. The employer must contribute an amount that equals your contribution. The federal government uses social security taxes to provide hospitalization insurance for people over 65, retirement income, survivor's benefits, and disability benefits.

▶ **Social Security Tax Withheld = Tax Rate × Gross Pay**

Example

Skills 30, 2 Application A Term Social Security

David Peralta's gross monthly salary is $2000. His earnings to date for the year total $16,000. What amount is deducted from David's pay this month for social security taxes?

Solution

1. Are earnings to date less than $42,600?
   Yes. $16,000 is less than $42,600.
2. Find the **social security tax withheld.**
   Tax Rate × Gross Pay
   7.15% × $2000.00 = $143.00 tax withheld

A. $178.75
B. $14.30

▶ **Self-Check** Cover the answers and complete the problems.

Find the social security tax withheld for this pay period.

A. Monthly salary: $2500.
Earnings to date: $31,000.

B. Weekly salary: $200.
Earnings to date: $8000.

Problems

	Tax Rate	x	Gross Pay	=	Social Security Tax Withheld
1.	7.15%	×	$1250	=	?
2.	7.15%	×	$ 600	=	?
3.	7.15%	×	$ 125	=	?
4.	7.15%	×	$1300	=	?
5.	7.15%	×	$1500	=	?

Use the F.I.C.A. tax rate of 7.15% of the first $42,600 to solve.

6. Curtis Griffith, mechanic. $9450 earned this year to date. $270 gross pay this week. How much deducted this week for social security?

7. Jackie Chandler, actuary. $1440 gross pay this check. $52,120 earned this year to date. How much deducted this pay period for social security?

8. Kenny Harris, a dock supervisor for Hainey Grain Co-op, is paid on a monthly basis. His gross pay this month is $1750.42. His earnings to date for this year are $15,753.78. How much is deducted from his paycheck this month for social security tax?

9. Alicia Johnson is an accountant for an electronics firm. She earns $28,080 a year and is paid on a semimonthly basis. How much is deducted per pay period for social security tax?

10. Elaine Marks was hired for the position of receptionist. She is to be paid on a monthly basis. How much will be deducted from each of Elaine's paychecks for social security?

RECEPTIONIST: $650 salary. Meet, greet clients. Park free. Great company. Call Charles Guidry 600-1325 for appointment.

Use the tables on pages 566–567 for federal withholding taxes.

11. Callie Shaw is employed by Madison Toy Manufacturing as a designer. She is married, earns $435 weekly, and claims no allowances. Her gross pay to date this year is $7981.35. How much is deducted from her paycheck this week for federal income and social security taxes?

12. Murray Tsongas earns $650 each week as a chief engineer for Putnam Laboratory. He is married and claims 2 allowances. How much is withheld from Murray's paycheck for the last week of December for federal income and social security taxes?

13. Pete Nichols was hired on January 2 for the supervisory position shown. He will be paid on a monthly basis. If Pete's social security deductions are 7.15% of the first $42,600, in which month will the last deduction be made?

MECHANICAL ENGINEER Supervise four engineers in special machine design. Good communicator. Downtown location. $54,000 salary.

Maintaining Your Skills

Look up the skills in parentheses if you need help or more practice.

Find the percentage. (Skill 30)

14. 7.15% of $480 15. $2\frac{1}{2}$% of 500 16. $15\frac{3}{5}$% of 80 17. $\frac{3}{4}$% of 20

Round to the place value indicated. (Skill 2)

Nearest ten 18. 219 19. 322.46 20. 3954.97 21. 23,981.98

Nearest hundred 22. 7928 23. 3764 24. 63,782 25. 147,366.20

2-5 Group Insurance

Objective: Compute the deduction for group insurance.

Many businesses offer **group insurance** plans to their employees. You can purchase group insurance for a lower cost than individual insurance. Businesses often pay part of the cost of the insurance. The remaining cost is deducted from your pay.

▸ **Deduction per Pay Period** = $\frac{\textbf{Total Annual Amount Paid by Employee}}{\textbf{Number of Pay Periods per Year}}$

Example

Skills 6, 11, 30 **Application** A **Term** Group insurance

Jane Wilson is a typist for Hilltop Corporation. She has family medical coverage through the group medical plan that Hilltop provides for its employees. The annual cost of Jane's family membership is $1200. The company pays 80% of the cost. How much is deducted from Jane's monthly paycheck for medical insurance?

Solution

1. Find the **percent paid by employee.**
   100% − 80% = 20%
2. Find the **total amount paid by employee.**
   $1200.00 × 20% = $240.00
3. Find the **deduction per pay period.**
   Total Amount Paid by Employee ÷ Number of Pay Periods per Year
   $240.00 ÷ 12 = $20.00 deducted per pay period

A. $25.00
B. $6.00

▸ **Self-Check** Cover the answers and complete the problems.

Find the deduction per pay period.

A. Annual cost of insurance: $1500.
Employee pays 20%.
12 pay periods.

B. Annual cost of insurance: $1040.
Employee pays 30%.
52 pay periods.

Problems

1. Lily Case, editor.
Annual group insurance costs $1200.
Company pays 40% of the cost.
How much does Lily pay semi-monthly?

2. Ken Miller, dance instructor.
Annual group insurance costs $890.
Employer pays 75% of the cost.
How much does Ken pay yearly?

3. Lee Ming's group medical insurance coverage costs $500 a year. The company pays 85% of the cost. How much is deducted each month from Lee's paycheck for medical insurance?

4. Nancy Goodson, a clerk at Neuman's Automotive Parts, earns $245.22 weekly. Her group medical insurance costs $524 a year, of which the company pays 60% of the costs. How much is deducted weekly from Nancy's paycheck for medical insurance?

5. Janet Glenn earns $294.78 a week as a radio dispatcher for the city. Because she is an employee of the city, the city pays 90% of the cost of any insurance coverage. Her family medical insurance costs $988 a year. How much is deducted each week for medical insurance?

6. Mike Sperl is an employment counselor for the city. He earns a biweekly salary of $670. His annual medical insurance costs $712.50 a year. The city pays 90% of the costs. How much is deducted each pay period from Mike's paycheck for medical insurance?

7. Alice LaVita is employed at Richter-Boyle Incorporated as a security consultant. Her annual salary is $18,600 and she is paid on a semimonthly basis. Her annual group medical coverage costs $810, of which the company pays 75%. How much is deducted each pay period from Alice's paycheck for medical coverage? After this deduction, what is her take-home pay?

8. Matt Ingels is a salesman for Western Paper Products, Inc. He is covered by travel insurance in addition to the basic medical coverage. Medical insurance costs $566.12 a year and travel insurance costs another $66.84 a year. The company pays 80% of all insurance costs. How much is deducted monthly for travel insurance? How much is deducted monthly for basic medical coverage?

9. Carla Cross has medical, dental, and term life insurance coverages through the company she works for. Medical coverage costs $636.80 a year, dental coverage costs $75.32 a year, and term life insurance costs $170.40 a year. The company pays 75% of the medical and 50% of the dental coverage. Term life insurance is entirely paid for by the company. What is the total amount deducted weekly for these coverages?

Maintaining Your Skills

Look up the skills in parentheses if you need help or more practice.

Divide. Round answers to the nearest hundredth. (Skill 11)

10. 600.48 ÷ 24 11. 16.97 ÷ 1.3 12. 0.00392 ÷ 0.30 13. 0.692 ÷ 0.008

Find the percentage. (Skill 30)

14. 8% of 240 15. 9.6% of 80 16. $5\frac{1}{4}$% of 1600 17. $8\frac{3}{10}$% of 95

2-6 Earnings Statement

Objective Compute the net pay per pay period.

You may have additional deductions taken from your gross pay for union dues, contributions to community funds, savings plans, and so on. The earnings statement attached to your paycheck lists all your deductions, your gross pay, and your **net pay** for the pay period.

▸ **Net Pay = Gross Pay − Total Deductions**

Example

Skills 5, 6 **Application** A **Term** Net pay

Andrew Morgan's gross weekly salary is $400. He is married and claims 2 allowances. The social security tax is 7.15% of the first $42,600. The state tax is $8.00 a week. Each week Andrew pays $8.25 for medical insurance and $1.50 for charity. Is Andrew's earnings statement correct?

DEPT.	EMPLOYEE	CHECK #		WEEK ENDING	GROSS PAY		NET PAY
04	Morgan, A	20566		9/17/--	$400.00		143.95
TAX DEDUCTIONS				PERSONAL DEDUCTIONS			
FIT	FICA	STATE	LOCAL	MEDICAL	UNION DUES	OTHERS	
15.60	28.20	8.00	---	8.25	----	1.50	

Solution

1. Find the **total deductions.**
   a. Federal withholding: (from table on page 567) $47.50
   b. Social security: 7.15% of $400.00 = ... 28.60
   c. State tax 8.00
   d. Medical insurance 8.25
   e. Charity 1.50

   Total $93.85

2. Find the **net pay.**

   Gross Pay − Total Deductions

   $400 − $93.85 = $306.15 net pay

$201.78

▸ **Self-Check** Cover the answer and complete the problem.

Jon Glik is single and claims 1 allowance. His gross weekly salary is $300. Each week he pays federal and social security taxes, $14.87 for medical insurance, and $20 for the credit union. What is his net pay?

Problems

Find the net pay.

1.

DEPT.	EMPLOYEE		CHECK #		WEEK ENDING	GROSS PAY		NET PAY
07	Vernon, J.		54601		8/24/--	$250.50		?
TAX DEDUCTIONS					PERSONAL DEDUCTIONS			
FIT	FICA	STATE		LOCAL	MEDICAL		UNION DUES	OTHERS
31.90	17.91	8.32		---	1.74		----	---

2.

DEPT.	EMPLOYEE		CHECK #		WEEK ENDING	GROSS PAY		NET PAY
12	Tanner, E.		11352		5/03/--	$304.20		?
TAX DEDUCTIONS					PERSONAL DEDUCTIONS			
FIT	FICA	STATE		LOCAL	MEDICAL		UNION DUES	OTHERS
37.70	21.75	8.05		1.58	2.31		5.00	2.00

Find the deductions and the net pay. The F.I.C.A. tax rate is 7.15% of the first $42,600. Use the tables on pages 566–567 for federal withholdings.

3. Betsy Cole, a harbor boat pilot, is married and claims 1 allowance.

DEPT.	EMPLOYEE		CHECK #		WEEK ENDING	GROSS PAY		NET PAY
A	Cole, B.		8002349		1/16/--	$455.00		?
TAX DEDUCTIONS					PERSONAL DEDUCTIONS			
FIT	FICA	STATE		LOCAL	MEDICAL		UNION DUES	OTHERS
?	?	12.18		2.69	3.40		12.00	---

4. Marge Ferrelli, an administrative assistant for the town hospital, is married and claims 1 allowance. Her state personal exemptions total $35.46 a week. The state tax rate is 2.5% of taxable income.

DEPT.	EMPLOYEE		CHECK #		WEEK ENDING	GROSS PAY		NET PAY
M	Ferrelli, M.		347528		11/2/--	$421.25		?
TAX DEDUCTIONS					PERSONAL DEDUCTIONS			
FIT	FICA	STATE		LOCAL	MEDICAL		UNION DUES	OTHERS
?	?	?		--	---		----	10.00

5. Jay Shorr is an economist and earns $22,000 a year. Jay is single and supports his mother. A state tax of 3% on his gross earnings is deducted from his weekly paycheck. In addition to federal income tax and social security tax, $4.50 is deducted for medical insurance.

DEPT.	EMPLOYEE		CHECK #		WEEK ENDING	GROSS PAY		NET PAY
15	Shorr, J.		91166		3/31/--	?		?
TAX DEDUCTIONS					PERSONAL DEDUCTIONS			
FIT	FICA	STATE		LOCAL	MEDICAL		UNION DUES	OTHERS
?	?	?		---	?		----	---

6. Cindy Marks is employed by Dardeen Associates in the Marketing Department. She is single and claims 1 allowance. Tax deductions include federal income tax, F.I.C.A., and a state tax of 2.5% on gross earnings. The company pays 70% of medical insurance of $988.

DEPT.	EMPLOYEE	CHECK #	WEEK ENDING	GROSS PAY	NET PAY
14	Marks, C.	10–422	10/31/--	$460.50	?

TAX DEDUCTIONS				PERSONAL DEDUCTIONS		
FIT	FICA	STATE	LOCAL	MEDICAL	UNION DUES	OTHERS
?	?	?	———	?	————	———

7. Kirk Glowaki, a shearing machine operator at Grayling Fabricators, earns $9.60 per hour for a regular 40-hour week. He is married and claims 6 allowances. The state tax is 3.5% of gross earnings. The local tax is $\frac{1}{2}$% of gross earnings. Kirk pays $16.20 a week for medical insurance and $5.00 a week for union dues.

DEPT.	EMPLOYEE	CHECK #	WEEK ENDING	GROSS PAY	NET PAY
SM	Glowaki, K.	4–942	4/12/--	?	?

TAX DEDUCTIONS				PERSONAL DEDUCTIONS		
FIT	FICA	STATE	LOCAL	MEDICAL	UNION DUES	OTHERS
?	?	?	?	?	?	———

*8. **A Brief Case** John Clark, fork lift driver at Sheraton Manufacturing, earns $8.00 an hour for a regular 40-hour week with time and a half for overtime. He is married and claims 4 allowances. The state tax rate is **25% of his federal income tax withholding.** Credit union deductions are $25 a week. Medical insurance costs $873.28 a year, of which the company pays 80% of the costs. What is John's net pay for a week in which he worked 48 hours?

Maintaining Your Skills

Look up the skills in parentheses if you need help or more practice.

Add. (Skill 5)

9. 16.002 + 161.320 + 342.117
10. 16 + 9.3 + 42.016
11. 429.9 + 76.107 + 3.05
12. 62.90 + 3.52 + 186.81
13. 0.005 + 0.319 + 0.223
14. 465.6 + 1.627 + 15.11
15. 42.60 + 187.90 + 5.42
16. 2041.42 + 106.90 + 22.84

Subtract. (Skill 6)

17. 571.21 − 16.96
18. 69.3 − 6.24
19. 62.39 − 86.14
20. 88.19 − 22.81

Reviewing the Basics

Skills

Write the number that is greater.

(Skill 1) **1.** $239 or $230.98 **2.** $194.45 or $200.45

Round to the nearest cent.

(Skill 2) **3.** $154.807 **4.** $25.209 **5.** $40.0953 **6.** $1.523

Solve. Round answers to the nearest cent.

(Skill 5) **7.** $48.72 + $54.59 **8.** $10.41 + $11.03 + $6.88

(Skill 6) **9.** $84.72 − $60.71 **10.** $8.89 − $0.62 **11.** $143.32 − $7.89

(Skill 11) **12.** $892.50 ÷ 12 **13.** $988 ÷ 52 **14.** $1147.27 ÷ 24

(Skill 30) **15.** 4.5% of $105.45 **16.** 3.5% of $404.10 **17.** 3.2% of $148.89

Applications

Use the formula TAX WITHHELD = TAX RATE × GROSS PAY to solve.

(Application A) **18.** Tax rate: 5%.
Weekly wage: $320.19.
Find the tax withheld.

19. Tax rate: 6.5%.
Monthly wage: $979.67.
Find the tax withheld.

(Application C) **20.** Marge Fuller is single, earns $253.86 weekly, and claims 2 allowances. Use the table to find the weekly federal income tax withheld.

Weekly Payroll Period
Single

Wages		Allowances		
At least	But less than	0	1	2
		Tax Withheld		
$210	$220	$27.90	$24.30	$20.90
220	230	29.80	26.20	22.50
230	240	31.70	28.10	24.50
240	250	33.60	30.00	26.30
250	260	35.50	31.90	28.20

Find the number of pay periods.

(Application K) **21.** Semimonthly for 1 year

22. Biweekly for 2 years

23. Monthly for 6 years

Terms

Write your own definition for each term.

24. Group insurance **25.** Income tax **26.** Net pay

27. Social security **28.** Graduated income tax **29.** Personal exemptions

Refer to your reference files at the back of the book if you need help.

CENTRAL DATA CO.

Application for Employment

NAME	LAST	FIRST	INITIAL	
PRESENT ADDRESS	STREET & NUMBER	TOWN OR CITY	STATE	ZIP
PERMANENT ADDRESS				
SOCIAL SEC. NO.		HOME TELEPHONE NO.	BUSINESS TELEPHONE NO.	

Position Applying For | Expected Salary

When you fill out some job applications, you may be asked to state your salary expectations. You can find out what current salary is being paid to employees in similar positions by checking available sources. Some reliable sources include the "want ad" section of your local newspaper, your local job service office, and the **Occupational Outlook Quarterly.**

The newspaper clipping shows the weekly salary paid for some positions in business. Use the clipping to answer the following questions.

THE JOB LEDGER 820-2253

TELLER TRNS . . . to $190
No exp necessary
CODING CLK . . . . to $175
Major ins co, exp helpful
COPY MACH TRN . to $165
Excl oppty. Will be sent to school, supvse copy center
PAYROLL ACCTNT to $275
Excptnl opporty for person with exp. top co and benefits
PAYROLL CLK . . . to $200
1 year payroll exp will do
TYPIST/FIGURES . to $185
Front office, growth opprty
ACCTNG CLKS . . $200-280
Several very fine openings
SWTCHBD TRN $190
Train with top co downtown
CLERICALS $135-170
Many varied local openings

1. Gary Jackson is applying for a switchboard trainee position. On his application he fills in a salary expectation of $200 per week. Does Gary's salary expectation exceed the advertised salary for a switchboard trainee?

2. Deborah Key is applying for an accounting clerk position. Her salary expectation is $220 per week. Does Deborah's salary expectation fall within the advertised salary range for an accounting clerk?

3. Barbara Chin is an experienced payroll accountant. She expects an annual salary of $18,000. Does Barbara's salary expectation exceed the advertised salary for a payroll accountant?

4. Morton Daley is applying for a position as payroll clerk. On his application he fills in an expected salary range of $12,000 to $14,000 yearly. Does Morton's salary range fall within the advertised salary for payroll clerks?

Unit Test

Lesson 2-1

1. Sherry Knight, a shipping supervisor, earns $443.60 a week. She is married and claims 2 allowances. How much is withheld from Sherry's weekly paycheck for federal income tax?

Weekly Payroll Period
Married

Wages		Allowances			
At least	But less than	0	1	2	3
		Tax Withheld			
$410	$420	$58.10	$53.90	$49.60	$46.00
420	430	60.30	56.10	51.80	47.70
430	440	62.50	58.30	54.00	49.80
440	450	64.70	60.50	56.20	52.00
450	460	66.90	62.70	58.40	54.20

Lesson 2-2

2. Wilma Huen earns $15,021 a year as a bank teller. The state income tax rate is 4.6% of taxable income. Her personal exemptions total $3700. How much is withheld each week from Wilma's gross pay for state income tax?

Lesson 2-3

3. Miles Reese, a commercial artist, earns an annual salary of $23,409.45. He is paid biweekly. His personal exemptions total $2000. How much is deducted each pay period from Miles's paycheck for state income tax?

STATE TAX

Taxable Wages	Tax Rate
First $2000	2%
Next $4000	3%
Next $4000	4.5%
Over $10,000	6%

Lesson 2-4

4. Harry Orwell, a medical technician, is paid $310.50 a week. His earnings to date this year total $9468.66. The social security tax rate is 7.15% of the first $42,600 earned. How much is deducted from Harry's paycheck this week for social security tax?

Lesson 2-5

5. Lisa Kimberly, a technical writer for Ace Electronics, earns $308.08 a week. Her medical insurance costs $669.76 a year, of which her company pays 75% of the costs. How much is deducted each week from Lisa's paycheck for medical insurance?

Lesson 2-6

6. Larry Alsales, a research assistant for Pinewood Chemical Industries, is married and claims no allowances. He earns $422.75 a week. The social security tax rate is 7.15% of the first $42,600 earned. The state tax is $4.95 a week. He has weekly deductions of $2.96 for medical insurance and $30 for payroll savings. Use the table above to find Larry's federal tax withheld. What is his net pay for a week?

Income Tax: A Simulation

If you are employed, some money is probably withheld by your employer from each of your paychecks for federal income tax. You must, by law, prepare an income tax return by April 15 of each year and send it to the Internal Revenue Service (IRS).

On an income tax return you report your adjusted gross income, which is the total of your wages, salaries, tips, interest, and other income. Based on your adjusted gross income, you use tax tables to figure out your tax liability, the amount of income tax you must pay. Some of your tax liability is already paid by your withholdings. If your tax liability is greater than your withholdings, you owe the IRS an amount called the balance due. If your tax liability is less than your withholdings, the IRS will return the extra money to you as a tax refund.

Each year your employer must send you a Wage and Tax Statement form, called a W-2 form. This form tells how much money you earned, how much was withheld for social security tax (F.I.C.A.), and how much was withheld for federal, state, and local income tax. You will receive copies of your W-2 form to send with your federal, state, and local income tax returns, and a copy to keep for your records. If you have earned interest on a bank account, the bank will send you a form showing the amount of interest.

Suppose that last year you had a part-time job as a clerk at a sporting goods store. Your W-2 form might look like this.

		2. Employer's ← State number		Copy C For employee's records.	
3. Employer's name SPORTSWORLD		4. Subtotal ☐ Correction ☐ Void ☐		State form number	
		5. Employer's identification number		City form number	
		6. Advance EIC payment		7.	
8. Employee's social security no. 123-45-6789	9. Federal income tax withheld 248.05	10. Wages, tips, other compensation 2255	11. FICA tax withheld 161.23	12. Total FICA wages	
13. Employee's name (First, middle, last)		14. Pension plan coverage? Yes/No NO	15.	16. FICA tips	
		18. State income tax withheld 53.40	19. State wages, tips, etc. 2255	20. Name of State	
17. Employee's address and Zip code		21. Local income tax withheld	22. Local wages, tips, etc.	23. Name of locality	
Wage and Tax Statement 19–					
Form W-2					

1. How much were you paid in wages and other compensation?

2. How much was withheld for federal income tax? for F.I.C.A.? for state income tax?

Preparing an Income Tax Return

Because your income was all from wages, salaries, or tips and you did not have more than $400 in interest and dividends, you are allowed to use the 1040A return, often called the short form. This return is shown on pages 85–86.

Use the information in the W-2 form on page 83. You also received $10.66 in interest on your savings account.

3. Assume you are single. Next to ''Filing Status'' which box should you check?

4. Lines 5 a–e refer to your exemptions. Exemptions are people you support, including yourself. (They are called withholding allowances when your net pay is calculated.) You get an extra exemption if you or your spouse is over 65 or blind. The more exemptions you have, the less tax you pay. Assume you are under 65 and not blind. Which boxes would you check in line 5a? in line 5b? In line 5e, what is the total number of exemptions you claim?

5. What amount would you write in line 6?

6. What amount would you write in line 7?

7. Because you had no dividends or unemployment compensation, you can proceed to line 10. Add the amounts in lines 6 and 7 to find your adjusted gross income.

8. Proceed to line 14. Line 14 will have the same amount as line 10 because 11 and 12 do not apply to you. On page 2, write the amount again in lines 15 and 17. What amount did you write?

9. What amount would you write in line 18? Find your taxable income and write it in line 19.

10. A single person whose income is $2,300 or less has no tax liability. What would you write in line 20? Skip lines 21 and 22 and write the same amount on line 23.

11. What amount would you write in lines 24a and 25?

12. Compare the amounts on lines 23 and 25. Which is larger? Would you have a refund or a balance due?

13. Which line would you write in, 26 or 27? What amount would you write?

To complete the tax return, you would sign your name, write the date, and attach a copy of your W-2 form. If you owe a balance due, you would also attach a check or money order for the amount you owe.

Department of the Treasury—Internal Revenue Service

Form 1040A US Individual Income Tax Return (0)

OMB No. 1545-0085

Step 1 Name and address

Use the IRS mailing label. If you don't have a label, print or type:

Your first name and initial (if joint return, also give spouse's name and initial) | Last name | Your social security no.

Present home address (number and street) | Spouse's social security no.

City, town or post office, State, and ZIP code

Presidential Election Campaign Fund

Do you want $1 to go to this fund? ☐ Yes ☐ No

If joint return, does your spouse want $1 to go to this fund? ☐ Yes ☐ No

Step 2 Filing status

(Check only one)

1 ☐ Single (See if you can use Form 1040EZ.)

2 ☐ Married filing joint return (even if only one had income)

3 ☐ Married filing separate return. Enter spouse's social security number above and spouse's full name here. ______

4 ☐ Head of household (with qualifying person). If the qualifying person is your unmarried child but not your dependent, write this child's name here. ______

Exemptions

Always check the exemption box labeled Yourself. Check other boxes if they apply.

5a ☐ Yourself ☐ 65 or over ☐ Blind

b ☐ Spouse ☐ 65 or over ☐ Blind

Write number of boxes checked on 5a and b ______

c First names of your dependent children who lived with you ______

Write number of children listed on 5c + ______

Attach Copy B of Form(s) W-2 here

d Other dependents:

1. Name	2. Relationship	3. Number of months lived in your home.	4. Did dependent have income of $1,000 or more?	5. Did you provide more than one-half of dependent's support?

Write number of other dependents listed on 5d + ______

e Total number of exemptions claimed.

Add numbers entered on lines above = ______

Step 3 Total income

6 Wages, salaries, tips, etc. (Attach Form(s) W-2.) 6 ______

7 Interest income. (If line 7 is over $400, also complete Schedule 1, Part I.) 7 ______

Attach check or money order here

8a Dividends. (If line 8a is over $400, also complete Schedule 1, Part II.) 8a ______

b Exclusion. See the instructions on page 14. 8b ______

c Subtract line 8b from line 8a. Write the result. 8c ______

9a Unemployment compensation (insurance), from Form(s) 1099-G. 9a ______

b Taxable amount, if any, from the worksheet on page 15 of the instructions. 9b ______

10 Add lines 6, 7, 8c, and 9b. Write the total. This is your total income. 10 ______

Step 4 Adjusted gross income

11a IRA deduction, from the worksheet on page 17. 11a ______

b Write IRA payments made in 1984 that you included on line 11a: ($.)

12 Deduction for a married couple when both work. Complete Schedule 1, Part III. 12 ______

13 Add lines 11a and 12. Write the total. These are your total adjustments. 13 ______

14 Subtract line 13 from line 10. Write the result. This is your adjusted gross income. 14 ______

Step 5
Taxable income

15 Write the amount from line 14. 15 ______ .

16 Allowable part of your charitable contributions, from the worksheet on page 19 of the instructions. 16 ______ .

17 Subtract line 16 from line 15. Write the result. 17 ______ .

18 Multiply $1,000 by the total number of exemptions claimed on line 5e. 18 ______ .

19 Subtract line 18 from line 17. Write the result. This is your taxable income. 19 ______ .

Step 6
Tax, credits, and payments

If You Want IRS to Figure Your Tax, See Page 19 of the Instructions.

20 Find the tax on the amount on line 19. Use the tax table, pages 29–34. 20 ______ .

21a Partial credit for political contributions. See page 20 of the instructions. 21a ______ .

b Credit for child and dependent care expenses. Complete Schedule 1, Part IV. 21b ______ .

22 Add lines 21a and 21b. Write the total. These are your total credits. 22 ______ .

23 Subtract line 22 from line 20. Write the result. This is your total tax. 23 ______ .

24a Total Federal income tax withheld. This should be shown in Box 9 of your W-2 form(s). (If line 6 is more than $35,700, see page 23 of the instructions.) 24a ______ .

b Earned income credit, from the worksheet on page 24 of the instructions. See page 23 of the instructions. 24b ______ .

25 Add lines 24a and 24b. Write the total. These are your total payments. 25 ______ .

Step 7
Refund or amount you owe

26 If line 25 is larger than line 23, subtract line 23 from line 25. Write the result. This is the **amount of your refund.** 26 ______ .

27 If line 23 is larger than line 25, subtract line 25 from line 23. Write the result. This is the **amount you owe.** Attach check or money order for full amount payable to "Internal Revenue Service." Write your social security number and "1983 Form 1040A" on it. 27 ______ .

Step 8
Sign your return

Under penalties of perjury, I declare that I have examined this return and accompanying schedules and statements, and to the best of my knowledge and belief, they are true, correct, and complete. Declaration of preparer (other than the taxpayer) is based on all information of which the preparer has any knowledge.

Your signature	Date	Your occupation
x		
Spouse's signature (if joint return both must sign)	Date	Spouse's occupation
x		
Paid preparer's signature	Date	Preparer's social security no.
x		
Firm's name (or yours, if self-employed)		Employer identification no.
Address and Zip code		Check if self-employed ☐

For **Privacy Act and Paperwork Reduction Act Notice,** see page 38.

Tax Tables

For line 20 of a 1040A return, you need to find your tax liability. To do this, you use tax tables like the one below. Tax tables are included in the instruction booklet that comes with the return. Your tax liability depends on your filing status, the number of exemptions you claim, and your adjusted gross income (line 14).

If you are a single taxpayer you would use the tables below. For example, suppose your adjusted gross income is $8624 and you claim one exemption. Notice that $8624 is between $8600 and $8650, so your tax liability is $887.

Use the tax table to find the tax liabilities for single taxpayers with the following adjusted gross incomes.

	14.	15.	16.	17.	18.
Adjusted gross income	$8058	$8775	$14,176	$14,503	$14,099

Tax Table

If 1040A, line 19 is—		And you are—				If 1040A, line 19 is—		And you are—			
At least	But less than	Single	Married filing jointly	Married filing separately	Head of a household	At least	But less than	Single	Married filing jointly	Married filing separately	Head of a household
		Your tax is—						Your tax is—			
8,000						**14,000**					
8,000	**8,050**	795	568	928	733	**14,000**	**14,050**	1,892	1,510	2,277	1,815
8,050	**8,100**	802	575	937	740	**14,050**	**14,100**	1,903	1,519	2,290	1,824
8,100	**8,150**	810	583	947	748	**14,100**	**14,150**	1,913	1,527	2,303	1,834
8,150	**8,200**	817	590	956	755	**14,150**	**14,200**	1,924	1,536	2,316	1,843
8,200	**8,250**	825	598	966	763	**14,200**	**14,250**	1,934	1,544	2,329	1,853
8,250	**8,300**	832	605	975	770	**14,250**	**14,300**	1,945	1,553	2,342	1,862
8,300	**8,350**	840	613	985	778	**14,300**	**14,350**	1,955	1,561	2,355	1,872
8,350	**8,400**	847	620	994	785	**14,350**	**14,400**	1,966	1,570	2,368	1,881
8,400	**8,450**	855	628	1,004	793	**14,400**	**14,450**	1,976	1,578	2,381	1,891
8,450	**8,500**	862	635	1,013	800	**14,450**	**14,500**	1,987	1,587	2,394	1,900
8,500	**8,550**	870	643	1,023	808	**14,500**	**14,550**	1,997	1,595	2,407	1,910
8,550	**8,600**	879	650	1,032	815	**14,550**	**14,600**	2,008	1,604	2,420	1,919
8,600	**8,650**	887	658	1,042	823	**14,600**	**14,650**	2,018	1,612	2,433	1,929
8,650	**8,700**	896	665	1,051	830	**14,650**	**14,700**	2,029	1,621	2,446	1,938
8,700	**8,750**	904	673	1,061	839	**14,700**	**14,750**	2,039	1,629	2,459	1,948
8,750	**8,800**	913	680	1,070	848	**14,750**	**14,800**	2,050	1,638	2,472	1,957
8,800	**8,850**	921	688	1,080	857	**14,800**	**14,850**	2,060	1,646	2,485	1,967
8,850	**8,900**	930	695	1,089	866	**14,850**	**14,900**	2,071	1,655	2,498	1,976
8,900	**8,950**	938	703	1,099	875	**14,900**	**14,950**	2,081	1,663	2,511	1,986
8,950	**9,000**	947	710	1,108	884	**14,950**	**15,000**	2,092	1,672	2,525	1,995

Calculating Your Tax

A few years later, you are working full time. You are single and have no dependents. Your only sources of income are your salary and the interest on your savings account. Here are your W-2 form and your statement of earnings from your savings bank.

	2. Employer's ← State number	Copy C For employee's records.	
3. Employer's name	4. Subtotal ☐ Correction ☐ Void ☐	State form number	
	5. Employer's identification number	City form number	
	6. Advance EIC payment	7.	

8. Employee's social security no.	9. Federal income tax withheld	10. Wages, tips, other compensation	11. FICA tax withheld	12. Total FICA wages
123-45-6789	2236.00	14820	1039.63	

13. Employee's name (First, middle, last)	14. Pension plan coverage? Yes/No	15. *	16. FICA tips
	YES		

	18. State income tax withheld	19. State wages, tips, etc.	20. Name of State
	705.90	14820	
17. Employee's address and Zip code	21. Local income tax withheld	22. Local wages, tips, etc.	23. Name of locality
Wage and Tax Statement 19–			
Form W-2			

HAMILTON SAVINGS BANK
HAMILTON, INDIANA

PLEASE RETAIN FOR YOUR RECORDS
STATEMENT OF EARNINGS FOR 19____

FOR YOUR PROTECTION WE ASK THAT YOU COMPARE THIS INFORMATION BELOW WITH YOUR RECORDS AND NOTIFY OUR AUDITING DEPARTMENT OF ANY DISCREPANCY.

IDENTIFYING NO.:	
ACCOUNT NUMBER	INTEREST
13579	41.25

AS OF THE CLOSE OF BUSINESS 12/31/—
YOUR CURRENT BALANCE WAS 824.36
YOUR PASSBOOK BALANCE WAS 824.36

IN ACCORDANCE WITH FEDERAL REGULATIONS, WE HAVE REPORTED INTEREST EARNED ON YOUR ACCOUNT(S) TO THE INTERNAL REVENUE SERVICE. THESE EARNINGS ARE CONSIDERED AS INTEREST FOR TAX REPORTING PURPOSES AND SHOULD BE USED DURING PREPARATION OF YOUR INCOME TAX RETURN.

Use the forms above, the return on pages 85–86, and the tax table on page 87 to answer these questions.

19. What is your salary? your interest income? your adjusted gross income?

20. How much was withheld for federal income tax? What is the your tax liability?

21. Will you have a tax refund or a balance due? How much?

Unit 3 Checking Accounts

A bank *checking account* allows you to pay for goods and services by *check* instead of cash. A check is a form of payment to another person. The bank deducts the sum from your account and pays it to the person named on the check. You keep a record of your *deposits* and the checks you have written in a *check register.* A monthly *bank statement* sent by the bank also records your deposits, withdrawals, and checks written. The amount of money in your check register and bank statement should agree, or *balance.* Some checking accounts earn interest.

Automatic teller machines can do all the routine functions that a bank teller does.

3-1 Deposits

Objective Compute the total checking account deposit.

A **deposit** is an amount of money that you put into a bank account. You use a deposit slip to record the amounts of currency, coins, and checks you deposit. To open a checking account, you must make a deposit.

▸ **Total Deposit = (Currency + Coins + Checks) − Cash Received**

Example

Skills 5, 6 **Application** A **Term** Deposit

Margaret Miller has a check for $145.52 and a check for $83.97. She would like to receive $30 in cash and deposit the rest of the money in her checking account. What is Margaret's total deposit?

Solution

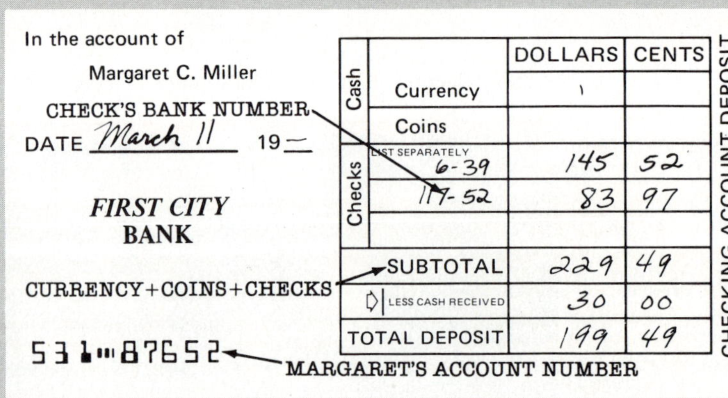

In the account of

Margaret C. Miller

CHECK'S BANK NUMBER

DATE *March 11* 19—

FIRST CITY BANK

CURRENCY+COINS+CHECKS

531⑈87652 ← MARGARET'S ACCOUNT NUMBER

		DOLLARS	CENTS
Cash	Currency		
	Coins		
Checks	LIST SEPARATELY 6-39	145	52
	117-52	83	97
	SUBTOTAL	229	49
	LESS CASH RECEIVED	30	00
	TOTAL DEPOSIT	199	49

CHECKING ACCOUNT DEPOSIT

A. $55.80
B. $35.00

▸ **Self-Check** Cover the answers and complete the problems.

A. ($40.00 + $0.80 + $15.00) − $0.00 = ? total deposit

B. ($25.00 + $30.00) − $20.00 = ? total deposit

Problems

Find the subtotal and total deposits.

	1.	2.	3.	4.	5.	6.
Currency	$50.00	$95.00			$200.00	$175.00
Coins	$ 1.80	$ 4.65				$ 10.00
Checks			$75.50	$165.26	$196.40	$496.73
			$13.80	$ 38.19	$ 31.55	$ 19.21
Subtotal	?	?	?	?	?	?
Less Cash Received			$20.00	$ 30.00	$ 25.00	$ 40.00
Total Deposit	?	?	?	?	?	?

Find the subtotal and total deposit.

			DOLLARS	CENTS
	CASH	CURRENCY		
		COINS		
	CHECKS	LIST SEPARATELY 14-2	37	48
		7-43	232	49
7.		SUBTOTAL		?
		▷| LESS CASH RECEIVED	75	50
8.		TOTAL DEPOSIT		?

			DOLLARS	CENTS
	CASH	CURRENCY	33	00
		COINS	2	49
	CHECKS	LIST SEPARATELY		
		120-18	377	59
9.		SUBTOTAL		?
		▷| LESS CASH RECEIVED		
10.		TOTAL DEPOSIT		?

11. Mary Meers has a paycheck for $265.89 and a refund check for $24.98. She would like to receive $37 in cash and deposit the remaining amount. What is Mary's total deposit?

12. Tom Moore deposited a check for $95.80 and a check for $101.17. He received $48 in cash. What was his total deposit?

13. Jack and Betty Greene deposited their paychecks for $501.44 and $618.25 and a check from their insurance company for $83.57. They received $150 in cash. What was their total deposit?

14. David Norton deposited his paycheck for $387.56, a refund check from a store for $19.78, and $54.63 in cash. What was David's total deposit?

15. Allison Scott deposits the following in her checking account: 3 five-dollar bills, 1 two-dollar bill, 14 one-dollar bills, 6 half dollars, 5 quarters, 17 dimes, 9 nickels, 16 pennies, and a check for $38.62. What is Allison's total deposit?

16. Morgan Ellis has a check for $177.40 and a check for $25.33. He would like to deposit the checks and receive 2 ten-dollar bills, 3 one-dollar bills, 4 quarters, and 10 dimes. What is Morgan's total deposit?

Maintaining Your Skills

Look up the skills in parentheses if you need help or more practice.

Add. (Skill 5)

17. $420.00 + $1.30 + $75.80
18. $200.00 + $0.65 + $171.85
19. $150.00 + $5.00 + $16.39
20. $496.74 + $123.75

Subtract. (Skill 6)

21. $450.00 − $30.00
22. $320.00 − $20.00
23. $347.50 − $25.00
24. $419.20 − $50.00
25. $619.23 − $45.00
26. $216.27 − $45.00

3-2 Writing Checks

Objective

Write a check.

After you have opened a checking account and made a deposit, you can write **checks.** A check directs a bank to deduct money from your checking account to make a payment. Your account must contain as much money as the amount of the check you are writing so that you do not **overdraw** your account.

Example

Skill 1 **Application** _ **Term** Check

Margaret Miller is buying a gift at Hansen's Department Store. The cost of the gift is \$35.48. Margaret is paying by check. How should Margaret write the check?

Solution

1. Write the date.
2. Write the name of the person or organization to whom payment will be made.
3. Write the amount of the check as a numeral.
4. Write the amount of the check in words with cents expressed as a fraction of a dollar.
5. Make a notation on the check to indicate its purpose.
6. Sign the check.

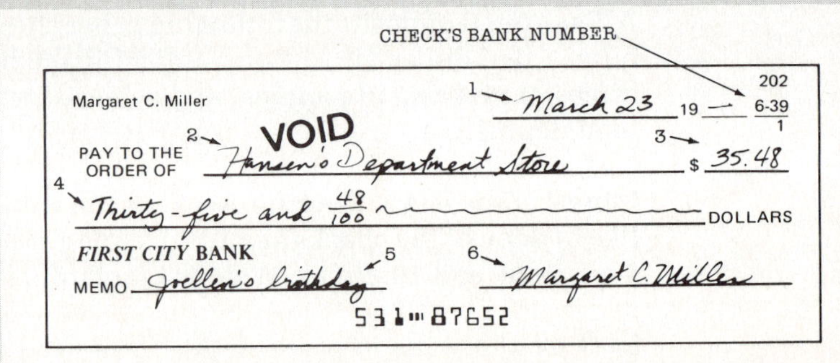

A. \$116.52
B. Two hundred fifty-three and $\frac{42}{100}$

▶ **Self-Check** Cover the answers and complete the problems.

A. Write one hundred sixteen and $\frac{52}{100}$ dollars as a numeral.

B. Write \$253.42 in words with cents expressed as a fraction of a dollar.

Problems

Write each amount in words as it would appear on a check.

1. \$10.50	2. \$375.00	3. \$55.83	4. \$9.72
5. \$94.06	6. \$29.00	7. \$4196.00	8. \$13,700.00
9. \$103.77	10. \$982.75	11. \$9456.86	12. \$44,381.15

13. Janet Smith wrote check number 311 to Midway Dental Clinic for $68.92. Did she write the amount in words correctly? If not, write the amount in words correctly.

14. For what purpose did Janet Smith write the check to the Midway Dental Clinic?

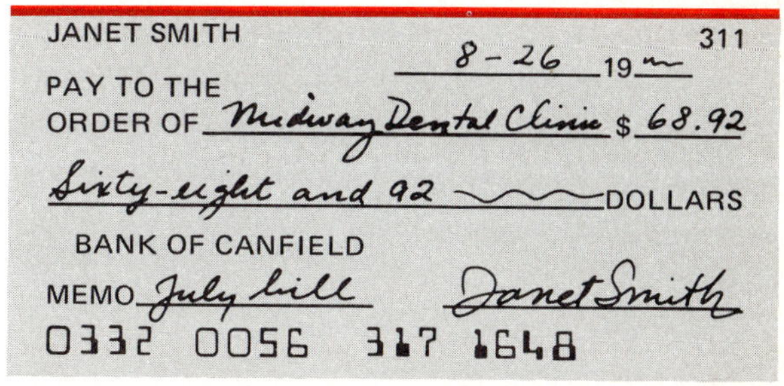
JANET SMITH 311
8-26 19__
PAY TO THE ORDER OF Midway Dental Clinic $ 68.92
Sixty-eight and 92 ~~ DOLLARS
BANK OF CANFIELD
MEMO July bill Janet Smith
0332 0056 317 1648

15. Pete Newel is paying for an auto repair bill with a check in the amount of $137.52. Did Pete write the amount of the check correctly, both as a numeral and in words?

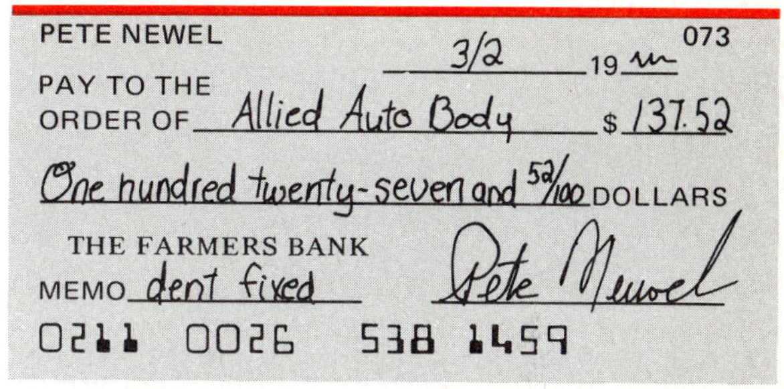
PETE NEWEL 073
3/2 19__
PAY TO THE ORDER OF Allied Auto Body $ 137.52
One hundred twenty-seven and 52/100 DOLLARS
THE FARMERS BANK
MEMO dent fixed Pete Newel
0211 0026 538 1459

16. What is the number of the check that Pete Newel used to pay for his auto repair bill?

17. Amy Lepon wrote check number 55 to Downtown Electric Company to pay her bill for the month of June. The bill totaled $36.88. Is the check made out in the correct amount?

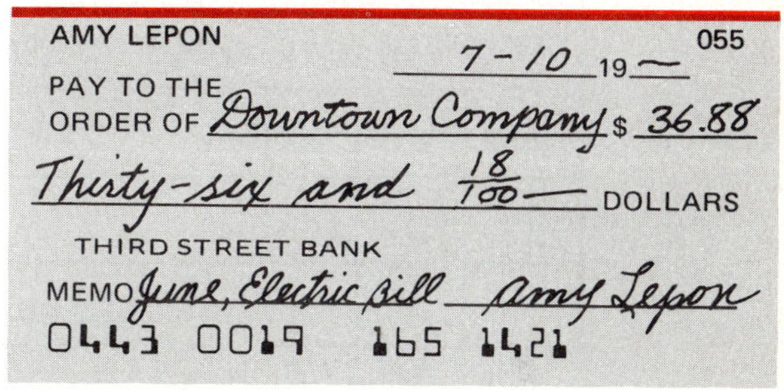
AMY LEPON 055
7-10 19__
PAY TO THE ORDER OF Downtown Company $ 36.88
Thirty-six and 18/100 DOLLARS
THIRD STREET BANK
MEMO June, Electric bill Amy Lepon
0443 0019 165 1421

18. Is the check Amy Lepon wrote made out to the correct company? If not, write the correct company name.

Maintaining Your Skills

Look up the skill in parentheses if you need help or more practice.

Write the number. (Skill 1)

19. twenty-seven and $\frac{15}{100}$ dollars

20. ninety-two and $\frac{50}{100}$ dollars

Write in word form with cents expressed as a fraction of a dollar. (Skill 1)

21. $18.35
22. $60.54
23. $345.00
24. $796.72
25. $8178.52
26. $14,718.47
27. $1610.25
28. $19,155.67

3-3 Check Registers

Objective Compute the balance in a check register.

You use a **check register** to keep a record of your deposits and the checks you write. The **balance** is the amount of money in your account. When you make a deposit, add the amount of the deposit to the balance. When you write a check, subtract the amount of the check from the balance.

▶ **New Balance = Previous Balance − Check Amount**
▶ **New Balance = Previous Balance + Deposit Amount**

Example

Skills 5, 6 **Application** _ **Term** Balance

Margaret Miller's checking account had a balance of $173.84. She wrote a check for $35.48 on March 23. On March 25, she made a deposit of $120.52. What is the new balance in Margaret's account?

Solution

CHECK NO	DATE	CHECKS ISSUED TO OR DESCRIPTION OF DEPOSIT	AMOUNT OF CHECK		✓	AMOUNT OF DEPOSIT		BALANCE	
		PREVIOUS BALANCE − CHECK AMOUNT BALANCE BROUGHT FORWARD →						173	84
202	3/23	Hansen's Dept. Store	35	48				138	36
	3/25	Deposit				120	52	258	88
		PREVIOUS BALANCE + DEPOSIT AMOUNT							

A. $111.90
B. $332.45

▶ **Self-Check** Cover the answers and complete the problems.

A. Balance $156.90
Amount of check − 45.00
Balance ?

B. Balance $212.45
Deposit + 120.00
Balance ?

Problems Find the new balance after each check or deposit.

	AMOUNT OF CHECK		✓	AMOUNT OF DEPOSIT		BALANCE	
	BALANCE BROUGHT FORWARD →					149	35
1.	16	72					
2.				126	84		
3.	241	47					

	AMOUNT OF CHECK		✓	AMOUNT OF DEPOSIT		BALANCE	
	BALANCE BROUGHT FORWARD →					423	86
4.	13	80					
5.	42	95					
6.				104	00		

7. Your balance is $79.65 on May 23. Deposit $256.90 on May 30. Write a $96.79 check on April 2. What is your new balance?

8. Your balance is $81.03 on April 5. Write a $21.78 check on April 9. Write a $35.62 check on April 11. What is your new balance?

9. Richard Stevens opened a new checking account by depositing his paycheck for $309.71. The check register shows his transactions since opening his account. What is his balance after each transaction?

CHECK NO.	DATE	CHECKS ISSUED TO OR DESCRIPTION OF DEPOSIT	AMOUNT OF CHECK		AMOUNT OF DEPOSIT		BALANCE	
		BALANCE BROUGHT FORWARD →					309	71
101	3/12	Connor's Clothiers	117	88				
102	3/18	Hammonds Jewelers	47	62				
103	3/20	Corner Food King	52	37				

10. Diane Russell's checkbook balance was $103.22 on October 5. Her check register shows her transactions since. What is Diane's balance after each transaction?

CHECK NO.	DATE	CHECKS ISSUED TO OR DESCRIPTION OF DEPOSIT	AMOUNT OF CHECK		AMOUNT OF DEPOSIT		BALANCE	
		BALANCE BROUGHT FORWARD →					103	22
571	10/6	Don's Sporting Goods	85	95				
	10/9	Deposit			231	40		
572	10/10	Dept. of Public Works	14	59				
573	10/19	Central Telephone	35	15				

11. Graham Olin's latest transactions are shown on the check register. Find his balance after each transaction.

CHECK NO.	DATE	CHECKS ISSUED TO OR DESCRIPTION OF DEPOSIT	AMOUNT OF CHECK		AMOUNT OF DEPOSIT		BALANCE	
		BALANCE BROUGHT FORWARD →					216	42
472	11/17	Kline Dept. Store	16	00				
473	11/18	Cash	100	00				
	11/20	Deposit			345	70		
474	11/21	Allen's Hardware	96	20				
	12/4	Deposit			250	00		

(Continue on next page.)

12. Tony Wallace opened a new checking account by depositing his IRS return of $245.70. The check register shows his transactions to date. What is his balance after each transaction?

CHECK NO.	DATE	CHECKS ISSUED TO OR DESCRIPTION OF DEPOSIT	AMOUNT OF CHECK		AMOUNT OF DEPOSIT		BALANCE	
		BALANCE BROUGHT FORWARD →					245	70
1	4/20	High School Pep Club	15	00				
2	4/25	Discount Sport Shop	21	63				
3	4/26	Racquet Club	12	50				
	5/1	Deposit			100	00		
4	5/3	Mr. Formal Rental	32	47				
5	5/3	City Florists	18	43				
6	5/4	Majestic Restaurant	23	45				
7	5/15	Herb Jones Ring Co.	71	64				

13. Helen McInness's latest transactions are shown on the check register. Find her balance after each transaction.

CHECK NO.	DATE	CHECKS ISSUED TO OR DESCRIPTION OF DEPOSIT	AMOUNT OF CHECK		AMOUNT OF DEPOSIT		BALANCE	
		BALANCE BROUGHT FORWARD →					418	92
916	8/20	National Bank	212	45				
	8/22	Deposit			475	00		
917	9/3	Federal Mortgage Co.	515	78				
918	9/4	Natural Gas Co.	71	62				
	9/5	Deposit			220	00		
	9/19	Deposit			165	00		
919	9/20	National Bank	212	45				
920	9/22	Cash	50	00				

Maintaining Your Skills

Look up the skills in parentheses if you need help or more practice.

Add. (Skill 5)

14. $214.75 + $135.50
15. $745.36 + $100.00
16. $491.78 + $312.50
17. $61.75 + $191.36
18. $1471.81 + $581.48
19. $671.80 + $234.65

Subtract. (Skill 6)

20. $479.23 − $112.60
21. $247.89 − $49.92
22. $361.85 − $7.47
23. $141.82 − $94.63
24. $2417.80 − $535.60
25. $971.82 − $214.91

3-4 Bank Statements

Objective Compute the present balance on a checking account bank statement.

When you have a checking account, you receive a **statement** and **canceled checks** from the bank each month. Canceled checks are the checks that the bank has paid by deducting money from your account. Your statement lists all your checks that the bank has paid and your deposits that the bank has recorded since your last statement. The statement may include a **service charge** for handling the account.

▶ **Present Balance** = **Previous Balance** + **Deposits Recorded** − **Checks Paid** − **Service Charge**

Example

Skills 5, 6 **Application** A **Term** Statement

Margaret Miller received her bank statement and canceled checks for March. She checks the statement. What is her present balance?

Solution

FIRST CITY **BANK**

Margaret Miller

Date Last Statement: 3/1/–

Balance Last Statement: 35.87

CLOSING DATE: 4/1/–

Checking Account Number: 531-87652

CHECKS AND OTHER CHARGES			DEPOSITS AND CREDITS		BALANCE
Date	Number	Amount	Date	Amount	
3/23	202	35.48	3/11	199.49	235.36
3/27	203	45.75	3/25	120.52	199.88
					320.40
	Service charge	.20			274.65
					274.45

Total of all deposits processed

Total of all checks processed

PREVIOUS BALANCE	DEPOSITS RECORDED	CHECKS PAID	SERVICE CHARGE	PRESENT BALANCE
$35.87	$320.01	$81.23	$0.20	$274.45

▶ **Self-Check** Cover the answers and complete the problems.

	Previous Balance	+	Deposits Recorded	−	Checks Paid	−	Service Charge	=	Present Balance
A.	$270.00	+	$150.00	−	$120.00	−	$2.50	=	?
B.	$175.50	+	$145.00	−	$211.40	−	$3.00	=	?

A. $297.50
B. $106.10

Problems

	1.	2.	3.	4.	5.	6.
Previous Balance	$ 50.10	$123.67	$ 49.07	$200.00	$ 101.99	$1513.11
Total Deposits	$240.00	$330.75	$401.00	$573.16	$1509.03	$ 412.46
Total Checks	$190.10	$398.17	$419.80	$699.44	$1598.86	$1809.52
Service Charge	$ 2.23	$ 1.25	$ 1.15	0	$ 3.75	$ 3.60
Present Balance	?	?	?	?	?	?

7. A portion of Janet Raines's bank statement is shown. Her previous balance was $183.42. What is her present balance?

CHECKS AND OTHER CHARGES			DEPOSITS AND CREDITS		BALANCE
DATE	NUMBER	AMOUNT	DATE	AMOUNT	
6/11	304	8.45	6/12	215.90	
6/15	305	11.02	6/19	215.90	
6/30	307	85.95	6/26	445.85	
Service charge		1.85			

8. A portion of William Powers's bank statement is shown below. His previous balance was $290.88. What is his present balance?

CHECKS AND OTHER CHARGES			DEPOSITS AND CREDITS		BALANCE
DATE	NUMBER	AMOUNT	DATE	AMOUNT	
11/2	1031	112.00	11/9	105.55	
11/5	1033	12.38	11/19	112.48	
11/20	1032	47.98	11/29	105.80	
No service charge					

9. A portion of Irene Brook's bank statement is shown. Her previous balance was $199.27. What is her present balance?

CHECKS AND OTHER CHARGES			DEPOSITS AND CREDITS		BALANCE
DATE	NUMBER	AMOUNT	DATE	AMOUNT	
1/18	386	147.50	1/20	564.50	
1/22	389	25.00	1/31	221.37	
1/30	387	417.75	2/12	164.22	
2/5	388	50.00			
Service charge		.50			

10. Robert North received his checking account statement and canceled checks for September. His previous balance was $247.82. What is his present balance?

CHECKS AND OTHER CHARGES			DEPOSITS AND CREDITS		BALANCE
DATE	NUMBER	AMOUNT	DATE	AMOUNT	
9/3	451	17.85	9/7	200.00	229.97
9/9	452	161.75	9/21	200.00	429.97
					268.22
					468.22
9/21	453	71.89			396.33
9/21	454	25.00			371.33
Service charge		2.50			368.83

PREVIOUS BALANCE	DEPOSITS RECORDED	CHECKS PAID	SERVICE CHARGE	PRESENT BALANCE
?	?	?	?	?

11. Wilma Baker received her checking account statement and canceled checks for June. Her previous balance was $421.84. What is her present balance?

CHECKS AND OTHER CHARGES			DEPOSITS AND CREDITS		BALANCE
DATE	NUMBER	AMOUNT	DATE	AMOUNT	
6/4	916	215.25	6/10	215.50	206.59
6/8	917	75.00			131.59
			6/24	215.50	347.09
6/20	918	117.85			229.24
6/21	919	192.35			36.89
					252.39
Service charge		3.00			249.39

PREVIOUS BALANCE	DEPOSITS RECORDED	CHECKS PAID	SERVICE CHARGE	PRESENT BALANCE
?	?	?	?	?

Maintaining Your Skills

Look up the skills in parentheses if you need help or more practice.

Add. (Skill 5)

12. $216.50 + $125.50 + $95.75

13. $345.70 + $217.40 + $231.50

14. $417.35 + $148.55 + $917.85 + $3.00

15. $23,960.45 + $1567 + $61.82

Subtract. (Skill 6)

16. $715.37 − 648.42

17. $584.31 − 63.49

18. $924.81 − 691.84

19. $1256.93 − 73.42

20. 162.82 − 78.93

21. 278.93 − 183.49

22. 263.890 − 23.461

23. 1941.862 − 216.499

3-5 Reconciling the Bank Statement

Objective Reconcile a check register and a bank statement.

When you receive your bank statement, you compare the canceled checks, the bank statement, and your check register to be sure they agree. You may find some **outstanding** checks and deposits that appear in your register but did not reach the bank in time to be processed and listed on your statement. You **reconcile** the statement to make sure that it agrees with your check register.

▶ **Adjusted Balance** = **Statement Balance** − **Outstanding Checks** + **Outstanding Deposits**

Example

Skills 5, 6 **Application** A **Term** Reconcile

Margaret Miller's check register balance is $283.63. She compared her statement, canceled checks, and check register. For each check and deposit listed on her statement, she placed a checkmark next to the information in her register. Margaret found these outstanding checks and deposits:

check #201 $61.52; check #204 $12.00; deposit $82.50

How does Margaret reconcile her statement?

Solution

RECONCILIATION STATEMENT

Check Register Balance $ 283.63

Service Charge − .20

$ 283.43 ← NEW BALANCE

Service charge must be entered in the register and subtracted from the balance.

If these balances agree, the records are correct.

Statement Balance $ 274.45

Outstanding Checks
#201 $61.52
#204 $12.00
− $ 73.52
$ 200.93

Outstanding Deposits
$82.50
+ $ 82.50

ADJUSTED BALANCE $ 283.43

A. $166.09
B. $849.12

▶ **Self-Check** Cover the answers and complete the problems.

A.		
Statement balance		$165.47
Outstanding checks	−	49.38
Outstanding deposits	+	50.00
Adjusted balance		?

B.		
Statement balance		$994.47
Outstanding checks	−	361.85
Outstanding deposits	+	216.50
Adjusted balance		?

Problems

Complete the table. Do the register and statement balances agree?

	1.	2.	3.	4.
Check Register Balance	\$488.15	\$285.51	\$1142.44	\$2585.45
Service Charge	\$ 1.20	\$ 2.10	\$ 1.10	0
NEW BALANCE	?	?	?	?
Statement Balance	\$620.41	\$407.56	\$1260.54	\$2124.00
Outstanding Checks	\$264.18	\$215.15	0	\$ 116.71
Outstanding Deposits	\$130.72	\$ 91.00	\$ 119.20	\$ 578.16
ADJUSTED BALANCE	?	?	?	?

5. After comparing her bank statement, canceled checks, and checkbook register, Jane Scott completes the reconciliation statement shown below. What is the adjusted balance? Do the register and statement balances agree?

RECONCILIATION STATEMENT

CHECK REGISTER BALANCE \$ 213.40

SERVICE CHARGE – 2.20

NEW BALANCE \$ 211.20

STATEMENT BALANCE \$204.20

OUTSTANDING CHECKS

#071 \$31.20

#080 \$24.20 – \$____

\$____

OUTSTANDING DEPOSITS

\$62.40

\+ \$____

ADJUSTED BALANCE \$____

** RECONCILIATION STATEMENT FOR YOUR CONVENIENCE

6. Charlie Tanner completes the reconciliation statement shown. What are the new and adjusted balances? Do they agree?

RECONCILIATION STATEMENT

CHECK REGISTER BALANCE \$ 309.43

SERVICE CHARGE – 1.85

NEW BALANCE \$____

STATEMENT BALANCE \$216.74

OUTSTANDING CHECKS

#609 \$29.41

– \$____

\$____

OUTSTANDING DEPOSITS

\$120.00

\+ \$____

ADJUSTED BALANCE \$____

** RECONCILIATION STATEMENT FOR YOUR CONVENIENCE

(Continue on next page.)

7. Colleen Pridemore received her bank statement and canceled checks for the period ending November 20. She compared the check register with the canceled checks and deposits listed on the statement, then placed a checkmark next to the items processed. She reconciles the bank statement.

a. What total amount does Colleen have in outstanding checks?
b. What total amount does she have in outstanding deposits?
c. What is her adjusted balance?
d. What is Colleen's new check register balance?
e. Do the register and adjusted balances agree?

CHECK NO.	DATE	CHECKS ISSUED TO OR DESCRIPTION OF DEPOSIT	AMOUNT OF CHECK	✓	AMOUNT OF DEPOSIT	BALANCE
		BALANCE BROUGHT FORWARD →				402.73
047	11/3	General Dept. Store	73.42	✓		329.31
	11/4	Deposit		✓	250.00	579.31
048	11/9	Central Tel. Co.	19.71	✓		559.60
049	11/15	Cash	60.00	✓		499.60
	11/20	Deposit			250.00	749.60
050	11/20	United Power Co.	23.19			726.41

DATE LAST STATEMENT: 10/20/--
BALANCE LAST STATEMENT: 402.73
83-30283
CLOSING DATE: 11/20/--

AND OTHER CHARGES DATE	NUMBER	AMOUNT	DEPOSITS AND CREDITS DATE	AMOUNT	BALANCE
11/5	047	73.42	11/4	250.00	652.73
11/15	048	19.71			579.31
11/15	049	60.00			559.60
Service charge		2.90			499.60
					496.70

PREVIOUS BALANCE	DEPOSITS RECORDED	CHECKS PAID	SERVICE CHARGE	PRESENT BALANCE
$402.73	$250.00	$153.13	$2.90	$496.70

CHECK REGISTER BALANCE	$ 726.41	STATEMENT BALANCE	$ 496.70
SERVICE CHARGE	− 2.90	OUTSTANDING CHECKS	
NEW BALANCE	$ ____	#50 $23.19	− $ ____
			$ ____
** RECONCILIATION STATEMENT FOR YOUR CONVENIENCE		OUTSTANDING DEPOSITS	
		$250.00	+ $ ____
		ADJUSTED BALANCE	$ ____

***8. A Brief Case** Jasper Henning received his bank statement and canceled checks for the month of May. He compared them against the items listed in his check register. He found the items shown below outstanding. Reconcile his bank statement.

CHECK NO.	DATE	CHECKS ISSUED TO OR DESCRIPTION OF DEPOSIT	AMOUNT OF CHECK	✓	AMOUNT OF DEPOSIT	BALANCE
		BALANCE BROUGHT FORWARD →				224 80
108	5/29	Dept. Motor Vehicles	19 00			205 80
	6/1	Deposit			405 80	611 60
109	6/4	McKinley Shoes	52 45			559 15
110	6/6	First Avenue Bank	299 00			260 15
	6/11	Deposit			300 50	560 65

DATE LAST STATEMENT 5/1/--
BALANCE LAST STATEMENT 1639.77
CLOSING DATE 6/1/--
NUMBER 30-92854

AND OTHER CHARGES			DEPOSITS AND CREDITS		BALANCE
DATE	NUMBER	AMOUNT	DATE	AMOUNT	
5/8	103	1.90	5/7	100.00	1739.77
5/11	105	151.03	5/11	100.00	1737.87
5/12	101	50.00	5/15	383.31	1686.84
5/12	102	69.95			1566.89
5/14	106	26.40			1540.49
5/25	107	1400.00			1923.80
5/27	104	299.00			523.80
Service charge — none					224.80

PREVIOUS BALANCE	DEPOSITS RECORDED	CHECKS PAID	SERVICE CHARGE	PRESENT BALANCE
$1639.77	$583.31	$1998.28	None	$224.80

CHECK REGISTER BALANCE	$ 560.65	STATEMENT BALANCE	$224.80
SERVICE CHARGE	— ______	OUTSTANDING CHECKS	
NEW BALANCE	$ ______	______	
		______	— $ ______
		______	$ ______
** RECONCILIATION STATEMENT FOR YOUR CONVENIENCE		OUTSTANDING DEPOSITS	
		______	+ $ ______
		______	
		ADJUSTED BALANCE	$ ______

Maintaining Your Skills

Look up the skill in parentheses if you need help or more practice.

Subtract. (Skill 6)

9. \$412.30 – \$1.25 **10.** \$219.63 – \$2.50 **11.** \$96.78 – \$3.00

12. \$349.82 – \$116.78 **13.** \$491.60 – \$247.83 **14.** \$127.80 – \$64.92

15. \$421.92 – \$250.00 **16.** \$216.91 – \$150.00 **17.** \$97.83 – \$51.47

Reviewing the Basics

Skills

Write the amount in words with cents expressed as a fraction of a dollar.

(Skill 1)

1. $56.39
2. $347.33
3. $1245.97
4. $13,692.00

Solve.

(Skill 5)

5. $72.08 + $49.95 + $80.64
6. $1380.53 + $75.48 + $1.49
7. $439.39 + $137.95
8. $4820.30 + $899.75

(Skill 6)

9. $516.73 − $75.00
10. $873.91 − $315.90
11. $237.85 − $215.73
12. $617.83 − $551.32

Applications

Use the formula to solve.

Total Deposit = (Currency + Coins + Checks) − Cash Received

(Application A)

13. Deposit slip for Jan Jones.
Check for $203.55.
Coins totaling $3.39.
Currency totaling $38.
What is the total deposit?

14. Deposit slip for Tim Barnes.
Check for $109.33.
Currency totaling $73.
Received a $50 bill.
What is the total deposit?

Terms

Match each term with its definition on the right.

15. Balance
16. Reconcile
17. Deposit
18. Statement
19. Check

a. to obtain agreement between two financial records by accounting for outstanding items

b. an amount of money that you put into an account

c. the amount of money in your account

d. a written order directing the bank to deduct money from your checking account to make a payment

e. items which did not reach the bank in time to be processed and listed in your statement

f. a record prepared by the bank listing all transactions the bank has recorded

Refer to your reference files at the back of the book if you need help.

Problem Solving
Restating the Problem

To solve a word problem, you may find it helpful to restate the problem in a different way. A word problem that appears very difficult or complicated can be restated and become simple to solve. Or, a word problem may involve a series of simpler problems that will lead to the solution of the original problem.

Example 1 A single elimination tournament is to be held among 62 contestants. Single elimination means that you are out of the tournament if you have 1 loss. How many matches will have to be played to determine a winner?

Solution Given that it is a single elimination tournament, every player but the winner will have to lose 1 match. Sixty-one contestants will have to lose 1 match; therefore, 61 matches will have to be held.

Example 2 A 24-inch bicycle wheel makes 5 revolutions per minute. A stone stuck in the treads will travel how many feet in 1 hour?

Solution Restate the problem to follow a series of simpler problems.

1. How many inches will the stone travel in 1 revolution?
   $C = \pi d$ or 3.1416×24 inches $= 75.3984$ inches
2. How many inches will the stone travel in 1 minute?
   5 rpm $\times$ 75.3984 inches per revolution $=$ 376.992 inches
3. How many inches will the stone travel in 1 hour?
   376.922 inches per minute $\times$ 60 minutes per hour $=$ 22,619.52 inches
4. How many feet will the stone travel in 1 hour?
   22,619.52 inches $\div$ 12 inches per foot $=$ 1884.96 feet

---

Problems

1. Your automobile's gas mileage is 32.5 miles per gallon. Last month your automobile was driven 2015 miles. A gallon of gasoline cost $1.19 last month. What did it cost to buy gasoline for your automobile last month?

2. Two cyclists start toward each other from points 35 miles apart. One cyclist travels 15 mph while the other cyclist travels 20 mph. A trained bird leaves the shoulder of 1 cyclist and travels to the shoulder of the other cyclist and back and forth until the cyclists meet. If the bird flies at 40 mph, how far will the bird fly?

3. Three men agree to share a hotel room and pay $30 when they check in. Later, the clerk finds that he should have charged them only $25. He sends the bellhop up with a $5 refund for the room, but because the bellhop cannot divide $5 evenly by 3, he returns only $3 to the men and keeps $2. Therefore, each man paid only $9 as his share of the room for a total of $27. The bellhop kept the other $2 for a total of $29. What happened to the extra $1?

Unit Test

Lesson 3-1

1. A portion of Cathy Cantrell's deposit slip is shown. What is her total deposit?

		DOLLARS	CENTS
CASH	CURRENCY		
	COINS	9	50
CHECKS	LIST SEPARATELY 9-14	62	80
	18-4	186	81
SUBTOTAL			
LESS CASH RECEIVED		80	00
TOTAL DEPOSIT			

Lesson 3-2

2. Milo Dinar wrote a check to Acton Electronics in the amount of $599.95 for the purchase of a microcomputer. The check number is 363. Write the amount of the check in words with cents expressed as a fraction of a dollar.

Lesson 3-3

3. A portion of Bill Burton's check register is shown listing his transactions since June 5. What is his new balance?

CHECK NO.	DATE	CHECKS ISSUED TO OR DESCRIPTION OF DEPOSIT	AMOUNT OF CHECK		AMOUNT OF DEPOSIT		BALANCE	
BALANCE BROUGHT FORWARD →							717	22
	6/5	Deposit			115	90		
082	6/9	Miller's Hardware	33	87				
083	6/18	Corner Drug	24	96				

Lesson 3-4

4. A portion of Maria Lee's bank statement is shown. Her balance from last month's statement was $593.43. What is her present balance?

CHECKS AND OTHER CHARGES			DEPOSITS AND CREDITS		BALANCE
DATE	*NUMBER*	*AMOUNT*	*DATE*	*AMOUNT*	
10/4	902	52.95	10/8	225.35	
10/4	904	12.21	10/22	304.55	
10/15	903	185.00			
Service charge		2.35			

Lesson 3-5

5. Maria Lee compares her checkbook register, canceled checks, and bank statement. She finds that she has check numbers 901 and 905 outstanding in the amounts of $86.35 and $39.57. She also has an outstanding deposit of $200. She reconciles the statement. What should her adjusted balance be?

Unit 4 Savings Accounts

▸ A *savings account* is used for savings only. You cannot write checks on this account. Instead of a check register, you keep track of your money with a *passbook*. Each time you deposit or withdraw money, your bank records the transaction in your passbook. Money deposited in a savings account earns *interest*. Interest is money earned on the funds, or *principal*, in your account. The more often the interest is computed, or *compounded*, the more money your account will earn.

A savings account may pay a higher interest rate than a checking account.

4-1 Deposits

Objective

Complete a savings account deposit slip and compute total deposit.

To open a savings account you must make a **deposit.** Each time you make a deposit, it is added to the balance of your account. You fill out on savings account deposit slips the cash and checks that you deposit.

▸ **Total Deposit = (Currency + Coins + Checks) − Cash Received**

Example

Skills 5, 6 **Application** A **Term** Deposit

Paul Bentley has \$17.00 in currency, \$2.23 in coins, and a check for \$18.24 to deposit in his savings account. How much will Paul deposit?

Solution

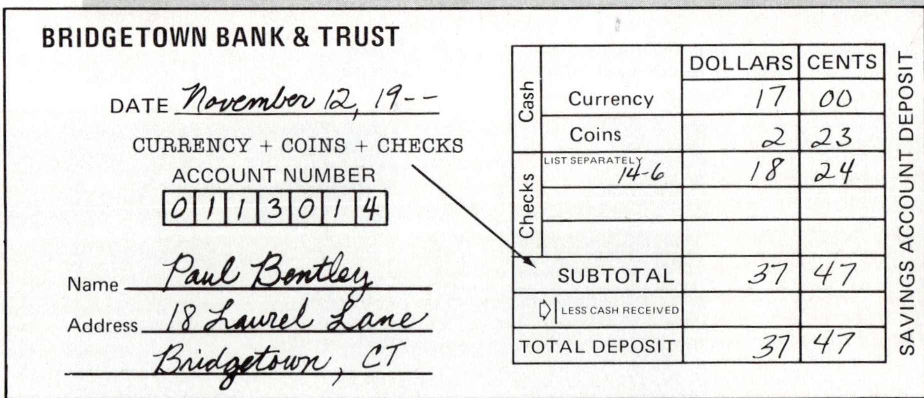

BRIDGETOWN BANK & TRUST

DATE *November 12, 19--*

CURRENCY + COINS + CHECKS

ACCOUNT NUMBER 0 1 1 3 0 1 4

Name *Paul Bentley*

Address *18 Laurel Lane*

Bridgetown, CT

		DOLLARS	CENTS
Cash	Currency	17	00
	Coins	2	23
Checks	LIST SEPARATELY 14-6	18	24
	SUBTOTAL	37	47
	LESS CASH RECEIVED		
	TOTAL DEPOSIT	37	47

SAVINGS ACCOUNT DEPOSIT

A. \$65.00
B. \$60.00

▸ **Self-Check** Cover the answers and complete the problems.

Find the total deposit.

A. Deposited checks for \$50 and \$35. Received \$20 in cash.

B. Deposited checks for \$70 and \$25. Received \$35 in cash.

Problems

Find the subtotal and total deposit.

	1.	2.	3.	4.	5.
Deposits	\$ 52.00 2.65 45.75	\$27.00 1.37 34.28	\$ 62.50 38.95	\$216.70 195.85	\$ 15.00 91.75 118.42
Subtotal	?	?	?	?	?
Less Cash Received	0	0	\$ 25.00	\$ 50.00	\$ 30.00
Total Deposit	?	?	?	?	?

6. Susan Foster.
Deposits $52 in cash.
Deposits $9.25 in coins.
What is the total deposit?

7. Norman Kaplan.
Deposits $4.99 in coins.
Deposits a check for $84.15.
What is the total deposit?

8. Louis Rodero.
Deposits a check for $125.18.
Deposits a check for $39.48.
Deposits $50 in cash.
What is the total deposit?

9. Barbara Miller.
Deposits a check for $214.63.
Deposits a check for $96.13.
Receives $40 in cash.
What is the total deposit?

10. Consuela Sanders fills out the savings deposit form shown. What is her total deposit?

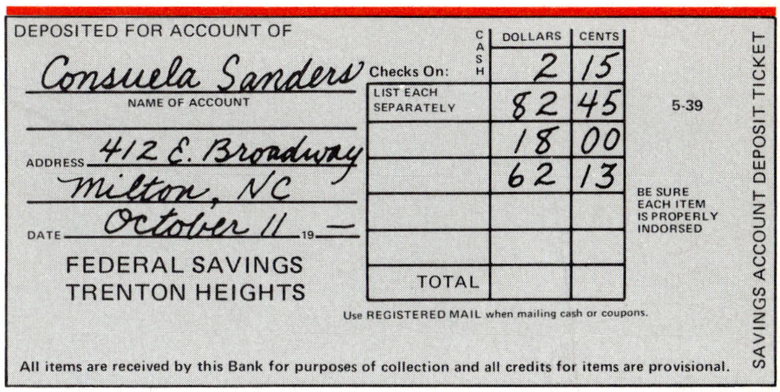
DEPOSITED FOR ACCOUNT OF

Consuela Sanders

NAME OF ACCOUNT

ADDRESS 412 E. Broadway

Milton, NC

DATE October 11 19—

FEDERAL SAVINGS
TRENTON HEIGHTS

	DOLLARS	CENTS
CASH	2	15
Checks On: LIST EACH SEPARATELY	82	45
	18	00
	62	13
TOTAL		

5-39

BE SURE EACH ITEM IS PROPERLY INDORSED

Use REGISTERED MAIL when mailing cash or coupons.

All items are received by this Bank for purposes of collection and all credits for items are provisional.

SAVINGS ACCOUNT DEPOSIT TICKET

11. Alfred Shields deposited a check for $112.28 and another check for $92.17 in his savings account. He received $80 in cash. What was his total deposit?

12. Desmond Green completed a savings account deposit slip on which he recorded checks for $153.18 and $11.99 for deposit. He received $65.17 in cash. What was Desmond's total deposit?

13. Darlene Spencer deposited 2 twenty-dollar bills, 1 ten-dollar bill, 10 quarters, 5 dimes, 17 pennies, and a check for $91.10 in her savings account. What was Darlene's total deposit?

14. Steve Calisi had 2 checks for $319.84 and $79.31. He would like to deposit the 2 checks and receive 6 twenty-dollar bills, 5 one-dollar bills, and 10 quarters. What is Steve's total deposit?

15. Arlene Doller would like to deposit 2 checks for $198.50 and $63.72. She would like to receive $15 in cash. What is Arlene's total deposit?

Maintaining Your Skills

Look up the skills in parentheses if you need help or more practice.

Add. (Skill 5)

16. $72.50 + 14.45
17. $16.50 + 9.22
18. $246.18 + 15.93
19. $196.43 + 216.91
20. $45.00 + 1.32 + 16.92

Subtract. (Skill 6)

21. $94.85 − 10.00
22. $197.75 − 55.50
23. $212.95 − 50.00
24. $573.85 − 65.00
25. $98.45 − 48.75

4-2 Withdrawals

Objective Complete a savings account withdrawal slip.

A **withdrawal** is a sum of money that you take out of your savings account. Your withdrawal is subtracted from the balance of your account. You fill out on withdrawal slips the money you are taking from your account.

Example

Skill 1 **Application** _ **Term** Withdrawal

Paul Bentley would like to withdraw $27 from his savings account. His account number is 0113014. How should he fill out the withdrawal slip?

Solution

1. Write the date of withdrawal.
2. Write the savings account number.
3. Write the amount withdrawn in words with cents expressed as a fraction of a dollar.
4. Write the amount withdrawn as a numeral.
5. Sign the withdrawal slip.

BRIDGETOWN BANK & TRUST

1 → DATE December 12, 19--

2 → 0 1 1 3 0 1 4 SAVINGS ACCOUNT NUMBER

3 → Twenty-seven and 00/100 DOLLARS

4 → $ 27 | 00 (dollars, cents)

5 → Paul Bentley SIGN HERE

WITHDRAWAL

A. Forty-six and $\frac{52}{100}$ dollars
B. $194.37

▸ **Self-Check** Cover the answers and complete the problems.

A. Write $46.52 in words with cents expressed as a fraction of a dollar.

B. Write one hundred ninety-four and $\frac{37}{100}$ dollars as a numeral.

Problems Write each amount in words as it would appear on a withdrawal slip.

1. $20.25	2. $18.90	3. $65.72	4. $91.05
5. $103.00	6. $215.64	7. $1496.17	8. $2715.75

For each of the following exercises, write (a) the account number, (b) the amount as a numeral, and (c) the amount in words.

9.

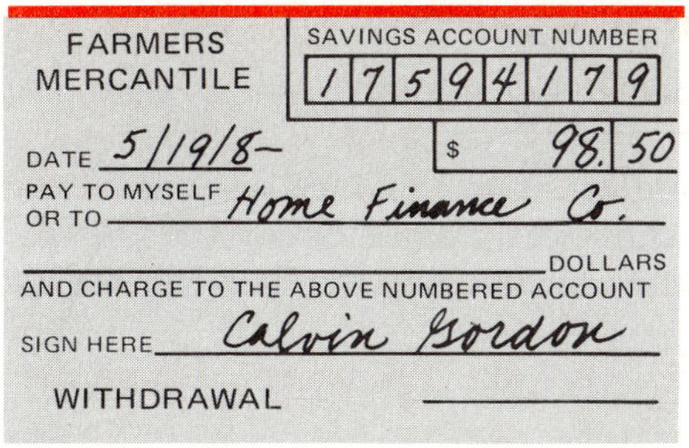
FARMERS MERCANTILE
SAVINGS ACCOUNT NUMBER 1 7 5 9 4 1 7 9
DATE 5/19/8– $ 98.50
PAY TO MYSELF OR TO Home Finance Co.
DOLLARS
AND CHARGE TO THE ABOVE NUMBERED ACCOUNT
SIGN HERE Calvin Gordon
WITHDRAWAL

10.

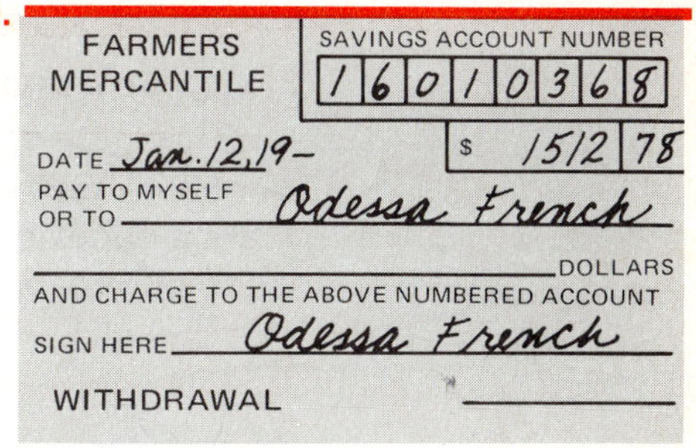
FARMERS MERCANTILE
SAVINGS ACCOUNT NUMBER 1 6 0 1 0 3 6 8
DATE Jan. 12, 19– $ 1512.78
PAY TO MYSELF OR TO Odessa French
DOLLARS
AND CHARGE TO THE ABOVE NUMBERED ACCOUNT
SIGN HERE Odessa French
WITHDRAWAL

11. Mike Moran's savings account number is 81-0-174927. He fills out a savings withdrawal slip for $281.92 to purchase a gift.

12. Carol Dilulo has been saving for a trip abroad. Her travel agent has arranged a trip to Europe that will cost $1560.98. Carol withdraws the amount from her savings account. Her account number is 13-122-541.

13. John King has been saving to buy a commemorative stamp for his stamp collection. He fills out the savings withdrawal slip shown.

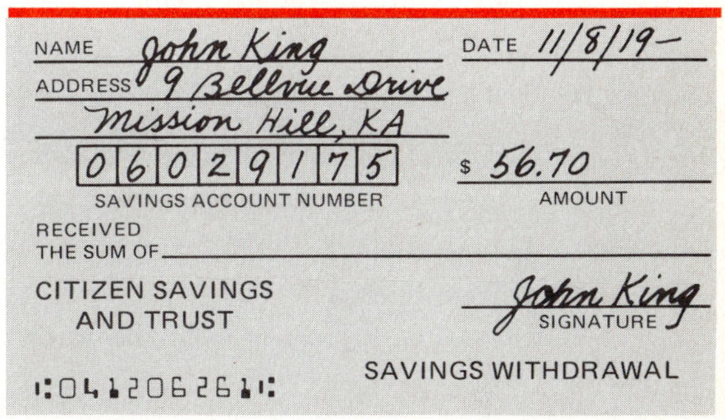
NAME John King DATE 11/8/19–
ADDRESS 9 Bellevue Drive
Mission Hill, KA
0 6 0 2 9 1 7 5
SAVINGS ACCOUNT NUMBER
$ 56.70
AMOUNT
RECEIVED THE SUM OF
CITIZEN SAVINGS AND TRUST
John King
SIGNATURE
SAVINGS WITHDRAWAL
⑆0412062611⑆

14. Robert Mercer has finally saved enough money to buy a motorcycle. The total purchase price is $1395.58. He fills out a withdrawal slip for the amount. His savings account number is 7086-5.

Maintaining Your Skills

Look up the skill in parentheses if you need help or more practice.

Write in words with cents expressed as a fraction of a dollar. (Skill 1)

15. $74.89 16. $192.43 17. $216.50 18. $17.21 19. $1241.80

Write as a numeral. (Skill 1)

20. ninety-four and $\frac{50}{100}$ dollars

21. five hundred sixty-two and $\frac{19}{100}$ dollars

22. three thousand two hundred fifteen and $\frac{74}{100}$ dollars

23. thirty-one thousand four hundred twenty-one and $\frac{19}{100}$ dollars

4-3 Passbooks

Objective

Compute the new balance in a savings account passbook.

Your bank may provide you with a savings account **passbook.** When you make a deposit or withdrawal, a bank teller records in your passbook the transaction, any interest earned, and the new balance.

▶ **New Balance = Previous Balance + Interest + Deposits − Withdrawals**

Example

Skills 5, 6 **Application** A **Term** Passbook

Amy Bentley has a passbook savings account. On April 1, she deposited $62.25 in her account. The bank teller recorded the transaction and the interest earned during the past savings period. What is the new balance in her account?

DEPOSITOR: Amy Bently
ACCOUNT NO.: 08- 021235-4

	DATE	INTEREST	DEPOSIT	WITH-DRAWAL	BALANCE
18	01-01	4.87			784.15
19	01-12		150.00		934.15
20	02-11		24.50		958.65
21	03-03			20.00	938.65
22	04-01	12.37			951.02
23	04-01		62.25		?
24					

Solution

Find the **new balance.**

Previous Balance	+	Interest	+	Deposit	−	Withdrawal		
$938.65	+	$12.37	+	$62.25	−	0	=	$1013.27 new balance

A. $780.00
B. $1087.50

▶ **Self-Check** Cover the answers and complete the problems.

Find the new balance.

A. Previous balance, $800; interest, $10; deposit, 0; withdrawal, $30.

B. Previous balance, $1050; interest, $12.50; deposit, $25; withdrawal, 0.

Problems

	1.	2.	3.	4.	5.
Previous Balance	$371.21	$542.18	$795.30	$2009.20	$351.72
Interest	$ 9.29	$ 26.58	$ 21.10	—	$ 17.23
Deposit	$ 50.75	—	—	—	—
Withdrawal	—	$125.00	—	$ 200.00	$110.80
New Balance	?	?	?	?	?

6. On November 20, Rose Mendez withdrew $240 from her savings account. What is the new balance in her account?

DEPOSITOR: Rose Mendez				
DATE	WITHDRAWAL	DEPOSIT	INTEREST	BALANCE
10/2				1215.98
11/15			12.28	1228.26
11/20	240.00			?

7. On January 16, Neal Gordon deposited $71.25 in his savings account. The teller also credited $6.89 in interest to his account. What is the new balance in his account?

DEPOSITOR: Neal Gordon				
DATE	WITHDRAWAL	DEPOSIT	INTEREST	BALANCE
1/01				500.00
1/05		135.15		635.15
1/16		71.25		
1/16			6.89	?

8. Find the balance for each date on this savings account passbook.

PASSBOOK ACCOUNT NUMBER 07-17615				
DATE	DEPOSIT	WITHDRAWAL	INTEREST	BALANCE
2/06	400.00			784.15
3/15	50.00			?
4/01			7.38	?
5/15		75.00		?
6/01	297.34			?
7/01			13.97	?
8/20	313.56			?
9/15		100.00		?
10/01			19.01	?
10/05		125.00		?
10/12		100.00		?
10/20	214.75			?
11/1		95.00		?

Maintaining Your Skills

Look up the skills in parentheses if you need help or more practice.

Add. (Skill 5)

9. $250.00 + $75.00
10. $417.40 + $10.50
11. $647.85 + $212.58
12. $412.93 + $7.45 + $150.00

Subtract. (Skill 6)

13. $850.00 − $45.00
14. $96.70 − $15.00
15. $212.45 − $50.00
16. $614.50 − $125.00
17. $1416.89 − $93.43
18. $11,461.82 − $597.98

4-4 Account Statements

Objective Compute the new balance on a savings account statement.

When you have a savings account, your bank may mail you a monthly or quarterly account **statement.** The account statement shows all deposits, withdrawals, and interest credited to your account.

▶ **New Balance = Previous Balance + Interest + Deposits − Withdrawals**

Example

Skills 5, 6 **Application** A **Term** Statement

Amy Bentley received this savings account statement. After checking her passbook and transactions to be be sure all items have been recorded correctly, she checks the calculations. What is the balance in in her account on April 1?

Depositor: Amy Bentley

DATE	WITHDRAWAL	DEPOSIT	INTEREST	BALANCE
1-12		150.00		934.15
2-11		24.50		958.65
3-03	20.00			938.65
4-01			12.37	951.02
4-01		62.25		1013.27

Previous Statement		This Statement	
Date	Balance	Date	Balance
01-01	$784.15	04-01	?

Solution

Find the **new balance.**

Previous Balance + Interest + Deposits − Withdrawals

$784.15 + $12.37 + $236.75 − $20.00 = $1013.27 new balance

$587.50

▶ **Self-Check** Cover the answer and complete the problem.
Previous balance, $400; interest, $7.50; deposits of $100 and $250; withdrawals of $70 and $100. What is the new balance?

Problems

	Previous Balance	+	Interest	+	Deposits	−	Withdrawals	=	New Balance
1.	$ 200.00	+	$11.00	+	$ 70.00	−	$ 50.00	=	?
2.	$ 375.00	+	$17.50	+	$225.00	−	$100.00	=	?
3.	$ 765.00	+	$15.25	+	$150.00 $150.00	−	$ 85.00	=	?
4.	$2475.00	+	$26.35	+	$455.00	−	$190.00 $150.00	=	?

5. Irene Karras.
Previous balance: $515.20.
Interest: $10.34.
Deposits: $125.00 and $474.90.
Withdrawal: $70.00.
What is her new balance?

6. James Lee.
Previous balance: $964.74.
Interest: $17.45.
Deposits: $225.00 and $187.50.
Withdrawals: $110.00 and $145.00.
What is his new balance?

7. Sylvia Roach.
Previous balance: $21,563.73.
Interest: $351.48.
Deposits: $713.47 and $663.47.
Withdrawals: $450.00 and $362.50.
What is her new balance?

8. Scott Jamison.
Previous balance: $1868.70.
Interest: $21.47.
Deposits: $237.45 and $474.90.
Withdrawals: $135.00 and $215.00.
What is his new balance?

9. Richard Greene checks the calculations on his savings account statement. What is the balance in his account on June 30?

DEPOSITOR: Richard Greene			ACC. NO. 020-79-98	
DATE	WITHDRAWAL	DEPOSIT	INTEREST	BALANCE
4/09	525.90			121.14
5/30		316.72		437.86
6/29	118.00			319.86
6/30			4.60	

PREVIOUS STATEMENT		THIS STATEMENT	
DATE	BALANCE	DATE	BALANCE
3/31	$647.04	6/30	?

10. Robert Britto received his savings account statement. What is the balance in his account on June 30?

NAME: Robert Britto		ACCOUNT NUMBER: 12-36-5000		
DATE	DEPOSIT	WITHDRAWAL	INTEREST	BALANCE
01-28	25.00			630.49
02-03	70.30			700.79
02-15			3.17	703.96
03-15			3.19	707.15
04-10		600.00		
04-15			.49	
05-01	425.50			
05-15			1.20	
06-15			2.42	
06-30				

PREVIOUS STATEMENT		THIS STATEMENT	
DATE	BALANCE	DATE	BALANCE
12-31	$605.49	06-30	?

(Continue on next page.)

11. Violet Wood received her savings account statement. What is the balance in her account on August 30?

DEPOSITOR: Violet Wood		ACC. NO. 10-275-190		
DATE	WITHDRAWAL	DEPOSIT	INTEREST	BALANCE
7/01		29.14		731.36
7/07		114.22		845.58
7/10	200.00			645.58
7/15		142.14	9.01	
7/22		36.42		
8/01	150.00			
8/07		142.14		
8/10			10.15	
8/30	225.00			

PREVIOUS STATEMENT		THIS STATEMENT	
DATE	BALANCE	DATE	BALANCE
6/30	$702.22	8/30	?

12. Nancy Logan received her savings account statement. What is the balance in her account on December 31?

DEPOSITOR: Nancy Logan		ACC. NO. 16-86-1453		
DATE	WITHDRAWAL	DEPOSIT	INTEREST	BALANCE
10/7	250.00			13,928.92
10/11		500.00		14,428.92
10/12	200.00			14,228.92
10/31			140.75	14,369.65
11/7	250.00			14,119.65
11/30			138.23	14,257.88
12/7	250.00			
12/15		425.00		
12/31			139.72	

PREVIOUS STATEMENT		THIS STATEMENT	
DATE	BALANCE	DATE	BALANCE
9/30	$14,178.92	12/31	?

Maintaining Your Skills

Look up the skills in parentheses if you need help or more practice.

Add. (Skill 5)

13. \$250.00 + \$2.50 + \$50.00

14. \$475 + \$9.50 + \$335

15. \$937.50 + \$9.38 + \$9.00

16. \$516.40 + \$649.31 + \$243.60

17. \$4781.90 + \$87.90 + \$427.24 + \$357.61

Subtract. (Skill 6)

18. \$7942.70 − \$1451.80

19. \$17,865.95 − \$13,591.49

20. \$218.49 − \$199.89

21. \$31,415.54 − \$27,645.91

22. \$5617.48 − \$1217.48

23. \$10,451.47 − \$315.93

4-5 Simple Interest

Objective

Compute the simple interest.

When you deposit money in a savings account, you are permitting the bank to use the money. The amount you earn for permitting the bank to use your money is called **interest.** The **principal** is the amount of money earning interest. The **annual interest rate** is the percent of the principal that you earn as interest based on one year. **Simple interest** is interest paid on the original principal.

▶ **Interest = Principal × Rate × Time**

Example

Skills 28, 14, 30 **Application** J **Term** Simple interest

Loretta Kiley deposited \$800 in a savings account at Presidential Cooperative Bank. The account pays an annual interest rate of $5\frac{1}{2}\%$. She made no other deposits or withdrawals. After 3 months the interest was calculated. How much simple interest did Loretta's money earn?

Solution

Find the **interest.**

Principal × Rate × Time

$\$800.00 \times 5\frac{1}{2}\% \times \frac{3}{12}$

$\$800.00 \times 0.055 \times 0.25 = \$11.00$ interest

A. \$6.50
B. \$11.25

▶ **Self-Check** Cover the answers and complete the problems.

A. Principal, \$200. Annual interest rate, $6\frac{1}{2}\%$. What is the interest after 6 months?

B. Principal, \$600. Annual interest rate, $7\frac{1}{2}\%$. What is the interest after 3 months?

Problems

	1.	2.	3.	4.	5.
Principal	\$640.00	\$810.00	\$926.00	\$1642.00	\$2850.48
Rate	× 0.06	× 0.06	× 0.12	× 0.10	× 0.065
	?	?	?	?	?
Time (Months)	× $\frac{3}{12}$	× $\frac{3}{12}$	× $\frac{6}{12}$	× $\frac{12}{12}$	× $\frac{3}{12}$
Interest	?	?	?	?	?

6. Cathy Taber's savings account.
$570 on deposit.
$5\frac{1}{2}\%$ annual interest rate.
Time is 3 months.
Find the interest.

7. Mike Bothen's savings account.
$829 on deposit.
$6\frac{1}{2}\%$ annual interest rate.
Time is 1 year.
Find the interest.

8. Vernon Blake's savings account.
$1000 on deposit.
$5\frac{3}{4}\%$ annual interest rate.
Time is 12 months.
Find the interest.

9. Robert Drew's savings account.
$782.78 on deposit.
$6\frac{1}{4}\%$ annual interest rate.
Time is 6 months.
Find the interest.

10. Mike Aieta deposited $450 in a new savings account at Hillside Savings and Loan Association. No other deposits or withdrawals were made. After 3 months the interest was computed at an annual interest rate of $5\frac{1}{2}\%$. How much simple interest did Mike's money earn?

11. Joyce Griffin deposited $392.90 in a new savings account at City Square Bank and Trust. She made no other deposits or withdrawals. After 6 months the interest was computed at an annual rate of $5\frac{3}{4}\%$. How much simple interest did Joyce's money earn?

12. On May 1, Ameil Tyler opened a savings account at County Savings Bank with a deposit of $986.55. No other deposits or withdrawals were made. How much simple interest did Ameil's money earn by July 31?

> **$ COUNTY SAVINGS BANK $**
>
> *Good News to our regular savings account customers!*
>
> $5\frac{1}{2}\%$
>
> * Interest computed quarterly and paid quarterly.
>
> * Effective May 1

13. Trevor Rattray deposited his $500 award in a savings account at Lakeshore Savings Bank on May 1. At the end of July interest was computed at an annual interest rate of $6\frac{1}{4}\%$. How much simple interest did Trevor's money earn?

14. On March 1, Sonia Neves deposited a refund check for $722.85 in a savings account at Central Bank and Trust Co. At the end of November interest was computed at an annual interest rate of 6%. How much simple interest did Sonia's money earn?

Maintaining Your Skills

Look up the skills in parentheses if you need help or more practice.

Change the fractions and mixed numbers to decimals. (Skill 14)

15. $\frac{3}{4}$ 16. $\frac{1}{2}$ 17. $\frac{1}{4}$ 18. $5\frac{1}{4}$ 19. $6\frac{1}{2}$ 20. $7\frac{3}{4}$

Write the percents as decimals. (Skill 28)

21. $5\frac{1}{2}\%$ 22. $6\frac{1}{4}\%$ 23. $4\frac{3}{4}\%$ 24. $7\frac{1}{8}\%$ 25. $6\frac{5}{8}\%$ 26. $5\frac{3}{8}\%$

4-6 Compound Interest

Objective Compute the compound interest and the amount.

Interest that you earn in a savings account during an interest period is added to your account. The new balance is used to calculate the interest for the next interest period. **Compound interest** is interest earned not only on the original principal but also on the interest earned during previous interest periods. The **amount** is the balance in your account at the end of an interest period.

▶ **Amount = Principal + Interest**

Example

Skills 28, 14, 30 **Application** K **Term** Compound interest

Harold Bauer deposited $1200 in a savings account that earns 6% interest compounded quarterly. He made no deposits or withdrawals. What was the amount in the account at the end of the second quarter?

Solution

Find the **amount** for each quarter.

First Quarter: Interest = Principal × Rate × Time

Interest = $1200 × 6% × $\frac{1}{4}$ = $18.00

Amount = Principal + Interest

Amount = $1200 + $18.00 = $1218

Second Quarter: Interest = $1218 × 6% × $\frac{1}{4}$ = $18.27

Amount = $1218 + $18.27 = $1236.27

$1081.60

▶ **Self-Check** Cover the answer and complete the problem.

$1000 is deposited at 8% compounded semiannually for 1 year. Find the amount after 1 year.

Problems

	1.	2.	3.	4.	5.
Principal	$600	$800	$1240	$12,564	$24,938
Annual Interest Rate	4%	6%	$8\frac{1}{2}$%	$7\frac{1}{2}$%	9%
Interest Period	quarterly	monthly	semiannually	quarterly	annually
First Period Interest	$ 6.00	$ 4.00	$ 52.70	?	?
Amount	$606.00	$804.00	?	?	?
Second Period Interest	?	?	?	?	?
Amount	?	?	?	?	?

6. Sally Crosby's savings account. Principal is $900. 5.6% interest compounded quarterly. What is the amount in the account at the end of the second quarter?

7. Julio Cruz's savings account. Principal is $600. 6% interest compounded quarterly. What is the amount in the account at the end of the third quarter?

8. Pete Vasilli deposited $680 in a new regular savings account that earns 5.5% interest compounded semiannually. He made no other deposits or withdrawals. What was the amount in the account at the end of 1 year?

9. Ann Martin deposited $1550 in a new credit-union savings account on the first of a quarter. The principal earns 6% interest compounded quarterly. She made no other deposits or withdrawals. What was the amount in her account at the end of 3 months?

10. Jane and Earl Donovan's savings account had a balance of $5950 on May 1. The account earns interest at a rate of 5.25% compounded monthly. What is the amount in their account at the end of August if no deposits or withdrawals were made during the period?

11. Angelo Larraga had $6220 on deposit at Suburban Savings Bank on July 1. The money earns interest at a rate of 5.5% compounded quarterly. What is the amount in the account on April 1 of the following year if no deposits or withdrawals were made?

12. The Woods opened a savings account with a deposit of $1500 on May 1. The account pays interest at 9% compounded semiannually. What amount will the Woods have in their account on November 1?

13. Alicia DeJute had $10,000 put into a trust fund for her on her 16th birthday. The trust fund pays $7\frac{1}{4}$% interest compounded annually. What will Alicia's trust fund be worth on her 21st birthday?

Maintaining Your Skills

Look up the skills in parentheses if you need help or more practice.

Write the percents as decimals. (Skill 28)

14. $5\frac{1}{4}$%   15. $7\frac{1}{2}$%   16. $6\frac{3}{4}$%   17. 4.8%   18. 6.35%

Find the percentage. (Skill 30)

19. $750 × 5%   20. $680 × $6\frac{1}{2}$%   21. $5640 × 9%

22. $11,480 × $7\frac{1}{2}$%   23. $18,640 × $5\frac{1}{2}$%   24. $34,200 × 4%

Write the fractions as decimals. Round to the nearest hundredth. (Skill 14)

25. $\frac{2}{10}$ 26. $\frac{2}{15}$ 27. $\frac{5}{7}$ 28. $\frac{3}{5}$

4-7 Compound Interest Tables

Objective Find compound interest with tables.

To compute compound interest quickly, you can use a compound interest table which shows the amount of $1.00 for many interest rates and interest periods. To use the table, you must know the total number of interest periods and the interest rate per period.

AMOUNT OF $1.00			
TOTAL INTEREST PERIODS	INTEREST RATE PER PERIOD		
	1.250%	1.375%	1.500%
1	1.01250	1.01375	1.01500
2	1.02515	1.02768	1.03022
3	1.03797	1.04182	1.04567
4	1.05094	1.05614	1.06136
5	1.06408	1.07066	1.07728
6	1.07738	1.08538	1.09344
7	1.09085	1.10031	1.10984
8	1.10448	1.11544	1.12649

▶ **Amount = Original Principal × Amount of $1.00**
▶ **Compound Interest = Amount − Original Principal**

Example

Skills 11, 8 **Application** C **Term** Compound interest

Bano Savings and Loan pays 5.5% interest compounded quarterly on regular savings accounts. Carol Whiteside deposited $2000 for 2 years. She made no other deposits or withdrawals. How much interest did Carol earn during the 2 years?

Solution

1. Find the **total interest periods.**
   2 years × 4 quarters per year = 8 total interest periods
2. Find the **interest rate per period.**
   5.5% ÷ 4 quarters = 1.375% interest rate per period
3. Find the **amount of $1.00** from the table.
   1.11544
4. Find the **amount.**
   Original Principal × Amount of $1.00
   $2000 × 1.11544 = $2230.88
5. Find the **compound interest.**
   Amount − Original Principal
   $2230.88 − $2000.00 = $230.88 compound interest

$5306.80

▶ **Self-Check** Cover the answer and complete the problem.

$5000 invested at 3% compounded semiannually for 2 years. Find the amount.

AMOUNT OF $1.00							
TOTAL INTEREST PERIODS	INTEREST RATE PER PERIOD						
	1.2500%	1.3750%	1.5000%	2.7500%	2.8750%	3.0000%	3.1250%
1	1.01250	1.01375	1.01500	1.02749	1.02875	1.03000	1.03125
2	1.02515	1.02768	1.03022	1.05575	1.05832	1.06090	1.06347
3	1.03797	1.04182	1.04567	1.08478	1.08875	1.09272	1.09671
4	1.05094	1.05614	1.06136	1.11462	1.12005	1.12550	1.13098
5	1.06408	1.07066	1.07728	1.14527	1.15225	1.15927	1.16632
6	1.07738	1.08538	1.09344	1.17676	1.18538	1.19405	1.20277
7	1.09085	1.10031	1.10984	1.20912	1.21946	1.22987	1.24036
8	1.10448	1.11544	1.12649	1.24237	1.25452	1.26677	1.27912
9	1.11829	1.13078	1.14339	1.27654	1.29059	1.30477	1.31909
10	1.13227	1.14632	1.16054	1.31165	1.32769	1.34391	1.36031
11	1.14642	1.16209	1.17795	1.34772	1.36586	1.38423	1.40282
12	1.16075	1.17806	1.19562	1.38478	1.40513	1.42576	1.44666
13	1.17526	1.19426	1.21355	1.42286	1.44553	1.46853	1.49187
14	1.18995	1.21068	1.23175	1.46199	1.48709	1.51258	1.53849
15	1.20482	1.22733	1.25023	1.50219	1.52984	1.55796	1.58657
16	1.21988	1.24421	1.26898	1.54350	1.57382	1.60470	1.63615
17	1.23513	1.26132	1.28802	1.58595	1.61907	1.65284	1.68728
18	1.25057	1.27866	1.30734	1.62956	1.66562	1.70243	1.74000
19	1.26620	1.29624	1.32695	1.67438	1.71351	1.75350	1.79438
20	1.28203	1.31406	1.34685	1.72042	1.76277	1.80611	1.85045
21	1.29806	1.33213	1.36706	1.76773	1.81345	1.86029	1.90828
22	1.31428	1.35045	1.38756	1.81635	1.86559	1.91610	1.96791
23	1.33071	1.36902	1.40838	1.86630	1.91922	1.97358	2.02941
24	1.34735	1.38784	1.42950	1.91762	1.97440	2.03279	2.09283

Problems

Use the compound interest table above to solve. Round answers to the nearest cent.

	1.	2.	3.	4.	5.
Principal	$800.00	$950.00	$1100.00	$1232.00	$2955.60
Annual Interest Rate	5.5%	6%	5%	5.75%	18%
Interest Periods per Year	4	2	4	2	12
Total Time	2 years	1 year	3 years	5 years	6 months
Amount	?	?	?	?	?
Compound Interest	?	?	?	?	?

6. Principal is $750.
5.5% annual interest rate.
Compounded quarterly for 1 year.
What is the compound interest?

7. Principal is $895.50.
6% annual interest rate.
Compounded quarterly for 2 years.
What is the compound interest?

8. National Savings and Trust pays 5.5% interest compounded semiannually on regular savings accounts. Ruth Kittle deposited $1500 in a regular savings account for 2 years. She made no other deposits or withdrawals during the period. How much money did Ruth's money earn?

9. Patty McCormick deposited $800 in a savings plan with the employees' credit union where she works. The credit union savings plan pays 6% interest compounded quarterly. If she makes no other deposits or withdrawals, how much interest will her money earn in 1 year?

10. Reliance Bank pays 6.25% interest compounded semiannually on special notice savings accounts. Roger Streling deposited $2198.70 in a special notice savings account for $2\frac{1}{2}$ years. No other deposits or withdrawals were made during the period. How much interest did Roger's money earn?

11. Union Savings and Loan pays 5% interest compounded quarterly on regular savings accounts. John and Wilma Lawrence had $2000 on deposit for 1 year. At the end of the year, they withdrew all their money and deposited $2000 at Volunteer Co-operative Bank. How much more did the $2000 earn at Volunteer Co-operative for 1 year?

VOLUNTEER CO-OPERATIVE BANK
announces new interest rate!

5.75%

on regular savings accounts.
INTEREST COMPOUNDED SEMIANNUALLY.
MINIMUM OF $10 DEPOSIT REQUIRED

12. Barbara Jones opened a savings account at National Trust Bank on March 1. National Trust pays 15% compounded monthly on money left on deposit for at least 6 months. Barbara opened her account with an initial deposit of $10,000. What is Barbara's $10,000 worth on October 1? What did Barbara's $10,000 earn by October 1?

13. Alistair McKenzie opened a savings account at Madison Savings and Loan on January 1 with a deposit of $900. Madison Savings and Loan pays 11.5% compounded quarterly. What will Alistair's $900 be worth on January 1 of the following year? What will Alistair's $900 have earned by January 1 of the following year?

14. Martha and Barry Drew deposited $1000 in a savings account in their child's name when their child was born. The $1000 earns interest at $6\frac{1}{4}$% compounded semiannually. What will the $1000 be worth on the child's 12th birthday?

Maintaining Your Skills

Look up the skills in parentheses if you need help or more practice.

Divide. (Skill 11)

15. 5% ÷ 4	16. 9% ÷ 12	17. 6% ÷ 2	18. 7% ÷ 1
19. 6.5% ÷ 2	20. 4.5% ÷ 4	21. 5.25% ÷ 2	22. 16.5% ÷ 12
23. $7\frac{1}{2}$% ÷ 4	24. $6\frac{1}{4}$% ÷ 2	25. $19\frac{1}{2}$% ÷ 12	26. $11\frac{1}{2}$% ÷ 4

4-8 Daily Compounding

Objective Find compound interest using tables.

Usually the more frequently interest is compounded, the more interest you will earn. Many banks offer savings accounts with daily compounding. When interest is compounded daily, it is computed each day and added to the account balance. The account will earn interest from the day of deposit to the day of withdrawal. A table can be used to calculate the amount and interest for daily compounding.

AMOUNT OF $1.00 AT 5.5% COMPOUNDED DAILY, 365-DAY YEAR			
DAY	AMOUNT	DAY	AMOUNT
21	1.00316	31	1.00468
22	1.00331	32	1.00483
23	1.00347	33	1.00498
24	1.00362	34	1.00513
25	1.00377	35	1.00528

▸ **Amount = Original Principal × Amount of $1.00**

▸ **Compound Interest = Amount − Original Principal**

Example

Skill 8 **Application** G **Term** Compound interest

On May 30, Diane Fisher deposited $1000 in a savings account that pays $5\frac{1}{2}$% interest compounded daily. On July 1, how much interest had been earned on the principal in Diane's account?

Solution

1. Find the **number of days.**
   from May 30 to July 1 = 32 days
2. Find the **amount.**
   Original Principal × Amount of $1.00
   $1000.00 × 1.00483 = $1004.83
3. Find the **compound interest.**
   Amount − Original Principal
   $1004.83 − $1000.00 = $4.83 compound interest

A. $5026.40
B. $26.40

▸ **Self-Check** Cover the answers and complete the problems.

$5000 deposited at 5.5% compounded daily for 35 days.

A. Find the amount.

B. Find the compound interest.

Problems

Use the daily compound interest table on page 568 to solve. Round answers to the nearest cent. Interest is 5.5% compounded daily.

	1.	2.	3.	4.	5.
Principal	$10,000	$800	$4200	$1600	$12,960
Number of days	25	11	50	10	40
Amount	?	?	?	?	?
Compound interest	?	?	?	?	?

6. Larry Gavin's savings account.
$785 on deposit for 90 days.
$5\frac{1}{2}$% annual interest rate.
Interest compounded daily.
What is the compound interest?

7. Andrea Kelly's savings account.
$2105 on deposit for 50 days.
$5\frac{1}{2}$% annual interest rate.
Interest compounded daily.
What is the compound interest?

8. On June 10, Cameil French deposited $1006.25 in a savings account that pays $5\frac{1}{2}$% interest compounded daily. How much interest will Cameil's money earn in 31 days?

9. Anson Williams has a savings account that earns $5\frac{1}{2}$% interest compounded daily. On May 5, the amount in his account was $1482.56. How much interest will the money earn in 26 days?

10. On April 11, Steve Adler had $2531.46 in his savings account. The account pays $5\frac{1}{2}$% interest compounded daily. How much interest will Steve's money earn by the end of June?

11. On August 23, Buddy Livers had $390.13 in his savings account at Camden Savings and Trust. The account earns $5\frac{1}{2}$% interest compounded daily. What will be the amount in his savings account when he closes his account on October 1?

12. Carrie Wright's savings account shows a balance of $1308.27 on March 1. The same day, she made a deposit of $375 to the account. The bank pays interest at a rate of $5\frac{1}{2}$% compounded daily. What will be the amount in Carrie's account on May 30?

Maintaining Your Skills

Look up the skill in parentheses if you need help or more practice.

Multiply. (Skill 8)

13. $2000 × 1.02131
14. $8000 × 1.00135
15. $600 × 1.00907
16. $1400 × 1.00392
17. $750 × 1.00392
18. $4160 × 1.01059
19. $8690 × 1.00347
20. $42,600 × 1.00407
21. $75 × 1.00196
22. $982 × 1.00301
23. $9172.45 × 1.01670
24. $861.97 × 1.00015

Reviewing the Basics

Skills

Write in words.

(Skill 1) **1.** \$359.90 **2.** \$74.91 **3.** \$4024.09 **4.** \$14,641.89

Solve. Round answers to the nearest cent.

(Skill 5) **5.** \$154.84 + \$17.51 **6.** \$1814.19 + \$43.70 + \$5.03

(Skill 6) **7.** \$718.59 − \$319.69 **8.** \$1200.00 − \$678.19

(Skill 8) **9.** 1.01250 × \$2000 **10.** 1.11643 × \$816.24

(Skill 11) **11.** 8.5 ÷ 4 **12.** 6.74 ÷ 4

Write as a decimal.

(Skill 14) **13.** $6\frac{1}{4}$ **14.** $5\frac{1}{2}$ **15.** $27\frac{3}{4}$ **16.** $12\frac{3}{8}$

(Skill 28) **17.** $6\frac{3}{4}\%$ **18.** $4\frac{7}{8}\%$ **19.** 39.95% **20.** 11.145%

Applications

Use the table on page 122 to find the amount of \$1.00.

(Application C) **21.** 3 interest periods
rate per period 1.5%

22. 2 interest periods
rate per period 1.25%

(Application G) Find the elapsed time in days.

23. From March 1 to May 31 **24.** From June 15 to July 31

(Application J) Write as a fraction of a year.

25. 3 months **26.** 6 months **27.** 9 months **28.** 18 months

(Application K) Find the total number of interest periods.

29. Quarterly for 2 years **30.** Semiannually for $1\frac{1}{2}$ years

Terms

Use each term in a sentence that explains the term.

31. Passbook **32.** Withdrawal **33.** Compound interest

34. Simple interest **35.** Deposit **36.** Statement

Refer to your reference files at the back of the book if you need help.

A Microcomputer Application
N.O.W. Accounts

To complete this microcomputer application, you will need the diskette *Microcomputer Applications for Business Mathematics,* which accompanies this textbook.

Select option 1, N.O.W. Accounts, from the menu. Input the transactions for a N.O.W. account in the following problems to find the interest earned and the new balance. Assume the statement period begins on the first day and ends on the last day of the month.

1. Here is a portion of Jim Martin's bank statement. The previous balance is $235.80.

CHECKS AND OTHER CHARGES			DEPOSITS AND CREDITS	
DATE	NUMBER	AMOUNT	DATE	AMOUNT
6/9	301	24.50	6/10	150.00
6/11	302	145.67	6/15	45.00
6/15	303	9.86		

2. Here is a portion of Janice Tim's bank statement. The previous balance is $198.68.

CHECKS AND OTHER CHARGES			DEPOSITS AND CREDITS	
DATE	NUMBER	AMOUNT	DATE	AMOUNT
7/12	101	54.50	7/15	250.00
7/14	103	284.90	7/29	65.00
7/24	105	38.78		

3. Here is a portion of Fong Lee's bank statement. The previous balance is $889.99.

CHECKS AND OTHER CHARGES			DEPOSITS AND CREDITS	
DATE	NUMBER	AMOUNT	DATE	AMOUNT
11/5	819	983.78	11/1	450.00
11/11	820	421.99	11/19	1095.00
11/28	823	18.23	11/28	120.50

4. Here is a portion of T. J. Wadsworth's bank statement. The previous balance is $976.41.

CHECKS AND OTHER CHARGES			DEPOSITS AND CREDITS	
DATE	NUMBER	AMOUNT	DATE	AMOUNT
3/1	235	763.91	3/5	321.98
3/6	238	21.80		
3/15	239	108.77		

5. Here is a portion of Sandra Bush's bank statement. The previous balance is $376.39.

CHECKS AND OTHER CHARGES			DEPOSITS AND CREDITS	
DATE	NUMBER	AMOUNT	DATE	AMOUNT
2/7	522	224.50	2/10	355.16
2/11	524	391.16	2/20	445.00
2/22	521	31.68	2/25	40.00
2/27	527	15.19		

Unit Test

Lesson 4-1

1. James Holmes deposited \$24.50 in currency and a check for \$148.91 in his savings account. What was his total deposit?

Lesson 4-2

2. Susan Alder has to withdraw \$128 from her savings account. Write the amount as a numeral and in words.

Lesson 4-3

3. On September 13, Fred Dorsey deposited \$121.87 in his savings account. What is the new balance?

DEPOSITOR: Fred Dorsey				
DATE	WITHDRAWAL	DEPOSIT	INTEREST	BALANCE
8-24 9-13		 \$121.87		\$415.80 ?

Lesson 4-4

4. Janet Allard's savings account statement shows the following: a previous balance of \$257.84, a withdrawal of \$108.20, total deposits of \$461.22, and interest of \$3.17. What is the new balance?

Lesson 4-5

5. Beverly Roy deposited \$790.20 in a savings account that pays an annual interest rate of $5\frac{3}{4}$%. She made no other deposits or withdrawals. How much simple interest did Beverly's money earn in 3 months?

Lesson 4-6

6. Alex Bennett deposited \$680.50 in a regular savings account that earns 5.5% interest compounded quarterly. He made no other deposits or withdrawals. What is the amount at the end of the second quarter?

Lesson 4-7

7. Union Savings Bank pays 6% interest compounded quarterly on regular savings accounts. Marcos Torres deposited \$1500 in a regular savings account for $1\frac{1}{2}$ years. He made no other deposits or withdrawals. Use the table to find the interest earned.

TOTAL INTEREST PERIODS	AMOUNT OF \$1.00 Rate per Period	
	1.375%	1.500%
4	1.05614	1.06136
5	1.07067	1.07728
6	1.08539	1.09344

Lesson 4-8

8. Mary Ellen Spencer deposited \$795.70 in a savings account that pays 5.5% interest compounded daily. How much interest will her money earn in 32 days?

AMOUNT OF \$1.00 at 5.5% Compounded Daily, 365-day Year			
Day	Amount	Day	Amount
11	1.00165	31	1.00468
12	1.00180	32	1.00483
13	1.00196	33	1.00498

Unit 5 Cash Purchases

▸ A *sales receipt* is a proof of purchase showing the cost of your purchase and the *sales tax,* where applicable. To find the best buy based on price, you can use the *unit price,* or the cost per unit of an item. Stores may also sell products at *sales prices,* which are lower than the regular selling price. The *markdown* is the difference between the regular selling price and the sale price.

Good buying decisions are based on all product information, not just the selling price or unit price.

5-1 Sales Tax

Objective

Compute the sales tax.

Many state, county, and city governments charge a **sales tax** on certain items or services that you buy. The sales tax rate is usually expressed as a percent.

▸ **Sales Tax = Sales Tax Rate × Total Selling Price**

Example

Skills 30, 2 **Application** A **Term** Sales tax

Al Fox bought a pair of shoes that had a selling price of $32 and a pair of boots that had a selling price of $68. The sales tax rate on these items is 4%. What is the sales tax on his purchases?

Solution

1. Find the **total selling price.**
   $32.00 + $68.00 = $100.00

2. Find the **sales tax.**
   Sales Tax Rate × Total Selling Price
   4% × $100.00 = $4.00 sales tax

A. $9.00
B. $47.25

▸ **Self-Check** Cover the answers and complete the problems.

Find the sales tax.

A. Total selling price: $300.
Sales tax rate: 3%.

B. Total selling price: $1050.
Sales tax rate: 4.5%.

Problems

Item	Sales Tax Rate	×	Total Selling Price	=	Sales Tax
1. Notions	2%	×	$ 1.19	=	?
2. Typewriter	5%	×	$ 94.50	=	?
3. Calculator	3%	×	$ 31.73	=	?
4. Record	4.5%	×	$ 9.98	=	?
5. Sewing machine	7%	×	$590.00	=	?
6. Meal	6.25%	×	$ 37.50	=	?

Find the sales tax.

7.

3.95	GROC
0.79	GROC
2.70	DELI
1.05	MEAT
1.55	GROC

Tax rate: 4%

8.

5.98	MUSIC
6.98	MUSIC
2.17	AUTO
3.75	GARDEN
1.58	AUTO
12.98	TOYS

Tax rate: 3%

9.

16 OZ HAMMER	6.79
1 LB #2 NAILS	3.34
#3 SCREWDRIVER	2.27

Tax rate: 5%

10. John Lopes purchases a $13.88 turbine ventilator, an adjustable base for $6.49, and a 12-foot extension cord for $3.49. The sales tax rate is 6%. What is the total selling price? What is the sales tax?

11. Tom Giovanni purchases a weed trimmer for $39, potting soil for $3.85, a 50-foot garden hose for $4.44, and a chaise lounge for $13.89. The sales tax rate is 4.5%. What is the sales tax?

12. Paula Guest purchases the 8-track tape player shown. The sales tax rate is 3.5%. What is the sales tax?

Holiday Special
8-track stereo tape player
Only $139.88

13. At Skyland's Department Store, Ted Yankovich purchases a record for $7.98, a dictionary for $11.50, a paperback book for $2.25, and a magazine for $1.00. The sales tax rate is 3%. What is the sales tax on Ted's purchases?

14. Yvonne Haddad purchases a $35 portable workbench, an $18.99 hand saw, and an extra blade for $2.45. The sales tax rate is 5.5%. What is the sales tax on Yvonne's purchases?

15. Wilma Smith purchases a $15.95 tapestry handbag, a pair of canvas shoes for $7.79, and four polyester knit tops at $5.75 each. The sales tax rate is 3.75%. What is the sales tax on Wilma's purchases?

Maintaining Your Skills

Look up the skills in parentheses if you need help or more practice.

Find the percentage. (Skill 30)

16. 5% of 29.9 17. $2\frac{1}{2}$% of $12,640 18. 140% of 80 19. $\frac{1}{2}$% of 90

Round to the place value indicated. (Skill 2)

Nearest thousand	20. 856,951	21. 2159	22. $44,329	23. 1,344,456
Nearest ten	24. 124	25. 45,729	26. 41	27. 1051
Nearest cent	28. $45.125	29. $189.4539	30. $456.337	31. $1.3776
Nearest tenth	32. 46.741	33. 458.284	34. 149.014	35. 432.863

5-2 Total Purchase Price

Objective Compute the total purchase price.

Most stores give you a **sales receipt** as proof of purchase. The sales receipt may be a handwritten sales slip or a cash register tape. The receipt shows the selling price of each item or service you purchased, the total selling price, any sales tax, and the total purchase price.

▶ **Total Purchase Price = Total Selling Price + Sales Tax**

Example

Skills 8, 5 **Application** A **Term** Sales receipt

Rich Bowman purchased the school supplies listed on the sales receipt. He checks the information on the receipt. The sales tax rate is 3%. What should the total purchase price be?

Solution

ALDEN BOOKSTORE: 5th and Main Street

QUANTITY	DESCRIPTION	PRICE	AMOUNT
1	School Organizer	1 19	1 19
2	50-Sheet Letter Pads	89	1 78
1	Ballpoint pen	63	63
SALES SLIP Customer's Copy		Subtotal	3 60
		Sales Tax	11
		Total	3 71

Labels: Number Purchased; Price per Item; Total Price; TOTAL SELLING PRICE; TOTAL PURCHASE PRICE

$627.30 ▶ **Self-Check** Cover the answer and complete the problem.

Stephen Morgan bought a stereo receiver priced at $245, 2 speakers priced at $95 each, and a cassette deck priced at $180. The sales tax rate is 2%. Find the total purchase price.

Problems

	Item	Total Selling Price	×	Sales Tax Rate	=	Sales Tax	Total Purchase Price
1.	Plant	$ 4.79	×	5%	=	$0.24	?
2.	Funnel	$ 1.05	×	4.5%	=	$0.05	?
3.	Garden hose	$12.74	×	4%	=	?	?

The sales tax rate is 4%. Find the total purchase price.

4.
```
0.39  GROC
1.19  GROC
3.72  MEAT
7.19  MEAT
0.65  GROC
1.09  DELI
0.67  GROC
```

5.
```
3.98  GROC
2.10  MEAT
3.05  DELI
0.69  GROC
0.23  MISC
1.00  DELI
1.09  GROC
```

6.
```
12V BATTERY  59.95
30 WT. OIL    0.79
30 WT. OIL    0.79
30 WT. OIL    0.79
30 WT. OIL    0.79
```

7. Sales slip for Paula McGrath.
Pen for 59¢.
Ruler for 19¢.
Sales tax of $0.05.
What is the total purchase price?

8. Sales slip for Norris Dean.
Calculator for $9.95.
2 batteries at 49¢ each.
Sales tax of $0.55.
What is the total purchase price?

9. Sales slip for Ben Krassler.
Tube of toothpaste for $1.79.
Bottle of shampoo for $2.98.
4 bars of soap at 41¢ each.
Sales tax rate of 3%.
What is the total purchase price?

10. Sales slip for Mary Jo Butler.
AM-FM radio for $14.98.
2 batteries at 55¢ each.
Jar of hand lotion for $1.65.
Sales tax rate of 4.5%.
What is the total purchase price?

11. Frank Gonzales purchases a $15.45 drill bit set and a $6.95 extension cord. The sales tax rate is 5%. What is the total purchase price of Frank's purchases?

12. Jane Webster purchases the items shown on the sales slip. The sales tax rate is 2.5%. What is the total purchase price of Jane's purchases?

```
4  CANDLES - $.59 EACH
1  ZIPPER - $.97 EACH
2  MUGS - $1.37 EACH
```

13. James Doyle purchases an antenna for $35.95, an antenna rotor for $29.95, and 25 m of wire at 32¢ per meter. The sales tax rate is 4.6%. What is the total purchase price?

14. Amanda Kinney purchases 4 cans of motor oil at 95¢ a can, 8 spark plugs at $1.25 per plug, and an oil filter for $4.99. The sales tax rate is 6.5%. What is the total purchase price of Amanda's purchases?

15. At Save-More Store, Ada Baldon purchases a 3-lb chuck roast at $1.59 per pound, 6 pounds of bananas at 4 pounds for $1.00, a $2.99-box of laundry detergent, a gallon of milk for $1.69, a 75¢-box of clothes pins, and 6 cans of condensed soup at 2 cans for 74¢. The sales tax rate is 3%. There is no sales tax charged on food items. What is the total purchase price of Ada's purchases?

(Continue on next page.)

16. Carl Chu purchases the items shown on the sales slip. The sales tax rate is 3.2%. There is no sales tax charged on items of clothing. What is the total purchase price of Carl's purchases?

```
1 CAST IRON SKILLET - $8.95
2  BLOUSES - $5.49 EACH
2 LP RECORDS - $6.98 EACH
1 WORK SLACKS - $11.95
1 PAIR TENNIS SHOES - $21.98
```

17. Maria Ramos purchases a suit for $69.99, a blouse for $25, a pair of shoes for $36, and a matching bracelet and necklace for $12.99 each. She pays a state tax of 3% and a local tax of 1%. There is no tax on clothing. What is the total sales tax on her purchases? What is the total purchase price?

18. Janet Gault purchased the items listed on the following sales receipt. The sales tax is 5%. Complete the sales receipt.

The Camera's Eye			
Quantity	**Description**	**Price**	**Amount**
1	35 mm Ishu camera	$289.95	?
1	camera case	$ 22.50	?
4	filters	$ 6.95	?
6	rolls of film	$ 3.69	?
1	telephoto lens	$ 89.00	?
		Subtotal	?
		Sales tax	?
		Total	?

Maintaining Your Skills

Look up the skills in parentheses if you need help or more practice.

Multiply. (Skill 8)

19. 3.6×1.2 **20.** 18.6×8.142 **21.** 0.91×0.001 **22.** 0.01×0.06

Add. (Skill 5)

23. $42.19 + 164.00 + 1.52$

24. $18.09 + 6.3 + 0.007$

25. $3.65 + 0.0014 + 142.1$

26. $103.216 + 87.942 + 7.888$

27. $342.86 + 100.50 + 0.014$

28. $114.6 + 7.05 + 460$

29. $83.2 + 94.8 + 260.05$

30. $73 + 1.15 + 90.8$

31. $375.85 + 261.8 + 194.9$

32. $316.015 + 413.02 + 122.811$

5-3 Unit Pricing

Objective Compute the unit price.

Grocery stores often give **unit price** information for their products. Shoppers can use this information to determine which size of a product is the better buy based solely on price. The unit price of an item is its cost per unit of measure or count, such as dollars per pound or cents per dozen.

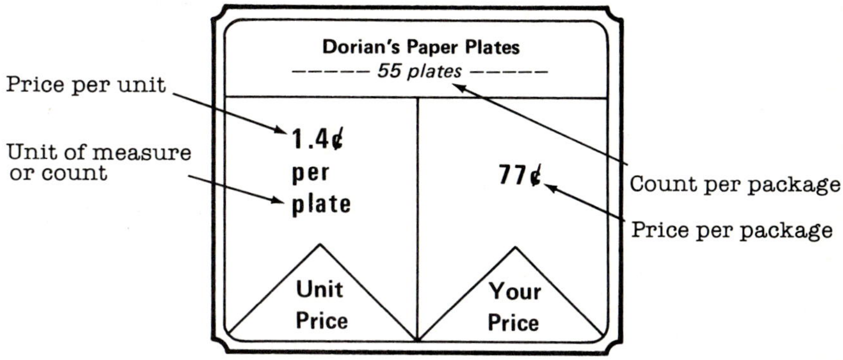

▶ **Unit Price** = $\frac{\textbf{Price per Item}}{\textbf{Measure or Count}}$

Example

Skills 11, 2 **Application** A **Term** Unit price

Amelia Velley purchased a 4-oz jar of honey for 89¢. What is the unit price of the jar of honey to the nearest tenth of a cent?

Solution

Find the **unit price**.

Price per Item ÷ Measure or Count

$0.89 ÷ 4 = $0.2225 = 22.25¢ = 22.3¢ unit price per ounce

A. 3.2¢
B. 0.63¢

▶ **Self-Check** Cover the answers and complete the problems.

A. A 28-oz can of tomatoes costs 90¢. Find the unit price per ounce to the nearest tenth of a cent.

B. A box of 120 coffee filters costs 75¢. Find the price per filter to the nearest hundredth of a cent.

Problems

Round to the nearest tenth of a cent.

	Price per Item	÷	Measure or Count	=	Unit Price
1.	74¢	÷	5 oz	=	? per oz
2.	$1.59	÷	40 oz	=	? per oz
3.	$5.44	÷	16 m	=	? per m
4.	79¢	÷	2 L	=	? per L

5. Sesame crackers.
2 boxes for $1.00.
What is the price per box?

6. Seedless limes.
6 limes for 39¢.
What is the price per lime?

7. Plastic food wrap.
200-foot roll.
Sells for 79¢.
What is the price per foot?

8. Room deodorizer.
2.5-oz bar.
Sells for $1.29.
What is the price per ounce?

9. Alice Mota recently purchased 4 tires for her automobile. The total purchase price was $121.24. What was the price per tire?

10. Travis Grady sees that green beans are on sale for 6 cans for $1.09. What is the price per can?

11. Pat Kruger purchases a 5-kg box of laundry detergent for $3.88. What is the price per kilogram?

12. A carton of ice cream costs $1.79. There are 64 ounces in the carton. What is the unit price?

13. Tom Baxter purchases a 16-oz bottle of hand lotion for 99¢. What is the unit price?

14. An 8-oz box of oatmeal is priced at 2 boxes for $1.12. What is the price per box? What is the unit price?

15. What is the unit price for each item in the advertisement?

a. Pizza
b. Paper towels
c. Two loaves of bread
d. Potatoes

Shoppers' Specials

PIZZA, ea . . . . 98¢
PAPER TOWELS, 2 rolls . . 98¢
RYE BREAD, 59¢ ea. . . . 2/98¢
POTATOES, 5 lb bag. . . . . 98¢

Maintaining Your Skills

Look up the skills in parentheses if you need help or more practice.

Divide. Round to the nearest hundredth. (Skill 11)

16. 170.60 ÷ 86 17. 12.5 ÷ 2.9 18. 26.813 ÷ 6.6 19. 0.75 ÷ 40

Round to the place value indicated. (Skill 2)

Nearest ten	20. 62.8	21. 19.6	22. 325.22	23. 1256.3
Nearest tenth	24. 22.86	25. 141.62	26. 2.558	27. 1041.672
Nearest hundredth	28. 4.781	29. 97.152	30. 120.125	

5-4 Finding the Better Buy

Objective

Find the better buy based on unit price.

Compare the unit prices of products when you're deciding which size to buy. The size with the lower unit price is the better buy. If the larger size of a product is the better buy, consider whether you will be able to use up the product before it spoils.

Example

Skills 11, 1 **Application** A **Term** Unit price

Delman's mozzarella cheese is sold in two sizes. The price of a 10-oz package is $0.79. The price of a 16-oz package is $1.19. Based on price alone, which package is the better buy?

Solution

1. Find the **unit price** for each item.
   Price per Item ÷ Measure or Count
   a. 10-oz package: $0.79 ÷ 10 = $0.079 = 7.9¢
   b. 16-oz package: $1.19 ÷ 16 = $0.0743 = 7.4¢
2. Find the **better buy.**
   Compare: 7.9¢ and 7.4¢ ← Better Buy

A. 20 oz. at 2.6¢ an ounce
B. 24 at 24.5¢ each

▶ **Self-Check** Cover the answers and complete the problems.

Find the better buy.

A. A 20-oz package costs 52¢. A 15-oz package costs 45¢.

B. A box of 12 diapers costs $2.99. A box of 24 diapers costs $5.89.

Problems

Find the unit price to the nearest tenth of a cent. Then determine the better buy based on price alone.

	Small Size	Large Size	Better Buy
1.	49¢ ÷ 16 oz = ?	65¢ ÷ 22 oz = ?	?
2.	45¢ ÷ 10 = ?	$4.70 ÷ 100 = ?	?
3.	$7.29 ÷ 14 oz = ?	$16.50 ÷ 30 oz. = ?	?

4. Magnetic clips.
   a. Box of 200 for $2.10.
   b. Box of 20 for 55¢.
   Which is the better buy?

5. Instant coffee.
   a. 10-oz jar for $5.75.
   b. 4-oz jar for $2.49.
   Which is the better buy?

6. Wanda Fong has a choice of two sizes of Plaster of Paris. The 5-lb box sells for $1.29 while a 1-lb box sells for 59¢. Which size is the better buy?

7. Bob Johnson wants to purchase some insect repellant. A 7-oz can sells for $2.39 while a 14-oz can sells for $3.59. Which size can is the better buy?

8. Linda Miller is shopping in the meat department of the grocery store and sees the ham shoulders advertised. Which ham shoulders are the better buy?

HOLIDAY SPECIAL
This week only

Ham Shoulders

Canadian Brand
7 pounds: $12.32
OR
Virginia Brand
5 pounds: $7.95

9. Bill Lake reads in the newspaper that his favorite morning cereal is on sale at two stores in his neighborhood. The Country Store is selling a 24-oz box for $1.08 while Food Mart is selling a 18-oz box for 79¢. Which size box is the better buy?

10. Peter Skaff wants to purchase some paint thinner. A 64-oz can costs $2.78 while a 16-oz can costs $1.09. Which is the better buy?

11. Jeanette Hurley's favorite shampoo is available in the following sizes: a 7-oz jar for $2.29, a 5-oz jar for $1.59, a 3-oz jar for $1.27, and a 1.7-oz jar for 89¢. What is the unit price of each? Which size jar is the best buy?

12. Jim Glaser is buying mint tea bags. The brand he prefers is sold in boxes of 16 bags for 47¢, 48 bags for 99¢, 100 bags for $1.69, and 150 bags for $1.99. Which size box is the best buy?

13. The Co-op Food Chain sells two sizes of a generic laundry detergent. The family size contains 10 pounds 11 ounces and costs $5.67. The regular size contains 1 pound 4 ounces and costs 75¢. What is the unit price of each? Which size is the better buy?

*14. **A Brief Case** Tania Pappas is shopping for the best buy of a brand of dog food. The dog food comes in the following sizes: a 2.27-kg bag for $1.19, a 4.54-kg bag for $2.09, and an 11.35-kg bag for $4.79. What is the unit price of each? Which size bag is the best buy?

Maintaining Your Skills

Look up the skills in parentheses if you need help or more practice.

Divide. Round answers to the nearest hundredth. (Skill 11)

15. $1.3\overline{)39.16}$ 16. $0.150\overline{)16.5}$ 17. $0.005\overline{)0.7905}$ 18. $16\overline{)429.13}$

Which number is greater? (Skill 1)

19. $1786 or $1689 20. $12,244 or $12,301 21. 4.9 or 9.4 22. $106 or $96

5-5 Markdown

Objective Compute the dollar amount of the markdown.

Stores often sell products at **sale prices,** which are lower than their regular selling prices. The **markdown,** or **discount,** is the amount of money that you save by purchasing a product at the sale price.

The **markdown rate,** or **discount rate,** of an item is its markdown expressed as a percent of its regular selling price. Businesses frequently advertise the markdown rate rather than the sale price.

- **Markdown = Regular Selling Price − Sale Price**
- **Markdown = Markdown Rate × Regular Selling Price**

Example 1

Skill **6** **Application** **A** **Term** **Markdown**

David Chien purchased a sofa at a sale price of $388.55. The regular selling price is $544.95. What was the markdown?

Solution

Find the **markdown.**

Regular Selling Price − Sale Price

$544.95 − $388.55 = $156.40 markdown

Example 2

Skills 30, 2 **Application A** **Term Markdown rate**

Artform Custom Shop is selling all summer furniture for 25% off the regular selling price. Joanna Lewis purchased furniture that regularly sold for $160. How much did she save by buying the furniture during the sale?

End of Season Sale!
This month only
Save 25% on all Summer Furniture
ARTFORM CUSTOM SHOP

Solution

Find the **markdown.**

Markdown Rate × Regular Selling Price

25% × $160.00 = $40.00 markdown

A. $80.00
B. $17.90

Self-Check Cover the answers and complete the problems.

A. The regular selling price of a computer is $399.95. The sale price is $319.95. Find the markdown.

B. A clock radio is on sale for 20% off the regular selling price of $89.50. Find the markdown.

Problems

	Item	Regular Selling Price	–	Sale Price	=	Markdown
1.	Dinnerware	$ 39.50	–	$ 24.50	=	?
2.	Scarf	$ 9.80	–	$ 4.90	=	?
3.	Rocking chair	$256.79	–	$199.95	=	?

	Item	Markdown Rate	×	Regular Selling Price	=	Markdown
4.	Handbag	50%	×	$ 78.50	=	?
5.	Notebook	20%	×	$ 1.15	=	?
6.	Motorcycle	10%	×	$2395.00	=	?
7.	Television	15%	×	$ 569.00	=	?

8. Pair of shoes.
Regularly sells for $24.95.
Sale price is $19.60
What is the markdown?

9. Station wagon.
Regularly sells for $7980.
Sale price is $6990.
What is the markdown?

10. A pair of dress shoes.
Regularly sells for $37.99.
Markdown rate is 30%.
What is the markdown?

11. Typewriter ribbons.
Regularly sell for $1.29 each.
Markdown rate is 35%.
What is the markdown?

12. Bargains on games usually occur after the holidays in January. Jane Koneski purchases a chess set that regularly sells for $14.95. The set is on sale for $8.99. What is the markdown?

13. April is usually a good month to purchase home improvement supplies. A sheet of cedar paneling at Wetzler Lumber Yard regularly sells for $18.50. Danny Anastasio purchases the paneling on sale for $14.95 a sheet. What is the markdown?

14. A 21″ color television which regularly sells for $795.95 is marked down 30% during Bartlett's May Sale. What is the markdown?

15. Beth Walsh sees a summer dress that regularly sells for $34.95 on sale at 40% off. How much can Beth save by purchasing the summer dress during the sale?

16. TBA Auto Parts holds a sale on car care products. All polishing cloths and cellulose sponges are marked down 35%. What is the markdown on polishing cloths that regularly sell for 80¢ each? What is the markdown on cellulose sponges that regularly sell for 95¢ each?

17. Tires are generally on sale during the month of September. Dolly Peterson purchased 4 of the glass-belted tires shown on sale. What was the total markdown on her purchase?

Ojai Tire Center

A78-13 Glass-belted

*****RADIALS*****

reg. $49.95 per tire

SALE PRICE $33.95 ea

18. Typewriter sales often occur in the month of June. Harlan Office Supplies has three electric typewriters Model 1181 remaining in stock. The Model 1181 regularly sells for $325 but is on sale for $279.95. What is the markdown?

19. Gift items, such as these shown, are frequently discounted after the holidays in December and January. What is the markdown for each item advertised?

CRYSTAL SERVING ACCESSORIES

15% off regular prices

Salt and pepper set, reg. $12.00
Sugar and Creamer, reg. $7.50
Handled mugs, set of 4, reg. $17.50
Jam jar with spoon, reg. $8.50
3 pc. salad set, reg. $18.00

20. Peter Matthews purchases a corduroy jacket during an April Coat Sale at Bay's Clothiers. All coats are discounted 25%. The corduroy jacket regularly sells for $79.59. What is the markdown?

21. January, May, July, and August are months in which many department stores hold "white sales." Use the advertisement to find the markdown on each item. Then compare the results with a markdown rate of 20% off the regular selling price on each item. Which results in a larger markdown?

AUGUST WHITE SALE

Full flat or fitted sheets
Reg. $5.99 . . . . 2/$9
Queen flat or fitted sheets
Reg. $8.99 . . . . 2/$15
King flat or fitted sheets
Reg. $10.99 . . . . 2/$19
Standard cases,
Reg. $5.99 . . . . $3.99 pr.
King cases,
Reg. $6.99 . . . . $4.99 pr.

Maintaining Your Skills

Look up the skills in parentheses if you need help or more practice.

Subtract. (Skill 6)

22. $100.14 - 6.25$ 23. $89.6 - 7.924$ 24. $8.6 - 4.22105$ 25. $64.588 - 62.824$

Find the percentage. (Skill 30)

26. 15% of $90 27. $20\frac{3}{5}$% of 160 28. $\frac{1}{4}$% of 600 29. 125% of 250

5-6 Sale Price

Objective Compute the sale price when markdown rate is known.

For items that are on sale, some stores advertise the amount of markdown and the regular selling price. If you know this information, you can calculate the sale price.

▶ **Sale Price = Regular Selling Price − Markdown**

Example

Skills **30, 6** **Application** **A** **Term** **Sale price**

Kay Obee is purchasing this cassette player-recorder at Highland's Hifi. What will Kay pay for the tape recorder?

SAVE 20%
Briarcliff stereo cassette player-recorder
Regular Price **$177.95**

Solution

1. Find the **markdown.**
   Markdown Rate × Regular Selling Price
   20% of $177.95 = $35.59
2. Find the **sale price.**
   Regular Selling Price − Markdown
   $177.95 − $35.59 = $142.36 sale price

A. $21.00
B. $141.75

▶ **Self-Check** Cover the answers and complete the problems.

A. The regular selling price of a sweater is $35. The markdown rate is 40%. Find the sale price.

B. The regular selling price of a bicycle is $189. The markdown rate is 25%. Find the sale price.

Problems

	Regular Selling Price	×	Markdown Rate	=	Markdown	Sale Price
1.	$ 27.00	×	10%	=	?	?
2.	$ 89.95	×	20%	=	?	?
3.	$ 8.99	×	15%	=	?	?
4.	$650.00	×	30%	=	?	?
5.	$230.00	×	35%	=	?	?

6. Ironstone dinnerware set. Regularly sells for $34.99. Markdown rate is 20%. What is the sale price?

7. Electric baseboard heater. Regularly sells for $19.79. Markdown rate is 15%. What is the sale price?

8. Lemmon's Hardware Store has marked down the Model 80 mini-bike 25% during its Spring Bike Sale. What is the sale price of the mini-bike?

TURDELL MINI-BIKE
Model 80

25% off

Regularly $299.59
Offer ends May 30

9. During an August Pre-school Sale, Casual Apparel, Inc., has marked down its line of dress shirts 20%. These shirts are regularly priced at $12.95 each. What is the sale price of these shirts?

10. The Wyoming Pizza Shack discounts 20% off the regular selling price on the purchase of 3 or more pizzas. What is the sale price on the purchase of 3 pizzas that regularly sell for $4.25, $3.95, and $4.50?

*11. **A Brief Case** Many manufacturers, stores, and service establishments offer customers special discounts through **coupons.** Ben Kaplan has 3 suits and 6 pairs of pants to be dry cleaned. Ben is normally charged $3.20 for each suit and $1.60 for each pair of pants to be cleaned at Sun Cleaners. He received the coupons shown below from Sun Cleaners and Star Cleaners, both in his neighborhood. Which establishment will charge less for Ben's dry cleaning?

STAR
With this coupon have any number of
2 PC MEN'S OR LADIES
SUITS **$2.39 ea.**
Coupon good with incoming orders only.
EXPIRES October 31

STAR
With this coupon have any number of
MEN'S OR LADIES
PANTS **$1.19 ea.**
Coupon good with incoming orders only.
EXPIRES October 31

Clip this coupon and SAVE
SUN CLEANERS

Present this coupon with incoming orders and receive

25% OFF

on all dry cleaning.
Offer good through Oct. 31

Maintaining Your Skills

Look up the skills in parentheses if you need help or more practice.

Find the percentage. Round answers to the nearest hundredth. (Skill 30)

12. 15% of $64.75
13. 4.5% of 325
14. 8% of 400
15. 25% of 62.5

Subtract. (Skill 6)

16. 572.814 − 469.008
17. 27. − 16.814
18. 0.014 − 0.0016
19. $159.26 − 152.18
20. $21.95 − 16.46

Reviewing the Basics

Skills

Write the number that is smaller.

(Skill 1) **1.** 13.4¢ or 14.2¢ **2.** 49.3¢ or 50.1¢ **3.** 16.25¢ or 16.3¢

(Skill 2) Round to the nearest cent.

4. \$11.1806 **5.** \$1.637 **6.** \$0.497 **7.** \$4.223

Solve.

(Skill 5) **8.** \$3.39 + \$1.57 + \$23.07 **9.** \$239.45 + \$11.10 + \$119.95

(Skill 6) **10.** \$76.40 − \$18.59 **11.** \$101.21 − \$32.53

(Skill 8) **12.** \$3.69 × 6 **13.** \$35.10 × 26 **14.** \$84.87 × \$0.046

(Skill 11) **15.** 85¢ ÷ 4 **16.** 75¢ ÷ 120 **17.** \$3.32 ÷ 16

(Skill 30) **18.** 15% of \$12.95 **19.** 2.2% of \$40 **20.** 6.5% of \$223.50

Applications

Use the formula Markdown = Markdown Rate × Regular Selling Price to find the markdown.

(Application A) **21.** Sewing machine.
Regular price \$369.89.
Markdown rate of 25%.

22. Alarm clock.
Regular price \$32.95.
Markdown rate of 15%.

Terms

Match each term with its definition on the right.

23. Sales tax

24. Sales receipt

25. Unit price

26. Markdown

27. Markdown rate

28. Sale price

a. the amount of money that is saved by buying an item at the sale price

b. the purchase price of an item after markdown

c. the total purchase price including sales tax

d. the markdown of an item expressed as a percent of the regular selling price

e. a cash register tape or sales slip showing the selling price of items purchased, any sales tax, and the total purchase price

f. the cost per unit measure or count of an item

g. a tax charged on the selling price of an item or service provided

Refer to your reference files at the back of the book if you need help.

Problem Solving Writing an Equation

A word problem can be translated into an equation that is solved by performing the same mathematical operation (adding, subtracting, multiplying, or dividing) to both sides. Solving the equation then leads to the solution of the problem.

To set up the equation, look for words in the word problem that explain which of the four mathematical operations to use.

Words	Symbols	Operation
the total; how many in all; the sum; plus	$+$	addition
the difference; how much more; how much smaller; minus	$-$	subtraction
the total for a number of equal items; the product	$\times$	multiplication
the number left over; the quotient	$\div$	division

Example 1 Find a number that when multiplied by 3 and diminished by 4 is 5 more than twice the number.

Solution Use the letter x to stand for the unknown number, and set up the equation.

$3x - 4 = 5 + 2x$

$3x = 9 + 2x$ (Add 4 to both sides.)

$x = 9$ (Subtract $2x$ from both sides.)

Example 2 In 40 hours at your regular rate of pay plus 3 hours of double-time (twice your regular rate of pay), you earn \$349.60. What is your regular rate of pay?

Solution Use the letter x to stand for your regular rate of pay.

$40x + 3(2x) = \$349.60$

$40x + 6x = \$349.60$

$46x = \$349.60$

$x = \$7.60$ (Divide each side by 46.)

Problems

1. One word processor can produce 17 documents per hour while a second word processor can produce 13 documents per hour. How many hours will it take to produce 315 documents?

2. A bottle and a cork cost \$1.10. The bottle costs \$1.00 more than the cork. How much does each cost?

3. An ant travels 33 cm in walking completely around the edge of a rectangle. If the rectangle is twice as long as it is wide, how long is each side?

Unit Test

Lesson 5-1

1. Glenda Brock purchases fabric for \$7.67, a spool of thread for 79¢, and a dress pattern for \$2.25. The sales tax rate is 7%. What is the sales tax on Glenda's purchases?

Lesson 5-2

2. Clifton Rice purchases 3 cans of corn at 46¢ a can, 2 pounds of green beans at 69¢ a pound, a \$1.49-box of laundry detergent, and a package of ground chuck for \$2.89. The sales tax rate is 3.5%. What is the total purchase price of his purchases?

Lesson 5-3

3. A brand of bacon which Sandy Blackman prefers is on sale at Foodway. What is the unit price of each package?

Shop and Save at

FOODWAY

This week's specials:

Country bacon

12-oz package for \$1.49

16-oz package for \$1.89

Lesson 5-4

4. Marty Tyree compares the prices for frozen turkey. A 7-lb package of frozen turkey is priced at \$9.00. A 5-lb package of frozen turkey is priced at \$6.25. Based on price alone, which is the better buy?

Lesson 5-5

5. Marcie Withers purchases a raincoat on sale for \$59.88. The regular selling price was \$65.95. What is the markdown on the raincoat?

6. Drake's Appliances regularly sells a Model K2504E 25″ console color television for \$695.95. For an August Clearance Sale, the Model K2504E is discounted 20% off the regular selling price. How much is saved by purchasing the television during the sale?

Lesson 5-6

7. Jeff Rivers purchases a record album that regularly sells for \$9.95. The album has been discounted 25%. What is the sale price of the album?

Cumulative Review (Skills/Applications) Units 1–5

Skills

Write the amount in words with cents expressed as a fraction of a dollar.

(Skill 1) **1.** \$67.59 **2.** \$471.43 **3.** \$1002.00

Write the number that is less.

4. 6.7¢ or 7.1¢ **5.** \$250 or \$197 **6.** \$4781.91 or \$4778.87

Round to the nearest cent.

(Skill 2) **7.** \$29.188 **8.** \$0.1447 **9.** \$128.393 **10.** \$2176.4191

Round to the nearest tenth of a cent.

11. 16.312¢ **12.** 7.371¢ **13.** 47.37¢ **14.** 96.177¢

Solve.

(Skill 4) **15.** \$69 − \$47 **16.** \$916 − \$749 **17.** \$1984 − \$1750

18. \$7629 − \$7351 **19.** \$9531 − \$8753 **20.** \$4178 − \$919

(Skill 5) **21.** \$221.00 + \$175.00 **22.** \$57.79 + \$12.99

23. \$329.48 + \$36.05 + \$8.00 **24.** \$4.76 + \$16.09 + \$0.89

(Skill 6) **25.** \$800.00 − \$600.00 **26.** \$125.83 − \$100.54

27. \$236.15 − \$73.24 **28.** \$761.80 − \$672.71 **29.** \$1325.60 − \$890.85

(Skill 8) **30.** \$9.15 × 40 **31.** \$8.50 × 42 **32.** \$8.25 × 37.5

33. \$0.89 × 0.607 **34.** \$0.15 × 0.003 **35.** \$2.70 × 0.05

(Skill 11) **36.** \$6.56 ÷ 16 **37.** \$1.56 ÷ 12 **38.** \$752.96 ÷ 52

39. \$13.02 ÷ 0.42 **40.** \$3.75 ÷ 0.25 **41.** \$129.48 ÷ 0.75

(Skill 30) **42.** 25% of \$275.95 **43.** 2.5% of \$4270

44. 37.5% of \$1240 **45.** 6.13% of \$21,600.50

46. $7\frac{3}{4}$% of \$4400 **47.** $8\frac{1}{2}$% of \$5967.75

(Skill 16) **48.** $\frac{1}{4} + \frac{1}{2}$ **49.** $2\frac{1}{2} + 1\frac{1}{2}$ **50.** $3\frac{3}{4} + 5\frac{1}{2}$

Write as a decimal.

(Skill 14) **51.** $\frac{1}{4}$ **52.** $\frac{1}{8}$ **53.** $25\frac{1}{2}$ **54.** $41\frac{3}{4}$

(Skill 28) **55.** 4.25% **56.** 27.20% **57.** $17\frac{1}{2}$% **58.** $35\frac{3}{4}$%

Applications Use the formula to find the markdown.

(Application A) Markdown = Markdown Rate × Regular Selling Price

59. Luggage cart.
Regular price is $29.95.
Markdown rate is 25%.

60. Patio furniture.
Regular price is $399.50.
Markdown rate is 30%.

61. Sports coat.
Regular price is $45.00.
Markdown rate is 15%.

62. Reading lamp.
Regular price is $25.50.
Markdown rate is 45%.

Use the table on page 568 to find the amount of $1.

63. 4 interest periods.
Rate per period is 3.125%.

64. 6 interest periods.
Rate per period is 3.000%.

65. 5 interest periods.
Rate per period is 2.875%.

66. 3 interest periods.
Rate per period is 1.500%.

(Application F) Find the elapsed time.

67. From 8:15 a.m. to 11:55 a.m.

68. From 1:15 p.m. to 5:20 p.m.

69. From 7:52 a.m. to 5:10 p.m.

70. From 8:17 a.m. to 5:10 p.m.

(Application G) Use the table on page 569 to find the elapsed time in days.

71. From May 1 to May 16

72. From June 4 to June 30

73. From April 15 to July 15

74. From March 10 to May 31

Write as a fraction of a year.

75. 2 months

76. 11 months

77. 10 months

78. 15 months

Find the number of occurrences.

79. Weekly for 2 years

80. Biweekly for 1 year

81. Monthly for $1\frac{1}{2}$ years

82. Semiannually for 6 years

Terms Write your own definition for each term.

83. Straight-time pay
84. Overtime pay
85. Commission
86. Income tax
87. Social security
88. Group insurance
89. Sales tax
90. Unit price
91. Markdown
92. Sale price
93. Deposit
94. Reconcile
95. Withdrawal
96. Simple interest
97. Compound interest
98. Time card
99. Passbook
100. Piecework

 Refer to your reference files at the back of the book if you need help.

Cumulative Review Test Units 1-5

Lesson 1-2

1. Mary Jenkins works as a computer operator and earns $8.00 an hour for a regular 40-hour week. Her overtime rate is $1\frac{1}{2}$ times her regular hourly rate. This week Mary worked her regular 40 hours plus $7\frac{3}{4}$ hours of overtime. What is her total pay?

Lesson 1-5

2. Alice Meeks is an accountant for National Foods. She earns an annual salary of $28,524 paid on a semimonthly basis. What is her semimonthly pay?

Lesson 1-6

3. T. J. Banks sells hospital equipment for a 7% straight commission. Her sales totaled $27,845 in September. What is her pay for the month?

Lesson 2-1

4. Thomas King, a shipping supervisor, earns $448.50 a week. He is married and claims 3 allowances. How much is withheld from Thomas's weekly paycheck for federal income tax?

WEEKLY Payroll Married					
Wages		Allowances			
At least	But less than	0	1	2	3
$410	$420	$58.10	$53.90	$49.60	$46.00
420	430	60.30	56.10	51.80	47.70
430	440	62.50	58.30	54.00	49.80
440	450	64.70	60.50	56.20	52.00
450	460	66.90	62.70	58.40	54.20
460	470	69.40	64.90	60.60	56.40
470	480	71.90	67.10	62.80	58.60
480	490	74.40	69.60	65.00	60.80
490	500	76.90	72.10	67.30	63.00
500	510	79.40	74.60	69.80	65.20

Lesson 2-2

5. Betty Adams earns $22,464 a year as a systems analyst. The state income tax rate is 4.6% of taxable income. Her personal exemptions total $3692. How much is withheld each week from Betty's gross pay for state income tax?

Lesson 2-4

6. Wilbur Clark, a medical technician, is paid $334.60 a week. His earnings to date this year total $9468.66. The social security tax rate is 7.15% of the first $42,600 earned. How much is deducted from Wilbur's paycheck this week for social security tax?

Lesson 2-6

7. John Miller, a research assistant for Master Plastics Industries, is married and claims no allowances. He earns $456.85 a week. The social security tax rate is 7.15% of the first $42,600 earned. The state tax is $9.13 a week. He has weekly deductions of $12 for medical insurance and $50 for payroll savings. Use the table above to find John's federal tax withheld. What is his net pay for a week?

Lesson 3-2

8. Bill Cooke wrote a check to World Electronics in the amount of $849.96 for a microcomputer. The check number is 363. Write the amount of the check in words with cents expressed as a fraction of a dollar.

Lesson 3-4

9. A portion of Tania Cole's bank statement is shown. Her balance from last month's statement was $624.82. What is her present balance?

CHECKS AND OTHER CHARGES			DEPOSITS AND CREDITS		BALANCE
DATE	*NUMBER*	*AMOUNT*	*DATE*	*AMOUNT*	
10/4	902	71.84	10/8	247.80	
10/4	904	19.63	10/22	371.40	
10/15	903	147.00			
Service charge		3.50			

Lesson 3-5

10. Tania Cole compares her checkbook register, canceled checks, and bank statement. She finds that she has check numbers 901 and 905 outstanding in the amounts of $91.93 and $45.92. She also has an outstanding deposit of $200. She reconciles the statement. What should her adjusted balance be?

Lesson 4-4

11. Martha Bingston's savings account statement shows the following: a previous balance of $313.71, a withdrawal of $141.83, total deposits of $318.35, and interest of $4.50. What is the new balance?

Lesson 4-5

12. Beverly Roy deposited $1471.50 in a savings account that pays an annual interest rate of $6\frac{1}{2}$%. She made no other deposits or withdrawals. How much simple interest did Beverly's money earn in 6 months?

Lesson 4-7

13. Union Savings Bank pays $5\frac{1}{2}$% interest compounded quarterly on regular savings accounts. Marcos Torres deposited $2400 in a regular savings account for $1\frac{1}{2}$ years. He made no other deposits or withdrawals. Use the table to find the interest earned.

TOTAL INTEREST PERIODS	AMOUNT OF $1.00 Rate per Period	
	1.375%	1.500%
4	1.05614	1.06136
5	1.07067	1.07728
6	1.08539	1.09344

Lesson 5-2

14. Michael Noll purchases 3 cans of soup at 67¢ a can, 2 pounds of green beans at 79¢ a pound, a $1.49 box of laundry detergent, and a package of chicken for $3.74. The sales tax rate is 5.5%. What is the total purchase price?

Lesson 5-4

15. Tammy Dillon compares the prices for frozen turkey. A 7-lb package of frozen turkey is priced at $10.50. A 5-lb package of frozen turkey is priced at $7.95. Based on price alone, which is the better buy?

Lesson 5-5

16. Conner Appliance regularly sells a Model K2504E 25″ console color television for $849.99. For an August clearance sale, the Model K2504E is discounted 30% off the regular selling price. How much is saved by purchasing the television during the sale?

Unit 6 Charge Accounts and Credit Cards

▸ A *charge account* allows you to "buy now and pay later." You pay for goods or services with a *credit card.* At the end of a month, the credit company bills you for all the items you purchased on credit. This account *statement* shows any *finance charges* you must pay if your previous bill was not paid in full. A finance charge is interest charged on the amount you owe. Finance charges are figured by the *previous-balance method,* the *unpaid-balance method,* or the *average daily balance method.*

Credit cards are an easy and popular way to pay for consumer goods.

6-1 Sales Receipts

Objective Compute the total purchase price.

When you make a purchase with a credit or charge card, the salesclerk prepares a **sales receipt.** The receipt shows your name and account number, the price of each item you purchased, the sales tax, and the total purchase price. You sign the sales receipt and receive a copy for your records.

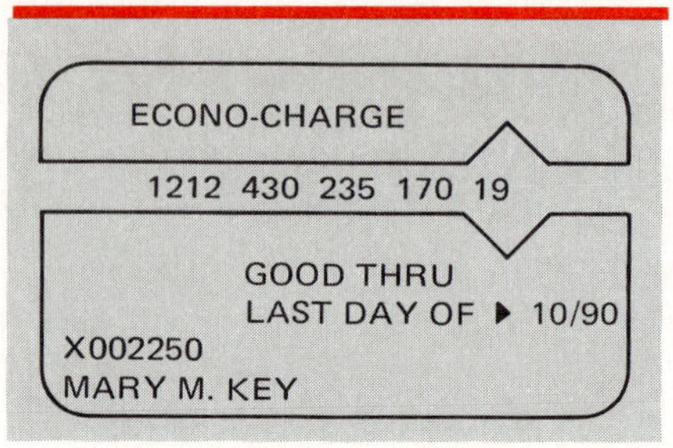

▶ **Total Purchase Price = Total Selling Price + Sales Tax**

Example

Skill 30 **Application** A **Term** Sales receipt

Kathryn Fairbanks purchased the items listed on the sales receipt. The sales tax rate is 4%. Is the total purchase price correct?

Solution

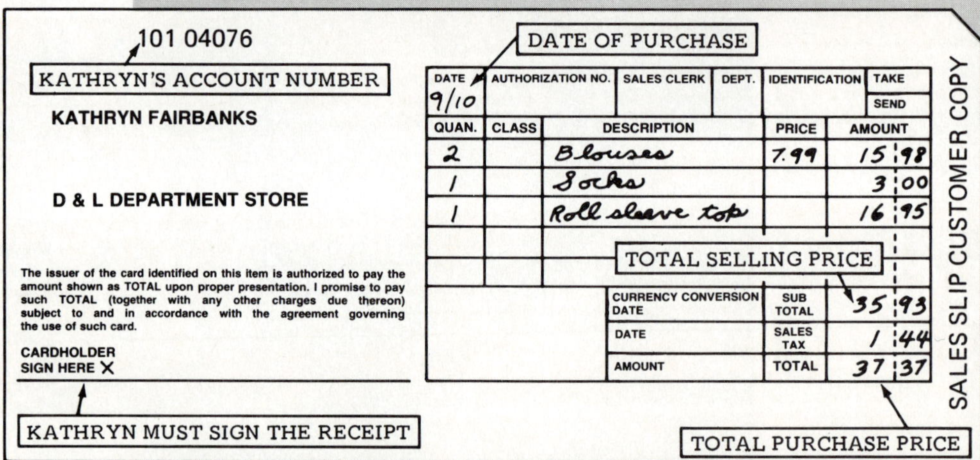

A. $6.00; $126.00
B. $4.50; $79.50

▶ **Self-Check** Cover the answers and complete the problems.

Find the sales tax and the total purchase price.

A. Cost of suit: $120.
Sales tax rate: 5%.

B. Price of tire: $75.
Sales tax rate: 6%.

Problems

	Sales Tax Rate	×	Total Selling Price	=	Sales Tax	Total Purchase Price
1.	2%	×	$ 34.00	=	?	?
2.	6%	×	$143.00	=	?	?
3.	5.5%	×	$127.00	=	?	?

Complete the sales receipt.

DATE	AUTH. NO.	IDENTIFICATION	CLERK	REG/DEPT	☑TAKE ☐SEND
3/14/—	42		JR		

QTY	CLASS	DESCRIPTION	PRICE	AMOUNT
1		Pressure Cooker		29.95
1		Dinnerware Set		34.95
4.			SUB TOTAL	?
	CUSTOMER SIGNATURE X Darcia Adams		TAX	4.22
5.	SALES SLIP		TOTAL	?

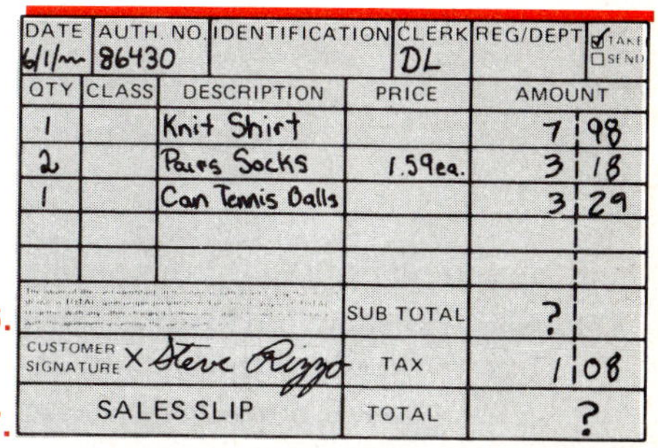

DATE	AUTH. NO.	IDENTIFICATION	CLERK	REG/DEPT	☑TAKE ☐SEND
6/1/—	86430		DL		

QTY	CLASS	DESCRIPTION	PRICE	AMOUNT
1		Knit Shirt		7.98
2		Pairs Socks	1.59 ea.	3.18
1		Can Tennis Balls		3.29
6.			SUB TOTAL	?
	CUSTOMER SIGNATURE X Steve Rizzo		TAX	1.08
7.	SALES SLIP		TOTAL	?

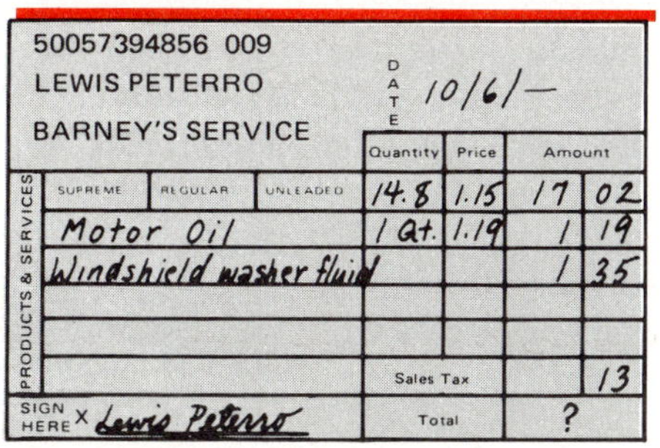

50057394856 009
LEWIS PETERRO
BARNEY'S SERVICE
DATE 10/6/—

PRODUCTS & SERVICES	Quantity	Price	Amount
SUPREME / REGULAR / UNLEADED	14.8	1.15	17.02
Motor Oil	1 Qt.	1.19	1.19
Windshield washer fluid			1.35
	Sales Tax		.13
8. SIGN HERE X Lewis Peterro	Total		?

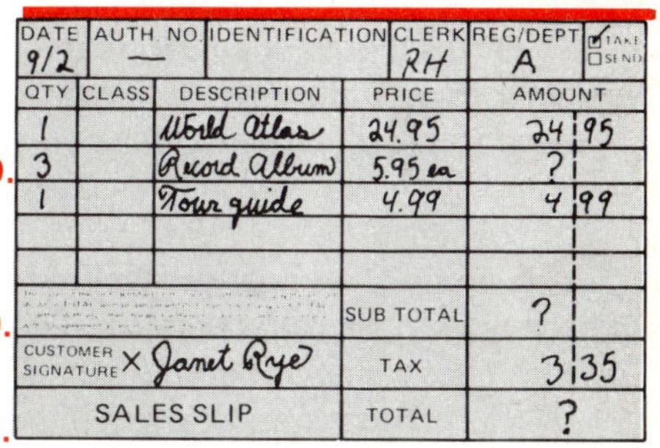

DATE	AUTH. NO.	IDENTIFICATION	CLERK	REG/DEPT	☑TAKE ☐SEND
9/2	—		RH	A	

QTY	CLASS	DESCRIPTION	PRICE	AMOUNT
1		World Atlas	24.95	24.95
9. 3		Record Album	5.95 ea	?
1		Tour guide	4.99	4.99
10.			SUB TOTAL	?
	CUSTOMER SIGNATURE X Janet Rye		TAX	3.35
11.	SALES SLIP		TOTAL	?

12. Sales slip for Tim Hall.
    Tune-up for $34.75.
    4 tires at $45.99 each.
    Headlight for $7.98.
    4 quarts of oil at $2.85 each.
    Sales tax is $11.90.
    What is the total purchase price?

13. Sales slip for George Eng.
    2 bags of lawn seed at $19.95 each.
    1 bag of topsoil for $6.95.
    Lawn spreader for $15.99.
    Lawn sprinkler for $24.95.
    Sales tax is $3.21.
    What is the total purchase price?

14. Carrie Wayne charges 215 gallons of home heating oil. The oil costs 92.4¢ per gallon. What is the total purchase price of Carrie's oil delivery?

15. Jarrad Jackson purchases a new watchband at Tillman's Jewelers. The watchband costs $12.95 plus a 4% sales tax. He charges the purchase on his bank charge card. What is the total purchase price?

16. Frank Robideau purchases a pair of track shoes at Hall Sports. The shoes cost $38.99 plus a sales tax of 5.5%. What is the total purchase price?

17. Henrietta Zimbler purchased a set of golf clubs at Sport Stores, Inc. The clubs cost $149.99 plus a sales tax of 6%. What is the total purchase price?

(Continue on next page.)

18. Quentin Gilbertson charged the following items on his bank charge card: a suit for $159.99, 3 shirts at $17.99 each, 2 ties at $12.50 each, and a pair of shoes for $89.99. There is a sales tax of 4%. What is the total purchase price?

19. Velma Gilkerson charged the following items on her department store charge card: a dress for $57.99, 2 skirts for $35.99 each, 2 blouses at $23.99 each, and 2 pair of shoes for $39.99 each. There is a sales tax of 6%. What is the total purchase price?

20. At Bargain City, Shirley Burke purchases a gallon of paint thinner for $3.98, 3 gallons of paint at $11.95 a gallon, 4 tubes of caulking at $1.79 each, 2 paint rollers at $2.59 each, and 2 paint brushes at $4.79 each. There is a sales tax of 6%. What is the total purchase price?

21. At Budget Mart, Fenton Hardee purchases a fielder's glove for $14.99, 3 pairs of athletic socks at $1.79 a pair, a softball bat for $9.95, and a baseball cap for $4.99. There is a sales tax of 2.5%. He charges the purchases on his bank charge card. What is the total purchase price?

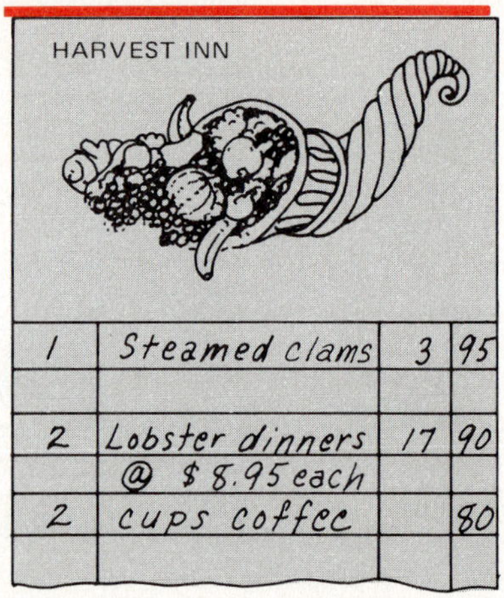

HARVEST INN

1	Steamed clams	3	95
2	Lobster dinners	17	90
	@ $ 8.95 each		
2	cups coffee		80

22. At the Harvest Inn, Barbara and James McAlister dined on clams and lobster. A portion of the check for the dinner is shown. There is an 8% meals tax. Jim charged the dinner on his bank charge card and added a tip of 15% on the cost of the meal before tax. What was the total cost of dining out?

Maintaining Your Skills

Look up the skills in parentheses if you need help or more practice.

Write the percents as decimals. (Skill 28)

23. 5% 24. 6% 25. 5.5% 26. 6.75% 27. 3.20%

Add. (Skill 5)

28. $44.65 + $12.55 + $45.78 29. $123.43 + $23.89 + $9.85

30. $56.75 + $3.65 + $424.45 31. $9.25 + $45.76 + $32.03 + $1.13

Find the percentage. (Skill 30)

32. 5% of 80 33. 8% of 32 34. 5.5% of 15 35. 3.2% of 87

36. 4.2% of $120 37. 3.5% of $99 38. 4.75% of $44.50

6-2 Account Statements

Objective Compute the new balance in a charge account.

With a credit card or charge account, you receive a monthly **statement** listing all transactions processed by the closing date for that month. If your previous bill was not paid in full, a **finance charge**is added. A finance charge is interest that is charged for delaying payment.

▶ **New Balance = Previous Balance + Finance Charge + New Purchases − (Payments + Credits)**

Example

Skills 6, 5 **Application** A **Term** Finance charge

Kathryn Fairbanks received this charge account statement. What is the new balance on October 2?

MAIL THIS PORTION WITH PAYMENT TO P.O. BOX 1027

D & L
DEPARTMENT STORE

☐ IF ADDRESS IS INCORRECT, PLEASE CHECK BOX AND CORRECT ON REVERSE SIDE

Kathryn Fairbanks
1234 MAIN ST.
TOLEDO, OH 43605

KATHRYN'S ACCOUNT NUMBER → ACCOUNT NUMBER 101-04076

NEW BALANCE 302.91

TOTAL MINIMUM AMOUNT NOW DUE 31.00

SEND PAYMENT TO REACH US BY 10/30/

DATE	DEPT NO	DESCRIPTION	PURCHASES	CREDITS	PAYMENTS	ITEM NO
09/07	162	SPORTSWEAR		23.44		0-06-18005
09/10	363	SPORTSWEAR	37.37			0-07-20015
09/17		PAYMENT—THANK YOU			50.00	0-06-47023
09/19	214	JR SPORTSWEAR		7.52		0-07-45005
09/19	53	DRESSES—APRONS		15.06		0-07-45006
09/27	34	COSMETICS	15.62			0-07-64004
09/29	1	LINENS—TOWELS	9.25			0-07-70012

$50 PAYMENT TOWARDS LAST MONTH'S BILL

LAST DATE TRANSACTIONS WERE PROCESSED

AMOUNT KATHRYN OWED LAST MONTH

AMOUNT KATHRYN NOW OWES

BILL CLOSING DATE	PREVIOUS BALANCE	FINANCE CHARGE	TOTAL PURCHASES THIS MONTH	CREDITS	PAYMENTS	NEW BALANCE
9/30	332.64	4.05	62.24	46.02	50.00	302.91

PAY NEW BALANCE IN FULL WITHIN 25 DAYS OF BILL CLOSING DATE TO AVOID FINANCE CHARGE NEXT MONTH.
FINANCE CHARGE, IF ANY, IS COMPUTED ON THE AVERAGE DAILY BALANCE OF $270.19 BY APPLYING MONTHLY PERIODIC RATES AS FOLLOWS:

MONTHLY PERIODIC RATE:	BALANCE	ANNUAL PERCENTAGE RATE:
1.50 %	FIRST $600	18%
1.00 %	OVER $600	12%

SUBJECT TO A MINIMUM FINANCE CHARGE OF $.50.

NOTICE: SEE REVERSE SIDE FOR IMPORTANT INFORMATION

ACCOUNT NUMBER 101-04076

MIN. ON

CREDIT FOR ITEMS RETURNED

AMOUNT PAST DUE

TOTAL MINIMUM AMOUNT DUE

Solution

Find the **new balance.**

Previous Balance	+	Finance Charge	+	New Purchases	−	(Payments + Credits)
$332.64	+	$4.05	+	$62.24	−	($50.00 + $46.02)
		$398.93			−	$96.02 = $302.91 new balance

A. $559.00
B. $176.17

Self-Check Cover the answers and complete the problems.

Find the new balance.

	Previous Balance	+	Finance Charge	+	New Purchases	−	Payments & Credits	=	New Balance
A.	$600.00	+	$9.00	+	$50.00	−	$100.00	=	?
B.	$197.77	+	$3.16	+	$17.74	−	$ 42.50	=	?

Problems

What is the new balance for the credit statements shown?

1.

BILLING DATE	PREVIOUS BALANCE	FINANCE CHARGE	NEW PURCHASES	PAYMENTS & CREDITS	NEW BALANCE
8/15/--	$400.00	$6.00	$120.00	$100.00	?

2.

BILLING DATE	PREVIOUS BALANCE	FINANCE CHARGE	NEW PURCHASES	PAYMENTS & CREDITS	NEW BALANCE
1/22/--	$500.75	$7.50	$142.00	$150.00	?

3.

BILLING DATE	PREVIOUS BALANCE	FINANCE CHARGE	NEW PURCHASES	PAYMENTS & CREDITS	NEW BALANCE
3/1/--	$550.75	$8.26	$57.50	$120.00	?

4.

BILLING DATE	PREVIOUS BALANCE	FINANCE CHARGE	NEW PURCHASES	PAYMENTS & CREDITS	NEW BALANCE
9/4/--	$123.23	$0.96	$20.36	$53.23	?

5.

BILLING DATE	PREVIOUS BALANCE	FINANCE CHARGE	NEW PURCHASES	PAYMENTS & CREDITS	NEW BALANCE
6/15/--	$375.19	0	$206.14	$375.19	?

6.

BILLING DATE	PREVIOUS BALANCE	FINANCE CHARGE	NEW PURCHASES	PAYMENTS & CREDITS	NEW BALANCE
8/1/--	$1394.21	$17.44	0	$350.00	?

7. What is the new balance for the credit statement shown?

PAYMENTS (P) & CREDITS (C) DEDUCTED—DATES
20.00 P 04-07
20.00 P 03-28

PREVIOUS BALANCE	PURCHASES ADDED
65.61	45.19
FINANCE CHARGE ADDED .38	NEW BALANCE ?
MINIMUM PAYMENT 20.00	MUST BE RECEIVED BY 05-15 TO AVOID ADDITIONAL FINANCE CHARGE

		ANNUAL PERCENTAGE RATE	PERIODIC MONTHLY RATE
TO	$500	18 %	1—1/2 %
OVER	$500	12 %	1 %

8. What is the new balance for the credit statement shown?

ACCOUNT NUMBER		BILLING DATE
064-173-388-62		03-10
PREVIOUS BALANCE 266 72	—PAYMENTS 66 72	RECEIVED ON 02-26
—CREDITS 14 59	FINANCE CHARGE 3 73	+ PURCHASES 4 13
+ INS. PREMIUMS	= NEW BALANCE ?	AMOUNT DUE 30 00
	NEXT BILLING 04-12	

		PERIODIC RATES	ANNUAL PERCENTAGE RATE(S)
UP TO	$500	1.50%	18.0%
OVER	$500	1.00%	12.0%
MINIMUM CHARGE		.50	

9. Monthly statement for Lila Key.
Credit Master bank card.
Previous balance of $107.35.
Payment of $20.
New purchases: $9.50, $21.73.
Finance charge of $1.61.
What is the new balance?

10. Monthly statement for Doyle Whitten.
Elliman Department Store charge card.
Previous balance of $34.80.
Payment and credit totaling $34.80.
Finance charge of $0.52.
No new purchases.
What is the new balance?

11. Monthly statement for Walt Dunn.
Department store charge card.
Previous balance of $457.19.
Payment of $50.
New purchases: $37.17, $29.88, $3.19.
Finance charge of $7.42.
What is the new balance?

12. Monthly statement for Sarah Rogers.
All-charge bank card.
Previous balance of $239.48.
Payment of $65.
New purchases: $177.42, $28.20.
Finance charge of $4.19.
What is the new balance?

13. Wanda Miller received this charge account statement. Find her new balance.

DEPT.	DESCRIPTION	CHARGES	PAYMENT/CREDIT	DATE	REF.#
109	Moderate Coordinates	121.78	31.78	1/25	6004
85	Mens Wear	35.98	117.48	1/25	7018
	PAYMENT		44.40	2/1	8014

BILLING DATE: 2/16

PREVIOUS BALANCE	PAYMENTS & CREDITS	UNPAID BALANCE	FINANCE CHARGE	NEW PURCHASES	NEW BALANCE
$185.42	?	— —	$2.40	?	?

Maintaining Your Skills

Look up the skills in parentheses if you need help or more practice.

Add. (Skill 5)

14. $442.60 + $25.10 + $48.90 + $16.50

15. $34.19 + $224.60 + $9.55 + $17.50

16. $94.52 + $123.17 + $9.20 + $127.63

17. $3.16 + $14.50 + $19.25 + $134.78

Subtract. (Skill 6)

18. $21.50 − $2.50

19. $121.65 − $11.35

20. $327.16 − $15.12

21. $137.42 − $39.58

22. $698.57 − $289.68

23. $1246.23 − $998.39

6-3 Finance Charge—Previous-Balance Method

Objective Compute the finance charge by the previous-balance method.

Some credit card companies use the **previous-balance** method to compute finance charges. They compute the finance charge based on the amount you owed on the closing date of your last statement. The **periodic rate** is the monthly finance charge rate.

▶ **Finance Charge = Periodic Rate × Previous Balance**

▶ **New Balance = Previous Balance + Finance Charge + New Purchases − (Payments + Credits)**

Example

Skills 30, 2 **Application** A **Term** Previous balance

Paul Simmons has a charge account with a store that charges 1.5% of the previous balance for the finance charge. A portion of Paul's statement is shown. He checks the computations. What is the new balance of Paul's account on October 5?

PREVIOUS BALANCE	CLOSING DATE THIS MONTH	CLOSING DATE LAST MONTH
64.00	October 5, 19--	September 6, 19--

TOTAL PURCHASES	PAYMENTS & CREDITS	FINANCE CHARGE	NEW BALANCE	MINIMUM PAYMENT
43.85	38.95	?	?	10.00

Solution

1. Find the **finance charge.**
   Periodic Rate × Previous Balance
   1.5% of $64.00 = $0.96

2. Find the **new balance.**
   Previous Balance + Finance Charge + New Purchases − (Payments + Credits)
   $64.00 + $0.96 + $43.85 − $38.95
   $108.81 − $38.95 = $69.86 new balance

A. $1.20
B. $1.182 = $1.18

▶ **Self-Check** Cover the answers and complete the problems.

The periodic rate is 1.5%. Find the finance charge.

A. Previous balance of $80.

B. Previous balance of $78.80.

Problems

Use the previous-balance method to find the finance charge. Round answers to the nearest cent.

	1.	2.	3.	4.	5.
Previous Balance	$40.00	$170.00	$142.00	$157.40	$387.52
Periodic Rate	× 1.5%	× 1.5%	× 1.6%	× 1.25%	× 1.75%
	?	?	?	?	?

The finance charge is 1.5% of the previous balance. Find the finance charge and the new balance for each statement shown.

6.

BILLING DATE	PREVIOUS BALANCE	FINANCE CHARGE	NEW PURCHASES	PAYMENTS & CREDITS	NEW BALANCE
4/11/--	$300.00	?	$100.00	$150.00	?

7.

BILLING DATE	PREVIOUS BALANCE	FINANCE CHARGE	NEW PURCHASES	PAYMENTS & CREDITS	NEW BALANCE
5/7/--	$190.00	?	$42.00	$25.00	?

8.

BILLING DATE	PREVIOUS BALANCE	FINANCE CHARGE	NEW PURCHASES	PAYMENTS & CREDITS	NEW BALANCE
3/1/--	$42.50	?	$12.50	$40.00	?

9.

BILLING DATE	PREVIOUS BALANCE	FINANCE CHARGE	NEW PURCHASES	PAYMENTS & CREDITS	NEW BALANCE
9/15/--	$357.00	?	$15.85	$80.00	?

10.

BILLING DATE	PREVIOUS BALANCE	FINANCE CHARGE	NEW PURCHASES	PAYMENTS & CREDITS	NEW BALANCE
12/12/--	$464.24	?	$122.92	$175.00	?

11.

BILLING DATE	PREVIOUS BALANCE	FINANCE CHARGE	NEW PURCHASES	PAYMENTS & CREDITS	NEW BALANCE
10/30/--	$1132.20	?	0	$350.00	?

Use the previous-balance method of computing finance charges to solve.

12. Foster Wheeling's charge account.
Previous balance of $216.
Periodic rate is 2.0%.
What is the finance charge?
New purchases totaling $32.50.
Payment and credit totaling $85.45.
What is the new balance?

13. Candice Webb's charge account.
Previous balance of $94.30.
Periodic rate is 0.75%.
What is the finance charge?
New purchases totaling $143.45.
Payments totaling $124.50.
What is the new balance?

(Continue on next page.)

14. Sue Delleles has a charge account at Cannon's Department Store where the finance charge is 1.5% of the previous balance. A portion of her account statement is shown. Find the finance charge and the new balance.

PREVIOUS BALANCE	CLOSING DATE THIS MONTH	CLOSING DATE LAST MONTH
$87.22	June 21, 19--	July 20, 19--

TOTAL PURCHASES	PAYMENTS & CREDITS	FINANCE CHARGE	NEW BALANCE	MINIMUM PAYMENT
$33.80	$45.82	?	?	$10.00

15. Bob Trummel has a charge account where the finance charge is 1.25% of the previous balance. A portion of his statement for May is shown. What is the total of his payments and credits? What is the finance charge? What is the new balance of Bob's account?

DATE	REFERENCE #	DEPT.	DESCRIPTION	PURCHASES	PAYMENTS	CREDITS
5-04	34029	03	Sport coat	85.00		
5-12	40085	06	Robe			29.00
5-23	29450	10	Payment		50.00	
5-23	37047	03	Men's slacks	58.82		
5-25	77460	04	Cologne	6.95		

PREVIOUS BALANCE	CLOSING DATE LAST MONTH	CLOSING DATE THIS MONTH
$130.35	May 31, 19--	April 30, 19--

TOTAL PURCHASES	PAYMENTS & CREDITS	FINANCE CHARGE	NEW BALANCE	MINIMUM PAYMENT
$150.77	?	?	?	$20.00

*16. **A Brief Case** Many credit account companies charge periodic rates that vary depending on the amount of the previous balance. Use the periodic rates on the statement shown to calculate the finance charge and the new balance according to the previous-balance method.

PREVIOUS BALANCE	−PAYMENTS	RECEIVED ON
779¦74	110¦00	12-09
−CREDITS	+ FINANCE CHARGE	+ PURCHASES
45¦63	?	264¦74
+INS. PREMIUM	=NEW BALANCE	AMOUNT DUE
¦	?	85¦00
PAY NEW BALANCE BY TO AVOID FINANCE CHARGE ▷	NEXT BILLING DATE 01-27	ANNUAL PERCENTAGE RATE(S)
BALANCES	PERIODIC RATE(S)	
UP TO $500	1.50%	18.0%
OVER $500	1.00%	12.0%
MINIMUM CHARGE	.50	

Maintaining Your Skills

Look up the skills in parentheses if you need help or more practice.

Round to the nearest hundredth. (Skill 2)

17. 37.126 18. 9.344 19. 8.5271 20. 15.8864 21. 3.5973

Round to the nearest cent. (Skill 2)

22. $1.245 23. $3.8512 24. $8.5255 25. $0.3345 26. $17.502

Find the percentage. (Skill 30)

27. 4% of 238 28. 1.75% of 888 29. 1.6% of 32.24 30. 1.75% of 465.5

6-4 Finance Charge— Unpaid-Balance Method

Objective Compute the finance charge by the unpaid-balance method.

Some companies use the **unpaid-balance** method of computing finance charges. They compute the finance charge based on that portion of the previous balance that you have not paid.

- ▶ **Unpaid Balance = Previous Balance − (Payments + Credits)**
- ▶ **Finance Charge = Periodic Rate × Unpaid Balance**
- ▶ **New Balance = Unpaid Balance + Finance Charge + New Purchases**

Example

Skill 30 Application A Term Unpaid balance

A portion of Barb Valdez's charge account statement is shown. The monthly finance charge is 1.5% of the unpaid balance. What is the new balance of Barb's account?

88	PAYMENT/Thank you		52.10		

BILLING DATE: 5/13

PREVIOUS BALANCE	PAYMENTS & CREDITS	UNPAID BALANCE	FINANCE CHARGE	NEW PURCHASES	NEW BALANCE
$112.10	$52.10	$60.00	$0.90	$97.45	?

Solution

1. Find the **unpaid balance.**
   Previous Balance − (Payments + Credits)
   $112.10 − $52.10 = $60.00
2. Find the **finance charge.**
   Periodic Rate × Unpaid Balance
   1.5% of $60.00 = $0.90
3. Find the **new balance.**
   Unpaid Balance + Finance Charge + New Purchases
   $60.00 + $0.90 + $97.45 = $158.35 new balance

A. $100; $1.50; $151.50
B. $320; $4.80; $384.80

▶ **Self-Check** Cover the answers and complete the problems.

The periodic rate is 1.5%. Find the unpaid balance, the finance charge, and the new balance.

	Previous Balance	Payments and Credits	New Purchases
A.	$300.00	$200.00	$50.00
B.	$420.00	$100.00	$60.00

Problems

For problems 1 through 8, use a periodic rate of 1.5% and the unpaid-balance method of computing the finance charge.

	(Previous Balance −	Payments + Credits	= Unpaid Balance)	+ Finance Charge	+ New Purchases	= New Balance
1.	($300.00 −	$100.00	= ?)	+ ?	+ $ 60.00	= ?
2.	($400.00 −	$150.00	= ?)	+ ?	+ $ 75.00	= ?
3.	($250.00 −	$ 50.00	= ?)	+ ?	+ $ 40.00	= ?
4.	($225.50 −	$ 55.50	= ?)	+ ?	+ $ 22.50	= ?
5.	($637.50 −	$ 57.50	= ?)	+ ?	+ $122.50	= ?
6.	($127.33 −	$100.00	= ?)	+ ?	+ $ 57.21	= ?
7.	($319.90 −	$319.90	= ?)	+ ?	+ $117.64	= ?
8.	($837.17 −	$145.00	= ?)	+ ?	+ $ 99.90	= ?

9. Sara Green's account statement.
Unpaid balance of $64.
Periodic rate is 1.5%.
What is the finance charge?
New purchases of $26.
What is the new balance?

10. Steve Corroll's account statement.
Unpaid balance of $12.60.
Periodic rate is 1.25%.
What is the finance charge?
New purchases of $41.25.
What is the new balance?

11. Elsie Buell's account statement.
Unpaid balance of $75.80.
Periodic rate is 1.6%.
What is the finance charge?
New purchases of $32.90.
What is the new balance?

12. Verda Lau's account statement.
Unpaid balance of $211.35.
Periodic rate is 1.75%.
What is the finance charge?
New purchases of $15.35.
What is the new balance?

13. A portion of Thomas Rockicki's charge account statement is shown. The finance charge is 2% of the unpaid balance. What is the new balance?

PREVIOUS BALANCE	PAYMENTS & CREDITS	UNPAID BALANCE	FINANCE CHARGE	NEW PURCHASES	NEW BALANCE
$219.92	$25.00	$194.92		$49.35	?

14. A portion of Lucille Minkwitz's charge account statement is shown. The finance charge is 1.25% of the unpaid balance. What is the new balance?

PREVIOUS BALANCE	PAYMENTS & CREDITS	UNPAID BALANCE	FINANCE CHARGE	NEW PURCHASES	NEW BALANCE
$173.48	$111.23			$37.46	?

15. Alice Ray has a charge account at Rowley's Department Store where the finance charge is 2% of the unpaid balance. Find the indicated amounts on the statement shown.

DEPT.	DESCRIPTION	CHARGES	PAYMENT/CREDIT	DATE	REF. #
55	Coats, Suits	47.88		3/2	6982
44	Drapes, Curtains	54.32		3/2	8640
32	Hosiery	15.35		3/2	9000
98	Daywear		11.45	3/5	7640
88	PAYMENT		50.00	3/9	500

BILLING DATE: 10/10

PREVIOUS BALANCE	PAYMENTS & CREDITS	UNPAID BALANCE	FINANCE CHARGE	NEW PURCHASES	NEW BALANCE
$254.79	?	?	?	?	?

16. Bill Baker has a charge account at Fieldman's Department Store where the finance charge is 1.75% of the unpaid balance. Find the indicated amounts on the statement shown.

DEPT.	DESCRIPTION	CHARGES	PAYMENT/CREDIT	DATE	REF. #
09	Housewares	12.31		9/12	9064
14	PAYMENT		50.00	9/13	A345
03	Sporting goods	39.83		9/26	11309
08	Sporting goods		4.89	10/02	3960

BILLING DATE: 10/10

PREVIOUS BALANCE	PAYMENTS & CREDITS	UNPAID BALANCE	FINANCE CHARGE	NEW PURCHASES	NEW BALANCE
$283.36	?	?	?	?	?

Maintaining Your Skills

Look up the skills in parentheses if you need help or more practice.

Add. (Skill 5)

17. $115.10 + $45.75 + $48.35 + $1.22

18. $4.50 + $27.62 + $18.91 + $134.72

Subtract. (Skill 6)

19. $329.24 − $87.30

20. $891.37 − $89.44

21. $443.99 − $212.92

22. $110.01 − $48.87

23. $78.05 − $5.54

24. $909.67 − $27.53

Find the percentage. (Skill 30)

25. 4% of 180

26. 9% of 50

27. 3.2% of 600

28. 5.7% of 290

29. 0.8% of 200

30. 1.15% of 397.2

31. 1.75% of 888

6-5 Finance Charge—Average Daily Balance (No new purchases included)

Objective

Compute the finance charge based on the average daily balance.

Many companies calculate the finance charge using the **average-daily-balance** method where no new purchases are included. The average daily balance is the average of the account balance at the end of each day of the billing period. For this method of computing finance charges, new purchases posted during the billing period are not included when figuring the balance at the end of the day.

$$\text{Average Daily Balance} = \frac{\text{Sum of Daily Balances}}{\text{Number of Days}}$$

Example

Skills 8, 11 **Application** G **Term** Average daily balance

A portion of Wilbur Marchal's credit card statement is shown.

REFERENCE	POSTING DATE	TRANSACTION DATE	DESCRIPTION	PURCHASES & ADVANCES	PAYMENTS & CREDITS
131809	9/05	8/24	Ski Shop	18.77	
265118	9/18		PAYMENT		54.83
407372	9/20	9/01	Delucca's	24.19	
329416	10/01	8/30	Brighton Service	30.99	

Billing Period	Prev. Balance	Periodic Rate	Average Daily Balance	FINANCE CHARGE
9/04 - 10/03	$174.83	2%	?	$2.91

Payments & Credits	Purchases & Advances	New Balance	Minimum Payment	Payment Due
$54.83	$73.95	?	$20.00	10/25

A finance charge was added to Wilbur's account balance because he did not pay his last bill in full. The finance charge was computed using the average daily balance where new purchases are not included. Only the payment of $54.83 was used to determine the average daily balance. What is the average daily balance?

Solution

1. Find the **sum of daily balances.**

Dates	Payment	End-of-Day Balance		Number of Days	Sum of Balances
9/4-17		$174.83	×	14	$2447.62
9/18	$54.83	$120.00	×	1	120.00
9/19-10/3		$120.00	×	15	1800.00
			Total	30	$4367.62

2. Find the **average daily balance.**

   Sum of Daily Balances ÷ Number of Days

   $4367.62 ÷ 30 = $145.5873 = $145.59 average daily balance

$4000
$300
$300; 19; $5700
30; $10,000
average daily balance = $333.33

▶ **Self-Check** Cover the answers and complete the problems.
Find the average daily balance.

Dates	Payment	End-of-Day Balance	×	Number of Days	=	Sum of Balance
9/9-9/18		$400	×	10	=	?
9/19	$100	$300	×	1	=	?
9/20-10/8		?	×	?	=	?
			Total	?		?

The finance charge is calculated by multiplying the periodic rate times the average daily balance.

▶ **Finance Charge = Periodic Rate × Average Daily Balance**

▶ **New Balance = Unpaid Balance + Finance Charge + New Purchases**

Example

Skills **8, 11** **Application** **G** **Term** **Average daily balance**

Wilbur checks the finance charge and the new balance. The finance charge is 2% of the average daily balance. What is the new balance?

Solution

1. Find the **unpaid balance.**

   Previous Balance − (Payments + Credits)

   $174.83 − $54.83 = $120.00

2. Find the **finance charge.**

   Periodic Rate × Average Daily Balance

   2% of $145.59 = $2.9118 = $2.91

3. Find the **new purchases.** $73.95

4. Find the **new balance.**

   Unpaid Balance + Finance Charge + New Purchases

   $120.00 + $2.91 + $73.95 = $196.86 new balance

A. $1.80
B. $141.80

▸ **Self-Check** Cover the answers and complete the problems.

The finance charge is 1.5% of the daily balance.

A. Find the finance charge. B. Find the new balance.

PREVIOUS BALANCE	TOTAL CHARGES	TOTAL PAYMENTS AND CREDITS	AVERAGE DAILY BALANCE	FINANCE CHARGE	NEW BALANCE
$150.00	$20.00	$30.00	$120.00	?	?

Problems

	Billing Periods	Payment	End-of-day Balance	Number of Days	Sum of Balances	
1.	6/01-6/15 6/16 6/17-6/30	 $50.00 	$84.00 34.00 34.00	15 1 14	$1260.00 34.00 ?	What is the average daily balance?
	TOTALS			?	?	
2.	7/15-8/2 8/3 8/4-8/14	 $100.00 	$300.00 200.00 200.00	? ? ?	? ? ?	What is the average daily balance?
	TOTALS			?	?	

3. Amy Bertelli has a bank charge card. Use the portion of the account statement shown to find the average daily balance, excluding new purchases.

REFERENCE	POSTING DATE	TRANSACTION DATE	DESCRIPTION	PURCHASES & ADVANCES	PAYMENTS & CREDITS
6646598	8/25	8/20	Al's Hardware	34.65	
7000507	8/28		PAYMENT RECEIVED		40.00

BILLING PERIOD	PREVIOUS BALANCE	PERIODIC RATE	AVERAGE DAILY BALANCE	FINANCE CHARGE
8/01 - 8/31	$240.00	1.5%	?	---------

4. The finance charge is computed using the average-daily-balance method where no new purchases are included. Find the finance charge and the new balance for the following statements.

BILLING PERIOD	PREVIOUS BALANCE	PERIODIC RATE	AVERAGE DAILY BALANCE	FINANCE CHARGE
2/3 - 3/2	$182.00	2%	$137.00	?

PAYMENTS & CREDITS	PURCHASES & ADVANCES	NEW BALANCE	MINIMUM PAYMENT	PAYMENT DUE
$90.00	0	?	$15.00	3/20

5.

BILLING PERIOD	PREVIOUS BALANCE	PERIODIC RATE	AVERAGE DAILY BALANCE	FINANCE CHARGE
11/15 - 12/14	$263.40	1.5%	$181.46	?

PAYMENTS & CREDITS	PURCHASES & ADVANCES	NEW BALANCE	MINIMUM PAYMENT	PAYMENT DUE
$163.40	$40.74	?	$20.00	1/2

6.

BILLING PERIOD	PREVIOUS BALANCE	PERIODIC RATE	AVERAGE DAILY BALANCE	FINANCE CHARGE
7/1 - 8/1	$74.50	1.25%	$45.66	?

PAYMENTS & CREDITS	PURCHASES & ADVANCES	NEW BALANCE	MINIMUM PAYMENT	PAYMENT DUE
$44.50	$23.95	?	$10.00	8/20

7. A portion of Celia Faust's account statement for March from Inbank Charge Company is shown. The finance charge is computed using the average-daily-balance method where new purchases are excluded. Find the average daily balance, the finance charge, and the new balance.

REFERENCE	POSTING DATE	TRANSACTION DATE	DESCRIPTION	PURCHASES & ADVANCES	PAYMENTS & CREDITS
450345	3/20		PAYMENT		34.56
458343	3/27	3/14	Aston Oil Co.	10.90	

BILLING PERIOD	PREVIOUS BALANCE	PERIODIC RATE	AVERAGE DAILY BALANCE	FINANCE CHARGE
3/4 - 4/3	$84.56	2%	?	?

PAYMENTS & CREDITS	PURCHASES & ADVANCES	NEW BALANCE	MINIMUM PAYMENT	PAYMENT DUE
$34.56	$70.90	?	$10.00	4/21

8. Darla Huff received this statement from Don's Department Store. Find the average daily balance, the finance charge, and the new balance.

REFERENCE	POSTING DATE	TRANSACTION DATE	DESCRIPTION	PURCHASES & ADVANCES	PAYMENTS & CREDITS
4500298	4/15	4/01	Tennis shorts	19.95	
1027485	4/11		PAYMENT		50.00
5473390	4/23	4/21	Paperback book	3.99	
1374655	4/25		PAYMENT		50.00

BILLING PERIOD	PREVIOUS BALANCE	PERIODIC RATE	AVERAGE DAILY BALANCE	FINANCE CHARGE
4/1 - 5/1	$198.00	1%	?	?

PAYMENTS & CREDITS	PURCHASES & ADVANCES	NEW BALANCE	MINIMUM PAYMENT	PAYMENT DUE
$100.00	$23.94	?	$25.00	6/30

(Continue on next page.)

9. Thomas Valli received this statement from Crowley Bank. Find the average daily balance, the finance charge, and the new balance.

REFERENCE	POSTING DATE	TRANSACTION DATE	DESCRIPTION	PURCHASES & ADVANCES	PAYMENTS & CREDITS
1616787	4/25	4/19	Ace Plumbing	59.43	
3945557	4/30		PAYMENT		95.55
3957475	5/5	5/1	Terr Hardware	22.79	
4000076	5/20		PAYMENT		50.00

BILLING PERIOD	PREVIOUS BALANCE	PERIODIC RATE	AVERAGE DAILY BALANCE	FINANCE CHARGE
4/21 - 5/20	$295.55	2%	?	?

PAYMENTS & CREDITS	PURCHASES & ADVANCES	NEW BALANCE	MINIMUM PAYMENT	PAYMENT DUE
$145.55	$82.22	?	$30.00	6/12

10. Nicki Louie received this statement from Credit Charge Company. Find the indicated amounts on the statement shown.

REFERENCE	POSTING DATE	TRANSACTION DATE	DESCRIPTION	PURCHASES & ADVANCES	PAYMENTS & CREDITS
31784	7/25		PAYMENT		30.00
103645	7/25	7/12	Men's shoes	34.99	
116748	7/30	7/28	LP record		7.98
345803	8/8		PAYMENT		30.00
57845	8/8	8/7	5-pc. cookware	24.99	

BILLING PERIOD	PREVIOUS BALANCE	PERIODIC RATE	AVERAGE DAILY BALANCE	FINANCE CHARGE
7/22 - 8/21	$449.26	1.75%	?	?

PAYMENTS & CREDITS	PURCHASES & ADVANCES	NEW BALANCE	MINIMUM PAYMENT	PAYMENT DUE
?	?	?	$30.00	9/11

***11. A Brief Case** Some companies compute the finance charge using the average-daily-balance method where new purchases are included when figuring the daily balances during the posting period.

REFERENCE	POSTING DATE	TRANSACTION DATE	DESCRIPTION	PURCHASES & ADVANCES	PAYMENTS & CREDITS
1-32734	12/10	12/8	Housewares	44.91	
2-44998	12/20		PAYMENT		63.00

BILLING PERIOD	PREVIOUS BALANCE	PERIODIC RATE	AVERAGE DAILY BALANCE	FINANCE CHARGE
12/1 - 12/31	$124.37	2%	$131.85	?

PAYMENTS & CREDITS	PURCHASES & ADVANCES	NEW BALANCE	MINIMUM PAYMENT	PAYMENT DUE
$63.00	$44.91	?	$20.00	1/21

a. Jeff Zdybek has a charge account where the finance charge is computed using the average-daily-balance method that includes new purchases. He checks to be sure the average daily balance is correct on the statement at the bottom of page 168. Find the finance charge and the new balance.

Date	Payments	Purchases	End-of-Day Balance		Number of Days	Sum of Balances
12/1-9			$124.37	×	9	$1119.33
12/10		$44.91	169.28	×	1	169.28
12/11-19			169.28	×	9	1523.52
12/20	$63.00		106.28	×	1	106.28
12/21-31			106.28	×	11	1169.08
						$4087.49

Average daily balance: $4087.49 ÷ 31 = $131.85

b. Find the average daily balance where new purchases are included, the finance charge, and the new balance.

REFERENCE	POSTING DATE	TRANSACTION DATE	DESCRIPTION	PURCHASES & ADVANCES	PAYMENTS & CREDITS
712719	2/12	2/7	Corner Drug	10.00	
745634	2/16		PAYMENT		30.00
745645	2/19	2/2	Touch Beauty Inc.	20.00	
894657	2/24	2/10	PA Sound Store	6.00	

BILLING PERIOD	PREVIOUS BALANCE	PERIODIC RATE	AVERAGE DAILY BALANCE	FINANCE CHARGE
2/1 - 2/28	$400.00	1.5%	?	?

PAYMENTS & CREDITS	PURCHASES & ADVANCES	NEW BALANCE	MINIMUM PAYMENT	PAYMENT DUE
$30.00	?	?	$30.00	3/15

Maintaining Your Skills

Look up the skills in parentheses if you need help or more practice.

Multiply. (Skill 8)

12. 12 × 300 **13.** 9 × 450 **14.** 15 × 37.5 **15.** 8 × 24.56

16. 6 × 125.5 **17.** 8 × 32.50 **18.** 65 × 40.8 **19.** 25 × 142.5

Divide. Round answers to the nearest hundredth. (Skill 10)

20. 3460 ÷ 30 **21.** 4800 ÷ 42 **22.** 1123.75 ÷ 31 **23.** 10,225.6 ÷ 28

Find the average. (Application Q)

24. 34, 53, 72, 28, 13 **25.** 124, 206, 158, 185, 185, 174

26. 235, 315, 264, 219, 44 **27.** 32.8, 43.31, 12.66, 24.89, 3.35

Reviewing the Basics

Skills

Round to the nearest cent.

(Skill 2) 1. $146.1378 2. $152.180 3. $36.4681 4. $4.796

Solve.

(Skill 5) 5. $135.23 + $21.77 + $1.98 6. $147.15 + $5.09 + $15.82 + $128.91

(Skill 6) 7. $389.40 − $292.72 8. $515.91 − $82.29 9. $27.01 − $4.37

(Skill 8) 10. $320 × 7 11. $432 × 15 12. $1310.79 × 20

(Skill 8) 13. $32.14 × 8 14. $198.78 × 16 15. $56.99 × 21

(Skill 11) 16. $510.60 ÷ 30 17. $1825.60 ÷ 28 18. $3773.40 ÷ 30

19. $472.75 ÷ 31 20. $4521.97 ÷ 31 21. $20,281.95 ÷ 31

(Skill 30) 22. 1.5% of 580 23. 1% of 345.2 24. 2% of 562.10

Applications

Find the elapsed time in days.

(Application G) 25. From June 4 to June 29 26. From April 3 to April 19

27. From May 19 to June 4 28. From July 23 to August 22

Find the average.

(Application Q) 29. 45, 32, 64, 51, 31, 55 30. 12.1, 14.6, 21.1, 15.3, 19.4

31. $45, $64, $55, $43, $46 32. 9.54, 6.23, 1.10, 3.54

Terms

Match each term with its definition on the right.

33. Sales receipt

34. Finance charge

35. Previous balance

36. Unpaid balance

37. Average daily balance

a. the average of the balances in a charge account at the end of each day in the billing period

b. the amount of money that remains unpaid in a charge account

c. a record of the total purchase which a charge card user signs each time the account is used

d. interest which a charge or credit card user pays for delaying payment

e. the monthly finance charge rate

f. the amount that a charge or credit card user owed at the close of the last billing period

Refer to your reference files at the back of the book if you need help.

The Changing Face of Business

Electronic funds transfer systems (EFTS) are changing the way that many consumers are conducting their financial affairs. EFTS were made possible through the introduction of computer technology. It may be possible in the near future for you to conduct all your financial transactions without ever setting foot in the office of the financial institution. This will mean greater convenience to you, the consumer.

A transaction involving an EFTS usually begins with a plastic card that has been magnetically coded to your bank. The card is used in conjunction with a personal identification number at some remote terminal. Once you are tied-in to your account, you can perform those services that you have paid for. For instance, you may be able to transfer funds from your checking account into your savings account, or vice versa. You may obtain cash at any time of the day from an automatic teller machine. Or, you can have your check verified and cashed at a supermarket. A few of the other services available through EFTS include automatic bill paying, automatic payroll service, and point-of-sale terminals.

Answer the following questions concerning EFTS.

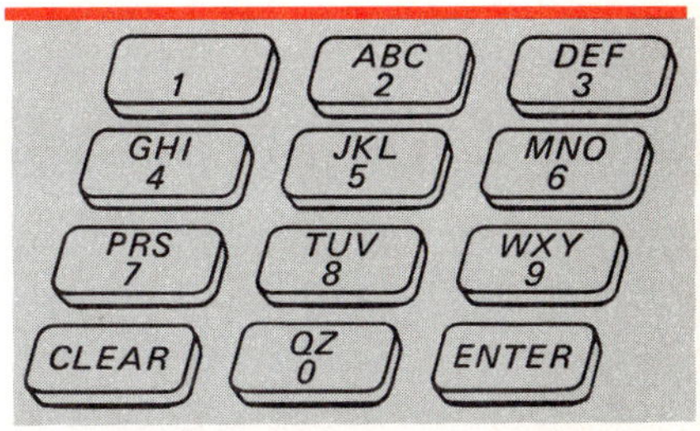

1. Janice Hayes would like to cash a check at the grocery store. She can verify her check at the verification machine. She inserts her bank card and enters her personal identification number. So that she can remember her number easily, she has associated her number with the name "PAUL." What is Janice's personal identification number?

2. Jeremy Reed is able to use his credit card to obtain travelers' checks at the airport from an automatic teller machine. The machine is only able to give amounts in multiples of $250. Jeremy will need at least $800. What amount in travelers' checks must he request?

3. Although there are advantages to using EFTS, there are a few disadvantages. One disadvantage of EFTS is the loss of confidentiality of the electronic transactions. Name other disadvantages which might be associated with the use of electronic funds transfer systems.

Unit Test

Lesson 6-1

1. Josie Ellis used her bank charge card to purchase a storage shed. The shed cost $205.95 plus a 6% sales tax. What was the total purchase price on the sales receipt?

Lesson 6-2

2. Find the new balance on the charge account statement.

PREVIOUS BALANCE	CLOSING DATE THIS MONTH	CLOSING DATE LAST MONTH
$41.57	November 13, 19— —	October 14, 19— —

TOTAL PURCHASES	PAYMENTS & CREDITS	FINANCE CHARGE	NEW BALANCE	MINIMUM PAYMENT
$37.28	0	$0.62	?	————

Lesson 6-3

3. Steve McKenzie has a charge account that charges 1.5% of the previous balance for a finance charge. His statement shows purchases totaling $48.14, a previous balance of $470, and a payment of $85. What is the new balance?

Lesson 6-4

4. A portion of Mary Lewis's account statement is shown. The finance charge is 1.5% of the unpaid balance. Find the unpaid balance, the finance charge, and the new balance.

BILLING DATE	PREVIOUS BALANCE	PAYMENTS AND CREDITS	UNPAID BALANCE	FINANCE CHARGE	NEW PURCHASES	NEW BALANCE
10/13	$314.52	$126.14	?	?	$124.15	

Lesson 6-5

5. Find the average daily balance where new purchases are not included when figuring the daily balances.

REFERENCE	POSTING DATE	TRANSACTION DATE	DESCRIPTION	PURCHASES & ADVANCES	PAYMENTS & CREDITS
13100	9/14	8/23	Wesford Theatre	37.00	
20046	9/14	8/27	Record Mart	24.65	
54545	9/17		PAYMENT		30.00

BILLING PERIOD	PREVIOUS BALANCE	PERIODIC RATE	AVERAGE DAILY BALANCE	FINANCE CHARGE
9/1 - 9/30	$140.00	1.5%	?	

PAYMENTS & CREDITS	PURCHASES & ADVANCES	NEW BALANCE	MINIMUM PAYMENT	PAYMENT DUE
$30.00	$61.65	——————	$20.00	10/21

6. In problem 5, find the finance charge and the new balance using the average-daily-balance method where new purchases are not included.

Unit 7 Loans

▸ A *loan* is money that you have borrowed and must repay with *interest*. Interest is the cost of borrowing money. A *single-payment loan* is repaid in one payment after a specified period of time. For a *discount loan*, you pay the interest first, before you receive the loan. An *installment loan* is a third type of loan. You pay a portion of the loan and a portion of the interest in several installments. Because all banks do not charge the same amount of interest for a sum of money, you should shop around for the best buy. Using the *annual percentage rate* is one way to compare the cost of borrowing money. If you repay a loan early you may receive a *refund* of part of the finance charge.

Before you can get a loan, you must fill out a credit application form. A credit investigation is made from the information about your income, bank accounts, and any credit that has been extended to you.

7-1 Single-Payment Loans

Objective Compute the maturity value and interest rate of a single-payment loan.

A **single-payment** loan is a loan that you repay with one payment after a specified period of time. The **maturity value** of the loan is the total amount you repay. It includes both the **principal** and the interest owed. The principal is the amount borrowed.

The **term** of a loan is the amount of time for which the loan is granted. A single-payment loan may be granted for a stated number of years, months, or days. When the term is a certain number of days, the lending agency may calculate interest in one of two ways. **Ordinary interest** is calculated by basing the length of time of the loan on a 360-day year. **Exact interest** is calculated by basing the length of time on a 365-day year.

▶ **Maturity Value = Principal + Interest Owed**

Example

Skills 28, 14 Application J Term Single-payment loan

Sara Vickers's bank granted her a single-payment loan of $1200 for 63 days at an interest rate of 12%. Her bank charges ordinary interest. What is the maturity value of Sara's loan?

Solution

1. Find the **interest owed.**
   Principal × Rate × Time
   $\$1200.00 \times 12\% \times \frac{63}{360} = \$25.20$
2. Find the **maturity value.**
   Principal + Interest Owed
   $1200.00 + $25.20 = $1225.20 maturity value

▶ **Self-Check** Cover the answers and complete the problems.

A. Compute the ordinary interest and the maturity value.
$\$300 \times 10\% \times 60/360 = ?$

B. Compute the exact interest and the maturity value.
$\$500 \times 12\% \times 73/365 = ?$

A. $5.00 interest, $305 maturity value
B. $12 interest, $512 maturity value

Problems

	1.	2.	3.	4.	5.
Principal	$600	$1200	$400	$2400	$1875
Annual Interest Rate	12%	15%	13%	12.5%	13.75%
Term	45 days	60 days	36 days	126 days	72 days
Ordinary Interest	$9.00	?	?	?	?
Maturity Value	?	?	?	?	?

6. Alice Stroud.
Single-payment loan of $800.
Interest rate of 14%.
108 days, ordinary interest.
What is the interest owed?
What is the maturity value?

7. Raymond Shaw.
Single-payment loan of $5000.
Interest rate of 10%.
73 days, exact interest.
What is the interest owed?
What is the maturity value?

8. Gordon Holliday's bank granted him a single-payment loan of $1500 for 144 days at an interest rate of 11%. His bank charges ordinary interest. What is the maturity value of Gordon's loan?

9. Celia Burke's bank granted her a single-payment loan of $2000 for 146 days at an interest rate of 15%. Her bank charges ordinary interest. What is the maturity value of Celia's loan?

10. Harry Dacey obtained a single-payment loan of $350 to pay a repair bill. He agreed to repay the loan in 219 days at an interest rate of 11.75% exact interest. What is the maturity value of his loan?

11. Kenneth Carey obtained a single-payment loan of $8400 to pay for fertilizer for his commercial orchard. He agreed to repay the loan in 270 days at an interest rate of 15.25% exact interest. What is the maturity value of his loan?

12. Manuel Rodriguez would like to borrow $1000 for 90 days. People's Finance Company charges an interest rate of 18% at ordinary interest. Mutual Finance Company also charges an interest rate of 18% but at exact interest. What would be the maturity value of Manuel's loan at People's Finance Company? What would be the maturity value of the loan at Mutual Finance Company?

13. Mary Bivins would like to obtain a single-payment loan of $1375. First Finance Company charges ordinary interest and will grant the loan for 180 days at an interest rate of 17.5%. Southside Loan Company will grant the loan for 9 months at an interest rate of 16.75%. Where should Mary obtain the loan to pay the lower maturity value?

Maintaining Your Skills

Look up the skills in parentheses if you need help or more practice.

Write the percents as decimals. (Skill 28)

14. 10% 15. 10.5% 16. 9% 17. 9.25% 18. 12.85%

Reduce the fractions to lowest terms. (Skill 12)

19. $\frac{30}{40}$ 20. $\frac{180}{360}$ 21. $\frac{90}{360}$ 22. $\frac{120}{360}$ 23. $\frac{73}{365}$

Change the fractions to decimals. (Skill 14)

24. $\frac{90}{360}$ 25. $\frac{45}{360}$ 26. $\frac{126}{360}$ 27. $\frac{292}{365}$ 28. $\frac{219}{365}$

7-2 Discount Loans

Objective

Compute the proceeds of a discount loan.

A **discount loan** is a loan in which the interest owed is paid in advance. When the loan is granted, the lender subtracts the amount of interest from the amount of the loan. The amount of money that you receive after the interest has been deducted is called the **proceeds.** Discount loans are generally short-term loans that are repaid over a specified number of days.

▶ **Proceeds = Principal − Interest Owed**

Example

Skills 28, 6 **Application** J **Term** Discount loan

Gerald Chin obtained a discount loan of $2400 at an interest rate of 14%. The loan was granted for 120 days. The bank charges ordinary interest. What proceeds did Gerald receive?

Solution

1. Find the **interest owed.**

   Principal × Rate × Time

   $\$2400 \times 14\% \times \frac{120}{360} = \$112$

2. Find the **proceeds.**

   Principal − Interest Owed

   $2400 − $112 = $2288 proceeds

A. $72 interest, $3528 proceeds
B. $84 interest, $4716 proceeds

▶ **Self-Check** Cover the answers and complete the problems.

Find the interest and the proceeds.

A. $\$3600 \times 15\% \times \frac{48}{360} = ?$

$3600 − ? = ?

B. $\$4800 \times 18\% \times \frac{35}{360} = ?$

$4800 − ? = ?

Problems

	1.	2.	3.	4.	5.
Principal	$280	$1000	$7200	$2520	$2400
Rate	15%	14%	18%	15%	$16\frac{1}{2}\%$
Term	90 days	108 days	80 days	30 days	85 days
Ordinary Interest	$10.50	?	?	?	?
Proceeds	?	?	?	?	?

6. Glenn Reed.
Discount loan of $2000.
16% annual interest rate.
Term of 63 days, ordinary interest.
What is the interest owed?
What are the proceeds?

7. Geneva Parker.
Discount loan of $1200.
14.5% annual interest rate.
Term of 45 days, ordinary interest.
What is the interest owed?
What are the proceeds?

8. Noreen Regan.
Discount loan of $1400.
13.5% annual interest rate.
Term of 73 days, exact interest.
What are the proceeds?

9. Alvin Tsoi.
Discount loan of $1500.
18.65% annual interest rate.
Term of 146 days, exact interest.
What are the proceeds?

10. Hilda Santos obtained a discount loan of $600 at an interest rate of 14%. The loan was granted for 54 days. The bank charges ordinary interest. What proceeds did Hilda receive?

11. Donald Trainor obtained a discount loan of $1200 at an interest rate of 16%. The loan was granted for 105 days. The bank charges ordinary interest. What proceeds did Donald receive?

12. Carl Dwyer obtained a discount loan of $1500 at an interest rate of 14.6%. He agreed to repay the loan in 225 days. What proceeds did Carl receive if the finance company charges exact interest?

13. Julie Burford was granted a discount loan of $8400 at 15.25% exact interest. She agreed to repay the loan in 60 days. What proceeds did she receive? How much did she pay the bank at the end of the 60 days?

Maintaining Your Skills

Look up the skills in parentheses if you need help or more practice.

Write the percents as decimals. (Skill 28)

14. 15% 15. 18.35% 16. 12.34% 17. 27.50% 18. 20.3%

Multiply. (Skill 8)

19. $800 \times 0.05 \times 0.25$ 20. $3600 \times 0.15 \times 0.30$ 21. $4500 \times 0.16 \times 0.75$

Subtract. (Skill 6)

22. $600.00 - 45.50$
23. $350.00 - 54.25$
24. $465.50 - 36.87$
25. $8550.00 - 243.76$

Reduce the fractions to lowest terms. (Skill 12)

26. $\frac{50}{360}$ 27. $\frac{110}{360}$ 28. $\frac{95}{365}$ 29. $\frac{80}{360}$

7-3 Installment Loans

Objective Compute the amount financed on an installment loan.

An **installment loan** is a loan that you repay in several equal payments over a specified period of time. Usually when you purchase an item with an installment loan, you must make a **down payment.** The down payment is a portion of the cash price of the item you are purchasing. The **amount financed** is the portion of the cash price that you owe after making the down payment.

▶ **Amount Financed = Cash Price − Down Payment**

Example

Skills 30, 28, 2 **Application** A **Term** Installment loan

Carla Metzler purchased a wall heating unit for $1035. She used the store's installment credit plan to pay for the unit. Carla made a down payment of 20% and financed the remaining amount. What amount did Carla finance?

Solution

1. Find the **down payment.**
   20% of $1035.00 = $207.00
2. Find the **amount financed.**
   Cash Price − Down Payment
   $1035.00 − $207.00 = $828.00 amount financed

A. $168 down payment $672 amount financed
B. $735 down payment $1715 amount financed

▶ **Self-Check** Cover the answers and complete the problems.

Find the down payment and the amount financed.

A. Sofa bed.
Cash price of $840.
20% down payment.

B. Oriental rug.
Cash price of $2450.
30% down payment.

Problems

	1.	2.	3.	4.	5.	6.
Cash Price	$975	$2940	$1330	$2510	$4362	$5890
Percent Down Payment	20%	30%	25%	28%	20%	25%
Down Payment	$195	?	?	?	?	?
Amount Financed	?	?	?	?	?	?

7. Cash price of $459.
20% down payment.
What amount is financed?

8. Cash price of $685.
28% down payment.
What amount is financed?

9. Cash price of $7549.50.
15% down payment.
What amount is financed?

10. Cash price of $2490.40.
25% down payment.
What amount is financed?

11. Les Grant purchased a dining room set for $2020. He used the store's credit plan. He made a 20% down payment. What amount did he finance?

12. Cheryl Morgan purchased a new set of radial tires for her car. The set of tires cost $179.50. Using the store's credit plan, she made a 40% down payment. What amount did she finance?

13. Cindy Hughs purchased a motorbike for $497.95. She made a down payment of 20% and financed the remaining amount using the store's installment plan. What amount did Cindy finance?

14. Amy and Eric Williams went on a two-week vacation at a total cost of $1278.95. They financed the trip through Sentinel Bank. They made a 25% down payment and financed the remaining amount on the installment plan. What amount did the Williams finance?

15. Jane and Milton Nash want to remodel their kitchen. They would like to finance part of the cost but do not want the amount financed to be more than $4000. The total cost of remodeling the kitchen is $5000. What percent of the total cost should their down payment be?

16. Barry Wells wants to purchase a car costing $5575. He will finance the car with an installment loan from the bank but would like to finance no more than $3000. What percent of the total cost of the car should his down payment be?

Maintaining Your Skills

Look up the skills in parentheses if you need help or more practice.

Write the percents as decimals. (Skill 28)

17. 22% 18. 35% 19. 15% 20. 40% 21. 28%

Find the percentage. (Skill 30)

22. 20% × 140 23. 30% × 225 24. 25% × 990 25. 15% × 795

Round to the nearest cent. (Skill 2)

26. $39.9428 27. $149.5764 28. $331.435 29. $18.9963

7-4 Installment Loans—Finance Charge

Objective Compute the finance charge on an installment loan.

When you obtain an installment loan, you must pay finance charges for the use of the money. Usually you repay the amount financed plus the finance charge in equal monthly payments. The amount of each monthly payment depends on the amount financed, the number of monthly payments, and the **annual percentage rate.** The annual percentage rate (APR) is an index showing the relative cost of borrowing money. Lenders are required by law to inform you of the finance charge and the APR.

APR	Term in Months	If You Finance. . . $200	$500	$1000	$1500
		Your **Monthly Payments** Are			
10%	6	34.31	85.78	171.56	257.34
	12	17.58	43.95	87.91	131.87
	18	12.01	30.02	60.05	90.08
	24	9.22	23.07	46.14	69.21
15%	6	34.80	87.01	174.03	261.05
	12	18.05	45.12	90.25	135.38
	18	12.47	31.19	62.38	93.57
	24	9.69	24.24	48.48	72.72
18%	6	35.10	87.76	175.52	263.28
	12	18.33	45.84	91.68	137.52
	18	12.76	31.90	63.80	95.70
	24	9.98	24.96	49.92	74.88

▶ **Total Amount Repaid = Number of Payments × Monthly Payment**
▶ **Finance Charge = Total Amount Repaid − Amount Financed**

Example

Skills 8, 6 **Application** C **Term** Annual percentage rate

Sandy Cole obtained an installment loan of $1000 to purchase a new washer and dryer. The annual percentage rate is 15%. She must repay the loan in 18 months. What is the finance charge?

Solution

1. Find the **monthly payment.** (Refer to the above table.)
   - **a.** APR of loan 15%
   - **b.** Term of loan . . . . 18 months
   - **c.** Amount financed $1000
   - Value from table $62.38
2. Find the **total amount repaid.**
   Number of Payments × Monthly Payment
   18 × $62.38 = $1122.84
3. Find the **finance charge.**
   Total Amount Repaid − Amount Financed
   $1122.84 − $1000.00 = $122.84 finance charge

▶ **Self-Check** Cover the answers and complete the problems.

A. (6 payments × $34.31) − $200 = ? finance charge
B. (24 payments × $74.88) − $1500 = ? finance charge

A. $5.86
B. $297.12

Problems

Use the table on page 180 to solve the following.

	1.	2.	3.	4.	5.
APR	10%	15%	18%	10%	15%
Term (Months)	12	18	24	6	12
Amount Financed	$200	$1000	$500	$1500	$1000
Monthly Payment	?	?	?	?	?
Total Repaid	?	?	?	?	?
Finance Charge	?	?	?	?	?

6. Pamela Smith.
   Installment loan of $1500.
   12 monthly payments.
   APR is 18%.
   What are the monthly payments?
   What is the finance charge?

7. Mark Kennedy.
   Installment loan of $1000.
   24 monthly payments.
   APR is 15%.
   What are the monthly payments?
   What is the finance charge?

8. Elmer Harding obtained an installment loan of $1500 to landscape the front of his house. The APR is 15%. The loan is to be repaid in 24 monthly payments. What is the finance charge?

9. Roger Brown obtained an installment loan of $1000 to pay for some new furniture. He agreed to repay the loan in 18 monthly payments at an APR of 10%. What is the finance charge?

10. Brian and Carol Collins obtained an installment loan of $1500. They obtained the loan at an APR of 10% for 12 months. What is the finance charge?

11. Frank and Della Ray are purchasing an electric stove with an installment loan. The electric stove sells for $699.95. They agree to make a down payment of 20% and to make 36 monthly payments of $20.24 each. What is the finance charge?

12. Jeff and Hazel Andrews would like to obtain an installment loan of $1500 to repair the gutters on their home. They can get the loan at an APR of 15% for 24 months or at an APR of 18% for 18 months. Which loan costs less? How much do the Andrews save by taking the loan that costs less?

13. Maria Galloway would like an installment loan of $500. National Savings and Loan will loan her the money at 15% for 12 months. Economy Finance Company will loan her the money at 18% for 24 months. Which loan costs less? How much will she save by taking the loan that costs less?

(Continue on next page.)

APR	Term in Months	If You Finance. . . . $1500	$2000	$2500	$3000
		Your **Monthly Payments** Are			
22%	6	266.28	355.04	443.80	532.56
	12	140.39	187.18	233.98	280.78
	18	98.59	131.45	164.32	197.18
	24	77.81	103.75	129.69	155.63
25%	6	268.54	358.05	447.57	537.08
	12	142.56	190.08	237.61	285.13
	18	100.78	134.38	167.97	201.57
	24	80.05	106.74	133.42	160.11
26%	6	269.29	359.06	448.82	538.59
	12	143.29	191.06	238.82	286.59
	18	101.52	135.36	169.20	203.05
	24	80.81	107.74	134.68	161.62

14. Eldon Hamilton obtained an installment loan of $2500 for a used car. He financed the purchase with a finance company and agreed to repay the loan in 24 monthly payments at an APR of 25%. What is the finance charge?

15. Candy Matson obtained an installment loan of $3000 on a used sailboat. She financed the purchase through the boat dealer and agreed to repay the loan in 24 monthly payments at an APR of 26%. What is the finance charge?

16. Susan and Gene Kharche would like to obtain an installment loan of $2000. They can get the loan at an APR of 22% for 24 months or at an APR of 25% for 18 months. Which loan costs less? How much do the Kharches save by taking the loan that costs less?

17. Adolfo and Aurora Ramirez would like an installment loan of $2500. City Loan will loan the money at 25% for 18 months. Economy Line Finance Company will loan the money at 26% for 24 months. Which loan costs less? How much will be saved by taking the loan that costs less?

18. Lucretia and Don Bohnert plan to buy a home computer that costs $1999.95. They have a down payment of $499.95 and will finance the remainder at an APR of 22% for 18 months. What is the finance charge?

19. Carl and Fay Saylor obtained an installment loan to purchase the garage advertised for $3295. They made a down payment of $1295. The bank has agreed to loan them the remainder of the money at an APR of 26% for 24 months. What is the finance charge?

Designed for you! 22′×22′ **$3295**

Maintaining Your Skills

Look up the skills in parentheses if you need help or more practice.

Multiply. (Skill 8)

20. 21.50 × 6
21. 26.61 × 18
22. 9.98 × 24
23. 286.59 × 12
24. 137.52 × 12
25. 95.70 × 24

Subtract. (Skill 6)

26. 231.42 − 150.00
27. 2378.19 − 2250.00
28. 6199.68 − 2456.21
29. 4430.52 − 4029.86

7-5 Determining the APR

Objective Use a table to find the annual percentage rate of a loan.

If you know the number of monthly payments and the finance charge per $100 of the amount financed, you can use a table to find the annual percentage rate of the loan. You can use the APR of loans to compare the relative cost of borrowing money.

APR	10.00%	10.25%	10.50%	10.75%	11.00%	11.25%	11.50%	11.75%	12.00%	12.25%	12.50%
TERM	FINANCE CHARGE PER $100 OF AMOUNT FINANCED										
6	2.94	3.01	3.08	3.16	3.23	3.31	3.38	3.45	3.53	3.60	3.68
12	5.50	5.64	5.78	5.92	6.06	6.20	6.34	6.48	6.62	6.76	6.90
18	8.10	8.31	8.52	8.73	8.93	9.14	9.35	9.56	9.77	9.98	10.19
24	10.75	11.02	11.30	11.58	11.86	12.14	12.42	12.70	12.98	13.26	13.54

▸ **Finance Charge per $100 = $100 × $\frac{\text{Finance Charge}}{\text{Amount Financed}}$**

Example

Skills 11, 2 **Application** C **Term** Annual percentage rate

Ralph Stangle obtained an installment loan of $400 to pay for a new television set. The finance charge is $27. Ralph agreed to repay the loan in 12 monthly payments. What is the annual percentage rate?

Solution

1. Find the **finance charge per $100.**
   $100 × (Finance Charge ÷ Amount Financed)
   $100 × ($27.00 ÷ $400.00)
   $100 × 0.0675 = $6.75 finance charge per $100
2. Find the **APR.** (Refer to the above table.)
   In the row for 12 payments, find the number closest to $6.75. It is $6.76. Read the APR at the top of the column. APR is 12.25%.

A. $3.31 finance charge per $100, 11.25% APR
B. $12.72 finance charge per $100, 11.75% APR

▸ **Self-Check** Cover the answers and complete the problems.

A. 6-month loan.
$100 × ($26.48 ÷ $800) = ?
APR = ?

B. 24-month loan.
$100 × ($108.12 ÷ $850) = ?
APR = ?

Problems

Complete the table. Use the APR table on page 570.

	1.	2.	3.	4.	5.
Finance Charge	$31.80	$435.54	$426.30	$1080	$804.30
Amount Financed	$900	$2100	$3000	$5400	$4200
Finance Charge per $100	$3.53	?	?	?	?
Number of Payments	6	24	18	36	30
Annual Percentage Rate	?	?	?	?	?

6. Stanley Bourne.
Installment loan of $1500.
Finance charge of $211.50.
24 monthly payments.
What is the APR?

7. Julia Correy.
Installment loan of $1900.
Finance charge of $669.75.
36 monthly payments.
What is the APR?

8. Oneta Stanley.
Installment loan of $600.
Finance charge of $124.32.
18 monthly payments.
What is the APR?

9. Thomas Stapleton.
Installment loan of $2400.
Finance charge of $406.80.
30 monthly payments.
What is the APR?

10. Helen Johnson obtained an installment loan of $970 to purchase a sofa bed. The finance charge is $169.52. Helen agreed to repay the loan in 24 monthly payments. What is the annual percentage rate?

11. Kenneth Larkin obtained an installment loan of $4500 to pay his daughter's college tuition. The finance charge is $1566.24. He agreed to repay the loan in 36 monthly payments. What is the APR?

12. Webster Arca obtained an installment loan of $280 to purchase 4 new tires for his car. The finance charge is $28.07. He agreed to repay the loan in 12 monthly payments. What is the APR?

13. Betty Naiman obtained an installment loan of $1800 to purchase a lawn and garden tractor. The finance charge is $414.54. She agreed to repay the loan in 24 monthly payments. What is the APR?

14. Mildred Guest wants to obtain an installment loan of $1960 to purchase the used car advertised. The bank has agreed to loan her the money. She must repay the loan in 24 months. The finance charge is $343.92. What is the APR of Mildred's loan?

SALE √√ $1960.00
1981 4-dr sedan, a/c, power equipment, one owner. No rust. Easy financing.

15. Jenny Mills needs an installment loan of $350 to purchase a VHS recorder. The store has agreed to loan her the money. She must repay the loan in 18 months. The finance charge is $58. What is the APR of her loan?

16. Larnsy and Carole Stout are buying a washer that costs $299.99 and a dryer that costs $249.99. To use the store's installment plan, they need a down payment of $49.98. They must make 18 monthly payments of $31.90 each. What is the APR on their installment loan?

17. Andrew Sever is buying a new furnace that costs $1860. The bank requires a down payment of 20% and 36 monthly payments of $54.50 each. What is the APR on Andrew's loan?

18. Jorge Flores is buying a new color television set that costs $629.95. To use the installment plan available at the department store, he must make a down payment of 25% and make 30 monthly payments of $19.59 each. What is the APR on Jorge's loan?

19. Larry Holland would like an installment loan of $5000 to be repaid in 36 months. Nathaniel Loan Company will grant the loan with a finance charge of $1215.60. City Finance Company will grant the loan with a finance charge of $1532.50. What is the APR on each loan?

20. Ty Chin would like an installment loan of $6000 to be repaid in 24 months. ABC Finance Company will grant the loan with a finance charge of $1488. Atco Financial Service will grant the loan with a finance charge of $1506. What is the APR on each loan?

21. Alice Gagne purchased a $245.95 computer chess game. She made a down payment of $99 and through the store's installment plan agreed to pay $25.83 a month for 6 months. What is the finance charge? What is the APR? What is the total amount paid for the chess game?

22. Geraldine Stachowiak purchased a home computer for $3249.29. She made a down payment of $649.29 and financed the remainder by agreeing to pay $128.55 per month for 24 months. What is the finance charge? What is the APR? What is the total amount paid for the computer?

Maintaining Your Skills

Look up the skills in parentheses if you need help or more practice.

Round to the nearest cent. (Skill 2)

23. $19.414 24. $12.1372 25. $40.2951 26. $18.6402

Divide. Round answers to the nearest hundredth. (Skills 10, 11)

27. $17\overline{)340}$ 28. $24\overline{)1080}$ 29. $47\overline{)2632}$ 30. $984\overline{)81672}$

31. $1.4\overline{)73.22}$ 32. $8.2\overline{)239.6}$ 33. $76.1\overline{)94.8}$ 34. $64.2\overline{)104}$

Multiply. (Skill 8)

35. 100 × 2.73 36. 10 × 0.4712 37. 100 × 0.0316 38. 100 × 0.91734

7-6 Refund of Finance Charge

Objective

Compute the amount of finance charge refunded for early repayment of a loan.

If you repay an installment loan before the final due date, you may be entitled to a refund of part of the finance charge. The refund is usually stated as a percent of the finance charge. The lender may determine the percent refund using a **rebate schedule,** which is a table of percent refunds.

TERM OF LOAN	NUMBER OF MONTHS LOAN HAS RUN													
	1	2	3	4	5	6	7	8	9	10	11	12	13	14
3	50.00	16.67	0											
6	71.43	47.62	28.57	14.29	4.76	0								
9	80.00	62.22	46.67	33.33	22.22	13.33	6.67	2.22	0					
12	84.62	70.51	57.69	46.15	35.90	26.92	19.23	12.82	7.69	3.85	1.28	0		
15	87.50	75.83	65.00	55.00	45.83	37.50	30.00	23.33	17.50	12.50	8.33	5.00	2.50	0.83
18	89.47	79.53	70.18	61.40	53.22	45.61	38.60	32.16	26.32	21.05	16.37	12.28	8.77	5.85

▶ **Refund = Percent Refund × Finance Charge**

Example

Skills 30, 2 Application C Term Rebate schedule

Lillian Bartaldo had a 15-month installment loan that she repaid after 7 months. The total finance charge on the loan was $112.50. How much was Lillian's refund?

Solution

1. Find the **percent refund.**
   Refer to the above table.
   a. Term of loan in months 15
   b. Number of months loan has run 7
   Value from table is 30.00%

2. Find the **refund.**
   Percent Refund × Finance Charge
   30% of $112.50 = $33.75 refund

A. 12.50%
B. $31.25

▶ **Self-Check** Cover the answers and complete the problems.

A 15-month loan has run for 10 months. The finance charge was $250.

A. Find the percent refund. **B.** Find the refund.

Problems

Use the table on page 186 to find the percent refund.

	1.	2.	3.	4.	5.	6.
Term of Loan (Months)	15	15	9	15	6	12
Finance Charge	$80.00	$130.00	$450.00	$617.80	$320.00	$316.18
Months Loan Has Run	12	4	5	13	4	6
Percent Refund	5.00%	55.00%	?	?	?	?
Amount of Refund	?	?	?	?	?	?

7. Eileen Fleming.
15-month installment loan.
$114.50 total finance charge.
Repaid loan in 12 months.
What is the refund?

8. Henry Hewett.
9-month installment loan.
$52.70 total finance charge.
Repaid loan in 8 months.
What is the refund?

9. Ed Bart had a 15-month installment loan that he repaid in 10 months. The total finance charge was $124. How much was Ed's refund?

10. Kent Eu had a 9-month installment loan that he repaid in 7 months. The total finance charge was $147. How much was Kent's refund?

11. Tina Meier took a 12-month installment loan to help pay for her new living room set. The total finance charge was $90. Tina repaid the loan in 2 months. How much was Tina's refund?

12. Irene Carter obtained an installment loan of $1200 from the bank. She agreed to pay $106.34 a month for 12 months. The finance charge totaled $76.08. How much of the finance charge can Irene save by repaying the loan in 10 months rather than 12 months?

13. Clyde Moore had an installment loan of $250 in which he agreed to pay $15.25 a month for 18 months. How much is the finance charge? How much can he save by repaying the loan in 14 months?

Maintaining Your Skills

Look up the skills in parentheses if you need help or more practice.

Round to the nearest hundredth. (Skill 2)

14. 27.214
15. 316.045
16. 191.394
17. 427.896

Find the percentage. (Skill 30)

18. 15% × 90
19. 12% × 160
20. 5.50% × 37
21. 18.25% × 460

Reviewing the Basics

Skills

Round to the nearest cent.

(Skill 2) 1. \$242.473 2. \$38.375 3. \$91.108 4. \$42.862

Solve.

(Skill 6) 5. \$749.24 − \$631.58 6. \$24.19 − \$7.20 7. \$1009.13 − \$937.89

(Skill 8) 8. \$17.00 × 18 9. \$15.22 × 12 10. \$142.43 × 48

(Skill 11) 11. \$28.00 ÷ \$540 12. \$32.25 ÷ \$240 13. \$241.34 ÷ \$1790

(Skill 30) 14. 76% of \$500 15. 35% of \$440 16. 62.5% of \$221.98

Write as a decimal.

(Skill 14) 17. $\frac{1}{4}$ 18. $\frac{72}{360}$ 19. $\frac{5}{12}$ 20. $\frac{225}{360}$ 21. $\frac{73}{365}$

(Skill 28) 22. 81% 23. 15% 24. 11.5% 25. 19.5% 26. 10.125%

Write as a percent.

(Skill 26) 27. 0.19 28. 0.25 29. 0.124 30. 0.265 31. 0.1145

32. 0.235 33. 0.345 34. 1.387 35. 2.054 36. 0.0051

Applications

Use the table to find the monthly payment.

(Application C) 37. \$500 financed for 12 months at 15% APR.

38. \$1500 financed for 6 months at 15% APR.

APR	Term in Months	If You Finance. . . \$200	\$500	\$1000	\$1500
		Your **Monthly Payments** Are			
15%	6	34.80	87.01	174.03	261.05
	12	18.05	45.12	90.25	135.38
	18	12.47	31.19	62.38	93.57
	24	9.69	24.24	48.48	72.72

Write as a fraction of a year.

(Application J) 39. 8 months 40. 7 months 41. 5 months 42. 3 months

43. 293 days (exact year) 44. 120 days (ordinary year)

45. 45 days (ordinary year) 46. 180 days (exact year)

Terms

Use each term in a sentence that explains the term.

47. Single-payment loan 48. Rebate schedule 49. Discount loan

50. Annual percentage rate 51. Installment loan 52. Interest rate

A Microcomputer Application Rebates

To complete this microcomputer application, you will need the diskette *Microcomputer Applications for Business Mathematics,* which accompanies this textbook.

Select option 2, Rebates, from the menu. Input the information in the following problems to find the rebate amounts.

1. Paul Sigafoose.
Loan for $2000.
Terms: $181.67 a month, 12 months.
Finance charge is $180.
Paid off with the 8th payment.
What is the rebate amount?

2. Isaac Walko.
Loan for $8000.
Terms: $386.67 a month, 24 months.
Finance charge is $1280.
Paid off with the 20th payment.
What is the rebate amount?

3. Dorothy Pilliod.
Loan for $1500.
Terms: $262.50 a month, 6 months.
Finance charge is $75.
Paid off with the 5th payment.
What is the rebate amount?

4. Pedro Jujan.
Loan for $12,400.
Terms: $434.86 a month, 3 years.
Finance charge is $3255.
Paid off with the 21st payment.
What is the rebate amount?

5. Used car.
Loan for $3500.
Terms: $237.32 a month, 18 months.
Finance charge is $771.75.
Paid off with the 15th payment.
What is the rebate amount?

6. Small sailboat.
Loan for $1065.
Terms: $96.63 a month, 12 months.
Finance charge is $94.57.
Paid off with the 4th payment.
What is the rebate amount?

7. Alice Miklovic obtained an installment loan of $974 to pay for a motorcycle. The finance charge is $136.84. She must make 12 payments of $92.57 each. The loan is paid off with the 8th payment. What is the rebate amount?

8. Sherman Ibarra obtained an installment loan of $1456 for energy-saving improvements to his home. The finance charge is $1019.60. He must make 60 payments of $41.26 each. The loan is paid off with the 42nd payment. What is the rebate amount?

9. A used motor home was financed with a $10,000 installment loan for 4 years. The finance charge is $4288.64. The monthly payment is $297.68. The loan is paid off with the 30th payment. What is the rebate amount?

10. A new sailboat was financed with a $14,600 installment loan for 5 years. The finance charge is $8853.40. The monthly payment is $390.89. The owner sold the boat after 2 years, thus paying off the loan with the 25th payment. What is the rebate amount?

Unit Test

Lesson 7-1

1. Steve Newly's bank granted him a single-payment loan of $1375 for 81 days at an annual interest rate of 14%. His bank charges ordinary interest. What is the maturity value of Steve's loan?

Lesson 7-2

2. Evelyn Turner obtained a discount loan of $1800 at an annual interest rate of 15.25%. The loan was granted for 150 days. The bank charges ordinary interest. What proceeds did Evelyn receive?

Lesson 7-3

3. Todd Barker purchased a stereo for $698 using the store's installment credit plan. Todd made a 20% down payment and financed the remaining amount. What amount did he finance?

Lesson 7-4

4. Susan York obtained an installment loan of $1500 to purchase a bedroom set. The annual percentage rate is 18%. She plans to repay the loan in 24 months. Use the table to find the finance charge.

APR	Term in Months	If You Finance. . . $200	$500	$1000	$1500
		Your **Monthly Payments** Are			
18%	6	35.10	87.76	175.52	263.28
	12	18.33	45.84	91.68	137.52
	18	12.76	31.90	63.80	95.70
	24	9.98	24.96	49.92	74.88

Lesson 7-5

5. Juan Granger obtained an installment loan of $825 to pay for a new refrigerator. The finance charge is $135. He agreed to make 24 payments of $40 each. Use the table to find the annual percentage rate.

FINANCE CHARGE PER $100 FINANCED			
NUMBER OF PAYMENTS	APR		
	14.50%	14.75%	15.00%
6	4.27	4.35	4.42
12	8.03	8.17	8.31
18	11.87	12.08	12.29
24	15.80	16.08	16.37

Lesson 7-6

6. Alice Howe repaid a 12-month installment loan after 4 months. The total finance charge was $177.52. Use the rebate table to find Alice's refund.

TERM OF LOAN	NUMBER OF MONTHS LOAN HAS RUN			
	1	2	3	4
3	50.00	16.67	0	0
6	71.43	47.62	28.57	14.29
9	80.00	62.22	46.67	33.33
12	84.62	70.51	57.69	46.15

Unit 8 Automobile Transportation

A *sticker price* on a new automobile shows the *base price* and the cost of *options,* and the *destination charge* for delivering the car from the factory to the dealer. The auto dealer may take less than the sticker price if you make an offer that is higher than the *dealer's cost.* If you buy a used car, there are *used-car guides* that show the approximate price you can expect to pay for a particular model. In addition to the cost of purchasing an automobile, your transportation costs include *insurance, maintenance,* fuel, and other operating costs.

The sticker price breaks down the cost of the car and its options, and lists the model's estimated miles per gallon. Gas and mileage will affect your yearly operating costs.

8-1 Purchasing a New Automobile

Objective Compute the sticker price of a new automobile.

Automobile manufacturers are required by law to place a sticker on the window of each new car to show all charges for the car. The **base price** is the price of the engine, chassis, and any other piece of standard equipment for a particular model. **Options** are extras for convenience, safety, or appearance, such as a radio, air conditioning, and tinted glass. The **destination charge** is the cost of shipping the car from the factory to the dealer. The **sticker price** is the total of the base price, options, and destination charge.

▶ **Sticker Price = Base Price + Options + Destination Charge**

Example

Skill 5 **Application** A **Term** Sticker price

Martin Suen is shopping for a compact car. A portion of the sticker for one that he is interested in is shown below. What is the sticker price for this compact?

LAMONT TRAVELER VN100		BASE PRICE: $5843.00
C/C	OPTIONS DESCRIPTION	LIST PRICE
TDY	Tan/Gold two-tone paint	161.25
Y76	High-back vinyl bench seats	289.45
F20	Air conditioning	637.50
G22	Oil pressure gauge	11.00
M83	Dual low-mounted mirrors	42.45
970	DESTINATION CHARGE	183.00

Solution

1. Find the **options.**
   $161.25 + $289.45 + $637.50 + $11.00 + $42.45 = $1141.65

2. Find the **sticker price.**
   Base Price + Options + Destination Charge
   $5843.00 + $1141.65 + $183.00 = $7167.65 sticker price

A. $6730
B. $12,245

▶ **Self-Check** Cover the answers and complete the problems.

Find the sticker price.

	Base Price	Options	Destination Charge
A.	$ 5394	$350; $610; $136	$240
B.	$10,400	$250; $500; $730; $90	$275

Problems

Find the sticker price.

	1.	2.	3.	4.	5.
Base Price	$6000	$8900	$6540	$12,654	$14,756
Options	1200	1400	960	1,765	2,764
Destination Charge	300	250	345	238	465
Sticker Price	?	?	?	?	?

6. Laurel Classic.
Base price is $6105.
Options: 405 V8 for $660.
Auto. trans. for $335.
A/C for $649.
Tinted glass for $215.
Destination charge is $243.
What is the sticker price?

7. DVM 310 subcompact.
Base price is $4425.95.
Options: GT package for $1398.
AM/FM stereo for $335.
Denim seats for $109.45.
Dlx. seat belts for $76.
Destination charge is $189.25.
What is the sticker price?

8. Vicki Hackworth is interested in a 4-door sedan that has a base price of $6639.95. Factory-installed options total $1236.65. The destination charge from its assembly plant in St. Louis is $152. What is the sticker price for this car?

9. Vaughn Cawvey sees a Voyager sedan that he is interested in buying. There is a 5% state sales tax on the purchase of the car. If Vaughn pays the sticker price, how much tax must he pay?

C/C	DESCRIPTION	WEIGHT	LIST PRICE
PB7	VOYAGER PB200 CUSTOM DELUXE	3164	5752.00
B89	STARLIGHT BLUE METALLIC		0.00
C8W	BENCH SEATS - VINYL		0.00
D34	AUTOMATIC TRANSMISSION		370.00
F26	255 CID V6 - 2BBL.	23	58.55
G11	AIR CONDITIONING	95	166.45
H51	TINTED GLASS		51.60
S77	POWER STEERING	41	131.95
V3W	ROUGH VINYL ROOF - WHITE		148.40
Y62	TIRES - G78 X 14 WHITE SIDEWALL		60.00
309	DESTINATION CHARGE		168.50

Maintaining Your Skills

Look up the skills in parentheses if you need help or more practice.

Add. (Skill 5)

10. 5450 + 895 + 560 + 345

11. 4857 + 1093 + 784 + 224 + 90

12. 17,453 + 1540 + 988

13. 6785 + 1982 + 302 + 65 + 40

14. 9987 + 656.50 + 321.40 + 174

15. 4875.25 + 326.43 + 659.46 + 345.50

8-2 Dealer's Cost

Objective Compute the dealer's cost of a new automobile.

Automobile dealers pay less than the prices on the sticker for both the basic automobile and the options. Consumer magazines often report the dealer's cost as a percent of the sticker price. You may save money when purchasing a new automobile by making an offer that is higher than the estimated dealer's cost but lower than the sticker price.

▶ **Dealer's Cost** = **Percent of Base Price** + **Percent of Options Price** + **Destination Charge**

Example

Skills 30, 2 **Application** A **Term** Dealer's cost

Linda and Phil Chapman want to purchase a new Viceroy. The car has a base price of $4352, options totaling $1387, and a destination charge of $160. They read in a consumer magazine that the dealer's cost for a Viceroy is about 80% of the base price and 77% of the options price. What should they estimate as the dealer's cost?

Solution

1. Find the **percent of base price.**
   80% of $4352.00 = $3481.60
2. Find the **percent of options price.**
   77% of $1387.00 = $1067.99
3. Find the **dealer's cost.**
   Percent of Base Price + Percent of Options Price + Destination Charge
   $3481.60 + $1067.99 + $160.00 = $4709.59 dealer's cost

$6400
1500
230
$8130

▶ **Self-Check** Cover the answers and complete the problems.

The dealer's cost is 80% of the base price of $8000 and 75% of the options price of $2000 plus a destination charge of $230. Find the dealer's cost.

Problems

	(Dealer's Percent	× Base Price)	+	(Dealer's Percent	× Options Price)	+	Destination Charge	=	Dealer's Cost
1.	(80%	× $ 9000)	+	(75%	× $1500)	+	$240	=	?
2.	(85%	× $ 7000)	+	(80%	× $1400)	+	$352	=	?
3.	(82%	× $12,000)	+	(78%	× $3400)	+	$465	=	?

4. MX-7 Diesel Sedan.
Base price is $9100.
Options total $740.
Destination charge is $249.
Dealer pays 80% of base price and 80% of options price.
What is the dealer's cost?

5. Federal Edition.
Base price is $6500.
Options total $1100.
Destination charge is $202.
Dealer pays 84% of base price and 75% of options price.
What is the dealer's cost?

6. Royal Senator.
Base price is $10,648.
Options total $2040.54.
Destination charge is $306.
Dealer pays 78% of base price and 77% of options price.
What is the dealer's cost?

7. Federman Series 680 SE.
Base price is $6985.80.
Options total $740.85.
Destination charge is $199.
Dealer pays 84% of base price and 78% of options price.
What is the dealer's cost?

8. Bob Kirkland is looking at a new Mathers RT-100 that has a base price of $5326.65, options totaling $1629.15, and a destination charge of $164. The dealer's cost is about 80% of the base price and 77% of the price of the options. What is the sticker price of the car? What should Bob estimate as the dealer's cost?

9. Allison Kisner is considering the purchase of a recreational van. She sees one that has a base price of $16,350 and options totaling $2160. The destination charge is $180. She estimates that the dealer's cost is 82% of the base price and 79% of the price of the options. If Allison offers the dealer $250 over the estimated dealer's cost, what is her offer?

10. John Estes offered an automobile dealer $400 over the estimated dealer's cost on a car that has a base price of $8200 and options totaling $2960. The dealer's cost is about 85% of the base price and 81% of the price of the options. The destination charge is $490. What was John's offer?

11. Connie O'Neal is ordering a new car built to her specifications direct from the factory. The base price is $12,874.50 and options total $1408.55. There is a destination charge of $258. The dealer's cost is about 81% of the base price and 88% of the price of the options. What should Connie offer the dealer if she wants the offer to be $250 over the estimated dealer's cost?

Maintaining Your Skills

Look up the skill in parentheses if you need help or more practice.

Find the percentage. Round answers to the nearest hundredth. (Skill 30)

12. 8% of 300
13. 15% of 460
14. 22% of 658
15. 24% of 865
16. 45% of 8900
17. 78% of 4400
18. 88% of 7650
19. 92% of 788.50
20. 56.5% of 975.5
21. 15.5% of 561.88
22. 87.25% of 501.6

8-3 Purchasing a Used Automobile

Objective Compute the average retail price of a used automobile.

Automobile dealers usually advertise used cars for prices that are higher than what they expect you to pay. **Used-car guides**, published monthly, give the average prices for cars that were purchased from dealers during the previous month. The information can help you make decisions about how much to pay for a used automobile.

▶ **Average Retail Price** = **Average Retail Value** + **Additional Options** − **Options Deductions** − **Mileage Deduction**

Example

Skills 5, 6 **Application** C **Term** Used-car guide

Av'g Trd-in		BODY TYPE	Model	Av'g Loan	Av'g Ret'l
PALAMINO – V8-AT-PS-AC					
2050	4	H'dtp 2Dr	I59	1850	2625
2150	4	H'dtp "S"	I59	1950	2750
2175	4	H'dtp Sprt	I59	1975	2775
25		Add Vinyl Top		25	50
75		Add Sky Roof		75	100
25		Add Tape Deck		25	50
25		Add AM/FM Stereo		25	50
50		Add AM/FM Ster/Tpe		50	75
25		Add Power Seats		25	50
25		Add Power Windows		25	50
25		Add Turbine Wheels		25	50
125		Deduct Man. Trans.		125	125
175		Deduct w/o Air Cond.		175	175

Justine Tyler would like to purchase a Palamino Spirit, model I59, that is advertised for $2995. It has an AM/FM stereo radio, no air conditioning, and has been driven 78,000 miles. The used-car guide indicates that $250 should be subtracted if the mileage is between 75,000 and 80,000 miles. What average retail price should Justine keep in mind when she makes an offer for the car?

Solution

Find the **average retail price.**

Average Retail Value	+	Additional Options	−	Options Deductions	−	Mileage Deduction
$2775	+	$50	−	$175	−	$250 = $2400 average retail price

$3550

▶ **Self-Check** Cover the answer and complete the problem.

Retail Value	Additional Options	Options Deductions	Mileage Deductions	Average Retail Price
$3375	$625	$325	$125	?

Problems

Use the above table to find the retail value and average retail price.

	Model	Retail Value	Vinyl Top	AM/FM Stereo	Power Seats	Power Windows	Man. Trans.	A/C	Av'g Retail Price
1.	2 dr	$2625	No	No	No	No	Yes	Yes	?
2.	H'dtp "S"	?	Yes	Yes	No	No	Yes	No	?
3.	Spirit	?	No	Yes	No	Yes	No	Yes	?

4. 6-year-old Elite compact.
Average retail value is $1150.
Add $30 for vinyl top.
Add $125 for air conditioning.
Deduct $65 for manual transmission.
What is the average retail price?

5. 2-year-old Cramer sedan.
Average retail value is $6625.
Add $175 for AM/FM stereo tape.
Deduct $450 for no air conditioning.
Deduct $325 for excessive mileage.
What is the average retail price?

6. Pilah Sanchez owns a 4-year-old Faust 4-door sedan that she wants to sell in order to purchase a new car. One used-car guide shows the average retail value of her car is $4950. She adds $50 for having a vinyl top, $175 for cassette player, $75 for power windows, and $25 for power locks. She deducts $475 for having no air conditioner. What is the average retail price she can ask as a selling price for her car?

Use the used-car guide shown to solve problems 7–9.

7. Jane Prince owns a Mystic Sport Coupe 2D. The car has AM/FM stereo, power door locks, power windows, cruise control, and a standard transmission. What is the average retail price of the car?

8. Bill Whitten owns a used Mystic Sport Coupe Z28 with air conditioning, an automatic transmission, AM/FM stereo tape deck, rear window defroster, and tilt steering wheel. Recently he was asked if he wanted to sell his car. What is the average retail price he should keep in mind when pricing his car?

Av'g Trd-in	BODY TYPE	Av'g Loan	Av'g Ret'l
MYSTIC V8-PS-PE			
8550	11 Spt Cpe 2D	7700	9575
9450	12 Berlinetta Cpe	8525	10575
10350	Spt Cpe Z28	9325	11525
500	**Add** T-Top	450	600
125	**Add** AM/FM Stereo	125	150
150	**Add** AM/FM Stereo/Tp	150	200
75	**Add** Power Door Locks	75	100
100	**Add** Power Windows	100	125
100	**Add** Power Seats	100	125
75	**Add** Rear Wind Defroster	75	100
75	**Add** Cruise Control	75	100
75	**Add** Tilt Strg Wheel	75	100
200	**Deduct** 4 Cyl. Engine	200	200
375	**Deduct** Manual Trans.	350	375
550	**Deduct** W/out Air Cond	500	550

9. Ellen Steingold wishes to trade in her car on the purchase of a new automobile. She now owns a used Mystic Berlinetta Coupe with a T-Top, AM/FM stereo, air conditioning, power windows, and power seat. She must deduct $500 from the trade-in value for high mileage. What is the average trade-in value of her car?

Maintaining Your Skills

Look up the skills in parentheses if you need help or more practice.

Add. (Skill 3)

10. 2125 + 4200 + 175 + 545

11. 5060 + 325 + 3950 + 125 + 675

12. 1675 + 25 + 35 + 55 + 225

13. 175 + 760 + 5470 + 65 + 25 + 10

Subtract. (Skill 4)

14. 3450 − 475

15. 7890 − 4530

16. 3205 − 125

17. 1625 − 325

18. 3375 − 240

19. 18,750 − 995

20. $4555 − $855

21. $8720 − $2308

8-4 Automobile Insurance

Objective Use tables to compute the annual premium for automobile insurance.

Liability insurance, which includes **bodily injury insurance** and **property damage insurance,** protects you against financial losses if your car is involved in an accident. Bodily injury limits of 25/100 mean that the insurance company will pay up to $25,000 to any one person injured and up to $100,000 if more than one person is injured. Property damage insurance protects you against financial loss if your automobile damages the property of others.

Comprehensive insurance on your automobile protects you from losses due to fire, vandalism, theft, and so on. **Collision insurance** will pay to repair the damage to your automobile if it was involved in an accident. Either policy may be written with a **deductible clause.** A $50-deductible clause means that you pay the first $50 of the repair bill. The **annual base premium** is determined by the amount of insurance you want, the **age group** of your car, and the **insurance-rating group.** The insurance-rating group depends on the size and value of your car.

The **annual premium** is the amount you pay each year for insurance coverage. Your annual premium depends on the **base premium** and your **driver-rating factor.** The base premium depends on the amount of coverage you want. The driver-rating factor depends on your age, marital status, the amount you drive each week, and so on. If several people drive your car, the highest driver-rating factor among those who drive your car is used to determine the annual premium.

Insurance companies use tables to determine your base premium.

LIABILITY PREMIUM FOR A PRIVATE PASSENGER AUTOMOBILE

PROPERTY DAMAGE LIMITS	BODILY INJURY LIMITS 25/50	25/100	50/100	100/200	100/300
$ 25,000	$108.00	$113.60	$120.00	$131.60	$134.00
50,000	111.20	116.80	123.20	135.20	137.20
100,000	114.40	120.00	126.40	138.40	140.40

PHYSICAL DAMAGE PREMIUM

COVERAGE	AGE GROUP	INSURANCE-RATING GROUP 10	11	12	13	14	15
Comprehensive $50 Deductible	A	$ 56.80	$ 69.60	$ 81.20	$ 88.80	$ 95.60	$103.20
	B	54.00	66.40	77.20	84.40	91.20	98.40
	C	51.20	63.20	73.20	80.00	86.80	93.60
	D	48.80	59.60	69.60	76.00	82.00	88.40
Collision $50 Deductible	A	$158.00	$170.80	$184.00	$202.80	$211.60	$220.00
	B	150.40	162.80	175.20	180.80	193.20	201.60
	C	142.80	154.80	166.40	171.60	183.60	191.60
	D	135.20	146.40	157.60	162.80	174.00	181.60

▸ **Annual Base Premium = Liability Premium + Comprehensive Premium + Collision Premium**

▸ **Annual Premium = Annual Base Premium × Driver-Rating Factor**

Example

Skills 5, 8 **Application** C **Term** Annual premium

Grace Malone is the principal operator of her car. Her driver-rating factor is 1.90. Her insurance includes 25/50 bodily injury and $25,000 property damage. Her car is in age group A and insurance-rating group 10 (A, 10). She has $50-deductible comprehensive and $50-deductible collision insurance. What is her annual base premium? What is her annual premium?

Solution

1. Find the **annual base premium.**

Liability Premium	+	Comprehensive Premium	+	Collision Premium		
$108.00	+	$56.80	+	$158.00	=	$322.80 annual base premium

2. Find the **annual premium.**

Annual Base Premium	×	Driver-Rating Factor		
$322.80	×	1.90	=	$613.32 annual premium

$120.00
63.20
154.80
$338.00 base premium
$845.00 annual premium

▸ **Self-Check** Cover the answers and complete the problems.

Find the annual base premium and the annual premium.

25/100 bodily injury and $100,000 property damage.
Car is in age group C and insurance-rating group 11 (C,11).
$50-deductible comprehensive and $50-deductible collision.
Driver-rating factor is 2.50.

Problems

1. Driver-rating factor is 1.20.
Age/rating group is A, 13.
Coverages: 50/100 bodily injury.
$25,000 property damage.
$50-deductible comprehensive.
$50-deductible collision.
What is the annual base premium?
What is the annual premium?

2. Driver-rating factor is 2.00.
Age/rating group is D, 10.
Coverages: 100/300 bodily injury.
$25,000 property damage.
$50-deductible comprehensive.
$50-deductible collision.
What is the annual base premium?
What is the annual premium?

3. Driver-rating factor is 2.50.
Age/rating group is D, 12.
Coverages: 25/100 bodily injury.
$100,000 property damage.
$50-deductible comprehensive.
$50-deductible collision.
What is the annual base premium?
What is the annual premium?

4. Driver-rating factor is 3.10.
Age/rating group is C, 13.
Coverages: 100/200 bodily injury.
$50,000 property damage.
$50-deductible comprehensive.
$50-deductible collision.
What is the annual base premium?
What is the annual premium?

(Continue on next page.)

5. Jane Heslinger uses her car primarily to run errands. She has $50-deductible comprehensive, $50-deductible collision, 50/100 bodily injury, and $25,000 property damage coverages. Her driver-rating factor is 1.00 and her car is classified A,14. What is Jane's annual base premium? What is her annual premium?

6. Arnold O'Malley uses his car primarily for business. He has $50-deductible comprehensive, $50-deductible collision, 50/100 bodily injury, and $50,000 property damage coverages. His driver-rating factor is 2.10 and his car is classified B,12. What is his annual base premium? What is his annual premium?

7. Lori Smith uses her car mainly for pleasure. She has $50-deductible comprehensive, $50-deductible collision, 100/300 bodily injury, and $100,000 property damage coverages. Her driver-rating factor is 1.85 and her car is classified C,12. What is Lori's annual base premium? What is her annual premium?

8. Joe Barbuto uses his car to deliver office supplies. He has $50-deductible comprehensive, $50-deductible collision, 100/200 bodily injury, and $50,000 property damage coverages. His driver-rating factor is 3.10 and his car is classified D,14. What is his annual base premium? What is his annual premium?

9. Robert Brass delivers magazines to retail outlets. His driver-rating factor is 2.15. His insurance coverage includes 50/100 bodily injury and $25,000 property damage. He has $50-deductible comprehensive and $50-deductible collision. His car is in age group C and insurance-rating group 10. What is his annual base premium? What is his annual premium?

10. Paula McGee-Randolph drives to and from work. Her driver-rating factor is 2.40. Her insurance coverage includes 100/300 bodily injury and $100,000 property damage. She has $50-deductible comprehensive and $50-deductible collision. Her car is in age group A and insurance-rating group 13. What is her annual base premium? What is her annual premium?

11. Sahib Jabara owns and operates Jabara's Imports, Inc. His driver-rating factor is 3.15. His insurance coverage includes 25/100 bodily injury and $50,000 property damage. He has $50-deductible comprehensive and $50-deductible collision. His convertible is in age group B and insurance-rating group 14. What is his annual base premium? What is his annual premium?

12. Eric Thomas uses his car to drive to and from work. His driver-rating factor is 2.85. His insurance coverage includes 25/50 bodily injury and $50,000 property damage. He has $50-deductible comprehensive and $50-deductible collision. His car is in age group D and insurance-rating group 11. What is his annual base premium? What is his annual premium?

13. Beth Obritter uses her car for delivering newspapers. Her insurance coverage includes 25/50 bodily injury and $25,000 property damage. She has $50-deductible comprehensive and no collision. Her car is classified D,10. Her rating factor is 1.90. What is her annual premium?

14. Mark Yoos drives a farm automobile. His rating factor is 0.90. His insurance coverage includes 50/100 bodily injury and $100,000 property damage. What is his annual premium?

15. Neil Mulvey delivers eggs to retail stores. His 12-year-old car is not worth very much so he doesn't carry comprehensive or collision insurance. He carries 50/100 bodily injury and $50,000 property damage insurance. His rating factor is 3.10. What is his annual premium?

16. Tom Raintree owns a car that is classified D,11. His driver-rating factor is 4.10. He has 100/200 bodily injury, $25,000 property damage, $50-deductible comprehensive, and $50-deductible collision coverages. What is Tom's annual premium?

17. Paula McGee owns and operates Paula's Flower Shop. She purchased a van with which to make deliveries. The van is classified A,15. She has the following coverages: $100,000 property damage, 100/300 bodily injury, $50-deductible comprehensive, and $50-deductible collision. Paula's driver-rating factor is 1.35. What is her annual premium?

18. Scott Williams uses his car mainly for pleasure driving. His driver-rating factor is now 3.90. He learns that his rating factor will be 2.00 after he is married next year. How much less will his annual premium be if he has $25,000 property damage and 50/100 bodily injury insurance coverages?

Maintaining Your Skills

Look up the skills in parentheses if you need help or more practice.

Add. (Skill 5)

19. 918.45 + 97.62 + 64.48 + 93.35

20. 39.55 + 1.82 + 24.51 + 953.70

21. 57.08 + 112.46 + 9.94 + 86.82

22. 380.88 + 75.44 + 549.45 + 6.80

23. 1.50 + 0.79 + 71.40 + 6.97 + 7.88

24. 30.93 + 83.30 + 37.83 + 0.9 + 1.67

Multiply. (Skill 8)

25. 1.25 × 88.88

26. 2.40 × 640

27. 7.85 × 99.40

28. 45.51 × 63.10

29. 555 × 2.3

30. 2022 × 6.35

31. 56.78 × 2.15

32. 458.80 × 4.10

33. 654.40 × 3.15

8-5 Operating and Maintaining an Automobile

Objective Compute the total cost per mile of operating and maintaining an automobile.

Many costs are involved in operating and maintaining an automobile. **Variable costs,** like gasoline and tires, increase as the number of miles you drive increases. **Fixed costs,** like automobile insurance, registration fees, and **depreciation,** remain about the same regardless of how many miles you drive. Depreciation is a decrease in the value of your car because of its age and condition.

▶ **Cost per Mile** $= \dfrac{\text{Annual Variable Cost} + \text{Annual Fixed Cost}}{\text{Number of Miles Driven}}$

Example

Skills 11, 2 **Application** A **Term** Depreciation

Ellen Ashley purchased an automobile for $1000 one year ago. She drove about 6000 miles during the year and kept a record of all her expenses. She estimates the car's present value at $800. What was the cost per mile for Ellen to operate her car last year?

VARIABLE COSTS		FIXED COSTS	
Gasoline	$670.00	Insurance	$298.92
Oil	15.00	License/Registration	30.00
Maintenance	82.85	Depreciation	
New tire	39.57	($1000 - $800)	200.00
	$807.42		$528.92

Solution

Find the **cost per mile.**

$$\left(\begin{matrix}\text{Annual}\\\text{Variable Cost}\end{matrix} + \begin{matrix}\text{Annual}\\\text{Fixed Cost}\end{matrix}\right) \div \begin{matrix}\text{Number of}\\\text{Miles Driven}\end{matrix}$$

$$(\$807.42 + \$528.92) \div 6000$$

$$\$1336.34 \div 6000 = \$0.2227 = 22¢ \text{ cost per mile}$$

A. $1500
$0.15/mile
B. $2400
$0.32/mile

▶ **Self-Check** Cover the answers and complete the problems.

Find the total cost and the cost per mile.

	Variable Cost	Fixed Cost	Miles Driven
A.	$ 800.00	$ 700.00	10,000
B.	$1085.50	$1314.50	7500

Problems

Solve. Round answers to the nearest cent.

	(Annual Variable Cost	+	Annual Fixed Cost	=	Total Annual Cost)	÷	Miles Driven	=	Cost per Mile
1.	($ 865.00	+	$ 935.00	=	?)	÷	9000	=	?
2.	($ 791.00	+	$1463.00	=	?)	÷	11,500	=	?
3.	($2205.00	+	$1411.20	=	?)	÷	14,700	=	?
4.	($ 965.77	+	$ 644.37	=	?)	÷	7800	=	?
5.	($2312.50	+	$2212.50	=	?)	÷	12,500	=	?

6. Nora Halston, student.
   Drove 7500 miles last year.
   Fixed costs totaled $875.
   Variable costs totaled $743.
   What was the total annual cost?
   What was the cost per mile?

7. Despina Nesson, salesperson.
   Drove 34,500 miles last year.
   Fixed costs totaled $1700.
   Variable costs totaled $3820.
   What was the total annual cost?
   What was the cost per mile?

8. Gary Olmstead, accountant.
   Drove 18,460 miles last year.
   Fixed costs totaled $1218.36.
   Variable costs totaled $2196.74.
   What was the cost per mile?

9. Karim Rashidi, sales manager.
   Drove 26,350 miles last year.
   Fixed costs totaled $544.31.
   Variable costs totaled $1906.24.
   What was the cost per mile?

10. Bea Pappas drove her car 10,000 miles last year. Her variable costs totaled $580.25. Her fixed costs totaled $1439. What was the cost per mile for Bea to operate her car?

11. Mabel Marcer drove her car 14,550 miles last year. Her variable costs totaled $1399.71. Her fixed costs totaled $2080.65. What was the cost per mile for Mabel to operate her car?

12. Paul Kite drove his car 14,500 miles last year. His variable costs totaled $996.60. His fixed costs totaled $1248.40. What was the cost per mile for Paul to operate his car?

13. Cain Baffin drove his car 23,650 miles last year. His variable costs totaled $1864.63. His fixed costs totaled $1364.67. What was the cost per mile for Cain to operate his car?

14. Pinckney Keil purchased an automobile for $9450 one year ago. He drove it 11,500 miles during the first year and kept a record of all his expenses. His variable costs were: gasoline, $513.40; oil, $15; parking, $115.71; and repairs, $112.29. His fixed costs were: insurance, $313; license, $76.75; and depreciation. He estimates the car's present value at $7560. What is his cost per mile?

(Continue on next page.)

15. Elska Yakobian purchased an automobile for $14,850 one year ago. She drove it 13,800 miles during the first year and kept a record of all her expenses. Her variable costs were: gasoline, $1396.92; oil, $17; parking, $103.74; and repairs, $204.24. Her fixed costs were: insurance, $543; license, $46.75; and depreciation. She estimates the car's present value at $11,880. What was her cost per mile?

16. Rhoda Moskowitz and three friends agreed that they would share equally in the costs of operating and maintaining a car. Last year the car was driven 10,540 miles. Fixed costs totaled $733.50 while variable costs totaled $1052.50. How much did it cost per mile to operate the car? How much did it cost per person?

17. Luke Breaker and two neighbors car-pool to and from work in Luke's subcompact. The three agreed that they would share equally in the cost of operating and maintaining the car. Last year, the car was driven 18,500 miles. Fixed costs totaled $327.54 while variable costs totaled $826.65. How much did it cost per mile to operate the car for the year? How much did it cost each person to car-pool for the year?

Oil changes	$ 22.00
Tune-up	43.00
Tire alignment	18.75
Insurance	315.00
Property tax	32.27
Registration	48.50
Loan interest	439.85
Depreciation	525.00
Gasoline	484.21

18. Alice Grantley kept records on the operation and maintenance of her car for the previous year. In the summary shown, which of the items listed are fixed costs? Which are variable costs?

19. Last year, Alice Grantley drove her car 8000 miles. What was the total cost to operate and maintain her car for the year? What was her cost per mile?

20. John and Erma Black each own the same model car. Each kept records of the operating and maintenance costs. Last year, John drove his car 14,500 miles and Erma drove hers 9880 miles. Both had fixed costs totaling $534.50. John's variable costs totaled $889 and Erma's totaled $647.30. Who spent more per mile to operate and maintain the car? How much more per mile?

Maintaining Your Skills

Look up the skills in parentheses if you need help or more practice.

Round to the nearest hundredth. (Skill 2)

21. 908.673 22. 12.746 23. 3.8055 24. 11.0951

25. 0.03411 26. 0.4633 27. 1.0934 28. 0.0719

Divide. Round answers to the nearest hundredth. (Skill 10)

29. 891 ÷ 400 30. 1400 ÷ 7000 31. 950 ÷ 8000 32. 4320 ÷ 12,000

33. 1775 ÷ 9800 34. 2099 ÷ 22,500 35. 4886 ÷ 12,400 36. 3065 ÷ 15,884

8-6 Renting an Automobile

Objective Compute the cost per mile of renting an automobile.

From time to time you may need to **rent** a car. Some automobile rental agencies charge a daily rate plus a rate per mile while others charge a daily rate with no mileage charge. In either case you pay for the gasoline used. If you are under 21 years of age, your parents may be required to pay for the insurance on the automobile. If you are 21 or over, the rental agency usually pays for the insurance. The insurance generally has a collision deductible clause which states that you will pay for a portion of any damage to the car if it is in an accident. You can obtain complete insurance coverage with a **collision waiver** by paying an additional charge per day.

▶ **Cost per Mile** = $\dfrac{\text{Total Cost}}{\text{Number of Miles Driven}}$

Example

Skills 11, 2, 5 **Application** A **Term** Lease

Carl Williams rented a compact car for 3 days at $22.95 per day plus 24¢ per mile. He purchased the collision waiver for $3.00 per day. Carl drove 468 miles and paid $25.11 for gasoline. What was the total cost of renting the car? What was the cost per mile to rent the car?

Solution

1. Find the **total cost.**
   - **a.** Daily cost: $22.95 × 3 = ... $ 68.85
   - **b.** Mileage cost: $0.24 × 468 = ... 112.32
   - **c.** Gasoline cost............................. 25.11
   - **d.** Collision waiver: $3.00 × 3 = ... 9.00
   - $215.28

2. Find the **cost per mile.**
   Total Cost ÷ Number of Miles Driven
   $215.28 ÷ 468 = $0.46 cost per mile

▶ **Self-Check** Cover the answers and complete the problems.

Daily cost:	$24.00 × 4
Mileage cost:	22¢ × 430
Gasoline cost:	$20.10

A. Find the total. **B.** Find the cost per mile.

A. $ 96.00
94.60
20.10
$210.70
B. $0.49

Problems

Solve. Find the cost per mile to the nearest cent.

	(Total Daily Cost	+	Total Mileage Cost	+	Gasoline Cost	=	Total Cost)	÷	Miles Driven	=	Cost per Mile
1.	($78.75	+	$144.00	+	$25.25	=	$248.00)	÷	800	=	?
2.	($97.65	+	$130.00	+	$40.35	=	?)	÷	1000	=	?
3.	($68.85	+	$ 84.60	+	$23.99	=	?)	÷	470	=	?
4.	($79.00	+	$158.00	+	$49.50	=	?)	÷	800	=	?

Solve. Find the cost per mile to the nearest cent. (No mileage charge)

	(Rental Cost	+	Gasoline Cost	=	Total Cost)	÷	Miles Driven	=	Cost per Mile
5.	($ 18.90	+	$ 8.40	=	?)	÷	300	=	?
6.	($ 47.99	+	$ 7.50	=	?)	÷	250	=	?
7.	($149.00	+	$10.70	=	?)	÷	245	=	?
8.	($219.97	+	$46.83	=	?)	÷	562	=	?
9.	($189.99	+	$27.53	=	?)	÷	640	=	?

10. Freda Marsh rented a subcompact.
Cost $24 a day for 4 days.
Drove 570 miles at 15¢ a mile.
Gasoline cost $18.
What was the total cost?
What was the cost per mile?

11. Denny Rochetti rented a sedan.
Cost $22.50 a day for 7 days.
Drove 2200 miles at 17¢ a mile.
Gasoline cost $108.70.
What was the total cost?
What was the cost per mile?

12. Roy Bardi rented a sedan.
Cost $29.96 a day for 5 days.
Gasoline cost $10.40.
Drove 242 miles.
What was the total cost?
What was the cost per mile?

13. Carol Chen rented a compact.
Cost $43.99 a day for 2 days.
Gasoline cost $5.31.
Drove 170 miles.
What was the total cost?
What was the cost per mile?

14. Abrah Bahta rented a sedan for 4 days at $43.95 a day plus 15¢ a mile. He purchased the collision waiver for $3.50 per day. He drove 420 miles and paid $17.10 for gasoline. What was the total cost of renting the car? What is the cost per mile?

15. Judy Wines rented a sedan for 2 days at $45.95 a day plus 18¢ a mile. She purchased the collision waiver for $3.50 per day. She drove 140 miles and paid $7.20 for gasoline. What was the total cost of renting the car? What was the cost per mile?

16. Ace Car Rentals has a weekly rate of $99 for subcompact cars and no mileage charge. A subcompact is driven 450 miles and uses $16.59 in gasoline. What is the cost per mile?

17. Helm Car Rentals has a weekly rate of $239.97 for station wagons and no mileage charge. A station wagon is driven 375 miles and uses $40.63 in gasoline. What is the cost per mile?

18. Ella Shirk is renting a compact car while her own car is being repaired. The compact rents for $129 a week with no mileage charge. She is going to rent the car for 2 weeks and expects to drive it about 260 miles. She estimates that the gasoline will cost about $12. What is the total cost of renting the car? What is the cost per mile?

19. Jessica White decides to rent a station wagon for her vacation. The station wagon rents for $30.75 a day plus 21¢ a mile. She would use the car for 8 days and would drive it about 2000 miles. She estimates that gasoline would cost about $180. What would be the total cost to rent the car? How much would it cost per mile?

20. Joe Anthem and Paula Green plan to rent a luxury sedan for the prom. They will need the car for 1 day and drive it about 100 miles. The car rents for $35.95 a day plus 23¢ a mile. They estimate that gasoline would cost about $11. If they share equally in the costs, how much would it cost each of them to rent the sedan?

21. The Bakers plan to fly to their vacation spot and then rent a car to tour the island. The vacation package they signed specifies that a sedan would be available for $45.95 a day with no charge for mileage. What will it cost the Bakers to rent the car for 5 days if they spend about $30 for gasoline and $4.50 a day for the collision waiver? What will it cost per mile if they drive about 400 miles?

*22. **A Brief Case** Alice Coopersmith is moving to a new apartment. She can rent a 14-foot panel truck for $30 a day plus 26¢ a mile or an 18-foot truck for $45 a day plus 29¢ a mile. It would take 4 trips to make the move in the 14-foot truck but only 3 trips in the larger truck. She estimates that gasoline would cost about $45 for either truck. A round trip to her new apartment is 60 miles. Regardless of the number of trips, Alice would need the truck for only 1 day. To save on expenses, which size truck should Alice rent? How much would it cost per mile to rent the truck?

Maintaining Your Skills

Look up the skills in parentheses if you need help or more practice.

Add. (Skill 5)

23. 2.94 + 85.67 + 26.73 + 478.29

24. 66.03 + 4.75 + 70.01 + 868.19

25. 24.64 + 79.899 + 65.080 + 1.419

26. 19.97 + 98.35 + 29.70 + 88.01 + 94.49

Divide. Round answers to the nearest hundredth. (Skill 11)

27. 965.20 ÷ 62

28. 684.26 ÷ 32.2

29. 502.00 ÷ 261.9

30. 701.83 ÷ 35.5

31. 381.19 ÷ 495

32. 107.69 ÷ 476

Reviewing the Basics

Skills

Round to the nearest cent.

(Skill 2) **1.** \$109.844591 **2.** \$0.3514 **3.** \$364.3091 **4.** \$0.1973

Solve. Round answers to the nearest cent.

(Skill 5) **5.** \$18.56 + \$79.09 **6.** \$354.21 + \$11.29 **7.** \$14 + \$13.57 + \$27.85

(Skill 6) **8.** \$47.42 − \$39.59 **9.** \$832.06 − \$759.98 **10.** \$18 − \$9.88

(Skill 8) **11.** \$92.50 × 3.15 **12.** \$34.25 × 6.5 **13.** \$148.55 × 1.95

14. \$132.65 × 4.10 **15.** \$324.1 × 0.95 **16.** \$110.45 × 2.50

(Skill 11) **17.** \$1095 ÷ 9500 **18.** \$2746 ÷ 8840 **19.** \$1742.35 ÷ 14,000

(Skill 30) **20.** 75% of \$4800 **21.** 81% of \$9180 **22.** 92% of \$3542.75

Applications

Use the table to answer the following.

(Application C) **23.** What is the average trade-in value of a G30 custom pickup, model A90, with a manual transmission?

24. What is the average retail value of a G30 deluxe pickup, model A95, with power steering, air conditioning, and four-wheel drive?

Av'g Trd-in	BODY TYPE Model	Av'g Loan	Av'g Ret'l
	G30 Series E- V8 1-Tn AT		
1825	Custom Wdbed A90	1650	2425
1875	Deluxe Wdbed A95	1700	2475
500	Add 4WD	450	600
150	Add air cond.	150	200
150	Add SS equip.	150	200
850	Add camper sp.	600	1000
50	Add power steer	50	75
125	Deduct man. trans	125	150

Terms

Use each term in one of the sentences.

25. Sticker price

26. Rent

27. Used-car guide

28. Deductible clause

29. Depreciation

a. You can find price information on used cars that were purchased from dealers during the previous month in a __?__ .

b. A decrease in the value of your car because of its age or condition is called __?__ .

c. If you do not own an automobile, but need to use one on occasions, you can __?__ from a rental agency.

d. The __?__ in your insurance policy states that you must pay a portion of any repair bill.

e. The total of the base price, options price, and any destination charge is the __?__ .

Refer to your reference files at the back of the book if you need help.

A Microcomputer Application
Annual Percentage Rate (APR)

To complete this microcomputer application, you will need the diskette *Microcomputer Applications for Business Mathematics,* which accompanies this textbook.

Select option 3, Annual Percentage Rate (APR), from the menu. Input the information in the following problems to find the APR for each automobile installment loan.

1. Paul Sigafoose.
   Loan for $2000.
   Terms: $181.67 a month, 12 months.
   What is the APR?

2. Isaac Walko.
   Loan for $8000.
   Terms: $386.67 a month, 24 months.
   What is the APR?

3. Dorothy Pilliod.
   Loan for $1500.
   Terms: $262.50 a month, 6 months.
   What is the APR?

4. Pedro Jujan.
   Loan for $12,400.
   Terms: $434.86 a month, 3 years.
   What is the APR?

5. Used car.
   Loan for $3500.
   Terms: $237.32 a month, 18 months.
   What is the APR?

6. Small sailboat.
   Loan for $1065.
   Terms: $96.63 a month, 12 months.
   What is the APR?

7. Alice Miklovic obtained an installment loan of $974 to pay for a motorcycle. She must make 12 payments of $92.57 each. What is the APR?

8. Sherman Ibarra obtained an installment loan of $1456 for energy-saving improvements to his home. He must make 60 payments of $41.26 each. What is the APR?

9. A used motor home was financed with a $10,000 installment loan for 4 years. The monthly payment is $279.68. What is the APR?

10. A new sailboat was financed with a $14,600 installment loan for 5 years. The monthly payment is $390.89. What is the APR?

11. Rosemary Brecht purchased a new car for $9475. She had a down payment of $1475 and financed the remainder with an installment loan for 4 years. The monthly payment is $241.67. What is the APR?

12. Tammy Newcomer purchased a new van for $15,972.19. She had a down payment of $2000 and financed the remainder with an installment loan for 5 years. The monthly payment is $337.71. What is the APR?

Unit Test

Lesson 8-1

1. Danilo Bigley sees a station wagon that has a base price of $12,707.75. Factory-installed options total $1498.66. There is a destination charge of $143.50. What is the sticker price of the station wagon?

Lesson 8-2

2. Bob Peterson is interested in a 4-wheel drive pickup that has a base price of $8035, options totaling $937.55, and a destination charge of $285. He has read that the dealer's cost is about 82% of the base price and 80% of the price of the options. What is the estimated dealer's cost that Bob should keep in mind when making an offer?

Lesson 8-3

3. Francis Guinta owns a used Sport Van model WC21 with automatic transmission, a V8 engine, AM/FM stereo, and no air conditioning. What is the average retail price that Francis should keep in mind when pricing his car for sale?

Av'g Trd-in	BODY TYPE	Model	Av'g Loan	Av'g Ret'l
STARCREST VAN C10, C20-V8-AT				
8250	Sport Van	WC11	8025	8900
8225	Panel Van	WC21	8025	8875
8350	Sport Van	WC21	8125	9000
100	Add AM/FM Stereo		100	150
75	Add Cruise Control		75	100
550	Deduct w/o Air Cond.		500	550

Lesson 8-4

4. Ruth Houghton uses her car to drive to and from work. Her insurance coverage includes 50/100 bodily injury and $25,000 property damage. Her driver-rating factor is 1.15. She has $50-deductible comprehensive and $100-deductible collision insurance coverages on her car. Her car is in age group B and rating group 7. What is her annual base premium? What is her annual premium?

Property Damage Limits	Bodily Injury Limits		
	25/50	25/100	50/100
$ 25,000	$103.20	$114.00	$119.20
$ 50,000	$107.20	$118.00	$123.20
$100,000	$110.00	$121.20	$126.00

Age Group	Insurance Rating Group 6	7	8
	Comprehensive $50 ded		
A	$32.80	$35.60	$39.60
B	$30.80	$33.20	$36.80
	Collision $100 ded		
A	$84.00	$83.20	$102.40
B	$72.80	$90.80	$108.40

Lesson 8-5

5. Nora Hovey drove her car 13,220 miles last year. Her records show that fixed costs totaled $980.20 and variable costs totaled $1230.39. How much did it cost per mile for Nora to operate her car last year?

Lesson 8-6

6. Jen-Shiang Hong flew to Toronto on vacation. While there, he rented a car for 3 days at $45.95 a day plus 18¢ a kilometer. He drove the car 630 km. Gasoline cost $38.40. How much did it cost per kilometer for him to rent the car?

Car Expenses: A Simulation

The costs of owning a car include the purchase price, financing, insurance, depreciation, gasoline, maintenance, and repairs. If you are about to buy a car, you might calculate the cost per month. If you drive to work, you may want to know the cost per day.

1. Suppose you plan to work 5 days per week, 50 weeks per year. How many working days are there per year?

2. If the round trip each day is 12 miles, how many miles will you drive to and from work in a year?

3. Your car gets 25 miles per gallon. How many gallons of gasoline will you use? At $1.25 per gallon, how much will it cost?

4. You expect to spend $38 for oil changes, $70 for a tune-up, and $23 for a wheel alignment. Depreciation will be $500. What is the total of these expenses?

5. The base premium for your automobile insurance is $192. Assume you are unmarried, under 21, and the principal operator. Use the table on page 213 to find your driver-rating factor. What will your annual insurance premium be?

6. What is the total annual cost for gasoline, maintenance, depreciation, and insurance?

7. Divide the total annual cost by the number of working days to find the cost per day.

Find the cost per day for each of these cases. Use your driver-rating factor from question 5.

	8.	9.	10.
Working days per year	245	250	240
Miles per day	30	26	8
Miles per gallon	30	18	40
Cost per gallon	$1.20	$1.18	$1.28
Maintenance costs	$120	$150	$95
Depreciation	$600	$750	$700
Base insurance premium	$160	$140	$125
Cost per day	?	?	?

Purchase Price and Financing

You have decided to buy a car. One choice you have to make is whether to buy a new car or a used car. A new car has a higher purchase price and greater depreciation. Comprehensive and collision insurance are more expensive for a new car. A used car can be expected to have lower gasoline mileage and higher maintenance and repair costs.

There are two cars you are considering: a new Volney and a used Jess. To help choose which one to buy, you decide to calculate how much it would cost per month to run each car for the next three years.

CODE	DESCRIPTION	LIST PRICE
PH2	VOLNEY 2-DOOR HATCHBACK	6550.00
P36	SUNSET RED METALLIC	120.00
R58	AM/FM STEREO TAPE DECK	375.00
SA4	ADJUSTABLE SEATS	250.00
AT2	AUTOMATIC TRANSMISSION	350.00
V31	VINYL ROOF — WHITE	125.00
829	DESTINATION CHARGE	200.00

11. A portion of the sticker for the Volney is shown. What is the total sticker price?

12. The dealer's cost for the Volney is about 90% of the base price and 85% of the price of the options. If the dealer accepts your offer of $200 over the dealer's cost, what would the Volney cost you, including destination charge and 6% sales tax?

13. You would make a $2000 down payment for the Volney and finance the rest of the cost at 16% for 36 months.

a. What is the amount financed?

b. Use the table on page 570 to find the finance charge. What is the total amount repaid? What is the cost per month?

14. The used-car guide entry for the Jess is shown. You are interested in a 2-door hardtop with a vinyl roof, AM/FM stereo with tape deck, and automatic transmission. If you bought the Jess for its average trade-in value, what would it cost, including 6% sales tax?

Av g Trd-in	BODY TYPE Model	Av'g Loan	Av'g Ret'l
JESS — Series C JB LT			
3875	H'dtp 2Dr	2750	4000
3980	H'dtp "S"	2800	4250
50	Add Vinyl Top	50	75
75	Add AM/FM/Tape	75	100
25	Add Power Seats	25	50
125	Add Auto. Trans.	125	125

15. You would make a $2000 down payment for the Jess and finance the rest of the cost at $18\frac{1}{2}$% for 36 months. Using the table on page 570 to find the finance charge, what is the cost per month?

Insurance

In order to choose an automobile insurance company, you have read articles in consumer and automobile magazines, talked with relatives and friends, and asked insurance companies about their coverage and premiums. The tables below show the base premiums of the company you have chosen.

Property Damage Limits	Bodily Injury Limits 25/100	50/100	100/200
25,000	$113.60	$120.00	$131.60
50,000	116.80	123.20	135.20
100,000	120.00	126.40	138.40

Age Group	Insurance Rating Group 10	11	12
	Comprehensive $50 Deductible		
A	$56.80	$69.60	$81.20
B	54.00	66.40	77.20
	Collision $50 Deductible		
A	$158.00	$170.80	$184.00
B	150.40	162.80	175.20

16. For either car you choose, you plan to have 50/100 bodily injury and $50,000 property damage coverage. What is the annual base premium?

17. You plan to have $50-deductible comprehensive and $50-deductible collision insurance coverage. The Volney is in age group A and rating group 11. What is the annual base premium?

18. The Jess is in age group B and rating group 10. What is the annual base premium for comprehensive and collision insurance?

19. Use the table below to find your driver-rating factor. Assume you are unmarried, under 21, and the principal operator. For each car, what is the total cost per month for insurance?

DRIVER-RATING FACTORS FOR AUTOMOBILE INSURANCE

Driver Classification	Multiple of Base Premium: Under 21 Years Occasional Operator	Under 21 Years Principal Operator	21–24 Years Occasional Operator	21–24 Years Principal Operator	25–29 Years Principal Operator
Unmarried youthful operators					
Females					
Pleasure	1.55	1.80	1.30	1.60	Classify as Adult
To and from work or business use	1.70	1.95	1.50	1.85	
Farm automobile	1.15	1.35	0.95	1.20	
Males					
Pleasure	2.30	3.90	1.70	2.90	1.65
To and from work or business use	2.50	4.10	1.95	3.10	1.90
Farm automobile	1.70	2.95	1.30	2.15	1.30

Maintenance, Depreciation, and Other Costs

A car depreciates in value most rapidly when it is new. You can estimate how a car will depreciate by looking at the values of older cars of the same model. Depreciation is often expressed as a percent of the purchase price.

Maintenance and repair costs tend to increase as a car ages. You can estimate these costs by talking with mechanics and with people who own the same model. Consumer and automobile magazines sometimes have articles comparing these costs for various models.

20. The Volney should lose 18% of its sticker price the first year, 15% of its sticker price the second year, and 12% the third year. What is the total percent? How much less will the Volney be worth in three years? What is the average depreciation per month?

21. The Jess can be expected to have a value of $3100 in three years. What is the depreciation per month?

22. For the Volney, you expect to spend $120 the first year, $450 the second year, and $500 the third year for maintenance and repairs. What is the average cost per month?

23. You expect the Jess to have maintenance and repair costs of $500 the first year, $600 the second year, and $700 the third year. It will also need tires for $250 plus 6% sales tax. What is the average cost per month?

24. You plan to drive 12,000 miles per year. Gasoline costs $1.25 per gallon. The Volney gets 35 miles per gallon. What is the cost of gasoline per month?

25. The Jess gets 28 miles per gallon. What is the cost of gasoline per month?

26. Make a table like this one to compare the costs of the two cars.

	Monthly Cost	Volney	Jess
Fixed Costs	Loan Payment	?	?
	Insurance	?	?
	Depreciation	?	?
	License and Registration	$47.20	$45.60
Variable Costs	Gasoline	?	?
	Maintenance, Repairs, Tires	?	?
	Total	?	?

Unit 9 Housing Costs

▸ Most people finance the cost of a home by taking out a *mortgage* loan. You pay a portion of the selling price and then finance the remaining amount. Most mortgages must be repaid in equal monthly installments that include the amount financed and the interest. Some monthly payments include a monthly payment for *real estate taxes.* Real estate taxes are paid to the city or county for schools, roads, and other expenses. Other housing costs include *homeowner's insurance* against fire, theft, and property damage, and costs for *utilities* such as heating fuel, water, and electricity.

A realtor assists a buyer in finding a suitable home. The realtor charges a percent commission on the purchase price of the house.

9-1 Mortgage Loans

Objective

Compute the mortgage loan amount.

When you purchase a home, you will probably make a down payment and finance the remaining portion of the selling price with a **mortgage loan** from a bank or savings and loan association. Generally, the down payment is between 10% and 40% of the selling price. The mortgage loan is usually repaid with interest in equal monthly payments. The mortgage gives the bank the right to sell the property if you fail to make the payments.

▶ **Mortgage Loan Amount = Selling Price − Down Payment**

Example

Skills 30, 4, 6 **Application** A **Term** Mortgage loan

Jane and Peter Ronfelt are considering the purchase of a new home for $62,500. A 15% down payment is required. What is the amount of the mortgage loan needed to finance the purchase?

Solution

1. Find the **down payment.**
   15% of $62,500 = $9375
2. Find the **mortgage loan amount.**
   Selling Price − Down Payment
   $62,500 − $9375 = $53,125 mortgage loan amount

A. $15,000
$45,000
B. $14,000
$26,000

▶ **Self-Check** Cover the answers and complete the problems.

Find the down payment and the amount of the mortgage.

A. $60,000 selling price, 25% down. B. $40,000 selling price, 35% down.

Problems

Buyer	Selling Price	Percent Down Payment	Down Payment	Mortgage Loan Amount
1. Eliza Connelly	$37,000	20%	$7400	?
2. Jeff Goldberg	$52,500	15%	?	?
3. Briget McClure	$93,800	25%	?	?
4. Manuel Freitas	$46,200	18%	?	?

5. John Rooney.
Home priced at $38,500.
20% down payment required.
What is the down payment?
What is the mortgage loan amount?

6. Steve and Alvira Green.
Condominium priced at $45,900.
25% down payment required.
What is the down payment?
What is the mortgage loan amount?

7. Don and Virginia Chin.
Mobile home priced at $17,300.
Made a 15% down payment.
What is the mortgage loan amount?

8. Cathy and Alan Bassett.
Home priced at $180,000.
Made a 28% down payment.
What is the mortgage loan amount?

9. John and Mary Angel decide that the home advertised is ideally suited to the needs of their family. They wish to make a 25% down payment and finance the remaining amount through their bank. What is the amount of their mortgage loan?

> A home with character, charm and personality, 4 bedrooms, 3 full baths, beautiful vaulted living room ceiling, deluxe kitchen, large master bedroom suite, big family room.
>
> **At $79,900 it's a steal.**

10. Alice and William Stoll offered $35,000 for a home that had been priced at $37,500. After some negotiating, the Stolls and the seller agreed on a selling price of $36,500. What is the amount of the mortgage loan if they make a 20% down payment?

11. Thelma Jackson offered $93,300 for a home that had been priced at $98,600. The seller agreed to the offer. A bank is willing to finance the purchase if she can make a down payment of 25%. What is the amount of her mortgage loan?

12. The Samsons have decided to purchase a duplex. They would use the rental income from one part of the duplex to help meet the mortgage payments. A selling price of $104,800 was agreed upon. What is the amount of the mortgage loan if a down payment of $22\frac{1}{2}\%$ is required?

13. Bob and Sue Ivanoski have saved $12,000 with which to make a down payment on their future home. Their bank has informed them that the minimum down payment required to obtain a mortgage loan is 15%. What is the most that they can spend for a home and expect to receive bank approval for their loan?

Maintaining Your Skills

Look up the skills in parentheses if you need help or more practice.

Find the percentage. (Skill 30)

14. 15% of 20,000
15. 25% of 24,000
16. 20% of 80,000
17. 40% of 85,000
18. 22% of 55,500
19. 9% of 125,500

Subtract. (Skill 4)

20. 20,000 − 12,400
21. 24,000 − 18,000
22. 80,000 − 16,000
23. 58,500 − 11,250
24. 21,000 − 14,000
25. 125,500 − 62,250

9-2 Monthly Payment and Total Interest

Objective

Compute the total interest charged.

Banks that make mortgage loans charge **interest** for the use of their money. The interest rate will vary from bank to bank, so it pays to shop around. If you know the annual interest rate, the amount of the loan, and the length of the loan, you can use a table to find the monthly payment, the total amount paid, and the total interest charged.

MONTHLY PAYMENT FOR A $1000 LOAN			
Annual Interest Rate	Length of Loan (Years)		
	20	25	30
10.00%	$ 9.66	$ 9.09	$ 8.78
10.50%	9.99	9.45	9.15
11.00%	10.33	9.81	9.53
11.50%	10.67	10.17	9.91
12.00%	11.02	10.54	10.29
12.50%	11.37	10.91	10.68
13.00%	11.72	11.28	11.07
13.50%	12.08	11.66	11.46

- **Monthly Payment** = $\frac{\text{Amount of Mortgage}}{\$1000}$ × **Monthly Payment for a $1000 Loan**
- **Amount Paid = Monthly Payment × Number of Payments**
- **Total Interest Charged = Amount Paid − Mortgage Amount**

Example

Skills 8, 6 **Application** C **Term** Interest

Sandra and Peter Miller have applied for a $40,000 mortgage loan at an annual interest rate of 12.50%. The loan is for a period of 30 years and will be paid in equal monthly payments that include interest. What is the total amount of interest charged?

Solution

1. Find the **monthly payment.** (Refer to the above table.)

$$\frac{\text{Amount of Mortgage}}{\$1000} \times \text{Monthly Payment for a \$1000 Loan}$$

$$\frac{\$40{,}000}{\$1000} \times \$10.68 = \$427.20$$

2. Find the **amount paid.**

Monthly Payment × Number of Months
$427.20 × (12 months × 30 years)
$427.20 × 360 = $153,792.00

3. Find the **total interest charged.**

Amount Paid − Mortgage Amount
$153,792.00 − $40,000.00 = $113,792.00 total interest

$676.80
$203,040
$143,040

▶ **Self-Check** Cover the answer and complete the problem.

Find the monthly payment, amount paid, and interest charged for a $60,000 mortgage loan at an annual interest rate of 13% for 25 years.

Problems

Use the table on page 218 to solve.

	1.	2.	3.	4.	5.
Mortgage	$40,000	$60,000	$85,000	$75,500	$195,000
Years	20	25	30	20	30
Rate	10.00%	12.00%	11.50%	13.00%	13.50%
Payment	?	?	?	?	?
Amount Paid	?	?	?	?	?
Total Interest	?	?	?	?	?

Use the table on page 569 to solve.

6. Jeffry and Lyn Egan.
$60,000 mortgage.
Terms: 11% for 25 years.
What is the monthly payment?
What is the total amount paid?
What is the total interest charged?

7. Sylvia and Raymond Duvall.
$50,000 mortgage.
Terms: 14% for 20 years.
What is the monthly payment?
What is the total amount paid?
What is the total interest charged?

8. Clara Koper.
$50,000 mortgage.
Terms: 10.5% for 20 years.
What is the total interest charged?

9. Karlis Krisjanis.
$80,000 mortgage.
Terms: 15% for 30 years.
What is the total interest charged?

10. Zeke and Abigail Stone have obtained a $40,000 mortgage loan at an annual interest rate of 12% for 25 years. What is the monthly payment? What is the total amount to be paid?

11. Charles and Sandy Novak reached an agreed upon price of $74,000 with the owner for the purchase of a house. They made a down payment of $14,000 and could finance the remaining amount in one of two ways: at 11.5% for 25 years or at 12% for 20 years. Which mortgage results in a larger amount of interest paid? How much greater?

12. How much can be saved in total interest by financing $40,000 at 14.5% for 15 years rather than 20 years?

Maintaining Your Skills

Look up the skills in parentheses if you need help or more practice.

Multiply. (Skill 8)

13. 24 × 20.50 14. 36 × 123.2 15. 12 × 332.40 16. 15 × 27.16

Subtract. (Skill 6)

17. 15,500 − 12,200 18. 42,461 − 19,320 19. 153,321.50 − 60,000

9-3 Closing Costs

Objective Compute the total closing costs.

At the time you sign the documents transferring ownership of your new home, you must pay any **closing costs** that the bank charges. The closing costs may include fees for the bank's lawyers, credit checks and title searches, taxes, and the preparation of the documents. Some banks charge a flat fee regardless of the amount of the loan. Some banks charge a percent of the amount of the loan. Other banks charge itemized fees at the closing.

▶ **Closing Costs = Sum of Bank Fees**

Example

Skills 30, 3 **Application** A **Term** Closing costs

Rose and Vincent Manuso were granted a $40,000 mortgage loan. When they sign the papers to purchase their new home, they will have to pay the closing costs shown. What is the total of the closing costs?

CREDIT REPORT: $35
LOAN ORIGINATION FEE: 2% of MORTGAGE
ABSTRACT OF TITLE: $60
ATTORNEY FEE: $100
DOCUMENTATION STAMP: 0.3% OF MORTGAGE
PROCESSING FEE: 2.23% OF MORTGAGE

Solution

Find the **sum of the bank fees.**

a. Credit report	$ 35
b. Loan origination: 2% of $40,000	800
c. Abstract of title	60
d. Attorney fee	100
e. Documentation stamp: 0.3% of $40,000	120
f. Processing fee: 2.23% of $40,000	892
Total closing costs	$2007

$35 + $1200 + $60 + $100 + $180 + $1338 = $2913

▶ **Self-Check** Cover the answer and complete the problem.

Use the list of closing costs in the above example to find the closing costs on a $60,000 mortgage.

Problems

Use the list of closing costs in the example above to solve.

1. Barry and Jana Roberts. Mortgage loan of $30,000. What is the total closing cost?

2. Jack and Ella Pritchard. Mortgage loan of $55,000. What is the total closing cost?

3. Joseph and René Murphy.
Mortgage loan of $78,000
What is the total closing cost?

4. Norman and Kristi Ming.
Mortgage loan of $53,900.
What is the total closing cost?

5. Rick and Marilyn Giovan are financing $28,700 at 11.5% for 30 years. They must pay these closing costs. What is the Giovans' total closing cost?

Appraisal fee	$35.00
Credit report	$25.00
Title search	$189.00
Service fee	1.5% of mortgage
Legal fees	$250.00

6. Rob and Judy Franklin have agreed to purchase a house for $46,500. Kenmore Savings and Loan Association is willing to lend the money at 12.25% for 25 years provided the Franklins can make a $10,000 down payment. The total closing cost is 4.25% of the amount of the mortgage loan. What is their closing cost?

7. Janet and Reed Atlein are interested in purchasing a $40,000 home. They plan to make a 25% down payment and finance the remaining amount through Peabody Savings Association. Peabody Savings has these closing costs: credit report, $25; appraisal report, $155; title insurance, $45; survey and photographs, $225; recording fee, $15; legal fees, $280; and property taxes, $289. If the loan is approved, how much cash will the Atleins need to secure the loan, including the down payment?

*8. **A Brief Case** Many states have usury laws that set a legal maximum interest rate that lending agencies can charge. When the trend in interest rates is above the maximum allowed, many lending agencies will add points to their basic mortgage charges. A "point" is 1% of the mortgage loan and is included in the closing costs. Five points is a one-time charge of 5% of the mortgage loan.

Dina and Jeremy Kane have agreed to purchase a $72,800 home. They plan to make a 25% down payment. Beacon Hill Bank is willing to finance the mortgage at 11.75% for 30 years plus 2 points. How much cash will the Kanes need to secure the loan, including the down payment?

Title search: $35
Appraisal fee: $40
Credit report: $5
Legal fee: $375
Property tax: $750
Fire insurance: $230

Maintaining Your Skills

Look up the skill in parentheses if you need help or more practice.

Find the percentage. (Skill 30)

9. 5% × 900
10. 4% × 860
11. 7% × 252
12. 2.4% × 780
13. 6.3% × 230
14. 4.2% × 30,000
15. 0.3% × 72,000
16. 0.15% × 45,100
17. 0.04% × 22,200
18. 8% × 815
19. 1.4% × 50,000
20. 0.61% × 83,300

9-4 The Monthly Payment

Objective Compute the monthly payment toward principal and the new principal.

Most mortgage loans are repaid in equal monthly payments. Each payment includes an amount for payment of interest and an amount for payment of the **principal** of the loan. The amount of interest is calculated using the simple interest formula. The amount of principal that you owe decreases with each payment that you make. The chart shows the interest and principal paid in the first 4 months of a $40,000 mortgage loan.

$40,000 MORTGAGE LOAN AT 12% FOR 25 YEARS				
PAYMENT NUMBER	MONTHLY PAYMENT	AMOUNT FOR INTEREST	AMOUNT FOR PRINCIPAL	NEW PRINCIPAL
1	$421.60	$400.00	$21.60	$39,978.40
2	$421.60	$399.78	$21.82	$39,956.58
3	$421.60	$399.57	$22.03	$39,934.55
4	$421.60	$399.35	$22.25	$39,912.30

▶ **Payment to Principal = Monthly Payment − Interest**

▶ **New Principal = Previous Balance − Payment to Principal**

Example

Skills 6, 8, 30 **Application** A **Term** Principal

Ralph and Lois Johnson obtained a 25-year, $40,000 mortgage loan from Second Bank and Trust. The interest rate is 12%. Their monthly payment is $421.60. For the first payment, what is the interest? What is the payment to principal? What is the new principal?

Solution

1. Find the **interest.**
   Principal × Rate × Time
   $\$40{,}000.00 \times 12\% \times \frac{1}{12} = \$400.00$
2. Find the **payment to principal.**
   Monthly Payment − Interest
   $421.60 − $400.00 = $21.60
3. Find the **new principal.**
   Previous Balance − Payment to Principal
   $40,000.00 − $21.60 = $39,978.40 new principal

▶ **Self-Check** Cover the answers and complete the problems.

In the example above the new principal is $39,978.40.
For the second payment find the:

A. Interest on $39,978.40. B. Payment to principal. C. New balance.

A. $399.78
B. $21.82
C. $39,956.58

Problems

	Mortgage Amount	Interest Rate	First Monthly Payment	Amount for Interest	Amount for Principal	New Principal
1.	$35,000	12%	$ 368.90	$350.00	$18.90	?
2.	$48,000	12%	$ 493.92	?	?	?
3.	$54,000	13%	$ 597.78	?	?	?
4.	$80,000	14%	$ 937.60	?	?	?
5.	$94,400	16%	$1269.68	?	?	?

6. May and Cleveland Foster.
   Mortgage loan of $36,000.
   Interest rate is 12%.
   Monthly payment is $396.72.
   How much of the first monthly payment is for interest?

7. Joyce and Walter Holmes.
   Mortgage loan of $54,000.
   Interest rate is 18%.
   Monthly payment is $814.32.
   How much of the first monthly payment is for interest?

8. Mary Bielarczyk.
   Mortgage loan of $48,600.
   Interest rate is 12.5%.
   Monthly payment is $530.25.
   How much of the first monthly payment is for interest?
   How much of the first payment is for principal?
   What is the new principal?

9. Norman Burris.
   Mortgage loan of $98,000.
   Interest rate is 15.5%.
   Monthly payment is $1293.60.
   How much of the first monthly payment is for interest?
   How much of the first payment is for principal?
   What is the new principal?

10. Charles and Agnes Lynn obtained a 25-year, $60,000 mortgage loan from Peoples Savings and Loan Association. The interest rate is 12%. Their monthly payment is $632.40. For the first payment, what is the interest? What is the payment to principal? What is the new principal?

11. Elbert McGuire obtained a 30-year, $80,000 mortgage loan from First Federal Savings and Loan Association. The interest rate is 15%. His monthly payment is $1012. For the first payment, what is the interest? What is the payment to principal? What is the new balance?

12. Amelia Pollom obtained a 20-year, $36,000 mortgage loan from Toledo Trust Company. The interest rate is 12.5%. Her monthly payment is $409.32. For the first payment, what is the interest? What is the payment to principal? What is the new balance?

13. Matthew Yunker obtained a 25-year, $88,500 mortgage loan from Swancreek Trust Company. The interest rate is 14.5%. His monthly payment is $1100. For the first payment, what is the interest? What is the payment to principal? What is the new balance?

Refer to the table on page 569 for problems 14–18.

14. Geraldine Contreras obtained a 25-year, $72,000 mortgage loan from the Miners and Merchants Deposit Company. The interest rate is 14.5%. What is her monthly payment? For the first payment, what is the interest? What is the payment to principal? What is the new balance?

15. Boyer Featherstone obtained a 15-year, $94,500 mortgage loan from Pacific Trust Company. The interest rate is 10.5%. What is the monthly payment? For the first payment, what is the interest? What is the payment to principal? What is the new balance?

16. Jack and Marian Jones purchased a home for $62,500. They made a 20% down payment and financed the remaining amount at 11% for 30 years. Their monthly payment is $476.50. How much of the first monthly payment is used to reduce the principal?

17. Will and Emily Tyson purchased the home advertised. They made a down payment of $14,300 and financed the remaining amount at 12.5% for 30 years. What is the monthly payment? What is the new principal after the first monthly payment?

Rated X! X-tra nice, X-tra large master bedroom with dressing room, X-tra sized screen porch, X-tra space for a garden. Split plan 3 bedroom, 2 bth, dbl. garage. X-cellent value too. $54,900.

18. Dottie and Hollis Johnson purchased a condominium for $57,600. They made a $9600 down payment and financed the remaining amount at 11.5% for 25 years. What is their monthly payment? How much of the first monthly payment is used to reduce the principal? How much of the second monthly payment is used to reduce the principal?

19. Julie Breen purchased a mobile home for $20,600. She made a down payment of $5000 and financed the remaining amount at 10.5% for 20 years. Her monthly payment is $155.84. She has made 118 payments to date. Use the portion of Julie's repayment schedule to find her remaining debt after the 120th payment.

PAYMENT NUMBER	MONTHLY PAYMENT	AMOUNT FOR INTEREST	AMOUNT FOR PRINCIPAL	NEW PRINCIPAL
115	$155.84	$103.78	$52.06	$11,808.59
116	$155.84	$103.33	$52.51	$11,756.08
117	$155.84	$102.87	$52.97	$11,703.11
118	$155.84	$102.40	$53.44	$11,649.67

Maintaining Your Skills

Look up the skills in parentheses if you need help or more practice.

Subtract. (Skill 6)

20. 48,000 − 29.46
21. 91,800 − 39.55
22. 24,400 − 23.12
23. 78,902 − 22.98
24. 18,185 − 45.11
25. 14,915 − 107.7

9-5 Real Estate Taxes

Objective Compute the assessed value and taxes of real estate.

When you own a home, you will have to pay city or county **real estate taxes.** The money collected is used to operate and maintain roads, parks, schools, government offices, and so on. The amount of real estate tax that you pay in one year depends on the **assessed value** of your property and the **tax rate.** The assessed value is found by multiplying the **market value** of your property by the **rate of assessment.** The market value is determined by an assessor hired by the municipality. The rate of assessment is a percent.

The tax rate is sometimes expressed in **mills** per dollar of valuation. A mill is $0.001. A tax rate of 80 mills per dollar is a tax rate of $80 per $1000 of assessed value. When working with mills, it is often convenient to express mills in dollars by dividing by 1000.

▸ **Assessed Value = Rate of Assessment × Market Value**

▸ **Real Estate Tax = Tax Rate × Assessed Value**

Example

Skills 8, 30 **Application** A **Term** **Real estate taxes**

The Lake County tax assessor stated that the market value of the Courtland Farm is $240,000. The rate of assessment in Lake County is 40% of market value. The tax rate is 83.42 mills. What is the real estate tax on the Courtland Farm?

Solution

1. Express the **tax rate** as a decimal.
   83.42 mills ÷ 1000 = 0.08342
2. Find the **assessed value.**
   Rate of Assessment × Market Value
   40% of $240,000 = $96,000.00
3. Find the **real estate tax.**
   Tax Rate × Assessed Value
   0.08342 × $96,000.00 = $8008.32 real estate tax

A. 0.0555
B. $32,000
C. $1776

▸ **Self-Check** Cover the answers and complete the problems.

The tax rate is 55.5 mills, market value is $80,000, and rate of assessment is 40%. Find the following:

A. The tax rate as a decimal.
B. The assessed value.
C. The real estate tax.

Problems

1. The M. L. Gavin's home.
   Market value is $42,000.
   Rate of assessment is 30%.
   What is the assessed value?

2. The Heinsons' condominium.
   Market value is $39,800.
   Rate of assessment is 50%.
   What is the assessed value?

3. The Hightower Farm.
   Market value is $112,000.
   Rate of assessment is 35%.
   Tax rate is $66 per $1000.
   What is the real estate tax?

4. The D'Limas' home.
   Market value is $76,000.
   Rate of assessment is 40%.
   Tax rate is 32.45 mills.
   What is the real estate tax?

5. Ali Mann's home is located in a community where the rate of assessment is 50% of market value. The tax rate is $35 per $1000 of assessed value. Her home has a market value of $82,000. What is its assessed value?

6. The rate of assessment in Wakefield County is 35%. The tax rate is $52.25 per $1000 of assessed value. What is the real estate tax on a piece of property that has a market value of $68,000?

7. Andy and Lu Simms live in a locality where the tax rate is 51.385 mills. The rate of assessment is 60%. The property that the Simms own has a market value of $73,500. What is their real estate tax for a year?

8. Bruce and Mary Davis own a home that has a market value of $54,000. The Davises live in an area where the rate of assessment is 100% and the tax rate is 28.535 mills. What is the annual real estate tax?

9. Timothy North owns a mobile home and the property on which it sits. His mobile home has a market value of $14,000 and his property has a market value of $23,000. Timothy lives in a county where the rate of assessment is 80% and the tax rate is $31.85 per $1000 of assessed value. What is his yearly real estate tax?

10. Kathy and Milo Bartlett received a tax statement which shows that their land has an assessed value of $4500 and their buildings have an assessed value of $12,125. The rate of assessment in their locality is 35%. What is the market value of their property?

Maintaining Your Skills

Look up the skills in parentheses if you need help or more practice.

Multiply. (Skill 8)

11. 27.3 × 68.1
12. 9.08 × 8.54
13. 17.3 × 3.37
14. 0.656 × 1.48
15. 73.82 × 33.9
16. 0.474 × 4256

Find the percentage. (Skill 30)

17. 40% × 80,000
18. 55% × 40,000
19. 35% × 15,500
20. 82% × 128,800
21. 28% × 52,710
22. 60% × 37,500

9-6 Homeowner's Insurance

Objective Compute the amount of coverage.

When you own a home, you will probably purchase **homeowner's insurance** to protect yourself and your home from losses due to fire, theft, and personal liability. Homeowner's insurance also includes **loss-of-use coverage** to pay for some of the expenses of living away from home while damage to your home is being repaired. **Personal liability** and **medical coverage** protect you from financial losses if someone is injured on your property.

To receive full payment for any loss up to the amount of the policy, you must insure your home for at least 80% of its replacement value. Some companies may require you to insure your home for 90% to 100% of its replacement value. The **replacement value** is the amount required to reconstruct your home, if destroyed. Insurance companies use the amount of coverage on your home to calculate the amount of coverage you must carry on your garage, personal property, and for loss of use. Many companies use these percents of the amount of coverage on the home for each type of protection.

COVERAGE	PERCENT OF COVERAGE
Personal Property	50%
Loss of Use	20%
Garage and Other Structures	10%

▶ **Amount of Coverage = Percent × Amount of Coverage on Home**

Example

Skill 30 Application A Term Homeowner's insurance

The replacement value of Ann and Mark Stile's home is estimated at $83,500. The Stiles have insured their home for 80% of its replacement value. According to the guidelines above, what is the amount of coverage on the Stiles' personal property?

Solution

1. Find the **amount of coverage on home.**
   80% of $83,500 = $66,800
2. Find the **amount of coverage on personal property.**
   Percent × Amount of Coverage on Home
   50% of $66,800 = $33,400 coverage on personal property

▶ **Self-Check** Cover the answers and complete the problems.

A home is insured for 90% of its replacement value of $80,000, or $72,000. Using the percents above, find the coverage for:

A. Personal property. B. Loss of use. C. Garage.

A. $36,000
B. $14,400
C. $7200

Problems

Use the table on page 227 to find the percent of coverage.

1. The Branches' home.
Replacement value is $50,000.
80% coverage on home.
What is the amount of insurance?

2. The Yees' duplex.
Replacement value is $75,000.
100% coverage on home.
What is the amount of insurance?

3. The Henriques' home.
Replacement value is $48,500.
90% coverage on home.
What is the amount of insurance?
What is the amount of coverage for personal property?

4. The Greenfields' home.
Replacement value is $122,000.
80% coverage on home.
What is the amount of insurance?
What is the amount of coverage for loss of use?

5. The Watkins family is purchasing a home that has a replacement value of $42,500. They wish to insure their new home for 90% of its replacement value. What is the amount of insurance on their home? What is the amount of coverage on their garage?

6. The Browns recently purchased a home that has a replacement value of $165,000. They insured their new home for 80% of its replacement value. What is the amount of insurance on their home? What is the amount of coverage for personal property? Loss of use? Garage?

The Jenson Insurance Company uses these guidelines for homeowner's insurance coverages. The policies must be written for 80% of a home's replacement value. Use the guidelines to solve the following.

JENSON INSURANCE COMPANY	
Coverage—Home 80% Insured	
Garage/Structures	10%
Personal Property	50%
Loss of Use	20%
Personal Liability	50%
Medical Payments	1%

7. The Timiltys own a home that has a replacement value of $52,000. They insure it for 80% of its replacement value. What is the amount of coverage on the Timilty's personal property?

8. Leon Casting has a wood-frame home that has a replacement value of $80,500. He has insured it for 80% of its replacement value. What is the amount of coverage on his home? What is the amount of coverage for each type of protection listed in the guidelines?

Maintaining Your Skills

Look up the skill in parentheses if you need help or more practice.

Find the percentage. (Skill 30)

9. 10% of $20,000
10. 80% of $90,000
11. 50% of 45,000
12. 15% of 120,000
13. 15% of 65,500
14. 90% of 244,000

9-7 Homeowner's Insurance Premium

Objective Compute the annual homeowner's insurance premium.

The amount of your homeowner's policy **premium** depends on the amount of insurance, the location of your property, and the type of construction of your home. Your **fire protection class** is a number that reflects the quality of fire protection available in your area.

ANNUAL PREMIUMS FOR A TYPICAL HOMEOWNER'S POLICY								
AMOUNT OF INSURANCE COVERAGE	BRICK/MASONRY VENEER				WOOD FRAME			
	FIRE PROTECTION CLASS				FIRE PROTECTION CLASS			
	1-6	7-8	9	10	1-6	7-8	9	10
$ 20,000	$126	$152	$161	$185	$136	$161	$173	$197
25,000	134	161	171	196	144	171	183	209
30,000	146	175	186	213	157	186	200	228
35,000	156	188	200	229	168	200	214	245
40,000	159	192	204	233	171	204	218	250
45,000	163	197	210	240	176	210	224	257
50,000	166	201	214	245	179	214	228	262
60,000	196	237	252	289	211	252	270	309
70,000	220	266	283	324	237	283	303	347
80,000	248	300	320	367	268	320	342	393
90,000	275	333	354	407	297	354	380	435
100,000	301	364	387	444	324	387	415	476
120,000	357	432	461	529	385	461	493	566
150,000	451	546	582	669	487	582	624	716
200,000	603	731	779	895	651	779	835	959

Example

Skill 30 **Application** C **Term** Premium

The replacement value of Lois Anderson's home is $50,000. Lois has insured her home for 80% of its replacement value. The home is of wood-frame construction and has been rated in fire protection class 4. What is Lois's annual premium?

Solution

1. Find the **amount of coverage.**
   80% of $50,000 = $40,000
2. Find the **annual premium.** (Refer to the above table.)
   $40,000 coverage, wood frame,
   fire protection class 4 . . . . $171.00 annual premium

$252

▶ **Self-Check** Cover the answer and complete the problem.

A home is insured for $60,000. It has a brick/masonry veneer and is in protection class 9. Find the premium.

Problems

Use the table on page 229 to find the annual premium.

1. Ron Fleming's wood-frame house.
$40,000 homeowner's policy.
Fire protection class 5.
What is the annual premium?

2. The Baldwins' brick home.
$60,000 homeowner's policy.
Fire protection class 9.
What is the annual premium?

3. The Gillmores' brick home.
Replacement value of $100,000.
Insured for 90%.
Fire protection class 7.
What is the annual premium?

4. The Meeks' wood-frame home.
Replacement value of $62,500.
Insured for 96%.
Fire protection class 2.
What is the annual premium?

5. The Wilhoyts own a wood-frame home in an area rated fire protection class 1. Their two-family home has a replacement value of $87,500 and is insured for 80%. What is their annual premium?

6. The McDonalds own a brick home that has a replacement value of $87,500. They purchased a homeowner's policy for 80% of its replacement value. The McDonalds live in an area rated fire protection class 3. What is their annual policy premium?

7. Marlene Gagne has insured her home for $50,000. The two-story, wood-frame house is located in an area rated fire protection class 10. What is her annual policy premium?

*8. **A Brief Case** When you obtain a mortgage loan, the bank may require you to include with your monthly principal and interest payment an amount equal to $\frac{1}{12}$ of the amounts needed to pay your real estate taxes and fire insurance premium. The bank places the money in an escrow account. When the real estate taxes and insurance premium are due, the bank uses the money from the escrow account to make the payments.

Lynn and Martin Tucker own a brick home with a market value and replacement value of $50,000. They insured their home for 100% of its replacement value. The Tuckers live in an area where the rate of assessment is 35%, the tax rate is 51.58 mills, and the fire protection is rated class 3. The Tuckers' monthly principal and interest payment is $463.19. How much are the Tuckers' monthly payments for principal, interest, real estate taxes, and insurance?

Maintaining Your Skills

Look up the skill in parentheses if you need help or more practice.

Find the percentage. (Skill 30)

9. 70% of 50,000
10. 80% of 120,000
11. 96% of 88,000
12. 50% of 30,000
13. 40% of 24,500
14. 15% of 31,300
15. 40% of 125,000
16. 20% of 145,500
17. 33.3% of 98,500

9-8 Other Housing Costs

Objective

Compute the total housing cost and compare it with suggested guidelines.

In addition to your monthly mortgage payment, real estate taxes, and insurance payment, you will have other expenses for utilities, maintenance, and home improvements. **Utilities costs** may include charges for electricity, gas, water, telephone, and heating fuel. The Federal Housing Administration (FHA) recommends that your total monthly housing cost be less than 35% of your monthly net pay.

Example

Skills 1, 5, 30 **Application** — **Term** Utilities cost

Margaret and Charles Hayashi have a combined monthly take-home pay of $2972. They keep a record of their monthly housing expenses. The list of expenses for May is shown. Were their housing costs for May within the FHA guidelines?

Housing Expenses for May

Mortgage payment	$508.24
Insurance ($306 ÷ 12)	25.50
Real estate taxes ($1335 ÷ 12)	111.25
Electricity	45.90
Heating fuel	54.20
Telephone	26.18
Water	16.20
Loan payment on Oven	50.00
Repair storm door	18.64

Solution

1. Find the **total monthly cost.**
   $508.24 + $25.50 + $111.25 + $45.90 + $54.20 + $26.18 + $16.20 + $50.00 + $18.64 = $856.11

2. Find the **recommended maximum.**
   35% of $2972.00 = $1040.20

3. Compare. Is **total monthly cost** less than **recommended maximum?**
   Is $856.11 less than $1040.20? . . . . . . Yes, they are within the guidelines.

A. Yes, 35% of $2500 = $875
B. No, 35% of $2000 = $700

▸ **Self-Check** Cover the answers and complete the problems.

Would the Hayashis be within the FHA recommended guidelines if they had a combined monthly take-home pay of:

A. $2500 B. $2000

Problems

	1.	2.	3.	4.	5.	6.
Monthly Net Pay	$1000	$2900	$1580	$1284	$1439	$3942
Recommended FHA Maximum	?	?	?	?	?	?

7. Theodore and Yvette Rowen.
Monthly net pay is $980.
Housing expenses for June:

Mortgage payment	$313.65
Insurance	18.50
Real estate taxes	57.00
Electricity	45.44
Telephone service	34.78
Loan payment	62.35

What is the total housing cost?
Is it within the FHA recommendation?

8. Patrick and Farra Peralta.
Monthly net pay is $1283.
Housing expenses for March:

Mortgage payment	$468.80
Insurance	19.75
Real estate taxes	82.47
Electricity	73.60
Telephone service	29.45
Water	12.20

What is the total housing cost?
Is it within the FHA recommendation?

9. Joshua and Ruth Boccho.
Monthly net pay is $4200.
Housing expenses for July:

Mortgage payment	$932.80
Insurance	28.50
Real estate taxes	144.67
Utilities	232.40
Repairs	47.00
Loan payment	74.67

What is the total housing cost?
Is it within the FHA recommendation?

10. Melanie Whitson.
Monthly net pay is $1650.
Housing expenses for January:

Mortgage payment	$319.68
Insurance	16.33
Real estate taxes	42.24
Utilities	138.46
Repair driveway	40.00

What is the total housing cost?
Is it within the FHA recommendation?

11. Peg Bayless recorded her housing expenses for the month of August: mortgage payment, $297.90; insurance, $27; taxes, $124; electricity, $94.40; phone service, $33.50; fuel, $88.25; water, $15.44; and repairs, $72.77. Her monthly take-home pay is $1990. What is her total monthly cost? Is it within the FHA recommendation?

12. Greg Kwan recorded his housing expenses for the month of December: mortgage payment of $248.36, $29.17 for insurance premium, $102.25 for real estate taxes, installment payment of $56.75 for refrigerator, $44.70 for electricity, $24.40 for telephone service, $76.76 for home heating oil, and $11.80 for water. His monthly take-home pay is $991.64. What is his total monthly housing cost? Is it within the FHA recommendation?

13. Edwin Enright has a monthly net income of $2100. He keeps a record of his monthly housing expenses. A list of expenses for January is shown. Were his housing costs for January within the FHA guidelines?

Housing Expenses for January

Mortgage payment	$372.45
Insurance	30.00
Real estate taxes	56.60
Electricity	75.30
Heating fuel	72.50
Telephone	19.44
Water	31.00
Loan payment	45.00

Water/sewer charges	$192.00
Electricity	540.00
Telephone service	205.32
Water heater	495.32
Repair storm windows	380.07
New air conditioner	459.68
Replace gutters	130.00
New lawn mower	288.90

14. Gary and Barbara Bardo have a combined monthly net income of $2450. Their records show that for last year they paid $4589.40 in mortgage payments, $283 for insurance premiums, and $1985.52 in real estate taxes. In addition, they had the expenses shown. What was their average monthly housing cost for last year? Was it within the FHA recommendation?

Use the table on page 218 to find the loan payment and the table on page 229 to find the insurance premium.

*15. **A Brief Case** Willis and Norma Sherman recently purchased a brick house for $50,000. They made a 20% down payment and financed the remaining amount at 12.50% for 30 years. The tax rate in their area is 51.07 mills and the rate of assessment is 35%. They were required to purchase a homeowner's insurance policy for the amount of the mortgage. The fire protection in the neighborhood is rated class 9. For the month of August, they recorded the following housing expenses: $59.23 for electricity, $43.85 for telephone service, $12.88 for water, and $54.65 to install additional insulation. The Shermans have a combined monthly net income of $2166.66. What is their monthly mortgage payment? What is the annual insurance premium? What was their total monthly housing cost for August? Is it within the FHA recommendation?

Maintaining Your Skills

Look up the skills in parentheses if you need help or more practice.

Which number is greater? (Skill 1)

16. 2019.8 or 2107.9
17. 7381.08 or 72,236.50
18. 534.26 or 534.81
19. 820.40 or 876.07
20. 62,879 or 65,396
21. 268.84 or 378.4

Add. (Skill 5)

22. 45.89 + 84.84 + 25.30 + 346.21
23. 553.26 + 74.29 + 14.4 + 7.39 + 24.39
24. 784.89 + 777.29 + 9.91 + 25.8
25. 327.86 + 668.2 + 11.88 + 9 + 104.02

Find the percentage. Round answers to the nearest hundredth. (Skill 30)

26. 35% of 4300
27. 40% of 3613
28. 42% of 8831
29. 35% of 5089.7
30. 30% of 986.44
31. 15% of 58,547
32. 0.3% × 72,000
33. 0.15% × 45,100
34. 0.04% × 22,200

Reviewing the Basics

Skills

Solve. Round answers to the nearest hundredth.

(Skill 3)	1. $735 + $325 + $58	2. $270 + $1035 + $349	3. $27,050 + $3498
(Skill 4)	4. $495 − $317	5. $96,000 − $41,500	6. $83,000 − $6774
(Skill 5)	7. $1951.73 + $1051.98	8. $617.54 + $483.31	9. $51.71 + $4.22
(Skill 6)	10. $123.12 − $10.07	11. $553.33 − $69.84	12. $431.12 − $279.94
(Skill 8)	13. 75.02 × 380	14. 1017.45 × 0.624	15. $3317 × 0.35
(Skill 30)	16. 25% × $94,000	17. 2.25% × $47,500	18. 35% × $3172.40

Applications

(Application C) Use the table to find the monthly payment.

19. The Jericho National Bank is willing to loan Sam Jones $40,000 at 12.50% for 25 years on the purchase of a home. What is the monthly payment?

20. The Saddlers are purchasing a $60,000 home. Everson Savings and Loan Association will finance the purchase at 13% for 30 years if the Saddlers can make a 20% down payment. What is the monthly payment?

MONTHLY PAYMENT FOR A $1000 LOAN

Annual Interest Rate	Length of Loan (Years) 20	25	30
10.00%	$ 9.66	$ 9.09	$ 8.78
10.50%	9.99	9.45	9.15
11.00%	10.33	9.81	9.53
11.50%	10.67	10.17	9.91
12.00%	11.02	10.54	10.29
12.50%	11.37	10.91	10.68
13.00%	11.72	11.28	11.07
13.50%	12.08	11.66	11.46

Terms

Match each term with its definition on the right.

21. Mortgage loan
22. Closing costs
23. Real estate taxes
24. Premium
25. Principal
26. Utilities costs

a. An amount paid for an insurance policy

b. A loan whereby the lender has the right to sell the property if payments are not made

c. Fees paid at the time documents are signed transferring ownership of a home

d. Charges for public services

e. An amount owed upon which interest charged is calculated

f. Fees collected on the ownership of property used to support the operation of government

Refer to your reference files at the back of the book if you need help.

A Microcomputer Application Mortgage Payments

To complete this microcomputer application, you will need the diskette *Microcomputer Applications for Business Mathematics,* which accompanies this textbook.

Select option 4, Mortgage Payments, from the menu. Input the information in the following problems to find the monthly payment and the total amount paid for each mortgage loan.

1. Ordella and Steve Jeremeas.
   Home selling for $36,400.
   Terms: $6400 down payment.
   a. Loan at 13% for 25 years.
   b. Loan at 13.25% for 25 years.

2. Sherman and Rose Ferrer.
   Home selling for $76,400.
   Terms: $7400 down payment.
   a. Loan at 14% for 25 years.
   b. Loan at 13.50% for 30 years.

3. Mobile home selling for $19,900.
   Terms: $1900 down payment.
   a. Loan at 14.25% for 15 years.
   b. Loan at 15.00% for 12 years.

4. Mobile home selling for $23,500.
   Terms: $400 down payment.
   a. Loan at 15.00% for 10 years.
   b. Loan at 14.50% for 12 years.

5. Estate for $240,000.
   Terms: 10% down payment.
   a. Loan at 12.75% for 30 years.
   b. Loan at 12.75% for 35 years.

6. Farmhouse for $175,000.
   Terms: 20% down payment.
   a. Loan at 16.25% for 25 years.
   b. Loan at 16.00% for 23 years.

7. Clarence Dudek plans to purchase a home for $46,900. He has an $8900 down payment. First Federal will give him a loan at 14.25% for 25 years. Baltimore Trust Company will give him a loan at 14.75% for 20 years. Find the monthly payment and total amount paid for each loan.

8. Marcia Hannaford plans to purchase a commercial property for $129,000. She has a $24,000 down payment. Hamilton Mortgage Company will give her a loan at 15.51% for 20 years. Grogan Investors will give her a loan at 14.92% for 18 years. Find the monthly payment and total amount paid for each loan.

9. Marlin Breno plans to purchase a duplex for $85,400. After a 15% down payment, he plans to get a 30-year loan. Find the monthly payment and total amount paid for a loan at:
   a. 14%
   b. 14.50%

10. Joyce Irons plans to purchase a condominium for $86,000. After a 20% down payment, she plans to finance the remainder for 20 years. Find the monthly payment and total amount paid for a 12% loan and a 12.25% loan.

Unit Test

Lesson 9-1

1. The Dennisons are purchasing a $75,000 home. What is the amount of the mortgage loan needed to finance the purchase if they wish to make a 15% down payment?

Lesson 9-2

2. The Cataldis have a $50,000 mortgage loan from Home Loan Bank at an interest rate of 11% for 25 years. Use the table on page 569 to find the monthly payment and the total interest charged.

Lesson 9-3

3. The Crawfords were granted a $55,000 mortgage loan. At the time of the closing, they paid the closing costs shown. What is the total of the closing costs?

Credit report: $50
Loan origination fee: 2.1% of loan
Abstract of title: $75
Attorney fees: $200
Property tax: $717.55

Lesson 9-4

4. The Loves obtained a 30-year, $65,000 mortgage loan at Swan Creek Bank at an interest rate of 13.50%. The monthly payment is $744.90. What is the new principal after the first monthly payment?

Lesson 9-5

5. The Dixon County tax assessor stated that the market value of the Sitlow estate is $725,000. The rate of assessment in Dixon County is 30% of the market value. The tax rate is 50.131 mills. What is the annual real estate tax on the Sitlow estate?

Lesson 9-6

6. The Wilsons have insured their home for 80% of its replacement value of $85,000. The insurance company states that the amount of coverage on their personal property is 50% of the amount of coverage on their home. What is the amount of coverage on the Wilsons' personal property?

Lesson 9-7

7. The replacement value of the Halls' townhouse is $60,000. They insured it for 100% of its replacement value. The townhouse is of masonry veneer and is rated in fire protection class 8. Use the table of annual premiums on page 229 to find their annual homeowner's policy premium.

Lesson 9-8

8. The Delacours have a combined monthly net income of $2955.80. Their housing expenses for June are listed below. Were their housing expenses for the month within the FHA recommendations?

Mortgage payment:	$600.75	Telephone:	$27.19
Insurance:	$ 33.35	Heating oil:	$81.21
Property taxes:	$166.67	Electricity:	$55.65
Water/sewer service:	$ 13.23	Furniture payment:	$48.85

Unit 10 Insurance and Investments

▸ *Health* and *life insurance* protect you and your dependents against financial losses in case of illness or death. *Whole life* and *endowment life* insurance offer a savings feature. *Certificates of deposit, stocks,* and *bonds* are other types of *investments,* which often earn more interest than money in a bank account. A certificate of deposit is purchased for specific amounts that remain on deposit in a bank for a specified period of time. A stock is a *share* in a company. The company may pay you a return on your investment in the form of *dividends.* A bond is money you loan to a company for a specified period. When the loan is due, you are paid the *face value* of the bond, which includes the interest on your original investment.

The main purpose of life insurance is to provide income for dependents.

10-1 Health Insurance Premiums

Objective

Compute health insurance premiums.

An accident or illness could cut off your income, wipe out your savings, and leave you in debt. To protect against overwhelming medical expenses, many people have **health insurance.** A **basic plan** includes hospital and surgical-medical insurance. A **comprehensive plan** includes hospital, surgical-medical, and major medical insurance.

You can get health insurance by joining a **group plan** where you work. Your employer may pay part or all of the premium. Health insurance companies also offer **nongroup plans** for people not enrolled in group plans. You can choose an individual plan that covers only yourself or, for a larger premium, a family plan that covers yourself, your spouse, and your children.

▸ **Employee's Percent = 100% − Employer's Percent**

▸ **Employee's Contribution = Total Premium × Employee's Percent**

Example

Skills <u>**11, 30**</u> **Application** <u>**K**</u> **Term** <u>**Health insurance**</u>

Stanley Stefanic is employed by Kindle Company. He has a family membership in the group comprehensive medical insurance program. The annual premium includes \$1225 for hospital insurance, \$386 for surgical-medical insurance, and \$71 for major medical insurance, a total of \$1682. Stanley's employer pays 70% of the total cost. His contribution is deducted monthly from his paycheck. What is Stanley's annual contribution? What is his monthly deduction?

Solution

1. Find the **employee's percent.**
   100% − 70% = 30%

2. Find the **employee's contribution.**
   Total Premium × Employee's Percent
   \$1682 × 30% = \$504.60

3. Find the **employee's monthly deduction.**
   Employee's Contribution ÷ 12
   \$504.60 ÷ 12 = \$42.05 monthly deduction

A. \$169.00
B. \$14.08

▸ **Self-Check** Cover the answers and complete the problems.

Annual premium includes \$602 for hospital, \$201 for surgical-medical, and \$42 for major medical. The company pays 80% of the cost.

A. Find the employee's total annual contribution.

B. Find the employee's monthly deduction.

Problems

Plan	Annual Premium for: Hospital	+	Surgical-Medical	+	Major Medical	=	Total Annual Premium	Employer's Percent	Employee's Annual Contribution
1. Single	$ 495	+	$142	+	$28	=	$665	60%	?
2. Single	$ 511	+	$158	+	$33	=	?	80%	?
3. Family	$1260	+	$411	+	$79	=	?	75%	?
4. Family	$1142	+	$380	+	$62	=	?	90%	?
5. Family	$1252	+	$398	+	$76	=	?	40%	?

6. Jack Noble.
   Annual health insurance premium:
   Hospital $496.
   Surgical-medical $142.
   Major medical $38.
   Employer pays 85% of cost.
   What is the total annual premium?
   How much does Jack pay annually?
   How much is deducted from his weekly paycheck?

7. Irene Ferrine.
   Annual health insurance premium:
   Hospital $1059.
   Surgical-medical $350.
   Major medical $69.
   Employer pays 65% of cost.
   What is the total annual premium?
   How much does Irene pay annually?
   How much is deducted from her semimonthly paycheck?

8. Susan Feldman has a family membership in the company's group medical insurance program. The total cost is $1817. Susan's employer pays 75% of the total cost. Her contribution is deducted biweekly from her paycheck. How much is her annual contribution? How much is her biweekly deduction?

9. Roger and Dolores Campbell are students at the state university. The comprehensive health insurance rate is $25.94 per month for each student. The premiums are paid quarterly (every three months). What does Roger's and Dolores's health insurance cost per quarter?

10. Employees of the General Manufacturing Company may sign up for a group vision care health insurance at a cost of $1.00 per semimonthly (twice a month) pay period. The insurance actually costs the company $60 per employee per year. What percent of the cost of the insurance does the company pay?

Maintaining Your Skills

Look up the skills in parentheses if you need help or more practice.

Divide. Round answers to the nearest hundredth. (Skill 11)

11. 10.42 ÷ 7.6 **12.** 2.918 ÷ 0.42 **13.** 0.00318 ÷ 0.06 **14.** 7.14 ÷ 26

Find the percentage. (Skill 30)

15. 55% × 235 **16.** 14.2% × 842 **17.** 13.65% × 2300 **18.** 4.81% × 3710

10-2 Health Insurance Benefits

Objective Compute the amount paid by the patient for medical care.

A comprehensive health insurance plan includes three kinds of insurance. **Hospital insurance** pays most of the cost of hospitalization, including a semiprivate room and most laboratory tests. **Surgical-medical** insurance pays your doctor's fee for surgery, up to a certain amount. **Major medical** insurance is designed to protect you against the other costs of a serious and expensive illness. Most major medical policies have a **deductible clause.** A $250 deductible clause means that you must pay the first $250 of the amount not covered by your hospital and surgical-medical insurance. Many major medical policies also have a **coinsurance clause.** A typical one states that you must pay 20% of the amount remaining after you pay the deductible. Your insurance company pays the other 80%.

▶ **Amount Paid by Patient = Deductible + Coinsurance Amount**

Example

Skills <u>4, 30</u> **Application** <u>A</u> **Term** <u>Coinsurance</u>

Mary Greene's comprehensive medical insurance includes hospital, surgical-medical, and major medical insurance. The major medical policy has a $250 deductible and a 20% coinsurance clause. When Mary had surgery, her total bill for hospital care and physician's fee was $5350. The hospital and surgical-medical provisions of her insurance policy covered $3950. What amount did Mary pay?

Solution

1. Find the **amount not paid by hospital and surgical-medical insurance.**
   $5350 − $3950 = $1400
2. Find the **amount subject to coinsurance.**
   $1400 − $250 deductible = $1150
3. Find the **coinsurance amount paid by patient.**
   20% × $1150 = $230
4. Find the **total amount paid by patient.**
   Deductible + Coinsurance Amount
   $250 + $230 = $480 paid by patient

A. $1350
B. $370

▶ **Self-Check** Cover the answers and complete the problems.

Bambi Jones had a total medical bill of $26,300. Her insurance covered $24,850. Her medical policy has a $100 deductible and a 20% coinsurance clause.

A. Find the amount subject to coinsurance.

B. Find the total amount Bambi had to pay.

Problems

	Total Medical Bill	Amount Not Paid by Hospital and Surgical-Medical Insurance	Major Medical Deductible	Amount Subject to Coinsurance	Coinsurance Rate	Amount Paid by Patient	
						Coinsurance	Total
1.	$ 14,500	$1100	$250	$850	20%	$170	?
2.	$ 62,040	$1970	$100	?	20%	?	?
3.	$ 8215	$ 25	$250	?	20%	?	?
4.	$140,000	$1050	$500	?	20%	?	?

5. Mary Lopes.
Total bill is $18,240.
Amount paid by hospital and surgical-medical insurance is $16,495.
Major medical has $250 deductible and 20% coinsurance clause.
How much is subject to coinsurance?
What is the total amount Mary pays?

6. Louis Carbo.
Total bill is $32,985.
Amount paid by hospital and surgical-medical insurance is $32,041.
Major medical has $500 deductible and 20% coinsurance clause.
How much is subject to coinsurance?
What is the total amount Louis pays?

7. Wanda Young has hospital and surgical-medical insurance but no major medical. Her total bill for hospital care and doctor's fee for a tonsillectomy was $1190. The hospital and surgical-medical provisions of her insurance policy covered $950 of the total. What did Wanda pay?

8. Rachel Orsini's comprehensive medical insurance includes hospital, surgical-medical, and major medical insurance. The major medical policy has a $250 deductible and a 20% coinsurance clause. When Rachel had surgery, her total bill for hospital care and physician's fee was $5972. The hospital and surgical-medical provisions of her insurance policy covered $3912. What amount did Rachel pay?

Maintaining Your Skills

Look up the skills in parentheses if you need help or more practice.

Find the percentage. Round answers to the nearest hundredth. (Skill 30)

9. 20% of $355
10. 31.4% of 858
11. $\frac{3}{4}$% of 86

Subtract. (Skill 4)

12. 942 − 865
13. 76,325 − 66,842
14. 1,250,000 − 942,876
15. 842 − 829
16. 42,321 − 12,966
17. 693,450 − 99,890

10-3 Term Life Insurance

Objective Use tables to compute the annual premium for term life insurance.

The main purpose of **life insurance** is to provide financial protection for your dependents in case of your death. **Term life insurance** is the least expensive form of life insurance that you can buy. You buy term insurance for a specified term such as 1 year, 5 years, or to a specified age. Unless you renew your policy at the end of each term, the insurance coverage ends.

If you should die during the term of the policy, your **beneficiary,** the person you name in the policy, will receive the **face value** of the policy. The face value is the amount of insurance coverage that you buy. The annual premium depends on your age at the time you buy the policy and the number of **units.** One unit of insurance has a face value of $1000. The annual premium for term life insurance usually increases with each new term.

ANNUAL PREMIUMS PER $1000 OF LIFE INSURANCE: 5-YR TERM		
AGE	MALE	FEMALE
18	$ 2.64	$ 2.25
20	2.66	2.27
25	3.07	2.39
30	4.29	3.85
35	5.61	4.95
45	11.11	10.65
55	23.60	22.01
65	42.77	35.42

▶ **Annual Premium = Number of Units Purchased × Premium per $1000**

Example

Skills 11, 8 **Application** C **Term** Term life insurance

John Corbett is 20 years old. He wants to purchase a $30,000, 5-year term life insurance policy. What is his annual premium?

Solution

1. Find the **number of units purchased.**
   $30,000 ÷ $1000 = 30
2. Find the **premium per $1000.**
   Refer to the above table. $2.66
3. Find the **annual premium.**
   Number of Units Purchased × Premium per $1000
   30 × $2.66 = $79.80 annual premium

A. $154.00
B. $3421.60

▶ **Self-Check** Cover the answers and complete the problems.

Use the above table to find the annual premium for a 5-year term policy.

A. $40,000, female, age 30.

B. $80,000, male, age 65.

Problems

Use the table on page 242 to answer the following.

	Insured	Age	Coverage	Number of Units	×	Annual Premium per $1000	=	Annual Premium
1.	Lisa Winslow	18	$ 10,000	10	×	$2.25	=	?
2.	Ken Reeves	35	$ 20,000	?	×	?	=	?
3.	Loni Ritter	55	$ 25,000	?	×	?	=	?
4.	Alex Barnes	25	$200,000	?	×	?	=	?

5. Jim Carew's insurance policy.
5-year term life insurance.
$45,000 coverage.
He is 25 years old.
What is his annual premium?

6. Tina Lead's insurance policy.
5-year term life insurance.
$80,000 coverage.
She is 45 years old.
What is her annual premium?

7. Janet Crosby wants to purchase a $35,000, 5-year term life insurance policy. She is 55 years old. What is her annual premium?

8. Sam and Mary Chapman have 1 child. Sam is a career counselor and Mary is a child psychologist. Sam is 30 years old and Mary is 25 years old. Both want to purchase $30,000, 5-year term life insurance policies. What is Mary's annual premium? What is Sam's annual premium?

9. Robert and Abigail Duncan have 2 children. Robert was 30 years old when he first purchased a $75,000, 5-year term life insurance policy. He will be 35 years old this year and his insurance agent has informed him that his policy comes up for renewal. What will be his annual premium if he renews his policy? What total amount did he pay during the previous 5-year term? What total amount will he pay for the next 5-year term?

Maintaining Your Skills

Look up the skills in parentheses if you need help or more practice.

Divide. Round answers to the nearest thousandth. (Skill 11)

10. 9.316 ÷ 6.12
11. 36.36 ÷ 0.14
12. 1.825 ÷ 0.05
13. 2.624 ÷ 1000
14. 43.40 ÷ 100
15. 3400 ÷ 1000
16. 65,550 ÷ 10,000
17. 79.85 ÷ 10
18. 9.62 ÷ 100

Multiply. Round answers to the nearest hundredth. (Skill 8)

19. 42.162 × 100
20. 0.45 × 0.06
21. 2.59 × 0.045
22. 32.6 × 12.914
23. 0.032 × 100
24. 1.47 × 1000
25. 689 × 1000
26. 360 × 2500
27. 3.62 × 0.10

10-4 Other Types of Life Insurance

Objective Use tables to compute annual premiums for three types of life insurance.

Whole life insurance offers financial protection for your entire life. The premium usually remains the same each year. In addition to the face value that your beneficiary will receive, whole life insurance has a **cash value** and a **loan value.** The cash value is the amount of money you will receive if you decide to cancel your policy. The loan value, usually the same amount as the cash value, is an amount that the insurance company will loan you if you request it. **Limited-payment life insurance** also offers lifetime protection. It has a cash value and a loan value. You pay premiums only for a specified number of years or until you reach a certain age.

Endowment life insurance is a combination of life insurance and a savings plan. The policy covers you for a specified number of years. If you live beyond that time, you receive the cash value of the policy.

ANNUAL PREMIUMS PER $1000 LIFE INSURANCE

Age M	Age F	TERM TO AGE 65	WHOLE LIFE	20-YEAR PAYMENT	20-YEAR ENDOWMENT
20	25	$ 7.01	$13.61	$25.16	$45.22
25	30	7.87	15.67	27.63	45.49
30	35	9.03	18.30	30.60	46.04
35	40	10.66	21.74	33.97	47.42
40	45	12.93	26.25	38.69	49.13
45	50	15.88	32.15	44.49	52.56
50	55	20.27	39.92	51.38	57.23
55	60	28.43	49.49	60.15	63.00

▶ **Annual Premium = Number of Units Purchased × Premium per $1000**

Example

Skills 11, 8 **Application** C **Term** Whole life insurance

Susan Derby is 45 years old. She wants to purchase a 20-year endowment policy valued at $25,000. What is her annual premium?

Solution

1. Find the **number of units purchased.**
   $25,000 ÷ $1000 = 25
2. Find the **premium per $1000.**
   Refer to the above table. $49.13
3. Find the **annual premium.**
   Number of Units Purchased × Premium per $1000
   25 × $49.13 = $1228.25 annual premium

A. $2745
B. $2371

▶ **Self-Check** Cover the answers and complete the problems.

Find the annual premium.

A. 30-year-old male. $150,000 whole life policy.

B. 40-year-old female. 20-year endowment policy of $50,000.

Problems

Use the table on page 244 to solve the following.

1. Ann Carroll's insurance policy. $50,000, 20-year payment life. She is 40 years old. What is her annual premium?

2. James Corbin's insurance policy. $35,000 term, to age 65. He is 20 years old. What is his annual premium?

3. Bill Dolby is 30 years old and wants to purchase a $40,000 life insurance policy. He is considering a 20-year, limited-payment life insurance policy. What is his annual premium?

4. Maxwell Webster is 40 years old. He wants to purchase a $35,000, 20-year endowment policy. What is his annual premium? How much will he pay in total premiums in the next 20 years?

5. Terry and Lucy Moore each purchase a $30,000 whole life insurance policy. Both are 25 years of age. What is their total annual premium? How much more is Terry's annual premium than his wife's?

6. Margaret Young purchased a $25,000, 20-year limited-payment life insurance policy at the age of 35. What is her annual premium? If Margaret dies at the age of 38, how much will her beneficiary receive?

Use the table on page 242 to find the annual premium per $1000.

*7. **A Brief Case** Rather than make 1 annual payment, many insurance companies will allow you to make 2, 4, or 12 smaller payments. Due to the additional expense of collecting and handling the payments several times a year, a small fee is charged. Many companies use the guidelines shown below.

Janice Allen is 20 years old and has purchased a $20,000, 5-year term life insurance policy. She wants to make quarterly payments. What are her quarterly payments? How much can she save in 1 year by paying the annual premium?

Optional Payment Plans Percent of Annual Premium	
Semiannual Premiums	= 50.5%
Quarterly Premiums	= 25.5%
Monthly Premiums	= 8.5%

Maintaining Your Skills

Look up the skills in parentheses if you need help or more practice.

Divide. Round answers to the nearest hundredth. (Skill 11)

8. 14.4 ÷ 0.032
9. 27.614 ÷ 15.62
10. 0.088 ÷ 1.9
11. 144.62 ÷ 12.1
12. 685 ÷ 25
13. 10,244 ÷ 120
14. 954 ÷ 9.5
15. 2500 ÷ 4.50

Multiply. (Skill 8)

16. 0.91 × 0.84
17. 6.81 × 8.1
18. 0.511 × 0.05
19. 0.671 × 1000
20. 724 × 12
21. 859 × 11
22. 2433 × 68
23. 15,875 × 25

10-5 Certificates of Deposit

Objective Use tables to compute interest on certificates of deposit.

One way to invest your money is to purchase a **certificate of deposit,** which earns interest at a higher rate than a regular savings account. You buy certificates for specific amounts such as $500 or $1000 and must leave the money on deposit for a specified time, ranging up to 10 years. You are penalized for early withdrawal. Most certificates of deposit earn interest compounded quarterly. Although banks use computers to calculate interest earned, you can use a table to compute the interest earned.

AMOUNT PER $1.00 INVESTED, SEMIANNUAL, QUARTERLY, AND DAILY COMPOUNDING

Annual	Interest Period—1 Year			Interest Period—4 Years		
Rate	Semiannual	Quarterly	Daily	Semiannual	Quarterly	Daily
9.75%	1.099877	1.101123	1.102397	1.463443	1.470089	1.476904
10.00%	1.102500	1.103813	1.105156	1.477455	1.484506	1.491744
10.25%	1.105127	1.106508	1.107921	1.491585	1.499055	1.506731
10.50%	1.107756	1.109207	1.110694	1.505833	1.513738	1.521870
10.75%	1.110389	1.111912	1.113473	1.520200	1.528556	1.537160
11.00%	1.113025	1.114621	1.116260	1.534687	1.543509	1.552605
11.25%	1.115664	1.117336	1.119053	1.549294	1.558600	1.568203
11.50%	1.118306	1.120055	1.121853	1.564023	1.573829	1.583959
11.75%	1.120952	1.122779	1.124660	1.578874	1.589197	1.599874
12.00%	1.123600	1.125509	1.127475	1.593848	1.604706	1.615947

▶ **Interest Earned = Amount − Original Principal**

Example

Skills 2, 6, 8 **Application** C **Term** Certificate of deposit

Richard O'Shea invests $3000 in a 1-year certificate of deposit (CD) that earns interest at an annual rate of 10.00% compounded quarterly. How much interest will he earn at the end of 1 year?

Solution

1. Find the **amount.**
   Original Principal × Amount per $1.00
   $3000.00 × 1.103813 = $3311.439 = $3311.44

2. Find the **interest earned.**
   Amount − Original Principal
   $3311.44 − $3000.00 = $311.44 interest earned

A. $925.31
B. $23,845.20

▶ **Self-Check** Cover the answers and complete the problems.

Find the interest earned.

A. $8000 CD for 1 year. 11.25% annual interest rate compounded semiannually.

B. $50,000 CD for 4 years. 9.75% annual interest rate compounded daily.

Problems

Use the table on page 246 to find the amount per $1.00 invested.

	Annual Rate	Interest Period	Principal	×	Amount per $1.00	=	Amount	Interest Earned
1.	11.00%	1 year quarterly	$ 2500	×	1.114621	=	?	?
2.	10.25%	4 years daily	$ 15,000	×	?	=	?	?
3.	12.00%	4 years semiannually	$ 1000	×	?	=	?	?
4.	9.75%	1 year quarterly	$120,000	×	?	=	?	?

5. $3000 certificate of deposit.
Interest period of 1 year.
10.75% annual interest rate.
Compounded quarterly.
What is the amount at maturity?
What is the interest earned?

6. $3500 certificate of deposit.
Interest period of 4 years.
11.50% annual interest rate.
Compounded semiannually.
What is the amount at maturity?
What is the interest earned?

7. Doris Calloway purchased a 1-year certificate of deposit for $500 that earns interest at a rate of 9.75% compounded daily. What is the amount of the certificate at maturity? What is the interest earned?

8. Matthew and Clara Fuller purchased a 1-year certificate of deposit for $8500. The Fullers' certificate earns interest at a rate of 11.75% compounded daily. What is the amount of the certificate at maturity? What is the interest earned?

9. The Bradleys have $7500 that they want to invest in a certificate of deposit. Granite Trust offers a 4-year certificate that earns interest at a rate of 11.50% compounded quarterly. Hancock Cooperative Bank offers a 4-year certificate that earns interest at a rate of 11.25% compounded daily. Which certificate earns the most interest? By how much?

Maintaining Your Skills

Look up the skills in parentheses if you need help or more practice.

Round to the nearest ten. (Skill 2)

10. 38 11. 122 12. 1043 13. 207 14. 1459 15. 496 16. 5399

Subtract. (Skill 6)

17. 0.093 − 0.042 18. 921.6283 − 236.84 19. 5.135 − 0.594

Multiply. (Skill 8)

20. 0.473 × 9 21. 122 × 1043 22. 16.39 × 84.76 23. 9.61 × 1000

10-6 Effective Annual Yield

Objective

Compute the effective annual yield.

Many banks advertise not only the annual interest rates, but also the **effective annual yields** of their certificates of deposit. The annual yield is the rate at which your money earns simple interest in one year. The interest earned at the annual rate is affected by the frequency of compounding. The effective annual yield is not affected, so you can use it to compare investments.

▶ **Effective Annual Yield** = $\frac{\textbf{Interest for One Year}}{\textbf{Principal}}$

Example

Skills 11, 26 **Application** A **Term** Annual yield

Donnel Johnson invested $5000 in a certificate of deposit for 3 years. The certificate earns interest at an annual rate of 10.50% compounded quarterly. What is the effective annual yield to the nearest hundredth of a percent?

Solution

1. Find the **interest for one year.** (Refer to the table on page 246.)
   Amount − Principal
   ($5000 × 1.109207) − $5000
   $5546.035 − $5000 = $546.035

2. Find the **effective annual yield.**
   Interest for One Year ÷ Principal
   $546.035 ÷ $5000.00 = 0.109207 = 10.9% effective annual yield

A. 9.99%
B. 12.55%

▶ **Self-Check** Cover the answers and complete the problems.

Find the effective annual yield. (Refer to the table on page 246.)

A. $10,000 certificate of deposit. 9.75% annual interest rate compounded semiannually.

B. $2500 certificate of deposit. 12% annual interest rate compounded quarterly.

Problems

Use the table on page 246 to find the amount per $1.00 invested.

	Annual Rate	Interest Period	Principal	×	Amount per $1.00	=	Amount	Interest Earned	Effective Annual Yield
1.	11.25%	1 year quarterly	$4100	×	1.117336	=	?	?	?
2.	12.00%	1 year quarterly	$6000	×	?	=	?	?	?

	Annual Rate	Interest Period	Principal ×	Amount per $1.00	= Amount	Interest Earned	Effective Annual Yield
3.	9.75%	1 year semiannual	$50,000 ×	?	= ?	?	?
4.	11.25%	1 year daily	$ 800 ×	?	= ?	?	?

5. $7000 certificate of deposit.
   9.75% annual interest rate compounded quarterly.
   4-year certificate.
   What is the effective annual yield?

6. $2500 certificate of deposit.
   10.75% annual interest rate compounded daily.
   $2\frac{1}{2}$-year certificate.
   What is the effective annual yield?

7. Ollie Durance invested $5000 in a certificate of deposit of 8 years at an annual interest rate of 10.75% compounded daily. What is the interest earned for 1 year? What is the effective annual yield?

8. Ellie Raphael invested $3750 in a 4-year certificate of deposit that earns interest at a rate of 11.25% compounded quarterly. What is the interest earned for 1 year? What is the effective annual yield?

The interest in the Golden Dollar Accounts is compounded quarterly.

GOLDEN DOLLAR ACCOUNTS		
Terms Available	Interest Rate	Minimum Deposit
1–$2\frac{1}{2}$ yr.	9.75%	$500
$2\frac{1}{2}$–4 yr.	10%	$1000
4–5 yr.	10.75%	$1500
6–7 yr.	11%	$2000
8–10 yr.	11.50%	$2500

9. Patricia Gibson invested $3000 in a Golden Dollar Account that matures in 6 years. What is the interest rate that the bank will give? What is the interest earned for 1 year? What is the effective annual yield?

10. Wallace Mitchell invested $5000 in a Golden Dollar Account that matures in 1 year. What is the interest rate that the bank will give? What is the interest earned at maturity? What is the effective annual yield?

11. Benjamin White invested $1000 in a Golden Dollar Account for 4 years. What is the interest rate that the bank will give? What is the interest earned at maturity? What is the effective annual yield?

Maintaining Your Skills

Look up the skills in parentheses if you need help or more practice.

Write as percents. Round answers to the nearest hundredth. (Skill 26)

12. 0.112741
13. 0.089146
14. 1.25642
15. 25.5
16. 2.0

Divide. (Skill 11)

17. 1500 ÷ 0.05
18. 28.42 ÷ 1.2
19. 679.2 ÷ 0.02
20. 3.60 ÷ 7.2

10-7 Stocks

Objective Compute the total cost of a stock investment.

You can invest your money in shares of **stocks.** Stock prices are published each day in the business sections of daily newspapers. Stock prices are usually quoted in eighths of a dollar.

Number of shares sold today, in hundreds

Highest and lowest price paid during the day

Difference between last sale today and last sale yesterday

Abbreviation of company name

Price paid for the last sale today

52-week High	52-week Low	Stock	Div.	Yld.	PE	Sales 100s	High	Low	Last	Net Chg.
$39\frac{3}{4}$	$29\frac{1}{8}$	Bastek	2.24	6.9	6	68	$32\frac{3}{4}$	$32\frac{1}{4}$	$32\frac{1}{4}$	$-\frac{3}{4}$
$18\frac{3}{8}$	$14\frac{1}{2}$	BCA Co	1.24	8.5	6	187	15	$14\frac{5}{8}$	$14\frac{5}{8}$	$-\frac{3}{8}$
25	13	Beta	.28	2.1	12	432	$13\frac{3}{4}$	$13\frac{1}{4}$	$13\frac{1}{4}$	$-\frac{5}{8}$
$12\frac{1}{4}$	$9\frac{1}{8}$	BF Low	1	9.5	...	352	$10\frac{3}{4}$	$10\frac{1}{2}$	$10\frac{1}{2}$	$-\frac{1}{8}$
42	33	Bigley	1.82	5.5	6	845	$33\frac{1}{4}$	$32\frac{1}{2}$	$33\frac{1}{8}$	$+\frac{1}{8}$

When you purchase a share of stock, you become a part owner of the corporation that issues the stock. You may receive a **stock certificate** as proof of ownership. The total amount you pay for the stock depends on the cost per share, the number of shares you purchase, and the **stock-broker's** commission.

▶ **Cost of Stock = Number of Shares × Cost per Share**

▶ **Total Paid = Cost of Stock + Commission**

Example

Skill 14 Application A Term Stocks

Julia Reinholt purchased 100 shares of stock at $\$39\frac{5}{8}$ per share. Her stockbroker charged a $35 commission. What is the total amount that Julia paid for the stock?

Solution

1. Find the **cost of stock.**
   Number of Shares × Cost per Share
   100 × $\$39\frac{5}{8}$
   100 × \$39.625 = \$3962.50

2. Find the **total paid.**
   Cost of Stock + Commission
   \$3962.50 + \$35.00 = \$3997.50 total paid

A. \$11,767
B. \$4296.50

▶ **Self-Check** Cover the answers and complete the problems.

Find the total paid.

A. 150 shares of stock at \$78 per share. Commission of \$67.

B. 500 shares of stock at $\$8\frac{1}{2}$ per share. Commission of \$46.50.

Problems

	Company	(Number of Shares	×	Cost per Share	=	Cost of Stock)	+	Commission	=	Total Paid
1.	AvGlo	(100	×	$7\frac{1}{8}$	=	$712.50)	+	$ 21.35	=	?
2.	Peasley	(1000	×	$30\frac{1}{2}$	=	?)	+	$150.25	=	?
3.	ICN Br	(500	×	$23\frac{1}{4}$	=	?)	+	$118.13	=	?
4.	HM Co	(255	×	$59\frac{5}{8}$	=	?)	+	$147.50	=	?

5. Carol Turrell.
500 shares of Air Genoa.
Cost per share: $26.50.
Commission is $132.40.
What is the cost of the stock?
What is the total paid?

6. Thomas Yee.
215 shares of Dianon Tire.
Cost per share: $40.125.
Commission is $40.15.
What is the cost of the stock?
What is the total paid?

7. Trudy Watson purchased 208 shares of KDI Petroleum at $10.75 per share. A broker's commission of $63.75 was charged. What was the cost of the stock? What was the total paid?

8. Thelma and Larry Mollier purchased 300 shares of Nielson Publishing at $9\frac{3}{8}$ per share and 150 shares of Ace Electronics at $18\frac{3}{4}$ per share. The broker charged a 1% commission. What was the total paid?

9. Dick Weiner wants to purchase either 100 shares of Paxtell Scientific at $38\frac{1}{8}$ per share or 350 shares of Callon Synergistics at $10\frac{7}{8}$ per share. Both companies show promising growth. The broker charges a 2% commission. Which purchase costs less? How much less?

10. Hamilton Investment Group charges a fee plus commission for stock transactions. Lillias May purchases 800 shares of stock at $11\frac{1}{4}$ per share through Hamilton Investment Group. What is the cost of the stock? What is the total paid?

HAMILTON INVESTMENT GROUP	
Sales	Commission Rate
Up to $4999	1%
$5000–$9999	1.25%
Over $10,000	1.5%
LOWEST FEES IN TOWN $49.98 plus commission	

11. Lisa Kittle purchases 500 shares of stock at $23\frac{5}{8}$ per share through Hamilton Investment Group. What is the total paid?

Maintaining Your Skills

Look up the skills in parentheses if you need help or more practice.

Change the fractions to decimals. (Skill 14)

12. $14\frac{7}{8}$ 13. $4\frac{1}{8}$ 14. $8\frac{3}{4}$ 15. $16\frac{3}{8}$ 16. $21\frac{1}{2}$

Multiply. (Skill 8)

17. $25.50 × 300 18. $52.75 × 242 19. $56.00 × 112 20. $32.10 × 68

10-8 Stock Dividends

Objective Compute the annual yield and annual dividend of a stock investment.

When you buy stocks, you may receive **dividends.** Dividends are your return on your investment. You may receive an amount specified by the corporation for each share of stock that you own. The **annual yield** is the dividend expressed as a percent of the cost of the stock. The higher the annual yield, the greater the return on your investment.

▶ **Annual Dividend = Annual Dividend per Share × Number of Shares**

▶ **Annual Yield** $= \dfrac{\textbf{Annual Dividend per Share}}{\textbf{Cost per Share}}$

Example

Skills 11, 26 **Application** A **Term** Dividends, annual yield

Jane Voy bought 60 shares of stock at $32.50 per share. The company paid annual dividends of $1.75 per share. What is the annual dividend? What is the annual yield to the nearest hundredth of a percent?

Solution

1. Find the **annual dividend.**
   Annual Dividend per Share × Number of Shares
   $1.75 × 60 = $105.00

2. Find the **annual yield.**
   Annual Dividend per Share ÷ Cost per Share
   $1.75 ÷ $32.50
   0.053846 = 5.38% annual yield

$24, 5%

▶ **Self-Check** Cover the answer and complete the problem.

Find the annual dividend and the annual yield.
120 shares at $4.00 cost per share. $0.20 annual dividend per share.

Problems

	Annual Dividend per Share	÷	Cost per Share	=	Annual Yield
1.	$3.20	÷	$60.00	=	?
2.	$0.32	÷	$ 5.50	=	?
3.	$4.80	÷	$40.00	=	?
4.	$5.10	÷	$80.00	=	?

Round the annual yield to the nearest hundredth of a percent.

5. Purchase price: $24.25 per share. Dividends were $1.50 per share. What is the annual yield?

6. Purchase price: $71.375 per share. Dividends were $3.50 per share. What is the annual yield?

7. Ada Thomas.
Owns 500 shares of MPD Systems.
Purchase price: $\$17\frac{3}{4}$ a share.
Dividends were $1.00 per share.
What were her annual dividends?
What is the annual yield?

8. William Ellet.
Owns 850 shares of Moran Funds.
Purchase price: $\$6\frac{5}{8}$ a share.
Dividends were $0.85 a share.
What were his annual dividends?
What is the annual yield?

9. Marjorie Raciti bought 150 shares of Apex Corporation stock at $9.00 per share. Last year the company paid annual dividends of $1.00 per share. What were her annual dividends? What is the annual yield?

10. Barry Monchick purchased 200 shares of Computech stock at $\$40\frac{1}{8}$ per share. Recently the company paid annual dividends of $2.35 per share. What were his annual dividends? What is the annual yield?

11. Gordon Halbertson owns 1000 shares of Baywater Energy stock, which he purchased at $\$3\frac{7}{8}$. Recently he read that the average selling price of his stock was $18. The company paid annual dividends of $0.90 per share last year. What is the annual yield on his stock? For an investor who purchased the stock at $18 per share, what is the annual yield?

12. The Krupps own 300 shares of Balwin Aviation, which they purchased at $17.50 per share. What was the dividend per share paid by the company last year? What were the Krupps' annual dividends? What is the annual yield?

52 Weeks High	Low	Stock	Div.	High	Low	Close	Net Chg.
$12\frac{1}{4}$	$7\frac{7}{8}$	ASTEX	.50	$8\frac{3}{4}$	$8\frac{5}{8}$	$8\frac{3}{4}$	$+\frac{1}{4}$
$24\frac{1}{4}$	$5\frac{3}{8}$	BANLINE		$23\frac{1}{8}$	$22\frac{5}{8}$	$22\frac{7}{8}$	....
$42\frac{3}{4}$	23	BALW AVI	.32	$42\frac{3}{4}$	$42\frac{1}{4}$	$42\frac{3}{4}$	$+\frac{3}{4}$
$3\frac{1}{8}$	$1\frac{7}{8}$	BYTE TK		$2\frac{1}{4}$	$2\frac{1}{4}$	$2\frac{1}{4}$	....

Use the table on page 246 to find the amount per $1.00 of compound interest.

*13. **A Brief Case** Walter Diaz purchased a 1-year, $5000 certificate of deposit that earns interest at a rate of 11.75% compounded daily. He also owns 175 shares of River Corporation, which he purchased at $\$24\frac{1}{8}$. Last year the company paid annual dividends of $1.75 per share. What is the interest earned on his certificate of deposit? What is the effective annual yield on the CD? What were his dividends on his stock? What is the annual yield? Which investment produced a greater annual yield?

Maintaining Your Skills

Look up the skills in parentheses if you need help or more practice.

Divide. Round answers to the nearest hundredth. (Skill 11)

14. 25.0 ÷ 200 15. 42.61 ÷ 30 16. 0.687 ÷ 0.008 17. 26.7 ÷ 1.1

Write the decimals as percents. (Skill 26)

18. 0.258 19. 0.007 20. 1.17 21. 13.42 22. 14.75

10-9 Selling Stocks

Objective

Compute the profit or loss from a stock sale.

When you sell your stocks, the sale can result in either a **profit** or a **loss.** If the amount you receive for the sale minus the sales commission is greater than the total amount you paid for the stocks, you have made a profit. If the amount you receive minus the sales commission is less than the total paid, your sale has resulted in a loss.

- **Net Sale = Amount of Sale − Commission**
- **Profit = Net Sale − Total Paid**
- **Loss = Total Paid − Net Sale**

Example

Skills 1, 6, 8 **Application** A **Term** Profit or loss

Jeff Montville paid a total of $2728.13 for 75 shares of stock. He sold the stock for $41.50 a share and paid a sales commission of $37.35. What is the profit or loss from the sale?

Solution

1. Find the **net sale.**
   Amount of Sale − Commission
   ($41.50 × 75) − $37.35
   $3112.50 − $37.35 = $3075.15
2. Is **net sale** greater than **total paid?**
   Is $3075.15 greater than $2728.13? . . . . Yes
3. Find the **profit.**
   Net Sale − Total Paid
   $3075.15 − $2728.13 = $347.02 profit

A. $174.50 profit
B. $5078.25 loss

Self-Check Cover the answers and complete the problems.

Find the profit or loss.

A. Paid $2600 for 40 shares. Sold for $70 per share. Commission of $25.50.

B. Paid $25,000 for 1000 shares. Sold for $20 per share. Commission of $78.25.

Problems

	Total Paid	(Selling Price per Share	×	Number of Shares)	− Commission	=	Net Sale	Amount of Profit or Loss
1.	$3000	($38	×	100)	− $42.75	=	?	$757.25 profit
2.	$4250	($16	×	200)	− $24.90	=	?	?
3.	$3925	($$50\frac{1}{4}$	×	100)	− $58.60	=	?	?
4.	$780	($$1\frac{1}{2}$	×	208)	− $16.45	=	?	?

5. Tom and Val McGuffey.
Bought 37 shares of ITA stock.
Paid a total of \$3812.75.
Sold at $\$92\frac{3}{4}$ per share.
Paid a \$95 sales commission.
What was the profit or loss?

6. Leslie and Carl Cooke.
Bought 115 shares of Pop Top stock.
Paid a total of \$9897.50.
Sold at $\$121\frac{7}{8}$ per share.
Paid a 2.5% sales commission.
What was the profit or loss?

7. Jay Hennessey owned 200 shares of Big N stock for which he paid a total of \$8047.50. He sold his stock at \$40.25 per share and paid a sales commission of \$140. What was the net amount of the sale? What was the profit or loss from the sale?

8. Bill and Wanda Brown sold 100 shares of United Tool stock at $\$50\frac{3}{4}$ per share and paid a \$50.75 sales commission. They had originally paid a total of \$4974.53 for the stock. What was the net amount of the sale? What was the profit or loss from the sale?

9. Sara Kennedy owned 200 shares of Tyler International stock for which she paid a total of \$3762.25. She sold the stock at \$22 per share and paid a 1.5% sales commission. What was the net amount of the sale? What was the profit or loss from the sale?

10. Ira and Joan Butler sold 850 shares of Consumer Research stock at $\$11\frac{5}{8}$ per share. They paid a 2.75% sales commission. They had originally paid a total of \$9946.06 for the stock. What was the profit or loss?

11. Eric Walton owned 80 shares of NewTech Computer stock for which he paid $\$34\frac{1}{2}$ per share plus a 2% commission. He sold at $\$38\frac{5}{8}$ and paid a 3% sales commission. What was the profit or loss from the sale?

12. The Tennysons purchased 450 shares of Software Design Study stock at $\$10\frac{3}{4}$ and sold 3 years later at $\$17\frac{7}{8}$. They paid a 1.25% sales commission when they purchased the stock and paid a 3.25% sales commission when they sold. What was the profit or loss from the sale?

Maintaining Your Skills

Look up the skills in parentheses if you need help or more practice.

Which number is greater? (Skill 1)

13. 321 or 321.4 14. \$28.19 or \$29.81 15. \$209.90 or \$209.89

Subtract. (Skill 6)

16. 86.15 − 86.10 17. 54.23 − 32.91 18. \$3254.50 − \$2198.49

Multiply. (Skill 8)

19. 81.2 × 0.003 20. 100 × 0.041 21. 12.058 × 6.8

10-10 Bonds

Objective Compute the annual interest and annual yield of a bond investment.

Many corporations and governments raise money by issuing **bonds.** When you invest in a bond, you do not become a part owner of the corporation. Instead, you are lending money to the corporation or government. In return for the loan, you are paid interest. On the date the bond matures, you will receive the **face value** of the bond. The face value is the amount printed on the bond.

Corporations usually issue bonds that mature in 10 to 30 years with face values that are multiples of $1000. Bonds may sell at a **discount** for less than the face value, or at a **premium** for more than the face value. The cost of a bond is usually a percent of the face value and is referred to as the **quoted price.** A price of 90 on a $1000 bond means that the bond sells for 90% of $1000, or $900. The interest that you receive from a bond is calculated on its face value.

- **Annual Interest = Interest Rate × Face Value**
- **Bond Cost = Percent × Face Value**
- **Annual Yield = $\dfrac{\text{Annual Interest}}{\text{Bond Cost}}$**

Example

Skills 30, 11 **Application** A **Term** Bonds

Celinda Vasquez purchased a $1000 bond at the quoted price of $79\frac{1}{2}$. The bond paid interest at a rate of $6\frac{1}{2}$%. What is the annual yield, to the nearest hundredth of a percent?

Solution

1. Find the **annual interest.**
   Interest Rate × Face Value
   $6\frac{1}{2}$% of $1000 = $65.00
2. Find the **bond cost.**
   Percent × Face Value
   $79\frac{1}{2}$% of $1000 = $795.00
3. Find the **annual yield.**
   Annual Interest ÷ Bond Cost
   $65.00 ÷ $795.00 = 0.08176 = 8.18% annual yield

A. $60.00
$905.00
6.63%
B. $725.00
$8200.00
8.84%

Self-Check Cover the answers and complete the problems.

Find the interest, bond cost, and annual yield.

A. $1000 bond at $90\frac{1}{2}$. Pays 6% interest.

B. $10,000 bond at 82. Pays $7\frac{1}{4}$% interest.

Problems

	Face Value of Bond	Quoted Price	Cost of Bond	Interest Rate	Annual Interest	Annual Yield
1.	\$10,000	94	\$9400	7%	\$700	?
2.	\$ 1000	$75\frac{1}{2}$	?	$6\frac{1}{4}$%	?	?
3.	\$10,000	$82\frac{3}{8}$	?	$9\frac{1}{2}$%	?	?
4.	\$ 1000	64.324	?	$8\frac{7}{8}$%	?	?

5. Jorge Gueva.
Purchased a \$1000 bond at $80\frac{1}{4}$.
Pays 5% annual interest.
What is the annual interest?
What was the cost of the bond?
What is the annual yield?

6. Sheila Kowalski.
Purchased a \$10,000 bond at $70\frac{3}{8}$.
Pays $7\frac{1}{4}$% annual interest.
What is the annual interest?
What was the cost of the bond?
What is the annual yield?

7. Betsy and Rollin Adams purchased a \$10,000 bond at a quoted price of $90\frac{3}{8}$. The bond pays annual interest at a rate of $6\frac{3}{4}$%. What is the annual interest earned? What is the cost of the bond? What is the annual yield?

8. The Grolier Housing Authority is offering \$5000 bonds that pay 8.4% annual interest. The quoted price of each bond is 94.128. Michael Gerrard wants to purchase 4 bonds through a broker who charges a 1.75% sales commission. What is the total cost of his purchase? What total interest will he earn yearly? What is the annual yield?

*9. **A Brief Case** Many newspapers give up-to-date bond information in the financial section. In the listing shown, the name of the corporation is given, then the annual interest rate the bond is paying, the year the bond matures, and so on. '05' means the bond matures in the year 2005.

Bonds		Cur. Yld.	Vol.	High	Low	Close
DICLO	$9\frac{3}{8}08$	13.	2	72	72	72
DYER	$10\frac{7}{8}09$	13.	10	$82\frac{1}{4}$	$82\frac{1}{4}$	$82\frac{1}{4}$
EANLO	$8\frac{3}{4}00$	13.	28	$70\frac{1}{2}$	70	70
EGG CO	$12\frac{1}{4}10$	13.	3	$93\frac{5}{8}$	$93\frac{5}{8}$	$93\frac{5}{8}$

Horace Dahl purchased three \$1000 bonds of the Dyer Corporation at the closing price for the day. He had to pay a broker's commission of 2.25%. What was the total cost of his purchase? What will be his total yearly earnings? In what year will the bond mature?

Maintaining Your Skills

Look up the skills in parentheses if you need help or more practice.

Find the percentage. (Skill 30)

10. $6\frac{1}{4}$% of \$700
11. $92\frac{1}{2}$% of \$15,000
12. $\frac{3}{4}$% of 40
13. 130% of 600
14. $17\frac{1}{2}$% of 34
15. 120% of 640
16. $2\frac{1}{4}$% of 82
17. 4.1% of 11,240

Divide. Round answers to the nearest hundredth. (Skill 11)

18. $2.12\overline{)46.80}$
19. $19.5\overline{)508}$
20. $0.004\overline{)86.28}$
21. $14.14\overline{)38.215}$

Reviewing the Basics

Skills

Round to the nearest cent.

(Skill 2) 1. \$72.125 2. \$1.381 3. \$7315.333 4. \$4.996

Solve. Round answers to the nearest cent.

(Skill 8) 5. \$1011.25 × 25 6. \$2.54 × 42 7. \$928.06 × 305

8. \$8.35 × 65 9. \$62.50 × 31 10. \$122.32 × 8

(Skill 11) 11. \$707.19 ÷ 3 12. \$34.20 ÷ 8 13. \$782.52 ÷ 12

(Skill 30) 14. 33.3% of \$3914.23 15. 5% of \$160 16. 19.2% of \$1923.17

Write as a decimal.

(Skill 14) 17. $\frac{3}{16}$ 18. $\frac{5}{8}$ 19. $1\frac{7}{8}$ 20. $\frac{7}{32}$ 21. $1\frac{3}{4}$ 22. $3\frac{3}{10}$

Write as a percent.

(Skill 26) 23. 0.009 24. 0.15 25. 0.153 26. 0.0375 27. 1.03

Applications

Use the table to find the annual premium.

(Application C) 28. Bob Colgate purchased a \$15,000 whole life insurance policy at the age of 35. What is his annual premium?

29. Claudine London purchased a 20-year, \$20,000 endowment policy at the age of 45. What is her annual premium?

ANNUAL PREMIUMS PER \$1000				
Age M	Age F	Whole Life	20-Year Payment	20-Year Endowment
35	40	\$21.74	\$33.97	\$47.42
40	45	26.25	38.69	49.13
45	50	32.15	44.49	52.56

Terms

Write your own definition for each term.

30. Term life insurance 31. Whole life insurance 32. Dividends

33. Certificate of deposit 34. Annual yield 35. Stocks

36. Profit or loss 37. Bonds 38. Health insurance

Refer to your reference files at the back of the book if you need help.

Problem Solving
Constructing a Table or Diagram

Constructing a table can be a good aid in solving some types of word problems. A table helps you produce data and display it in an organized way. Patterns often emerge when the data is organized. Once you have established a pattern, you can easily fill in missing information.

Example Each year, before the U.S. Supreme Court opens, each of the 9 justices shakes hands with every other justice. How many handshakes are involved in this ceremony?

Solution Construct a diagram or table until you find a pattern. Remember that when 2 justices meet, only 1 handshake is involved.

	Number of Justices	Handshakes	
Warren and John	2	1	
Warren, John, and Sandra	3	3	+ 2
Warren, John, Sandra, and William	4	6	+ 3
Warren, John, Sandra, William, and George	5	10	+ 4

At this point, a pattern is established. How many handshakes will be added to 10 if a sixth justice shakes hands with the others? The answer is 5. Each of the 5 justices already in the group will have to shake hands with the sixth justice, for a total of 15 handshakes. The table can be filled in with the number of handshakes involved for justices 7, 8, and 9, using the pattern established.

Number of Justices	Handshakes	
5	10	
6	15	+ 5
7	21	
8	28	
9	36	

Problems

1. Lt. Moore is a recruitment officer for the Marines. For each high school graduate that he recruits, he gets 5 points; for each college graduate, he gets 11 points. He earned 94 points last week by recruiting 14 high school or college graduates. How many of each type did he recruit?

2. I counted 7 cycle riders and 19 cycle wheels go past my house Saturday morning. How many bicycles and how many tricycles passed my house?

Unit Test

Lesson 10-1

1. Lanny Cargill is employed by Dental Associates. The annual premium for a comprehensive medical insurance program is $1400. His employer pays 70% of the cost. How much does Lanny pay?

Lesson 10-2

2. Sue Goetz has a medical bill of $16,000. Her hospital and surgical-medical insurance did not pay for $2000 of the total bill. Her major medical insurance has a $250 deductible and a 20% coinsurance clause. How much did Sue pay?

Lesson 10-3

3. Wally Williams purchases a $60,000 5-year term insurance policy. The annual base premium is $11.11 per $1000. What is his annual premium?

Lesson 10-4

4. Hosea Gomez purchased a whole life insurance policy for $35,000. The annual base premium is $18.30 per $1000. What is his annual premium?

Lesson 10-5

5. Gloria Steele invested $5000 in a 1-year certificate of deposit. The certificate earns interest at a rate of 10.25% compounded quarterly. How much interest will she earn on the date of maturity?

AMOUNT PER $1.00 INVESTED, COMPOUNDED QUARTERLY

Annual Rate	Interest Period 1 Year	4 Years
10.00%	1.103813	1.484506
10.25%	1.106508	1.499055
10.50%	1.109207	1.513738

Lesson 10-6

6. Henry Stubs invested $1500 in a certificate of deposit that earns interest at a rate of 10% and matures after 4 years. Interest is compounded quarterly. What is the annual yield?

Lesson 10-7

7. Josephine Sills purchased 200 shares of Nautilus Brothers stock at $\$8\frac{3}{8}$. She was charged a $25 commission. What was the total amount paid for the purchase?

Lesson 10-8

8. Donna Woo owns 120 shares of Petrie International for which she paid $\$25\frac{7}{8}$ per share. If the company pays dividends of $1.25 per share, what is the annual yield to the nearest hundredth of a percent?

Lesson 10-9

9. Jimmy Jones owns 52 shares of Cantrell stock for which he paid a total of $3462.23. He sold the stock for $64 per share and paid a sales commission of $23.24. What was his profit or loss on the investment?

Lesson 10-10

10. Donald Hollander purchased a $1000 bond at the quoted price of $85\frac{3}{4}$. The bond pays interest at a rate of $5\frac{3}{4}\%$. What is the annual yield to the nearest hundredth of a percent?

Unit 11

Recordkeeping

▸ *Recordkeeping* is a way for you to manage your money. By keeping track of your monthly *expenditures,* you can find out how you have spent your money and how much money you need to live on. Your next step is to prepare a *budget sheet* showing your monthly expenses. The purpose of a budget is to allow you to compare how much money you are spending with how much money you are earning.

The first step in planning a budget is to set realistic goals for yourself.

11-1 Average Monthly Expenditure

Objective

Compute the average monthly expenditure.

Your first step toward managing your money is to keep an accurate record of how you spend your money. Use a note pad to record any **expenditures** on the day you make them. At the end of the month, group them and total them. By keeping a record of your expenditures, you will be able to examine your spending habits.

▶ **Average Monthly Expenditure** = $\dfrac{\textbf{Sum of Monthly Expenditures}}{\textbf{Number of Months}}$

Example

Skills 5, 11 **Applications** A, Q **Term** Expenditures

The Talberts keep records of their expenditures. They want to know how much they spend each month, on the average. Here are their records for 3 months. What is their average monthly expenditure?

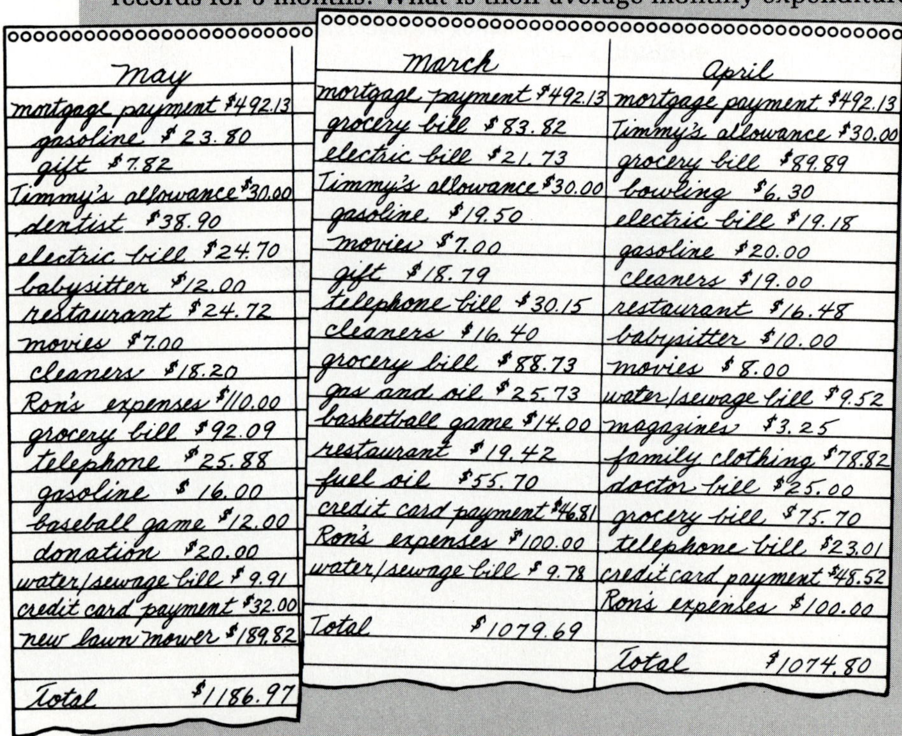

May

mortgage payment $492.13
gasoline $23.80
gift $7.82
Timmy's allowance $30.00
dentist $38.90
electric bill $24.70
babysitter $12.00
restaurant $24.72
movies $7.00
cleaners $18.20
Ron's expenses $110.00
grocery bill $92.09
telephone $25.88
gasoline $16.00
baseball game $12.00
donation $20.00
water/sewage bill $9.91
credit card payment $32.00
new lawn mower $189.82
Total $1186.97

March

mortgage payment $492.13
grocery bill $83.82
electric bill $21.73
Timmy's allowance $30.00
gasoline $19.50
movies $7.00
gift $18.79
telephone bill $30.15
cleaners $16.40
grocery bill $88.73
gas and oil $25.73
basketball game $14.00
restaurant $19.42
fuel oil $55.70
credit card payment $46.81
Ron's expenses $100.00
water/sewage bill $9.78
Total $1079.69

April

mortgage payment $492.13
Timmy's allowance $30.00
grocery bill $89.89
bowling $6.30
electric bill $19.18
gasoline $20.00
cleaners $19.00
restaurant $16.48
babysitter $10.00
movies $8.00
water/sewage bill $9.52
magazines $3.25
family clothing $78.82
doctor bill $25.00
grocery bill $75.70
telephone bill $23.01
credit card payment $48.52
Ron's expenses $100.00
Total $1074.80

Solution

Find the **average monthly expenditure.**

Sum of Monthly Expenditures	÷	Number of Months
\$1079.69 + \$1074.80 + \$1186.97	÷	3
\$3341.46	÷	3 = \$1113.82 average monthly expenditure

A. $400.00
B. $933.88

▸ **Self-Check** Cover the answers and complete the problems.

Find the average monthly expenditure.

A. January, $460; February, $320; March, $420.

B. May, $975.50; June, $899.50; July, $925; August, $935.52.

Problems

Find the average monthly expenditure.

	1.	2.	3.	4.	5.
January	$800	$456.50	$ 792.45	$ 774.56	$1038.39
February	750	337.25	1098.60	908.29	868.94
March	790	578.75	961.54	694.67	985.76
April	810	601.50	896.96	1074.22	1113.92
May	840	444.75	1111.55	739.05	1004.54
Total	?	?	?	?	?
Average	?	?	?	?	?

Use the Talberts' records of monthly expenditures on page 262 to answer the following.

6. What is the Talberts' average monthly expenditure for groceries?

7. Household costs include amounts for electric bills, telephone bills, water and sewage bills, home heating oil bills, laundry bills, and so on. What is their average monthly expenditure for household expenses?

8. Entertainment expenses include amounts for dining out, movies, and recreation. What is their average monthly expenditure for entertainment?

9. What do the Talberts pay each month to repay their mortgage loan?

10. Transportation costs include amounts for gasoline, oil, repairs, and so on. What is their average monthly expenditure for transportation costs?

11. How much do the Talberts put into savings each month, on the average?

12. Expenditures for personal spending include amounts for clothing, gifts, credit card payments, allowances, pocket money, and so on. What is their average monthly expenditure for personal spending?

13. The Talberts' records for June, July, and August show that their monthly expenditures totaled $1204.43, $1022.77, and $982.59, respectively. What is their average monthly expenditure for the past 6 months?

Maintaining Your Skills

Look up the skills in parentheses if you need help or more practice.

Add. (Skill 5)

14. $342.10 + $557.92

15. $17.91 + $19.22 + $88.88

Find the average. (Application Q)

16. 427, 515, 422, 501

17. 63.3, 45.8, 19.9

11-2 Preparing a Budget Sheet

Objective Use records of past expenditures to prepare a monthly budget.

If you have records of your past expenditures, you can use them to prepare a **budget sheet** outlining your total monthly expenses. **Living expenses** vary from month to month and include amounts for food, utility bills, pocket money, and so on. **Fixed expenses** are expenses that do not vary from one month to the next. **Annual expenses,** such as insurance premiums and real estate taxes, occur once a year.

▸ **Total Monthly Expenses** = **Monthly Living Expenses** + **Monthly Fixed Expenses** + **Monthly Share of Annual Expenses**

Example

Skills 5, 11 **Application** A **Term** Budget sheet

The Pereiras use records of their past expenditures to complete the budget sheet below. What is the total of their monthly expenses?

A MONEY MANAGER FOR *Juan & Inez Pereira* DATE *9/16/–*

MONTHLY LIVING EXPENSES	
Food/Grocery Bill	$*248.75*
Household Expenses	
Electricity	$*35.20*
Heating Fuel	$*43.54*
Telephone	$*36.95*
Water	$*8.62*
Garbage/Sewer Fee	$
Other *Cable TV*	$*25.00*
________	$
Transportation	
Gasoline/Oil	$*62.53*
Parking	$
Tolls	$
Commuting	$
Other *Repairs*	$*12.17*
Personal Spending	
Clothing	$*59.20*
Credit Payments	$*38.41*
Newspapers, gifts, etc.	$*6.25*
Pocket Money	$*41.59*
Entertainment	
Movies/Theater	$*15.00*
Sporting Events	$*12.00*
Recreation	$
Dining Out	$*33.58*
TOTAL	$*678.79*

MONTHLY FIXED EXPENSES	
Rent/Mortgage Payment (Rent circled)	$*200.00*
Car Payment	$*92.50*
Other Installments	
Appliances	$
Furniture	$*23.72*
Regular Savings	$*40.00*
EMERGENCY FUND	$*30.00*
TOTAL	$*386.22*

ANNUAL EXPENSES	
Life Insurance	$*92.40*
Home Insurance	$
Car Insurance	$*169.40*
Real Estate Taxes	$
Car Registration	$*12.00*
Pledges/Contributions	$*85.00*
Other ________	$
Total	$*358.80*
MONTHLY SHARE *(Divide by 12)	$*29.90*

MONTHLY BALANCE SHEET	
NET INCOME (Total Budget)	$
Living Expenses:	$*678.79*
Fixed Expenses:	$*386.22*
Annual Expenses:	$*29.90*
TOTAL MONTHLY EXPENSES	$
BALANCE	$

Solution

Find the **total monthly expenses.**

Monthly Living Expenses	+	Monthly Fixed Expenses	+	Monthly Share of Annual Expenses
$678.79	+	$386.22	+	$29.90 = $1094.91 total monthly expenses

A. $992.00
B. $639.00

▶ **Self-Check** Cover the answers and complete the problems.

Find the total monthly expenses.

A. Living, $600; fixed, $350; share of annual, $42.

B. Living, $372.50; fixed, $238.25; share of annual, $28.25.

Problems

Rhonda Carlin is a landscaper. Tom Carlin is a part-time real estate agent. They complete the budget sheet shown using records of their past expenditures.

Use the budget sheet to answer the following.

1. What is the total of the Carlins' monthly living expenses?
2. What is the total of their monthly fixed expenses?
3. What is the total of their annual expenses?
4. What must be set aside each month for annual expenses?
5. What is their total monthly expenditure?
6. Do the Carlins live within their monthly net income?
7. What individual expenses would be difficult for the Carlins to cut back on?

A MONEY MANAGER FOR Tom & Rhonda Carlin DATE 3/20/–

MONTHLY LIVING EXPENSES		MONTHLY FIXED EXPENSES	
Food/Grocery Bill	$209.75	Rent/Mortgage Payment (Mortgage circled)	$386.20
Household Expenses		Car Payment	$
Electricity	$38.11	Other Installments	
Heating Fuel	$31.45	Appliances	$32.16
Telephone	$35.80	Furniture	$
Water	$13.12	Regular Savings	$
Garbage/Sewer Fee	$	EMERGENCY FUND	$15.00
Other ______	$	TOTAL	$
______	$	**ANNUAL EXPENSES**	
Transportation		Life Insurance	$79.43
Gasoline/Oil	$94.87	Home Insurance	$89.45
Parking	$	Car Insurance	$175.60
Tolls	$30.00	Real Estate Taxes	$784.32
Commuting	$	Car Registration	$12.00
Other ______	$	Pledges/Contributions	$87.40
Personal Spending		Other ______	$
Clothing	$20.00	Total	$
Credit Payments	$46.83	MONTHLY SHARE *(Divide by 12)	$
Newspapers, gifts, etc.	$6.50	**MONTHLY BALANCE SHEET**	
Pocket Money	$30.00	NET INCOME (Total Budget)	$1075.35
Entertainment		Living Expenses:	$
Movies/Theater	$6.50	Fixed Expenses:	$
Sporting Events	$	Annual Expenses:	$
Recreation	$	TOTAL MONTHLY EXPENSES	$
Dining Out	$26.39	BALANCE	$
TOTAL	$		

(Continue on next page.)

Manny and Lisa Webb completed the budget sheet shown using records of their past expenditures.

Use the budget sheet to answer the following.

8. What is the total of the Webbs' monthly living expenses?

9. What is the total of their monthly fixed expenses?

10. What is the total of their annual expenses?

11. What must be set aside each month for annual expenses?

12. What is their total monthly expenditure?

13. Do the Webbs live within their monthly net income?

14. What individual expenses would be difficult for the Webbs to cut back on?

A MONEY MANAGER FOR *Manny & Lisa Webb* DATE *12/15/–*

MONTHLY LIVING EXPENSES	
Food/Grocery Bill	$323.08
Household Expenses	
Electricity	$41.78
Heating Fuel	$59.09
Telephone	$28.19
Water	$
Garbage/Sewer Fee	$15.20
Other ________	$
________	$
Transportation	
Gasoline/Oil	$50.53
Parking	$
Tolls	$
Commuting	$33.00
Other ________	$27.16
Personal Spending	
Clothing	$98.73
Credit Payments	$30.52
Newspapers, gifts, etc.	$7.10
Pocket Money	$18.89
Entertainment	
Movies/Theater	$15.26
Sporting Events	$7.09
Recreation	$9.25
Dining Out	$118.61
TOTAL	$

MONTHLY FIXED EXPENSES	
Rent/Mortgage Payment (Mortgage circled)	$564.81
Car Payment	$
Other Installments	
Appliances	$32.40
Furniture	$114.27
Regular Savings	$100.00
EMERGENCY FUND	$50.00
TOTAL	$

ANNUAL EXPENSES	
Life Insurance	$96.82
Home Insurance	$138.30
Car Insurance	$215.75
Real Estate Taxes	$967.61
Car Registration	$20.00
Pledges/Contributions	$125.00
Other ________	$
Total	$
MONTHLY SHARE *(Divide by 12)	$

MONTHLY BALANCE SHEET	
NET INCOME (Total Budget)	$1881.83
Living Expenses:	$
Fixed Expenses:	$
Annual Expenses:	$
TOTAL MONTHLY EXPENSES	$
BALANCE	$

Maintaining Your Skills

Look up the skills in parentheses if you need help or more practice.

Add. (Skill 5)

15. 42.10 + 36.5 + 22.8 + 48.3

16. 561.4 + 444.8 + 921.7 + 86.3

17. \$4.17 + \$3.29 + \$8.70 + \$3.21

18. \$15.91 + \$154.60 + \$3.72 + \$13.88

19. 576.012 + 371.01 + 224.153

20. 4.261 + 0.199 + 7.003 + 0.92

Divide. (Skill 11)

21. 342 ÷ 6

22. 857 ÷ 3

23. 757.2 ÷ 12

24. \$801 ÷ 2

25. \$1013.04 ÷ 12

26. \$376.24 ÷ 4

27. \$251.30 ÷ 2

28. \$8271 ÷ 4

11-3 Using a Budget

Objective

Compare amount budgeted to actual expenditure and personal spending to "typical spending."

Once you have completed a budget sheet outlining your past expenditures, you can use it to plan for future spendings. You may want to prepare a monthly **expense summary** to compare the amounts that you spend to the amounts that you budgeted. When you draft a budget, you should include an **emergency fund** to provide for unpredictable expenses, such as medical bills and repair bills.

Example

Skills 1, 6 **Application** ______ **Term** Emergency fund

The Crosbys kept accurate records of their expenditures. At the end of July, they prepared an expense summary. They had planned to spend $186.40 on groceries. They actually spent $163.90. How much more or less did they spend on groceries than they had budgeted for?

Solution

1. Compare.
   Is **amount spent** more or less than **amount budgeted?**
   Is $163.90 more or less than $186.40? . . . . Less
2. Find the **difference.**
   $186.40 − $163.90 = $22.50 less than amount budgeted

A. $5.00 less
B. $32.14 more

▶ **Self-Check** Cover the answers and complete the problems.

Find how much more or less the amount spent is than the amount budgeted.

A. Budgeted $142.25, spent $137.25.
B. Budgeted $462.21, spent $494.35.

Problems

1. Ed Wilson's groceries for May.
   Budgeted $254.34.
   Actually spent $223.58.
   Did he spend more or less than budgeted?
2. Ken Day's gas purchases for August.
   Budgeted $78.34.
   Actually spent $89.95.
   Did he spend more or less than budgeted?
3. Lila Mann's water bill for June.
   Budgeted $12.48.
   Actually spent $13.09.
   How much more or less did she spend than budgeted?
4. Mimi King's credit payment for June.
   Budgeted $38.43.
   Actually spent $64.30.
   How much more or less did she spend than budgeted?

Here is the Crosbys' expense summary for the month of July. They want to compare what they had budgeted to what they actually spent.

Expenditures for the Month of July		
Expenses	Amount Budgeted	Actual Amount Spent
Food	$186.40	$163.90
Household		
Electric bill	29.56	48.62
Telephone bill	18.44	24.56
Heating oil	53.63	23.90
Water bill	11.47	10.94
Cable TV bill	25.00	25.00
Transportation		
Gasoline purchases	115.57	103.45
Commuter ticket	35.00	35.00
Personal		
Clothing	13.19	38.00
Credit card payment	56.08	42.74
Newspapers, Gifts	7.35	13.29
Pocket money	50.00	75.00
Entertainment		
Movies	7.29	12.00
Dining out	88.28	46.73
Fixed		
Rent	235.00	235.00
Savings	400.00	400.00
Emergency fund	75.00	75.00
Renter's insurance premium	26.30	26.30

Use the Crosbys' expense summary to answer the following.

5. Which household expenses for the month were more than the amount budgeted?

6. Did they spend more or less than the amount budgeted for household expenses for the month? By how much?

7. How much did they budget for transportation costs? Were the amounts spent for transportation during the month more or less than the amount budgeted? By how much?

8. Which personal expenses did they spend more on than they had budgeted?

9. Were their total personal expenditures for the month more or less than the amount budgeted? By how much?

10. Were their total entertainment expenditures for the month more or less than the amount budgeted? By how much?

11. Were there any monthly fixed expenses for which the Crosbys spent more than the amount budgeted?

12. What annual expenses occurred during the month? Was the amount set aside for annual expenses more or less than the amount actually spent for annual expenses? By how much?

13. What was the Crosbys' total expenditure for the month of July? Was this amount more or less than the amount they had originally budgeted? By how much?

14. Your total monthly expenditure will vary from month to month. During the winter months, your home heating bills may "push" your total monthly expenditure over the amount budgeted. In some months, spending will be less. Name some factors that might affect your spending for specific months.

Here is the Morrisons' expense summary for December. They want to compare what they had budgeted to what they actually spent.

Use the Morrisons' expense summary to answer the following.

Expenditures for the Month of December		
Expenses	**Amount Budgeted**	**Actual Amount Spent**
Food	$325.00	$324.18
Household		
Electric	75.00	85.50
Telephone	25.00	17.92
Heating fuel	80.00	95.50
Water bill	30.00	30.00
Repairs	40.00	60.00
Transportation		
Gasoline/oil	75.00	64.20
Parking	25.00	24.00
Personal		
Clothing	100.00	95.40
Credit payment	50.00	35.00
Gifts	30.00	0
Pocket money	40.00	50.00
Entertainment		
Movies	30.00	20.00
Sports event	20.00	10.00
Dining out	100.00	88.50
Fixed Expenses		
Mortgage payment	315.00	315.00
Car payment	140.00	140.00
Savings	100.00	50.00
Home insurance	15.00	18.00

15. Which household expenses for the month were more than the amount budgeted?

16. Did they spend more or less than the amount budgeted for household expenses for the month? By how much?

17. How much did they budget for transportation costs? Were the amounts spent for transportation more or less than the amount budgeted? By how much?

18. Which personal expenses were more than the amount budgeted?

19. Were their total personal expenditures for the month more or less than the amount budgeted? By how much?

20. Were their total entertainment expenditures for the month more or less than the amount budgeted?

21. Were there any monthly fixed expenses for which the Morrisons spent more than the amount budgeted?

22. What was the Morrisons' total expenditure for the month of December? Was this amount more or less than the amount they had originally budgeted? By how much?

Maintaining Your Skills

Look up the skills in parentheses if you need help or more practice.

Which number is greater? (Skill 1)

23. 823.16 or 822.56 24. 64.31 or 58.30 25. 147.22 or 139.22

Subtract. (Skill 6)

26. 472.1 − 384.1 27. 386.91 − 352.81 28. 92.104 − 85.132

29. $900.00 − $357.62 30. $846.47 − $832.48 31. $3219.27 − $1942.21

Reviewing the Basics

Skills

Which number is greater?

(Skill 1) **1.** \$35.78 or \$41.09 **2.** \$16.29 or \$15.99 **3.** \$113.54 or \$98.45

Solve. Round answers to the nearest cent.

(Skill 5) **4.** \$24.72 + \$18.89 + \$145.43 + \$33.90 + \$76.77

5. \$55.83 + \$225 + \$41.09 + \$10.25 + \$89.82 + \$60.28

6. \$1.23 + \$2.25 + \$5.19 + \$0.92 + \$1.03 + \$0.23

(Skill 6) **7.** \$232.34 − \$89.56 **8.** \$171.06 − \$89.32 **9.** \$403.08 − \$289.39

(Skill 11) **10.** \$539.84 ÷ 7 **11.** \$597.06 ÷ 6 **12.** \$1105.26 ÷ 12

Write as a percent. Round to the nearest tenth of a percent.

(Skill 26) **13.** 0.18049 **14.** 0.65389 **15.** 0.08452 **16.** 0.0134

Applications

Find the average. Round answers to the nearest cent.

(Application Q) **17.** \$24.70, \$18.20, \$7.00, \$235.50, \$68.73

18. \$139.34, \$80, \$112.25, \$9.74, \$335.09, \$335.16

19. \$250, \$39.94, \$102.50, \$33.65, \$354.65, \$92.10, \$66.62, \$90

20. \$11.34, \$32.45, \$65.34, \$45, \$30.03, \$64.98, \$250, \$250, \$29.98

Terms

Use each term correctly in the paragraphs below. Then rewrite the paragraphs.

21. Budget sheet **22.** Annual expenses **23.** Monthly fixed expenses

24. Emergency fund **25.** Living expenses **26.** Expenditures

A __?__ is used to summarize your spending habits and aid in planning for future spendings. Your budget is based on your take-home pay. To determine how you spend your money, keep records of your monthly __?__ for three months to a year.

Real estate taxes, insurance payments, pledges, and contributions are examples of __?__ . Mortgage loan payments, rent, installment payments, and regular savings deposits are examples of __?__ . Amounts for food, clothing, pocket money, and utility bills are examples of __?__ . For your budget to succeed, an __?__ is necessary to provide for unpredictable expenses such as medical bills and repair bills.

Refer to your reference files at the end of the book if you need help.

Inside Economics

The Consumer Price Index (CPI) is a measure of the average change in prices of a certain number of goods and services. The year 1967 is used as the base year and the CPI for 1967 is set at 100. This means that a commodity which cost \$100 in 1967 and that has a CPI of 159 today would cost about \$159 today. The table below shows the nationwide CPIs of some commodities for a recent year. You can get an idea of the average cost of a commodity in 1967 by multiplying the current cost times 100 and dividing by the current CPI for the item.

$$\text{Cost in 1967} = \frac{100 \times \text{Current Cost}}{\text{Current CPI}}$$

Indicators for September	
Commodity	**CPI**
Fuel oil	496.5
Gasoline	369.8
Rent	108.9
Home	229.8
Groceries	295.3
Dining out	334.4
New car	208.1
Clothing	183.0

If you purchase a bag of groceries for \$45.90 and the current CPI for groceries is 295.3, what would you likely have paid for the same bag of groceries in 1967?

$$\text{Cost in 1967} = \frac{100 \times \$45.90}{295.3} = \$15.54$$

Use the table above to answer the following.

1. Thelma Ritter purchased \$13.50 in gasoline for her car. What could she have expected to pay for the same purchase in 1967?

2. Carl Cross purchased a standard-sized car for \$9525. What could he have expected to pay for a similar car in 1967?

3. Linda Stidham sees a house for sale for \$55,900. What would the same house have sold for in 1967?

4. Curtis Mann paid \$5.50 for a haircut when the CPI for personal care services was 167.2. What could he have expected to pay in 1967?

5. Alex Cunningham paid \$4.50 per movie ticket when the CPI for recreational services was 191.1. What might he have expected to pay for a ticket in 1967?

6. In a recent year it cost \$7.50 for a dry cleaning service and the CPI for this service is 164.0. What would it likely have cost to purchase the same service in 1967?

7. Gail Harris purchased a toy red wagon for \$26.95. Her father told her that he had purchased a similar wagon for her in 1967 for \$13.95. What is the CPI for toys when Gail made her purchase?

Unit Test

Lesson 11-1

1. Jackie Omar keeps records of her family's expenses. Total monthly expenses for September, October, and November were \$982.50, \$954.79, and \$997.16, respectively. What was their average monthly expenditure for these months?

A MONEY MANAGER FOR *Liz & David Stein* DATE *11/27/–*

MONTHLY LIVING EXPENSES	
Food/Grocery Bill	\$*310.94*
Household Expenses	
Electricity	\$*44.00*
Heating Fuel	\$
Telephone	\$*36.19*
Water	\$
Garbage/Sewer Fee	\$
Other *Child care*	\$*31.25*
______	\$
Transportation	
Gasoline/Oil	\$*78.16*
Parking	\$*15.20*
Tolls	\$
Commuting	\$
Other ______	\$
Personal Spending	
Clothing	\$*74.04*
Credit Payments	\$*48.11*
Newspapers, gifts, etc.	\$*7.81*
Pocket Money	\$*52.00*
Entertainment	
Movies/Theater	\$*18.75*
Sporting Events	\$
Recreation	\$*15.00*
Dining Out	\$*41.57*
TOTAL	\$

MONTHLY FIXED EXPENSES	
(Rent)/Mortgage Payment	\$*250.00*
Car Payment	\$*115.63*
Other Installments	
Appliances	\$
Furniture	\$*29.65*
Regular Savings	\$*50.00*
EMERGENCY FUND	\$*20.00*
TOTAL	\$

ANNUAL EXPENSES	
Life Insurance	\$*115.50*
Home Insurance	\$
Car Insurance	\$*211.72*
Real Estate Taxes	\$
Car Registration	\$*15.00*
Pledges/Contributions	\$*12.50*
Other *Property Tax*	\$*102.60*
Total	\$
MONTHLY SHARE *(Divide by 12)	\$

MONTHLY BALANCE SHEET	
NET INCOME (Total Budget)	\$*1278.29*
Living Expenses:	\$
Fixed Expenses:	\$
Annual Expenses:	\$
TOTAL MONTHLY EXPENSES	\$
BALANCE	\$

Complete the budget sheet. Use it to answer the following.

Lesson 11-2

2. What amount is set aside each month for annual expenses? What is the Steins' total monthly expenditure?

Lesson 11-3

3. The Steins had planned to spend \$93.36 on transportation costs. They actually spent \$103.52. How much more or less did they spend on transportation expenses than they originally planned?

Cumulative Review (Skills/Applications) Units 6–11

Skills

Find which number is greater.

(Skill 1) 1. 3.1 or 3.01 2. 14.73 or 14.721 3. 43.892 or 43.984

Round to the nearest cent.

(Skill 2) 4. $23.432 5. $143.584 6. $1.384 7. $59.996

Solve. Round to the nearest cent.

(Skill 3) 8. $23 + $46 9. $64 + $39 10. $61 + $76 + $43

11. $536 + $93 + $560 12. $644 + $32 + $272

(Skill 4) 13. $56 − $53 14. $127 − $88 15. $730 − $533

16. $8033 − $2083 17. $7872 − $3680 18. $7711 − $3799

(Skill 5) 19. $66.91 + $51.28 20. $37.44 + $9.80 21. $93.68 + $117.94

22. $67.45 + $40.49 23. $424.37 + $338.95 24. $449.25 + $627.40

(Skill 6) 25. $88.49 − $71.18 26. $54.58 − $13.26 27. $38.90 − $3.51

28. $35.48 − $29.89 29. $567.72 − $78.66 30. $1657.10 − $335.60

(Skill 8) 31. $428.37 × 25 32. $1.57 × 85 33. $41.12 × 354

34. $86.35 × 9.4 35. $9.05 × 7.4 36. $440.43 × 0.36

(Skill 11) 37. $96.89 ÷ 52 38. $8.85 ÷ 12 39. $581.37 ÷ 17.7

40. $813.92 ÷ 4.07 41. $66,742 ÷ 14.79 42. $1124.40 ÷ 88.2

Write as a decimal.

(Skill 14) 43. $\frac{5}{16}$ 44. $\frac{11}{32}$ 45. $\frac{135}{360}$ 46. $\frac{292}{365}$

Write as a percent to the nearest tenth of a percent.

(Skill 26) 47. 0.726 48. 0.085 49. 0.6829 50. 0.06138

Write as a decimal.

(Skill 28) 51. 25% 52. 12.3% 53. 108% 54. 4.4%

Solve. Round to the nearest cent.

(Skill 30) 55. 52% of $832 56. 35% of $15.81

57. 7.3% of $183 58. 6.2% of $73.89

59. 25% of $1.44 60. 0.83% of $240

61. 0.05% of $19.50 62. 0.25% of $60.25

Applications

Use the table to answer the following.

Monthly Payments $500 Loan at 18% APR	
Term in Months	Monthly Payment
6	$87.76
12	45.84
18	31.90
24	24.96

(Application C) 63. What is the monthly payment on a $500 loan for a term of 12 months?

64. What is the total amount repaid if the loan is repaid in 18 months?

65. For what term is the monthly payment lowest?

66. For what term is the total amount repaid lowest?

Use the table on page 569 to find the elapsed time in days.

(Application G) 67. From April 3 to April 25.

68. From May 8 to August 19.

69. From March 30 to July 12.

70. From July 12 to March 30.

71. From November 17 to March 16.

72. From August 29 to April 13.

Write as a fraction of a year.

(Application J) 73. 3 months

74. 8 months

75. 72 days (ordinary year)

76. 90 days (ordinary year)

77. 146 days (exact year)

Find the mean. Round answers to the nearest cent.

(Application Q) 78. $78.99, $65.44, $63.19, $72.35

79. $6.95, $1.69, $5.09, $8.34, $2

80. $350, $969.24, $205, $65.90, $83

81. $3.07, $5.93, $7, $7.50, $3, $5.65, $3.48, $9.04

Terms

Write your own definition for each term.

82. Sales receipt
83. Finance charge
84. Previous balance
85. Unpaid balance
86. Rebate schedule
87. Discount loan
88. Installment loan
89. Interest rate
90. Sticker price
91. Rent
92. Deductible clause
93. Depreciation
94. Mortgage loan
95. Closing costs
96. Premium
97. Principal
98. Term life insurance
99. Whole life insurance
100. Bonds
101. Stocks
102. Budget sheet
103. Fixed expenses
104. Emergency fund
105. Living expenses

Refer to your reference files at the back of the book if you need help.

Cumulative Review Test Units 6–11

Lesson 6-3

1. Alice Sloan has a charge account that charges 1.5% of the previous balance for a finance charge. Her statement shows purchases totaling $98.15, a previous balance of $364, and a payment of $90. What is the new balance?

Lesson 6-4

2. Matt Biehl has a charge account that charges 1.5% of the unpaid balance. His statement shows purchases totaling $32.22, a previous balance of $284, and a payment of $44. Find the finance charge and the new balance.

Lesson 6-5, 6-6

3. Find the average daily balance where new purchases are not included when figuring the daily balances. Then find the finance charge and the new balance.

REFERENCE	POSTING DATE	TRANSACTION DATE	DESCRIPTION	PURCHASES & ADVANCES	PAYMENTS & CREDITS
13100	9/14	8/23	Wesford Clothes	104.50	
20046	9/14	8/27	Record Mart	35.60	
54545	9/17		PAYMENT		30.00

BILLING PERIOD	PREVIOUS BALANCE	PERIODIC RATE	AVERAGE DAILY BALANCE	FINANCE CHARGE
9/1 - 9/30	$245.20	1.5%	?	?

PAYMENTS & CREDITS	PURCHASES & ADVANCES	NEW BALANCE	MINIMUM PAYMENT	PAYMENT DUE
$30.00	$140.10	?	$20.00	10/21

Lesson 7-1

4. Diane Morris's bank granted her a single-payment loan of $4450 for 75 days at an annual interest rate of 18%. Her bank charges ordinary interest. What is the maturity value of Diane's loan?

Lesson 7-4

5. Steve Howe obtained an installment loan of $1000 to purchase a stereo set. The annual percentage rate is 18%. He plans to repay the loan in 18 months. Use the table to find the finance charge.

APR	Term in Months	If You Finance. . . $200	$500	$1000	$1500
		Your **Monthly Payments** Are			
18%	6	35.10	87.76	175.52	263.28
	12	18.33	45.84	91.68	137.52
	18	12.76	31.90	63.80	95.70
	24	9.98	24.96	49.92	74.88

Lesson 8-2

6. Nora Peters is interested in a compact car that has a base price of $6085, options totaling $2147.42, and a destination charge of $180. She has read that the dealer's cost is about 82% of the base price and 80% of the price of the options. What is the estimated dealer's cost?

Lesson 8-4

7. Ralph Oberski uses his car to drive to and from work. His insurance coverage includes 25/100 bodily injury and $50,000 property damage. His driver-rating factor is 4.10. He has $50-deductible comprehensive and $100-deductible collision insurance coverages on his car. His car is in age group B and rating group 8. What is his annual premium?

Property Damage Limits	Bodily Injury Limits		
	25/50	25/100	50/100
$ 25,000	$103.20	$114.00	$119.20
$ 50,000	$107.20	$118.00	$123.20
$100,000	$110.00	$121.20	$126.00

Age Group	Insurance Rating Group 6	7	8
	Comprehensive $50 ded		
A	$32.80	$35.60	$39.60
B	$30.80	$33.20	$36.80
	Collision $100 ded		
A	$84.00	$83.20	$102.40
B	$72.80	$90.80	$108.40

Lesson 9-4

8. The Masters obtained a 25-year, $60,000 mortgage loan at First Federal Bank at an interest rate of 14.50%. Find the monthly payment. What is the new principal after the first monthly payment? Use the monthly payment table on page 569.

Lesson 9-5

9. The Franklin County tax assessor stated that the market value of the Lumm estate is $925,000. The rate of assessment in Franklin County is 35% of the market value. The tax rate is 48.15 mills. What is the annual real estate tax on the Lumm estate?

Lesson 10-1

10. Lisa Yunker is employed by P.S.I. Associates. The annual premium for comprehensive medical insurance is $1500. Her employer pays 60% of the cost. How much does Lisa pay?

Lesson 10-4

11. Trudy Kaminski purchased a whole life insurance policy with a face value of $50,000. The annual base premium is $21.74 per $1000. What is her annual premium?

Lesson 10-6

12. Jose Gomez invested $6500 in a certificate of deposit that earns interest at a rate of 10.25% and matures after 4 years. Interest is compounded quarterly. What is the annual yield?

AMOUNT PER $1.00 INVESTED COMPOUNDED QUARTERLY

Annual Rate	Interest Period 1 Year	4 Years
10.00%	1.103813	1.258577
10.25%	1.106508	1.271224
10.50%	1.109207	1.283998

Lesson 10-7

13. Alton Gordon purchased 500 shares of Keeners stock at $\$15\frac{5}{8}$. He was charged a $40 commission. What was the total amount paid for the purchase?

Lesson 11-1

14. Martha Duvall keeps records of family expenses. Total monthly expenses for September, October, and November were $1026.60, $1352.75, and $1128.25, respectively. What was their average monthly expenditure for these months?

Buying a Condominium: A Simulation

You have decided to purchase a condominium. You are chief systems programmer for Datacrunch, Inc., and your gross pay is $35,000 a year. You have $12,000 in a savings account.

Before you look at condominiums, you decide to figure out how much you can afford to spend.

Your bank suggests that you spend not more than 28% of your gross pay on principal, interest, and taxes.

Most condominium management companies in your area want a down payment of 20% of the selling price. You decide that you can use $10,000 of your savings account as a down payment.

Write your name on the Datacrunch badge. Use this worksheet to figure out how much you can afford to spend for your condominium.

Amount you can afford to spend per year on principal, interest, and taxes	
Amount you can afford to spend per month on principal, interest, and taxes	
Maximum selling price on which you can afford to make a 20% down payment	

DATACRUNCH, INC.

SYSTEMS DIVISION

NAME ___________

Employee's picture

Stonehill Savings Bank

DATE	DEPOSITS	INTEREST	WITHDRAWALS	BALANCE
JAN 28 –		47.71		11500.00
JAN 28 –	500.00			12000.00

2. Employer's ← State number

Copy C For employee's records.

3. Employer's name DATACRUNCH, INC.	4. Subtotal ☐ Correction ☐ Void ☐	State form number
	5. Employer's identification number	City form number
	6. Advance EIC payment	7.

8. Employee's social security no.	9. Federal income tax withheld	10. Wages, tips, other compensation	11. FICA tax withheld	12. Total FICA wages
123-45-6789	9484.88	35000	2502.50	

13. Employee's name (First, middle, last)	14. Pension plan coverage? Yes/No YES	15. •	16. FICA tips
	18. State income tax withheld 1658.16	19. State wages, tips, etc. 35000	20. Name of State
17. Employee's address and Zip code	21. Local income tax withheld	22. Local wages, tips, etc.	23. Name of locality
Wage and Tax Statement 19–			

Form W-2

Which Condominium?

After looking at several condominiums, you narrow your selection to two, one in the city and one in a suburb. The advertising booklet for each condominium gives the base price and the additional charges for extra features.

Townview

LUXURY AND CONVENIENCE IN SPRINGFIELD

One Bedroom Unit $42,000
790 sq ft

Fireplace: brick	$2500
Refrigerator with ice maker	460
Microwave oven: built in	800
Shag carpeting: 72 sq yd @ $5.50 per sq yd extra cost	

Country Squires Estates

Your Own Country Estate in Stafford Valley

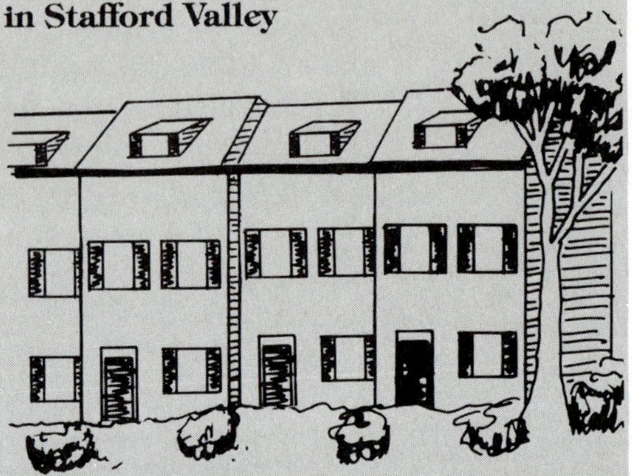

One Bedroom Unit $46,000
850 sq ft

Fireplace: stone	$2300
Refrigerator with ice maker	390
Microwave oven: built in	780
Shag carpeting: 81 sq yd @ $3.75 per sq yd extra cost	

Choose one of these condominiums. You will keep the same condominium for the rest of this simulation. You decide to purchase all the extra features listed in the advertising booklet. Use the prices in the booklet to find the total cost of your condominium.

Condominium you have chosen	
Base price	
Fireplace	
Refrigerator	
Microwave oven	
Carpeting: _____ sq yd × $_____	
Total cost of condominium	

Your Net Worth

After you choose the condominium you want, you fill out a mortgage application. The lender will check to make sure that you have as much money as you say you do, and that you have paid your debts on time.

The application also asks for a statement of assets and liabilities. You have $12,000 in a savings account and $450 in a checking account. You own a car worth $6000, on which you still owe $2000. Your life insurance has a net cash value of $2000. You have personal property, such as furniture and jewelry, with a total value of $4350. You owe $180 on your credit card.

Use this information to fill out the loan application.

RESIDENTIAL LOAN APPLICATION

BORROWER

Name	
Address	
Employer	
Position or Title	
Gross Annual Income	

STATEMENT OF ASSETS AND LIABILITIES

ASSETS		LIABILITIES	
Description	Cash or Market Value	Description	Unpaid Balance
Cash		Installment Debts	
Checking Account		Real Estate Loans	
Savings Account		Automobile Loans	
Automobile		Other Debts	
Life Insurance Net Cash Value			
Furniture and Personal Property			
Total Assets		Total Liabilities	
NET WORTH (Total Assets minus Total Liabilities)			

Bank Charges and Taxes

Now that you have selected the condominium you would like to buy, you will have to calculate what it will cost you each month.

Complete this worksheet to determine the amount you must finance.

Condominium you have chosen	
Total price	
Down payment (20% of total price)	
Amount financed	

Both development companies arrange financing. The mortgage for Townview is 13% for 25 years. The mortgage for Country Squires is $13\frac{1}{2}$% for 30 years.

Use the mortgage table (below left) and this worksheet to find your monthly payment.

Monthly payment per thousand dollars	
Amount financed in thousands of dollars	
Monthly payment for principal and interest	

Monthly Mortgage Payments Principal and Interest per $1000 Financed

Interest Rate %	15 Years	20 Years	25 Years	30 Years
$12\frac{1}{2}$	12.33	11.37	10.91	10.68
$12\frac{3}{4}$	12.49	11.54	11.09	10.87
13	12.66	11.72	11.28	11.07
$13\frac{1}{4}$	12.82	11.89	11.47	11.26
$13\frac{1}{2}$	12.99	12.08	11.66	11.46

Property Tax Rates

Town	Assessment Rate	Tax Rate per $1000
Sparta	63%	$109.40
Springfield	80%	136.00
Springton	75%	87.25
Stafford Valley	50%	166.50
Strawberry	100%	91.50

Use the property tax table (above right) and this worksheet to figure your taxes.

Total value of your condominium	
Assessment rate	
Assessed value	
Tax rate	
Taxes per year	
Taxes per month	
Total monthly cost for principal, interest, and taxes	

Is this amount less than 28% of your gross income?

Yes ☐ No ☐

Fees, Insurance, and Utilities

Your condominium management company charges you an annual fee. This fee covers the cost of trash pickup, snow removal, landscaping, maintenance, management costs, and insurance on the buildings.

The Townview fee is $0.60 per square foot per year. The Country Squires fee is $0.75 per square foot per year.

Use this worksheet to calculate your condominium fee.

Your condominium	
Fee per year: $_____ × _____ sq ft	
Fee per month	

The condominium fee includes insurance on the buildings. You have to get separate insurance on your furniture and other possessions. Your insurance company has a basic annual rate of $75. There are extra charges for insurance on certain valuable items.

Use these worksheets to calculate the cost of your insurance.

Item	Value	Annual Rate per $100	Annual Rate
camera	$200	$1.65	
bicycle	150	10.00	
jewelry	50	1.60	

Basic rate per year	
Extra charges per year	
Total rate per year	
Insurance cost per month	

Townview uses electricity for heating, cooking, and lighting. It is estimated that the unit will use 8750 kW•h per year. The cost of electricity is $0.06 per kW•h.

Country Squires uses gas for heating and cooking. The estimated annual consumption of gas is 745 cubic feet at $0.45 per cubic foot. The unit would also use 2400 kW•h of electricity per year at $0.07 per kW•h.

Use this worksheet to find your utilities costs.

Electricity: _____ kW · h × $_____ per kW · h	
Gas: _____ cu ft × $_____ per cu ft	
Utilities cost per year	
Utilities cost per month	
Total monthly cost for fee, insurance, and utilities	

Use this worksheet to summarize your results so far.

Your condominium	
Total cost	
Your net worth	
Monthly cost for principal, interest, and taxes	
Monthly cost for fees, insurance, and utilities	
Total monthly cost	

A common rule of thumb is that your monthly housing cost should be less than 35% of your monthly net pay. Your net pay is your gross pay minus deductions for taxes and F.I.C.A. Shown below is the stub attached to your monthly pay check.

Use the stub and this worksheet to calculate the monthly net pay.

DATACRUNCH, INC.

EARNINGS	AMOUNT		DEDUCTIONS			YEAR TO DATE		
REGULAR	2916	67	FICA TAX	208	54	GROSS PAY	2916	67
			FED. TAX	729	16	FICA TAX	208	54
			STATE TAX	138	18	FED. TAX	729	16
						STATE TAX	138	18
						FOR PERIOD ENDING		
						JANUARY 31, 19--		
						NET PAY		

Monthly net pay = $ — $	
35% of monthly net pay	

Is your total monthly housing cost less than 35% of your monthly net pay? yes ☐ no ☐

The Bottom Line

If you had another chance to decide about buying a condominium, what might you do differently? Why?

Part 3 Business Mathematics

▸ In Part 3, you will learn how to apply business mathematics to actual business practices. Eleven different departments of one business are covered in Part 3. Each department has special functions and uses specific business mathematics principles. You will learn firsthand how mathematics is used in each department and how each department contributes to the entire operation of the business.

▸ The following definitions provide an overview of the activities involved in the operation of the departments within a business. The lessons within each unit explain in detail how business mathematics is used to carry out these activities.

Personnel The administrative activities of hiring staff, developing wage and salary scales, and providing benefits for employees within your business.

Production The activities involved in manufacturing and packaging the items that your business sells.

Purchasing The activities involved in buying goods at discount prices from your suppliers.

Sales The activities involved in determining the selling price of your product in the marketplace.

Marketing 1. The activities concerned with researching buyer preference. 2. The activities concerned with communicating information about your product to prospective buyers.

Warehousing and Distribution The activities of storing, shipping, and maintaining accurate inventory records of your products.

Services 1. The activities related to using buildings, equipment, or utilities from other businesses for a fee. 2. The activities related to hiring employees of another business to do work for your business.

Accounting The activities related to monitoring normal expenses of your business, such as payroll and depreciation.

Accounting Records The activities of maintaining and analyzing documents that show the income, expenses, and value of your business.

Financial The activities concerned with managing your business's money, including paying taxes, borrowing, and investing.

Information Management The activities of organizing and analyzing data about your business through the use of a computer.

Unit 12 Personnel

▸ The Personnel Department of your business fills openings by *recruiting* new employees through advertising and interviews. The department also handles *wage and salary scales* for each position and *employee benefits* such as health and life insurance and paid vacations.

Part of the cost of running a business goes to employee benefits. Employee benefits add real dollars to your paycheck.

12-1 Hiring New Employees

Objective

Compute the cost of recruiting new employees.

To fill openings in your business, you may **recruit** new employees. The cost of recruiting includes advertising fees, interviewing expenses such as travel expenses, and hiring expenses such as moving expenses. In addition, you may pay an employment agency to locate job candidates.

Total Recruiting Cost = Advertising Expense + Interviewing Expense + Hiring Expense

Example

Skill 5 Application A Term Recruiting cost

The Cherry Hill Furniture Company is searching for a person to head the marketing department. The Personnel Department placed advertisements for a total cost of $1890 and employed the Worldwide Executive Search Company to locate candidates. Worldwide recommended George Trask, Susan Hollis, and Robert Sandora. Cherry Hill paid the candidates' travel expenses for interviews.

George Trask	Total Travel Expenses: $193
Susan Hollis	Total Travel Expenses: $362
Robert Sandora	Total Travel Expenses: $451

After the interviews, Cherry Hill hired Susan Hollis at an annual salary of $42,500. Cherry Hill paid these expenses to hire her:

Moving Expenses	$1800
Sale of Home (Real Estate Broker's Fee)	$4320
Worldwide's Fee (25% of Susan's first year salary)	$10,625

What was the total expense of recruiting Susan Hollis?

Solution

Find the **total recruiting cost.**

Advertising Expense + Interviewing Expense + Hiring Expense

$1890 + ($193 + $362 + $451) + ($1800 + $4320 + $10,625)

$1890 + $1006 + $16,745 = $19,641 total recruiting cost

A. $4720
B. $4160

Self-Check Cover the answers and complete the problems.

Find the total recruiting cost.

A. Advertising expenses, $1470; interviewing expenses, $800 and $750; hiring expenses, $200 and $1500.

B. Advertising expenses, $1250; interviewing expenses, $375 and $965; hiring expenses, $175 and $1395.

Problems

	Advertising Cost	+	Interviewing Cost	+	Hiring Cost	=	Total Recruiting Cost
1.	$ 131.70	+	$ 34.00	+	$1608.00	=	?
2.	$ 726.80	+	$ 14.30	+	$ 941.70	=	?
3.	$ 645.55	+	$327.86	+	$1782.00	=	?
4.	$ 230.75	+	$148.82	+	$2633.33	=	?
5.	$1491.83	+	$565.00	+	$2147.95	=	?

6. Cahill's Stores recruits head buyer.
Advertising cost: $647.50.
Interviewing cost: $215.60.
Hiring cost: $1471.50.
What is the total recruiting cost?

7. Magno Company recruits chief engineer.
Advertising cost: $561.85.
Interviewing cost: $151.73.
Hiring cost: $2516.75.
What is the total recruiting cost?

8. Essex Data Communications Corporation hired Herbert Stanhope as its new production coordinator at an annual salary of $18,952.

Advertising Costs: $463.72
Interviewing Expenses: Herbert Stanhope $72.50, Michelle Carlberg $48.65
Top Skills Agency Fee: 20% of first year's salary

What was the total cost of hiring the production coordinator?

9. Western Hills Investments, Ltd., hired the Financial Management Search Service to locate candidates for the position of Vice President, Trust Funds. The agency's fee is 25% of the first year's salary if one of its candidates is hired. Western Hills also ran several advertisements at a total cost of $1695.96. Western Hills interviewed three people:

John Gillespie	Alicia Cordoba	Diane Maxwell
Applied through agency	Answered advertisement	Applied through agency
Travel costs: $221.20	Travel cost: $104.56	Travel costs: $143.30

Western Hills hired Alicia Cordoba at an annual salary of $37,500. They paid her moving expenses of $816.75 and her real estate broker's fee of 8% to sell her $51,000 condominium. What was the total recruiting cost?

Maintaining Your Skills

Look up the skill in parentheses if you need help or more practice.

Add. (Skill 5)

10. $220 + $80 + $1790
11. $570 + $240 + $3450
12. $2150 + $1480 + $21,750
13. $1240 + $970 + $15,410
14. $467 + $124 + $1193
15. $983 + $485 + $1894
16. $1993 + $1056 + $12,491
17. $1956 + $358 + $1427

12-2 Administering Wages and Salaries

Objective Compute the new salary after merit increase and cost-of-living adjustment.

Your business may have **wage and salary scales** for the positions in the company. You can use the information to compare various jobs or to estimate the cost of giving employees **cost-of-living adjustments** or **merit increases.** A cost-of-living adjustment is a raise in your salary to help you keep up with inflation. A merit increase is a raise in your salary to reward you for the quality of your work.

▶ **New Salary = Present Salary + Cost-of-Living Adjustment + Merit Increase**

Example

Skill 30 Application C Term Salary scale

Calvin Daily is a computer operator, level 3, for Tech Services. The board of directors of Tech Services voted to give all employees a cost-of-living adjustment of 8.1%. In addition, Calvin was awarded a merit increase of 3.6% for outstanding work during the year. What will Calvin's salary be for the coming year?

JOB LEVEL	SECRETARY	COMPUTER OPERATOR	COMPUTER TECHNICIAN	COMPUTER PROGRAMMER
1	$ 8000	$10,815	$11,100	$15,150
2	9500	11,350	12,300	18,000
3	11,180	12,690	13,200	20,400
4	13,500	14,150	14,240	26,060

Solution

1. Find the **cost-of-living adjustment.**
   8.1% of $12,690.00 = $1027.89
2. Find the **merit increase.**
   3.6% of $12,690.00 = $456.84
3. Find the **new salary.**
   Present Salary + Cost-of-Living Adjustment + Merit Increase
   $12,690.00 + $1027.89 + $456.84 = $14,174.73 new salary

$8,480

▶ **Self-Check** Cover the answer and complete the problem.

A secretary, level 1, receives a 5% cost-of-living adjustment and a 1% merit increase. Find the new salary.

Problems

	Present Salary	+	Cost-of-Living Adjustment	+	Merit Increase	=	New Salary
1.	$12,300	+	$ 861	+	$615	=	?
2.	$13,500	+	$ 810	+	$540	=	?
3.	$15,150	+	$1136	+	$530	=	?
4.	$14,150	+	$ 919	+	$772	=	?
5.	$13,500	+	$ 402	+	$291	=	?

Level	Credit Correspondent	Senior Credit Correspondent
6	$11,456	$12,030
8	11,972	12,620
10	12,510	13,250

6. Joellen Dimatto.
Credit correspondent, level 6.
Cost-of-living adjustment 6.5%.
What is her new salary?

7. Carl Porter.
Credit correspondent, level 10.
Cost-of-living adjustment 5%.
Merit increase 4.8%.
What is his new salary?

8. Esther Middleton.
Senior credit correspondent, level 8.
Merit increase 8.2%.
What is her new salary?

9. Joseph Carlisle, a credit correspondent, level 8, received a 6% cost-of-living adjustment and a 4.5% merit increase. What is his new salary?

Refer to the table on page 288 for problem 10.

10. Mary Joseph is a computer technician, level 4. She receives a 4.8% cost-of-living adjustment and a 9% merit increase. What is her new salary?

11. Tomas Smit is a senior credit correspondent, level 10. He receives a 3.7% cost-of-living adjustment and a 1.3% merit increase. What is Tomas's new salary?

12. Frank Andrews works 40 hours a week for the Miracle-Clean Commercial Laundry. He earns $5.66 an hour.
    a. What is Frank's annual gross pay?
    b. If Frank receives a 6.8% merit increase, what would be his new annual gross pay?

Maintaining Your Skills

Look up the skills in parentheses if you have difficulty with the problems.

Find the present salary by using the table on page 288. (Application C)

13. Computer programmer, level 2
14. Computer operator, level 1
15. Secretary, level 2

Find the percentage. (Skill 30)

16. 5.5% × $11,350
17. 3.2% × $11,100
18. 10.6% × $13,200
19. 1.5% × $26,060
20. 2.9% × $12,030
21. 4.3% × $14,240

12-3 Employee Benefits

Objective

Compute the rate of employee benefits based on annual gross pay.

Your business may offer several **employee benefits.** Employee benefits include health and dental insurance, life insurance, pensions, paid vacations, unemployment insurance, and disability insurance. The total of the benefits may be figured as a percent of annual gross pay.

▶ **Rate of Benefits** $= \dfrac{\textbf{Total Benefits}}{\textbf{Annual Gross Pay}}$

Example

Skills 30, 31 **Application** A **Term** Employee benefits

The Personnel Department of the Glen Rock Printing Company is preparing annual reports on employee benefits. Kim Sonoyama's total annual benefits are what percent of her annual salary?

KIM SONOYAMA — Annual salary: $16,640.00 — Weekly salary: $320.00

Benefits	
Vacation: 2 weeks × $320.00	$ 640.00
Holidays: 8 days × ($320.00 ÷ 5)	512.00
Health Insurance: 12 months × $76.32	915.84
Dental Insurance: 12 months × $11.05	132.60
Disability Insurance: 3.6% of $16,640.00	599.04
Unemployment Insurance: 4.1% of $16,640.00	682.24
Social Security: 7.15% of $16,640.00	1189.76
Total	$4671.48

Solution

Find the **rate of benefits.**

Total Benefits ÷ Annual Gross Pay

$4671.48 ÷ $16,640.00 = 0.2807 = 28% rate of benefits

20%

▶ **Self-Check** Cover the answer and complete the problem.

Annual salary is $25,000. Total benefits are $5000. Find the rate of benefits.

Problems

Round answers to the nearest percent.

	Total Benefits	÷	Annual Gross Pay	=	Rate of Benefits
1.	$2197	÷	$ 8788	=	?
2.	$5745	÷	$19,180	=	?

	Total Benefits	÷	Annual Gross Pay	=	Rate of Benefits
3.	$3385	÷	$12,194	=	?
4.	$1364	÷	$ 7436	=	?
5.	$4868	÷	$16,500	=	?

6. Retail clerk's annual salary $9048.
Benefits:
  Vacation $348
  Holidays $278.40
  Health insurance $1044
  Unemployment insurance $370.97
  Social security $646.93
  Compensation insurance $325.73
What are the total benefits?
What is the rate of benefits?

7. Technician's annual salary $16,432.
Benefits:
  Vacation $632
  Holidays $505.60
  Term life insurance $43.85
  Unemployment insurance $673.71
  Social security $1174.89
  Compensation insurance $591.55
What are the total benefits?
What is the rate of benefits?

8. **a.** Complete the benefits chart for the Fort Perry Motel employees.
b. What is the rate of benefits for each employee?

Position	Annual Wage	Three-Week Vacation	2.8% Compensation Insurance	4.9% Unemployment Insurance	7.15% Social Security	Total Benefits
Manager	$14,800	?	?	?	?	?
Desk clerk	$ 9048	?	?	?	?	?
Laundry worker	$ 8008	?	?	?	?	?
Housekeeper	$ 7280	?	?	?	?	?

9. **a.** Complete the benefits chart for Dr. Warren's dental office staff.
b. What is the rate of benefits for each employee?
c. How much does Dr. Warren pay in benefits for her 3 employees?
d. How much more would it cost Dr. Warren to give each employee a 3-week vacation?

Position	Annual Wage	Two-Week Vacation	3.1% Compensation Insurance	3.8% Unemployment Insurance	7.15% Social Security	Total Benefits
Lab technician	$18,600	?	?	?	?	?
Hygienist	$12,500	?	?	?	?	?
Receptionist	$11,336	?	?	?	?	?

Maintaining Your Skills

Look up the skills in parentheses if you need help or more practice.

Find the percentage. (Skill 30)

10. 3.6% × $15,000   11. 4.1% × $15,000   12. 7.15% × $8900

Find the rate. Round answers to the nearest hundredth. (Skill 31)

13. $1625 ÷ $32,500   14. $1176 ÷ $16,800   15. $425 ÷ $8500

16. $2648 ÷ $26,480   17. $3560 ÷ $49,795   18. $1787 ÷ $7148

12-4 Travel Expenses

Objective Compute the total business-travel expenses.

If you travel for your business, you will probably be reimbursed, or paid back, for all authorized expenses during your trip. **Travel expenses** usually include transportation, lodging, and meals.

▶ **Total Travel Expense = Cost of Transportation + Cost of Lodging + Cost of Meals + Additional Costs**

Example

Skills 8, 5 **Application** A **Term** Travel expenses

The Personnel Department of the Mainland Insurance Company will reimburse Sherry Jacoby for attending a 2-day professional conference. Sherry drove to the conference, so Mainland will pay her $0.20 per mile. Sherry's expenses included:

Tolls: $0.90
Hotel: $37.54 per night for 1 night
Conference Registration: $27.50
Meals (including tips):

Friday luncheon	$ 6.25	Saturday	breakfast	$4.15
dinner	$11.50		luncheon	$5.70
			dinner	$8.85

Mileage: 280 miles round trip

What is the total cost of sending Sherry to the conference?

Solution

1. Find the **cost of transportation.**
   (280 × $0.20) + $0.90
   $56.00 + $0.90 . . . . $ 56.90
2. Find the **cost of lodging**. . . . . $ 37.54
3. Find the **cost of meals.**
   $6.25 + $11.50 + $4.15 + $5.70 + $8.85 . . . . $ 36.45
4. Find the **additional costs.**
   Conference registration . . . . $ 27.50

Total Travel Expense $158.39

$425

▶ **Self-Check** Cover the answer and complete the problem.

Find the total travel expense.
Cost of transportation, $250; cost of lodging, $50; cost of meals, $50; additional costs, $75.

Problems

Complete the table for these business trips.

	1.	2.	3.	4.	5.	6.
Name	L. Herman	J. Larson	A. Nguyen	J. Rotholz	S. Vancott	H. Burns
Miles Traveled	74	96	135	283	796	416
Cost at $0.20/mile	$14.80	?	?	?	?	?
Meals	$31.50	$18.45	$54.07	$147.30	$327.53	$248.97
Hotel Room	$38.20	0	$69.18	$172.62	$526.88	$243.30
Total Expenses	?	?	?	?	?	?

7. Justine Arega.
   Air fare: $521.36.
   Hotel: 3 nights at $35.74 each.
   Meals: $78.44.
   Entertainment: $47.50.
   What is the total travel expense?

8. Milton Jones.
   Train: $115.82.
   Hotel: 2 nights at $42.65 each.
   Meals: $59.60.
   Conference registration: $65.00.
   What is the total travel expense?

9. Ann Ellis is a consultant for Keller Business Machines, Inc. This month her travel expenses included airplane fares of $144.80 and $236.72, 192 miles of driving at $0.17 per mile, taxicab fares of $6.90 and $5.25, and meals totaling $59.84. What is the total travel expense for the month?

10. Roy Kondo, a trouble-shooter for Dot Plastics Corporation, flew to Mexico City to repair a heat sealing machine. Air fare was $327.54. Roy rented a car for 3 days for $18.20 a day plus $0.15 a mile. He drove a total of 66 miles. Roy's hotel bill was $42.20 a night for 2 nights. His meals cost $6.20, $9.14, $3.47, $4.80, $11.50, $3.47, and $5.12. What is Roy's total travel expense?

Maintaining Your Skills

Look up the skills in parentheses if you need help or more practice.

Multiply. (Skill 8)

11. 350 × $0.20
12. 900 × $0.30
13. 740 × $0.25
14. 180 × $0.15
15. 540 × $0.22
16. 450 × $0.19
17. 422 × $0.21
18. 986 × $0.17
19. 1842 × $0.17
20. 8612 × $0.23
21. 10,642 × $0.16
22. 12,463 × $0.13

Add. (Skill 5)

23. $24 + $32 + $16
24. $54 + $63 + $21
25. $112 + $86 + $72
26. $32 + $93 + $46

12-5 Employee Training

Objective Compute the total employee training costs.

Your business may pay the expenses involved in training employees. Your company may send you to special job-related programs or may offer special training programs within the company. Expenses for training during regular work hours include the cost of **release time.** When you are granted release time, you are paid your regular wages or salary while you are away from your job.

▶ **Total Training Cost** = **Cost of Release Time** + **Cost of Instruction** + **Additional Costs**

Example

Skills 8, 5 **Application** A **Term** Release time

The Lion Manufacturing Company chose 10 employees to attend a training program in the company. The employees were paid their regular wages while attending the three-day program. Their combined wages amounted to $404.62 per day. The chief quality control engineer, who earns $105.77 per day, was the course instructor. Refreshments were brought in twice a day at a cost of $21 per day. Supplies and equipment for the program amounted to $25 per person. What was the total cost for the training program?

Solution

1. Find the **cost of release time.**
   3 days × $404.62 = $1213.86
2. Find the **cost of instruction.**
   3 days × $105.77 = $317.31
3. Find the **additional costs.**
   (3 × $21) + (10 × $25)
   $63 + $250 = $313.00
4. Find the **total training cost.**
   Cost of Release Time + Cost of Instruction + Additional Costs
   $1213.86 + $317.31 + $313.00 = $1844.17 total training cost

$1100

▶ **Self-Check** Cover the answer and complete the problem.

Find the total training cost.

Cost of release time, $950; cost of instruction, $50; additional costs, $100.

Problems

	1.	2.	3.	4.	5.	6.
Number of Days	1	1	4	3	2	5
Daily Cost of Release Time	$592	$728	$4243	$1201	$34	$65
Daily Cost of Instruction	$ 80	$128	$ 258	$ 111	$75	$90
Daily Cost of Supplies	$ 46	$103	$ 120	$ 109	$10	$24
Total Training Cost	?	?	?	?	?	?

7. 3 attorneys.
Attend 2-day seminar.
$312 per day total for release time.
$87 per day total for instruction.
$6 per day total for supplies.
What is the total training cost?

8. 4 sales representatives.
Attend 3-day workshop.
$290 per day total for release time.
$95 per person for instruction.
$347 per person total travel expense.
What is the total training cost?

9. The Data Processing Department is sending 8 programmers to a 2-day seminar on a new version of BASIC. Three of the programmers earn $84 per day, while the other 5 earn $126 per day. Registration costs $250 per person. Materials cost $25 per person. What will the 2-day seminar cost the Data Processing Department?

10. The Wilson Company is sending their 5-person accounting department to a 2-day conference to learn a new, automated accounting procedure. One accountant earns $124.81 daily, two earn $110.15 daily, and two earn $83.81 daily. What will the 2-day conference cost the Wilson Company in lost productivity for the 5 accountants?

11. Ten employees of the Solar Manufacturing Company will undergo a 5-day sales training program. Their hourly wage rate averaged $8.93 per hour for an 8-hour day. The sales manager, at a daily salary of $184.19, conducted the 5-day session. Refreshments cost $24.95 per day. Sales kits were provided for each of the 10 employees at a cost of $47.95 each. What is the total cost of the sales training program?

Maintaining Your Skills

Look up the skills in parentheses if you need help or more practice.

Multiply. (Skill 8)

12. 4 × $125 13. 5 × $65 14. 10 × $15 15. 3 × $864 16. 2 × $193

Divide. Round to the nearest cent. (Skill 10)

17. $27,300 ÷ 52 18. $27,300 ÷ 260 19. $17,992 ÷ 260

Add. (Skill 5)

20. $1420 + $200 + $150 21. $764 + $135 + $37

22. $215 + $68 + $37 23. $4275 + $576 + $59

Reviewing the Basics

Skills

Solve.

(Skill 5) **1.** \$519.82 + \$643.71 **2.** \$16,043.25 + \$135.79 + \$1073.22

(Skill 8) **3.** \$598.03 × 6 **4.** 685 × \$0.19

(Skill 10) **5.** \$64,168 ÷ 52 **6.** \$16,968 ÷ 210

(Skill 30) **7.** 7.2% of \$26,380 **8.** 4.6% of \$8426.00

Round to the nearest tenth of a percent.

(Skill 31) **9.** n% of \$22,804 = \$6508 **10.** n% of \$13,020 = \$3214.50

Application

Use the table to answer.

(Application C) **11.** What is the base salary for a marketing assistant with 2 years of experience and a Bachelor's degree?

12. What is the base salary for a marketing assistant with 1 year of experience and a Master's degree?

Marketing Assistant—Base Salary

Years Experience	Degree: Bachelor's	Degree: Master's
0	\$20,384	\$22,880
1	22,014	24,710
2	23,744	26,668
3	25,676	28,824

Terms

Use each term in one of the sentences.

13. Employee benefits

14. Release time

15. Recruiting costs

16. Salary scale

17. Travel expenses

a. The Personnel Department may use a __?__ to compare the salaries of various jobs.

b. __?__ may include the cost of advertising in the newspaper.

c. Paid vacations and holidays are __?__ that most businesses give their employees.

d. When training sessions are held during working hours, the cost of __?__ is one of the expenses.

e. If a job requires travel, the employee may be reimbursed for __?__ which include transportation, lodging, and meals.

f. A __?__ is a raise in an employee's salary.

Refer to your reference files at the back of the book if you need help.

Problem Solving   Trial and Error

One way to solve a problem that seems to defy solution is the trial and error method. Guess a number that could be the solution and then check it out. Continue guessing and checking until you find the correct answer.

Example 1 You need 17 lbs of fertilizer. You can buy 3 lbs for \$2.29 or 5 lbs for \$3.25. What do you buy to obtain at least 17 lbs at the lowest cost?

Solution Guess and check until you find the lowest cost combination.

	Amount			
	5 lbs.	3 lbs.	Cost	Total
1.	4	0	4 × \$3.25	\$13.00
2.	3	1	(3 × \$3.25) + (1 × \$2.29)	12.04
3.	2	3	(2 × \$3.25) + (3 × \$2.29)	13.37
4.	1	4	(1 × \$3.25) + (4 × \$2.29)	12.41
5.	0	6	6 × \$2.29	13.74

The lowest cost combination is three 5 lbs. and one 3 lb. at a cost of \$12.04.

Example 2 Bob's father is five times as old as Bob is. Five years from now, he will be only three times as old. How old is Bob?

Solution Guess and check.

Current Age		Five Years from Now		
Bob's Age	Father's Age	Bob's Age	Father's Age	Check
3	15	8	20	No
4	20	9	25	No
5	25	10	30	Yes

Therefore, Bob is now 5 years of age.

Problems

1. The planet Kroton has only creatures with 3 legs (triads) and 4 legs (tetrads). Astronaut Alice Browning could not bear to look at these ugly creatures, so she kept her eyes on the ground. On her first day on Kroton she counted 31 legs. How many triads and tetrads did she meet?

2. It costs 14¢ to mail a post card and 22¢ to mail a letter. Victor Yee wrote to 13 friends and spent \$2.38 for postage. How many letters and how many post cards did he write?

3. Sandra Miller earns \$7.00 per hour Monday through Friday and \$10.50 per hour on weekends. One week she worked 49 hours and earned \$367.50. How many hours did she work on the weekend?

Unit Test

Lesson 12-1

1. Data Tech Corporation hired Eloise Floyd for a new clerk-typist position at an annual salary of $12,000. The hiring costs included:

   Advertising: $285
   Interviews: Mark Craig, $25; Eloise Floyd, $30
   Top Skills Agency fee: 20% of first year's salary

   What was the total cost of hiring Eloise?

Lesson 12-2

2. Ray Houlson is a statistician, level 2, for the Lifelong Insurance Corporation. Ray received a 7.2% cost-of-living increase and a 4.5% salary increase. What will Ray's new salary be?

Level	Statistician
1	$12,650
2	$14,770
3	$16,960

Lesson 12-3

3. Sonia Graham works for the Hollis Electrical Corporation. She earns $11,400 per year. Hollis provides these benefits for Sonia. To the nearest percent, what is the rate of benefits?

Vacation: 2 weeks
Holidays: 11 days
Compensation Insurance: 3.6%
Unemployment Insurance: 4.9%
Social Security: 7.15%

Lesson 12-4

4. Robert Hosang attended a professional conference last month. The Personnel Department reimbursed Robert for these expenses:

   Air fare: $273.80
   Hotel: 2 nights at $32.50 each
   Meals: $62.55
   Registration: $25.00

   What was the total travel expense?

Lesson 12-5

5. The Personnel Department of the Corbette Company is sending 4 of its employees to a local 2-day training program. The 4 employees' combined wages total $280 per day. The company paid the registration fee of $80 for each person. The company also paid for the employees' lunches for the 2 days at a total cost of $53.72. What was the total cost for the Corbette Company to send their employees to the training program?

Applying for a Job: A Simulation

Suppose you have decided to get a summer job to earn extra money. To decide what kind of work you might like to do, you talk to friends, family, teachers, and others. There are many factors to consider. Here are some of them.

Compensation
hourly pay
benefits
overtime pay
tips
commission

Transportation
location
public transportation
need for car
cost

Interest
interesting work
work related to career goals

Hours
regular hours
night work
weekend work
holiday work
overtime work

Conditions
duties
indoor work
outdoor work
clean, quiet environment

In comparing the wages of different jobs, consider costs too. For example, if you have to spend a lot of money for gasoline to drive to and from a job, you may end up with less money to spend than if you had a lower-paying job closer to home.

1. Look at the list of factors above. You may consider other factors important. List the ten factors that are most important to you, with the most important one first. (If you have never had a job, you may want to discuss the factors with someone who has.)

Calculate the spendable income for each job.

	Hourly Rate	Hours per Week	Gross Pay	Deductions	Net Pay	Transportation Cost	Spendable Income
2.	$3.50	40	$140	$10	$130	$ 5	?
3.	3.75	38	?	16.50	?	0	?
4.	3.90	40	?	12.50	?	10	?
5.	4.25	25	?	8.25	?	22	?

Job Interviews

Usually an employer interviews many applicants for a job opening. It is important that you dress neatly, speak clearly, and be able to answer the interviewer's questions. Many people are nervous before and during an interview. It is helpful to know what type of questions an interviewer might ask. Here are some of the most common questions. Write your answers to these questions.

6. What kind of work are you most interested in? Why?

7. What are your future vocational or educational plans?

8. What do you know about our company? Why do you think you might like to work for us?

9. What jobs have you held? What did you learn from them?

10. What courses in school did you like best? Least? Why?

11. What is your average grade in school? Do you think you have done the best academic work of which you are capable? If not, why not?

12. In what school activities have you participated? Which did you like most? Why? Have you been an officer in any activity?

13. How do you spend your spare time? What are your hobbies?

14. Do you prefer working with others or by yourself?

15. What is your major strength? What is your major weakness?

16. What have you done that shows initiative and willingness to work?

17. What qualifications do you have that make you feel you will be successful in your field?

An interview is an opportunity for you to ask questions about the job and the company. You may want to ask about pay, benefits, duties, tuition credit, opportunities for advancement, or other factors.

18. Find a want ad for a job that interests you. Write five specific questions you might ask if you were being interviewed for the job.

If there is time, you and someone in your class might take turns interviewing each other.

Job Application

When you apply for a job you will be asked to fill out a job application form. You should be prepared to complete all items on the form. If you do not have a social security number, you should apply for one. Be sure to receive permission to use the names of your personal character references.

Here is a sample of a job application form. Write down your answers as if you were filling out the form.

Application for Employment

Miss
Mr.
Mrs.
Ms. ____

(Last) *(First)* *(Middle initial)*

Present Address ____

(Street) *(City)* *(State)* *(Zip)*

Date ____________ Telephone No. ____________ Social Security No. ____________

In case of emergency, notify ____

(Name) *(Telephone No.)*

____

(Address)

EDUCATIONAL BACKGROUND

	Name and Address of School	Dates Attended		Date Graduated	Major Area of Study
		From	To		
Elementary School					
High School					
College					
Other					

Check business courses taken in school:

- ☐ Typewriting
- ☐ Shorthand
- ☐ Recordkeeping
- ☐ Accounting
- ☐ Office Practice
- ☐ General Business
- ☐ Business Law
- ☐ Business Math
- ☐ Consumer Economics
- ☐ Office Machines

Other ____________

Typing Speed ____________ WPM ____________

List extracurricular activities in school: ______

______

______

EMPLOYMENT: List most recent employment first

Name and Address of Employer	Dates Employed		Position	Reason for Leaving
	From	To		

REFERENCES (Do not list relatives)

Name	Position or Title	Address	Telephone
1.			
2.			
3.			

MISCELLANEOUS

Hours available for interview: ______

Number of days per week you wish to work: Minimum ______ Maximum ______

Are you willing to work evenings? ______ Weekends? ______

Signature ______

Unit 13 Production

▸ The Production Department of your business makes, or *manufactures*, and packages the products you sell. The costs of manufacturing an item include both materials and labor. The income from your sales must cover your manufacturing costs for you to break even. Sales beyond the *break-even point* result in profits. You may use *quality controls* to check for defective items during manufacturing. *Time studies* can tell you how much time is needed to make one item.

Because it is impractical to test every product, quality control checks are made on a representative sample.

13-1 Manufacturing

Objective Compute the prime cost of manufacturing an item.

The cost of manufacturing an item depends, in part, on the **direct material cost** and the **direct labor cost.** The direct material cost is the cost of the goods that you use to produce the item. The direct labor cost includes the wages paid to the employees who make the item. The **prime cost** is the total of the direct material cost and the direct labor cost. The prime cost is frequently expressed on a per-unit basis.

▶ **Prime Cost per Item** = **Direct Material Cost per Item** + **Direct Labor Cost per Item**

Example

Skills 11, 2 **Application** A **Term** Prime cost

Andia, Inc., produces aluminum shelving brackets. The machine operator stamps 20 brackets from each strip of aluminum. Each strip costs $0.60. The operator can stamp 1200 brackets per hour. The direct labor charge is $18 per hour. To the nearest tenth of a cent, what is the prime cost of manufacturing a bracket?

Solution

1. The **direct material cost per item** = $0.60 ÷ 20 = $0.03
2. The **direct labor cost per item** = $18.00 ÷ 1200 = $0.015
3. Find the **prime cost per item.**
   Direct Material Cost + Direct Labor Cost
   $0.03 + $0.015 = $0.045 or 4.5¢ cost per item

A. $0.07
B. $0.32

▶ **Self-Check** Cover the answers and complete the problems.

A. 10 brackets are made from a strip of metal costing $0.70. What is the direct material cost per item?

B. The direct material cost per item is $0.07. The direct labor cost per item is $0.25. Find the prime cost per item.

Problems

Round to the nearest tenth of a cent.

	Cost per Strip	Pieces per Strip	Direct Material Cost per Piece	Labor Cost per Hour	Pieces per Hour	Direct Labor Cost per Piece	Prime Cost
1.	$0.90	18	$0.05	$ 8.80	2200	$0.004	?
2.	$0.55	22	$0.025	$10.50	1500	?	?
3.	$0.68	17	?	$18.96	3160	?	?

Round to the nearest tenth of a cent.

4. 24 boxes per sheet of cardboard.
Cost: $3.60 per sheet.
Direct labor charge: $8.40 per hour.
240 boxes cut per hour.
What is the prime cost per box?

5. 4600 washers per steel strip.
Cost: $6.90 per strip.
Direct labor charge: $12.65 per hour.
12,650 washers punched per hour.
What is the prime cost per washer?

6. Donnely and Company manufactures plastic shower curtains. Each roll of printed plastic yields 112 curtains. Each roll costs $9.60. The direct labor charge is $10.20 per hour. The machine operator can cut and fold 85 curtains per hour. What is the prime cost of manufacturing 1 curtain?

7. Harbor Metalcraft Company manufactures aluminum products. One of their machines stamps small fans for hairdryers from strips of aluminum. Each strip of aluminum costs $0.75. The machine stamps 36 fans from each strip. The machine operator produces 12 fans per minute. The direct labor cost is $10.80 per hour. What is the prime cost of manufacturing 1 fan?

8. Rita DeVaney is a machine operator at Tech Manufacturing. She stamps 25 corner supports from a sheet of heavy gauge tin. Each sheet of tin costs $0.625. Rita's machine stamps out 1 corner support every 4 seconds. The direct labor cost is $8.60 per hour. What is the prime cost of manufacturing 1 corner support?

9. A strip of steel makes 39 gusset plates. The cost per strip is $0.80. The direct labor cost is $9.45 per hour. 180 plates are stamped per hour. What is the prime cost of manufacturing 1 gusset plate?

10. A strip of aluminum makes 20 electrical box housings. The aluminum costs $0.90 per strip. The direct labor cost is $8.75 per hour. One housing is stamped every 5 seconds. What is the prime cost of manufacturing 1 electrical box housing?

Maintaining Your Skills

Look up the skills in parentheses if you need help or more practice.

Round to the nearest tenth of a cent. (Skill 2)

11. $0.14546	12. $0.02186	13. $0.00145
14. 15.89¢	15. 6.1102¢	16. 1.095¢

Divide. Round answers to the nearest tenth of a cent. (Skill 11)

17. $0.70 ÷ 5	18. $0.87 ÷ 6	19. $0.85 ÷ 6	20. $0.97 ÷ 8
21. $1.24 ÷ 12	22. $1.45 ÷ 6	23. $12.60 ÷ 900	24. $8.25 ÷ 250
25. $14.56 ÷ 1200	26. $8.45 ÷ 900	27. $9.18 ÷ 600	28. $10.56 ÷ 1120

13-2 Break-even Analysis

Objective Compute the break-even point in the number of manufactured units.

Your business can prepare a **break-even analysis** to determine how many units of a product must be made and sold to cover production expenses. The **break-even point** is the point at which income from sales equals the cost of production. Units sold after that point will result in a **profit** for your business.

To calculate the break-even point, you must know the **total fixed costs,** the **variable costs per unit,** and the selling price per unit. Fixed costs include rent, salaries, and other costs that are not changed by the number of units produced. Variable costs include the cost of raw materials, the cost of packaging, and any other costs that vary directly with the number of units produced.

$$\text{Break-even Point in Units} = \frac{\text{Total Fixed Costs}}{\text{Selling Price per Unit} - \text{Variable Costs per Unit}}$$

Example

Skills 6, 11 **Application** A **Term** Break-even point

Wilson Electronics manufactures light bulbs. They plan to manufacture 750,000 special candleglow bulbs to be sold at $0.39 each. The fixed costs are estimated to be $140,250. Variable costs are $0.14 per unit. How many bulbs must be sold for Wilson Electronics to break even?

Solution

Find the **break-even point in units.**

Total Fixed Costs ÷ (Selling Price per Unit − Variable Costs per Unit)

$140,250 ÷ ($0.39 − $0.14)

$140,250 ÷ $0.25 = 561,000 break-even point in units

Self-Check Cover the answers and complete the problems.

A. 150,000
B. 950,000

Find the break-even point in units.

A. Selling price is $1.29.
Variable cost per unit is $0.29.
Total fixed costs are $150,000.

B. Selling price is $0.49.
Variable cost per unit is $0.39.
Total fixed costs are $95,000.

Problems

	Total Fixed Costs	÷	(Selling Price per Unit	−	Variable Costs per Unit)	=	Break-even Point in Units
1.	$ 102,000	÷	($ 2.49	−	$ 2.19)	=	?
2.	$ 630,000	÷	($ 3.79	−	$ 3.09)	=	?
3.	$ 382,500	÷	($ 12.98	−	$ 12.13)	=	?
4.	$30,660,065	÷	($237.98	−	$127.89)	=	?
5.	$ 95,000	÷	($ 0.79	−	$ 0.55)	=	?
6.	$ 65,000	÷	($ 9.99	−	$ 8.74)	=	?

7. Style-Rite table lamps.
Fixed costs: $535,800.
Selling price per lamp: $57.95.
Variable cost per lamp: $31.16.
What is the break-even point in units?

8. Weekend Fun volley balls.
Fixed costs: $705,600.
Selling price per volley ball: $14.79.
Variable cost per volley ball: $8.91.
What is the break-even point in units?

9. Waxco presses $33\frac{1}{3}$ r.p.m. stereo records. The fixed costs total $1,666,500. The selling price per record is $6.95. The variable cost per record is $5.14. What is the break-even point in number of records?

10. The Custom Car Company manufactures a limited number of luxury automobiles. The fixed costs of the operation total $33,552,000. The selling price per custom car is $42,750. The variable cost per custom car is $25,974. What is the break-even point in number of automobiles?

11. Cav-No-More toothpaste is manufactured by National Food Corporation. They have total fixed costs of $3,135,000. The variable cost per tube is 24¢. The selling price per tube of Cav-No-More toothpaste is 79¢. What is the break-even point in tubes of Cav-No-More toothpaste?

12. Tom and Judy Peters plan to produce and market monogrammed T-shirts. They anticipate selling each T-shirt for $4.95. Their variable costs per T-shirt include $1.19 for supplies and $1.00 per shirt for direct labor. The fixed costs of the operation total $27,600. How many monogrammed T-shirts do the Peters need to sell to break even?

Maintaining Your Skills

Look up the skills in parentheses if you need help or more practice.

Subtract. (Skill 6)

13. $0.49 − $0.24
14. $0.24 − $0.11
15. $0.74 − $0.39
16. $1.19 − $0.73
17. $4.99 − $2.22
18. $44.49 − $25.62

Divide. Round answers to the nearest whole number. (Skill 11)

19. $135,000 ÷ $1
20. $200,000 ÷ $0.50
21. $175,000 ÷ $2.50

13-3 Quality Control

Objective

Compute the percent of defective goods and determine if the process is in or out of control.

If your business deals with mass production, you need a **quality control** inspector to check the items that are manufactured. The inspector may examine a specified number of items. If the size of the item is incorrect or if the item is broken or damaged, it is classified as **defective.** The inspector computes the percent of the sample that is defective and plots the percent on a **quality control chart.** The quality control chart shows the percent of defective products that is allowable. If the actual percent of defective items is greater than the percent allowable, the process is said to be "out of control." Production must then be stopped and corrected.

▶ **Percent Defective** = $\frac{\textbf{Number Defective}}{\textbf{Total Number Checked}}$

Example

Skills 1, 31 **Application** N **Term** Quality control

Chris Hingham is a quality control inspector for Andia, Inc. At 9 A.M. Chris checked 50 stamped brackets produced during the previous hour and found 3 defective brackets. The process is in control if 4% or less of the sample is defective. Is the process in control or out of control? Plot the percent defective on a quality control chart.

Solution

1. Find the **percent defective.**
   Number Defective ÷ Total Number Checked
   3 ÷ 50 = 0.06 = 6% defective
   Process is out of control.
2. Plot the **percent defective** on a quality control chart.

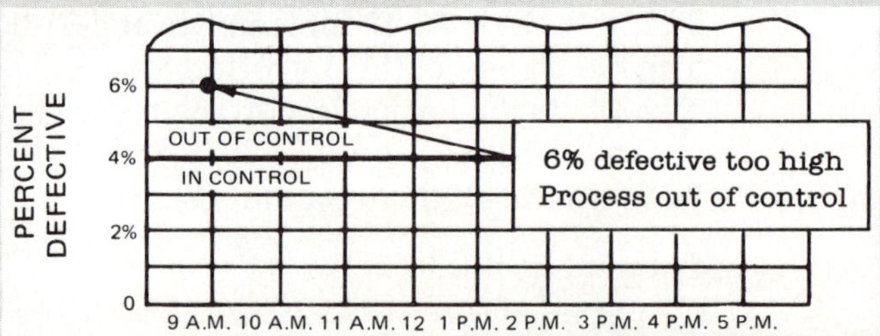

A. 4%
B. 8%

▶ **Self-Check** Cover the answers and complete the problems.

A. Number defective is 8.
Number checked is 200.
What is the percent defective?

B. Number defective is 2.
Number checked is 25.
What is the percent defective?

Problems

If more than 4% of the sample is defective, the process is out of control.

	Number Defective	÷	Number Checked	=	Percent Defective	In or Out of Control?
1.	1	÷	50	=	?	?
2.	2	÷	50	=	?	?
3.	0	÷	50	=	?	?
4.	4	÷	50	=	?	?
5.	6	÷	50	=	?	?
6.	5	÷	50	=	?	?

7. Cotton shirts.
250 in sample.
17 defective.
In control if 6% or less defective.
Is the process in or out of control?

8. Camper's cooking kit.
120 in sample.
7 defective.
In control if 5% or less defective.
Is the process in or out of control?

9. Lauri Cheung is a quality control inspector for Cityscene Records, Inc. The process is in control if 8% or less of each sample is defective. Lauri checked a sample of 25 records and found 4 defective records. What percent of the sample is defective? Is the process in or out of control?

10. Marvin Keller, a quality control inspector for Household Supply, checked a sample of 230 ballpoint pens. Fourteen of the pens were defective. If more than 6% of the sample is defective, the process is out of control. What percent of the sample is defective? Is the process in or out of control?

For Exercises 11 and 12, draw a quality control chart, compute the percent defective in each sample, and plot the percents on the chart.

11. Every 2 hours, May Litton checks samples from a punch press. Each sample contains 40 pieces. The operation is in control if 5% or less of each sample is defective. May found the following:

Time	9 A.M.	11 A.M.	1 P.M.	3 P.M.	5 P.M.
Number Defective	1	0	3	2	4

12. Every hour, 10 television sets are taken off the production line at United Communications and checked to see if they are defective. If more than 10% are defective, production is out of control. The number of defective sets for the 9 A.M. to 5 P.M. checks are:

Time	9 A.M.	10 A.M.	11 A.M.	12 NOON	1 P.M.	2 P.M.	3 P.M.	4 P.M.	5 P.M.
Number Defective	0	1	0	2	1	0	1	3	2

(Continue on next page.)

13. Every 3 hours, Lou Cleary pulls 20 bicycle rims off the assembly line and checks their diameters. If more than 8% of the rims are found to be defective, the assembly line is shut down. Determine whether the process is in or out of control at each time check. Draw a quality control chart.

Time	9:30 A.M.	12:30 P.M.	3:30 A.M.	Day
Number Defective	0	1	1	Monday
	1	0	0	Tuesday
	1	2	2	Wednesday
	0	0	1	Thursday
	1	3	2	Friday

14. Monica Peters randomly checks 60 transistors 10 times a day. If more than 5% are defective, the production is stopped and corrections are made. Complete the table and draw a quality control chart.

Checkpoint No.	1	2	3	4	5	6	7	8	9	10
Number Defective	1	2	1	0	2	1	3	4	1	0
Percent Defective	?	?	?	?	?	?	?	?	?	?

15. Victor Young checks microchip production at different times during the day. Victor will check a different number each time. If more than 4% are defective, he will close down production and have corrections made. The microchips are checked electronically. Complete the table and draw a quality control chart.

Time	9:40 A.M.	10:15 A.M.	11:50 A.M.	1:30 P.M.	3:30 P.M.	4:30 P.M.
Number Checked	70	40	85	100	30	50
Number Defective	3	1	3	4	2	1
Percent Defective	?	?	?	?	?	?

Maintaining Your Skills

Look up the skills in parentheses if you need help or more practice.

Find the rate. Round answers to the nearest tenth of a percent. (Skill 31)

16. 3 is what percent of 50?

17. 5 is what percent of 100?

18. 4 is what percent of 60?

19. 2 is what percent of 75?

20. 1 is what percent of 40?

21. 4 is what percent of 125?

22. 5 is what percent of 130?

23. 3 is what percent of 90?

Which number is greater? (Skill 1)

24. 4.6% or 5%

25. 6% or 7.2%

26. 3.6% or 4%

27. 5% or 6.7%

28. 4.3% or 7%

29. 4% or 0%

30. 8% or 8.8%

31. 7.1% or 7.2%

13-4 Time-study—Number of Units

Objective Use time-study results to compute the time spent on each task.

Your business may conduct a **time-study** to determine how long a particular job should take. A time-study involves watching an employee complete a job, recording the time required for each task, and calculating the average time for each task. You can use the averages to determine how many units a worker can produce in a fixed period of time.

$$\text{Number of Units} = \frac{\text{Actual Time Worked}}{\text{Average Time Required per Unit}}$$

Example

Skills 5, 11 **Application** Q **Term** Time-study

I-Pak did a time-study of Tina Walter's job as a carton packer. Tina's times were recorded. The averages were calculated.

$$\frac{\text{Sum of Times}}{\text{Number of Observations}}$$

TASK	OBSERVATIONS IN SECONDS #1	#2	#3	#4	#5	AVERAGE TIME
PICK UP CARTON	4.0	5.5	4.0	4.5	4.0	4.4 sec
FILL CARTON	12.0	13.0	14.5	12.5	11.0	12.6 sec
APPLY GLUE	14.5	12.5	13.0	13.5	13.5	13.4 sec
CLOSE CARTON	5.0	4.3	4.5	4.2	5.0	4.6 sec
REMOVE FILLED CARTON	6.0	6.0	6.0	6.0	6.0	6.0 sec

If Tina gets a 10-minute break each hour, how many cartons can she fill per hour?

Solution

1. Find the **average time required per unit.**
   4.4 + 12.6 + 13.4 + 4.6 + 6.0 = 41 seconds
2. Find the **actual time worked** per hour.
   (60 − 10) minutes × 60 seconds = 3000 seconds
3. Find the **number of units** per hour.
   Actual Time Worked ÷ Average Time Required per Unit
   3000 ÷ 41 = 73.17 = 73 units per hour

A. 11
B. 15

Self-Check Cover the answers and complete the problems.

Find the number of units per hour.

A. Average time is 5 minutes per unit. 55 minutes worked per hour.

B. Average time is $\frac{1}{2}$ hour per unit. $7\frac{1}{2}$ hours worked per day.

Problems

	Actual Time Worked	÷	Average Time per Unit	=	Number of Units
1.	50 minutes	÷	3.54 seconds	=	?
2.	55 minutes	÷	23.44 seconds	=	?
3.	45 minutes	÷	53.76 seconds	=	?
4.	52.5 minutes	÷	74.20 seconds	=	?

5. Towel drying a car.
Average time is 10.96 minutes per car.
45 minutes worked per hour.
How many cars can be dried per hour?

6. Filling a container.
Average time is 8.06 minutes per container.
55 minutes worked per hour.
How many containers can be filled per hour?

7. Bob Jessup prepared a time-study to determine the average time required to use the check approval machine in the Quality Food Store.

	Time in Seconds				
Task	**#1**	**#2**	**#3**	**#4**	**#5**
Put check in machine	1.0	2.0	1.5	2.5	1.5
Type in code	17.0	11.5	16.0	15.5	14.5
Approval stamped	2.0	1.5	1.5	1.0	2.0
Remove check	1.5	1.0	2.0	1.5	1.5

a. What is the average time required for each task?
b. What is the average time required to have a check approved?
c. How many checks can be approved at one machine in one hour?

8. The Blackmer Supply Company prepared a time-study of Laura Hadley's job as an assembler. How many units can Laura complete in 1 hour if she takes one 5-minute break during the hour?

	Time in Seconds				
Task	**#1**	**#2**	**#3**	**#4**	**#5**
Move water pump to filter	3.6	4.1	3.8	3.7	3.6
Insert filter coil	5.1	5.2	4.9	5.2	5.1
Affix gasket	5.2	4.8	5.1	5.0	5.2
Hand thread bolts	5.1	5.3	5.3	5.2	5.4
Machine tighten bolts	5.2	5.0	5.3	5.4	5.3

Maintaining Your Skills

Look up the skills in parentheses if you need help or more practice.

Add. (Skill 5)

9. 4.4 + 4.6 + 4.1 + 4.2 + 4.6
10. 2.3 + 2.5 + 1.8 + 1.7 + 1.7
11. 16.8 + 11.4 + 9.8 + 14.5 + 11.8
12. 5.3 + 4.9 + 5.2 + 5.1 + 4.9

Divide. Round answers to the nearest whole number. (Skill 11)

13. 3000 ÷ 4.5
14. 3300 ÷ 2.4
15. 3600 ÷ 15.2
16. 8 ÷ 0.25

13-5 Time-study—Percent of Time

Objective

Use time-study results to compute the percent of time spent on each task.

You can use a time-study to determine what percent of an employee's time is spent on various activities during a work day.

▶ **Percent of Time Spent on Activity** = $\dfrac{\text{Time Spent on Activity}}{\text{Total Time}}$

Example

Skills 5, 31 **Application** A **Term** Time-study

Steve Mazer is a mail clerk in a large office. A time-study showed that he spent his time on these activities. What percent of his time does Steve spend picking up and delivering mail?

Activity	Hours
Sorting mail	3.0
Picking up and delivering mail	2.5
Talking with employees	1.0
Taking coffee breaks	0.5
Making special deliveries	1.0
Total	8.0

Solution

Find the **percent of time spent on activity.**

Time Spent on Activity ÷ Total Time

2.5 ÷ 8 = 0.3125 = 31.25% of time spent on activity

A. 37.5%
B. 12.5%

▶ **Self-Check** Cover the answers and complete the problems.

A. What percent of his time does Steve spend sorting mail in the example above?

B. What percent of his time does Steve spend making special deliveries in the example above?

Problems

	Activity	Hours	÷	Total Time	=	Percent of Time
1.	Picking up furniture	2.50	÷	8	=	?
2.	Delivering furniture	4.25	÷	8	=	?
3.	Lunch	1.00	÷	8	=	?
4.	Returning truck	0.25	÷	8	=	?

5. Herb Palmer, store clerk.
Unpacks merchandise 3.5 hours per day.
Works 8 hours per day.
What percent of his day is spent unpacking merchandise?

6. Joy Delacruz, realtor.
Shows homes 4.5 hours per day.
Works 8 hours per day.
What percent of her day is spent showing homes?

7. Sara Hynes, receptionist.
Answers telephones 3.2 hours per day.
Works 7.5 hours per day.
What percent of her day is spent answering telephones?

8. Joseph Ng, credit correspondent.
Records payments 2.4 hours per day.
Works 7.25 hours per day.
What percent of his day is spent recording payments?

9. George Kloman is a product planner for the Capitol Products Corporation. George works 7.5 hours per day. He spends about 1.75 hours each day making calls to obtain cost information for the components of the products on which he is working. What percent of George's day is spent obtaining cost information?

10. Naomi Harrison is a baker at Lowello's Restaurant. She spends 4 hours each day baking bread. A time-study recorded her activities. What percent of the 4 hours does Naomi spend on each activity?

Activity	Hours
Gathering ingredients	0.5
Mixing dough	1.6
Placing dough in pans and putting in oven	0.25
Removing bread from oven and taking out of pans	0.15
Cleaning up work area	0.8
Waiting time	0.7
	4.0

11. Carl Stranger is a waiter at the Cityview Café. He works 8 hours each day. A time-study showed these figures. What percent of Carl's work day is spent on each of the activities?

Activity	Hours
Setting up tables	0.5
Taking orders	1.1
Waiting for food to be prepared	1.2
Carrying food to tables	1.4
Serving food	1.75
Clearing tables	1.3
Totaling checks	0.25
Taking breaks	0.5
	8.0

Maintaining Your Skills

Look up the skill in parentheses if you need help or more practice.

Add. (Skill 5)

12. 2.2 + 1.6 + 3.3 + 0.9

13. 1.8 + 3.4 + 1.6 + 1.2

14. 4.18 + 1.63 + 1.17 + 1.02

15. 4.6 + 31.8 + 3.6

16. 4.24 + 16.48 + 11.72 + 7.56

17. 16.48 + 31.8 + 75.62 + 19.5

13-6 Packaging

Objective Compute the dimensions of packaging cartons.

The production of your company's merchandise ends with **packaging,** placing the product in a container for shipment. The container eases handling and prevents breaking. The package may also identify the product and show it attractively. The size of the package depends on the size of each finished product.

▸ **Dimensions: Length, Width, Height**

Example

Skills 16, 20 **Application** — **Term** Packaging

The Universal Company manufactures silicone waterproofing that is sold in bottles with a diameter of $1\frac{3}{4}$ inches and a height of $4\frac{1}{2}$ inches. Universal plans to package 12 bottles to a carton for shipping. The carton is made of $\frac{1}{8}$-inch thick corrugated cardboard with $\frac{1}{16}$-inch cardboard spacers. What are the dimensions of the package that Universal needs?

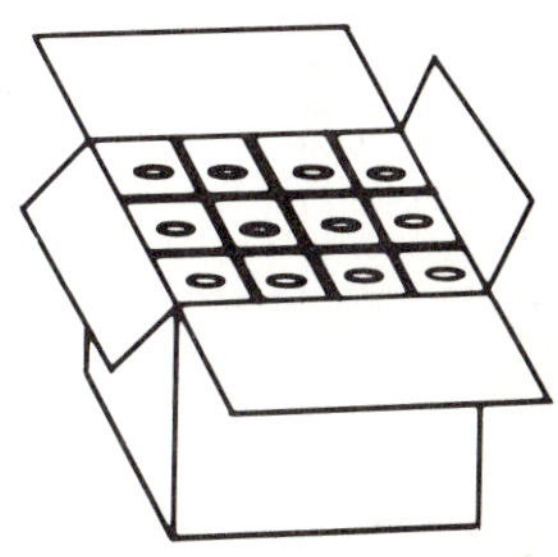

Solution

1. Find the **length.**

4 bottles × $1\frac{3}{4}$ inches in diameter	7	inches
3 partitions × $\frac{1}{16}$ inch wide	$\frac{3}{16}$	inches
2 ends × $\frac{1}{8}$ inch wide = $\frac{2}{8}$	$\frac{4}{16}$	inch
	$7\frac{7}{16}$	inches

2. Find the **width.**

3 bottles × $1\frac{3}{4}$ inches in diameter = $5\frac{1}{4}$	$5\frac{2}{8}$	inches
2 partitions × $\frac{1}{16}$ inch wide = $\frac{2}{16}$	$\frac{1}{8}$	inch
2 ends × $\frac{1}{8}$ inch wide = $\frac{2}{8}$	$\frac{2}{8}$	inch
	$5\frac{5}{8}$	inches

3. Find the **height.**

1 bottle × $4\frac{1}{2}$ inches = $4\frac{1}{2}$	$4\frac{2}{4}$	inches
2 top flaps × $\frac{1}{8}$ inch = $\frac{2}{8}$	$\frac{1}{4}$	inch
2 bottom flaps × $\frac{1}{8}$ inch = $\frac{2}{8}$	$\frac{1}{4}$	inch
	5	inches

4. Find the **dimensions.**
Length: $7\frac{7}{16}$ in.; Width: $5\frac{5}{8}$ in.; Height: 5 in.

A. $8\frac{1}{2}''$
B. $12\frac{1}{2}''$

▸ **Self-Check** Cover the answers and complete the problems.

A. Four packages, each 2 inches wide, next to each other. Each end is $\frac{1}{4}$-inch cardboard. How wide is the carton?

B. Six packages, each 2 inches long, next to each other. Each end is $\frac{1}{4}$-inch cardboard. How long is the carton?

Problems

Find the dimensions of the carton.

Height:

1. Top and bottom of 0.2 cm = ?
2. Height = ?
3. Total height = ?

Width:

4. 3 packages, each 1.5 cm wide = ?
5. 2 inserts, each 0.1 cm = ?
6. 2 ends, each 0.2 cm = ?
7. Total width = ?

Length:

8. 4 packages, each 2.5 cm long = ?
9. 3 inserts, each 0.1 cm = ?
10. 2 ends, each 0.2 cm = ?
11. Total length = ?

12. Dimensions?

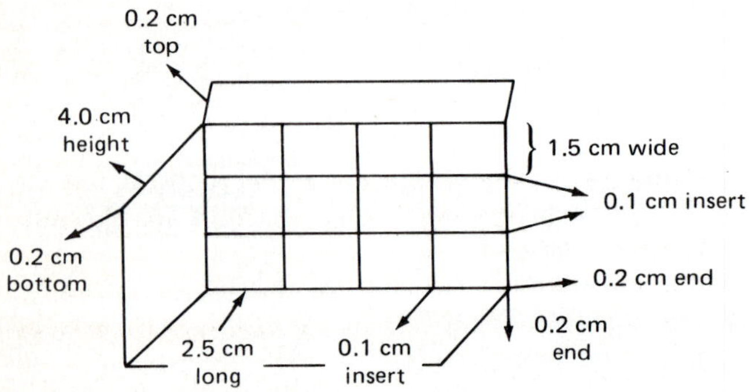

Draw a sketch for each carton in Exercises 13-17.

13. The Quick Copy Corporation packages 2 of these copiers per carton. One copier is placed on top of the other with a $\frac{3}{4}$-inch thick piece of cardboard between them. The carton is made of $\frac{5}{16}$-inch thick corrugated cardboard. What are the dimensions of the carton in which the copiers are packed?

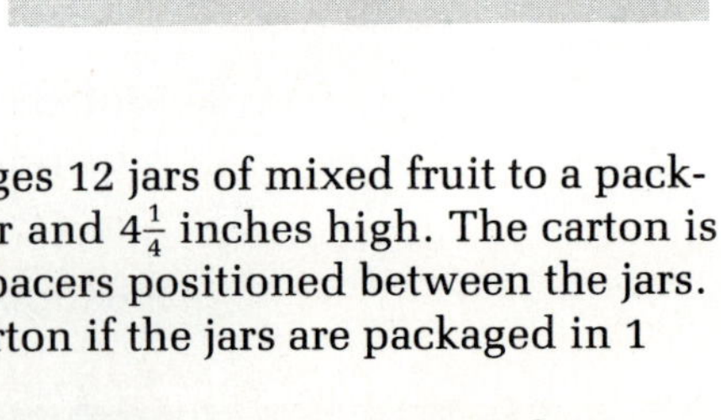

14. The Best Buy Foods Company packages 12 jars of mixed fruit to a package. Each jar is $2\frac{1}{2}$ inches in diameter and $4\frac{1}{4}$ inches high. The carton is $\frac{1}{8}$ inch thick and has $\frac{1}{16}$-inch thick spacers positioned between the jars.
    a. What are the dimensions of the carton if the jars are packaged in 1 row with 12 jars in the row?
    b. What are the dimensions of the carton if the jars are arranged in 2 rows with 6 jars in each row?

15. Time-rite Corporation packages 12 of their clock-radios in 1 carton. The carton is made of 0.4-cm thick corrugated cardboard. Spacers between the clock-radios are 0.2 cm thick.

a. What are the dimensions of the carton if the clock-radios are arranged in 2 rows of 6 clock-radios?

b. What are the dimensions of the carton if the clock-radios are arranged in 3 rows of 4 clock-radios?

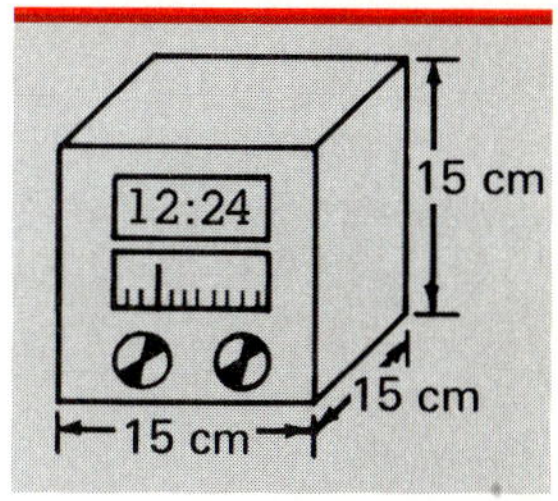

16. These colonial wall lamps are packaged 6 to a carton. The carton is constructed of $\frac{1}{4}$-inch thick cardboard. Cardboard spacers $\frac{1}{8}$-inch thick are positioned between the lamps.

a. What are the dimensions of the carton if the lamps are arranged standing up as shown, in 2 rows of 3 lamps?

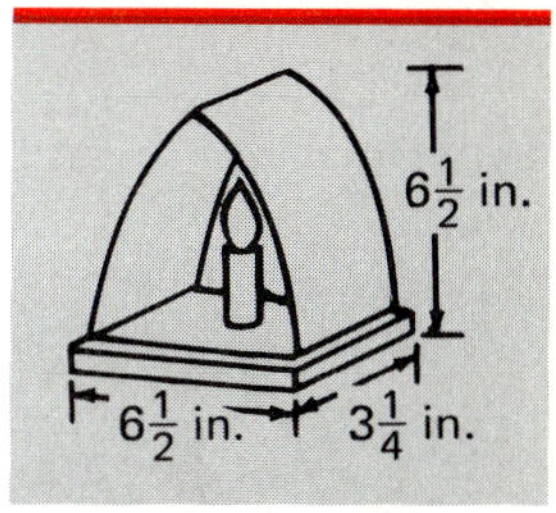

b. What are the dimensions of the carton if the lamps are arranged in 2 layers of 1 lamp by 3 lamps each, with $\frac{1}{2}$-inch spacers between layers.

17. These boxed electronic claculators are packaged 24 to a carton. The carton is made of 0.3-cm cardboard. There are no spacers.

a. What are the dimensions of the carton if the boxes are arranged standing up as shown, in 2 rows of 12 calculators?

b. What other arrangements could be used to package the calculators? Draw sketches and give the dimensions of the cartons.

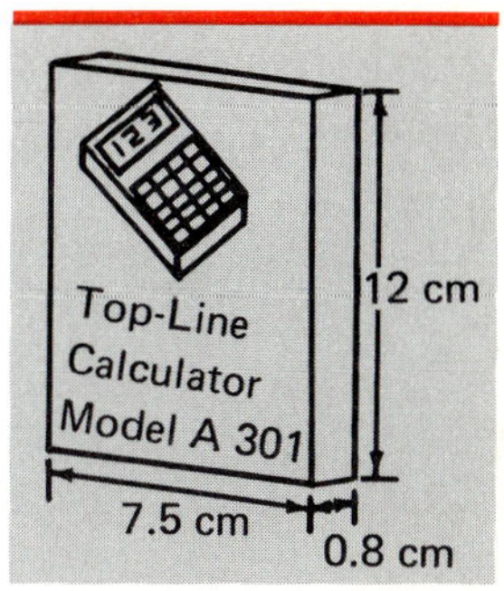

Maintaining Your Skills

Look up the skills in parentheses if you need help or more practice.

Add. (Skill 16)

18. $\frac{1}{2} + \frac{1}{4}$ 19. $\frac{1}{4} + \frac{3}{16}$ 20. $\frac{1}{2} + \frac{3}{16}$ 21. $\frac{1}{8} + \frac{5}{16}$ 22. $\frac{1}{2} + \frac{1}{8}$

23. $\frac{3}{4} + \frac{1}{16}$ 24. $\frac{1}{4} + \frac{3}{8} + \frac{1}{8}$ 25. $\frac{1}{2} + \frac{1}{16} + \frac{3}{16}$ 26. $\frac{3}{4} + \frac{1}{8} + \frac{3}{16}$ 27. $\frac{1}{2} + \frac{3}{8} + \frac{5}{16}$

Multiply. (Skill 20)

28. $4 \times \frac{1}{2}$ 29. $3 \times \frac{1}{8}$ 30. $5 \times \frac{3}{16}$ 31. $2 \times \frac{3}{8}$ 32. $4 \times \frac{3}{4}$

33. $6 \times \frac{1}{8}$ 34. $5 \times \frac{1}{4}$ 35. $7 \times \frac{1}{2}$ 36. $6 \times \frac{3}{16}$ 37. $8 \times \frac{3}{4}$

38. $6 \times \frac{5}{8}$ 39. $4 \times \frac{13}{16}$ 40. $2 \times \frac{1}{32}$ 41. $11 \times \frac{7}{8}$ 42. $4 \times \frac{3}{32}$

Reviewing the Basics

Skills

Round to the nearest tenth of a cent.

(Skill 2) 1. \$0.0498 2. \$1.05872 3. \$17.6548

Solve.

(Skill 5) 4. $4.8 + 7.2 + 9.5 + 3.8$ 5. $12.1 + 9.3 + 14.7 + 11.2 + 7.6$

(Skill 6) 6. \$0.73 − \$0.36 7. \$21.08 − \$15.74 8. \$146.73 − \$91.98

(Skill 11) 9. \$20.40 ÷ 1700 10. \$99,546 ÷ \$0.47 11. \$3,022,600 ÷ \$1.19

(Skill 16) 12. $\frac{1}{4} + \frac{3}{16} + 5$ 13. $7\frac{3}{4} + \frac{5}{16} + \frac{1}{2}$ 14. $2\frac{3}{8} + \frac{1}{2} + 3$

(Skill 20) 15. $6 \times \frac{1}{8}$ 16. $3 \times \frac{7}{16}$ 17. $4 \times 5\frac{3}{4}$

(Skill 31) 18. 7 is what percent of 20? 19. 3.5 is what percent of 14?

Applications

(Application N) Use the graph to answer.

20. What was the income for February?

21. What was the income for April?

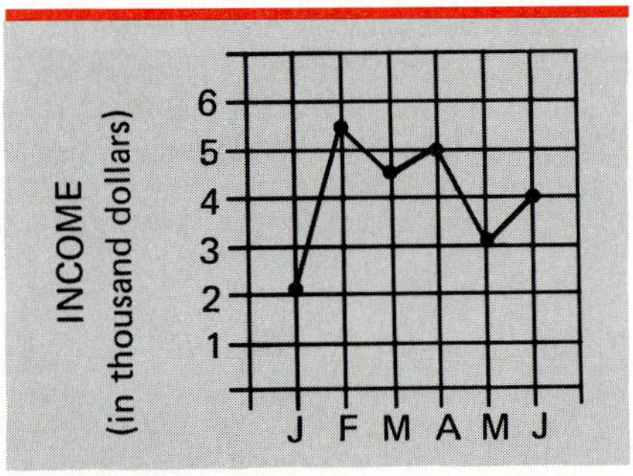

(Application Q) Find the mean.

22. 6.5, 8.2, 4.3, 7.0

23. 17.0, 14.5, 13.2, 16.5, 15.0

Terms

Match each term with its definition on the right.

24. Prime cost
25. Break-even point
26. Quality control
27. Time-study
28. Packaging

a. periodic inspection of the items you manufacture
b. the total of the direct labor cost and the direct material cost
c. the cost of wages paid to employees
d. the point at which income from sales equals the cost of production
e. determines how long a particular job should take
f. the final step in the production of your company's product

Refer to your reference files at the back of the book if you need help.

A Microcomputer Activity
Production Quantity

To complete this microcomputer application, you will need the diskette *Microcomputer Applications for Business Mathematics,* which accompanies this textbook.

Select option 5, Production Quantity, from the menu. Input the information in the following problems to find the failure rate and the production quantity needed.

1. Number defective for day: 6.
   Number ordered: 20,000.
   What is the failure rate?
   What quantity is needed?

2. Number defective for day: 15.
   Number ordered: 120,000.
   What is the failure rate?
   What quantity is needed?

3. Number defective for day: 3.
   Number ordered: 140,000.
   What is the failure rate?
   What quantity is needed?

4. Number defective for day: 2.
   Number ordered: 76,500.
   What is the failure rate?
   What quantity is needed?

5. Number defective for day: 7.
   Number ordered: 68,000.
   What is the failure rate?
   What quantity is needed?

6. Number defective for day: 9.
   Number ordered: 100,000.
   What is the failure rate?
   What quantity is needed?

7. Lee Mander found that 9 parts were defective and that 12,000 were ordered. What is the failure rate? How many are needed?

8. Sun Bing Lee found that 19 bolts were defective for an order of 8000. What is the failure rate? How many are needed?

9. The Quality Control Department of Acme Stamping Plant checks for defective parts. In a production run of #20 machine screws, 3 were defective. The initial order is for 6000 machine screws. What is the failure rate? How many are needed to fill the order?

10. Rowley Manufacturing produces plastic switch plate covers. A one-day sampling showed 4 defective plates. The current order is for 12,000 plates. What is the failure rate? How many are needed to fill the order?

11. A quality control inspector had these records for 1 week.

Day	Mon	Tues	Wed	Thur	Fri
Number defective	9	7	7	6	8
Number ordered	9000	9000	9000	9000	9000

Compute the failure rate for each day. How many are needed each day?

Unit Test

Lesson 13-1

1. Iron-Beam Steel produces cable from scrap steel that it buys for $192 per ton. The company produces 4 rolls of cable from each ton. The direct labor charge is $25.55 an hour. The employees produce 2 rolls of cable per hour. To the nearest tenth of a cent, what is the prime cost of manufacturing 1 roll of cable?

Lesson 13-2

2. Eastside Publishing Company plans to produce books on home repairs to be sold at $12 each. The fixed costs are estimated at $201,096. The variable costs are $8.12 per book. What is the estimated break-even point in units?

Lesson 13-3

3. A quality control inspector for the Contain-It Products Company found 13 out of 65 glass bottles with cracks during the 5 A.M. inspection. The process is in control if the percent defective is 5% or less. Is the process in control or out of control? Plot the percent defective on a quality control chart.

Lesson 13-4

4. Drew Hospital made a time-study of its outpatient pharmacy. The hospital recorded the times that Montro Parker spent unpacking cartons of bandages.

	Observations in Seconds				
Task	**#1**	**#2**	**#3**	**#4**	**#5**
Pick up carton	5.0	5.5	4.0	5.5	4.5
Open carton	4.0	4.5	5.5	4.0	5.0
Remove boxes	13.0	14.0	12.5	13.5	14.5
Discard carton	5.0	5.5	5.0	5.0	5.5

If Montro takes a 5-minute break each hour, how many cartons can he unpack in 1 hour?

Lesson 13-5

5. Georgia Summers is a paralegal for the law firm of Hayes, Ferrante, and Philips. Georgia prepared a time-study of her activities during the day. What percent of her time does Georgia spend reviewing cases?

Activity	Hours
Interviewing clients	2
Processing records	1
Reviewing cases	2
Lunch	1
Case research	2

Lesson 13-6

6. Al-Can ships its industrial containers in wooden boxes. The containers are arranged in 1 layer. Each box is made of $\frac{1}{8}$-inch thick wood with $\frac{1}{8}$-inch thick wood spacers. Each container is 6 inches high and $3\frac{1}{2}$ inches in diameter. What are the dimensions of each wooden box?

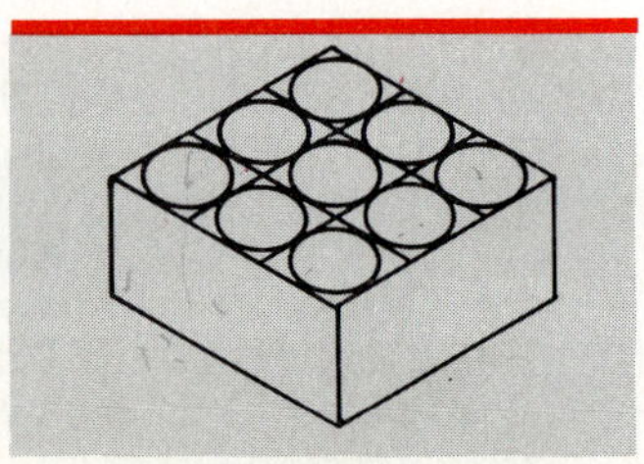

Unit 14 Purchasing

▶ The Purchasing Department orders merchandise for your business from suppliers. Most suppliers sell goods to businesses at a *discount,* or a reduction in price. The lower your *net price,* the price you actually pay for an item, the higher your profit when you sell the item. Suppliers offer three kinds of discounts: *trade, chain,* and *cash.*

A buyer for a clothing store tries to get the newest fashions at the best price from a supplier.

14-1 Trade Discounts

Objective

Compute the trade discount and the net price.

Most suppliers provide a catalogue from which your business can choose items to purchase. The catalogue includes a description and the **list price**, or **catalogue price**, of each item that the supplier sells. The list price is generally the price at which you may sell the item. You usually purchase the item at a **trade discount**, which is a discount from the list price. The trade discount is often expressed as a percent. The **net price** is the price you actually pay for the item.

▸ **Trade Discount = Trade-discount Rate × List Price**

▸ **Net Price = List Price − Trade Discount**

Example

Skills 30, 6 **Application** A **Term** Trade discount

Southland Hi-Fi is purchasing Karan stereo receivers, Model 9200, from a wholesaler. The list price in the wholesaler's catalogue is \$560. Southland receives a 30% trade discount. What is the net price of each receiver?

Solution

1. Find the **trade discount.**
   Trade-discount Rate × List Price
   30% × \$560 = \$168.00

2. Find the **net price.**
   List Price − Trade Discount
   \$560 − \$168 = \$392.00 net price

A. \$105
B. \$315

▸ **Self-Check** Cover the answers and complete the problems.

A. List price: \$420.
Trade discount rate: 25%.
Find the trade discount.

B. List price: \$420.
Trade discount: \$105.
Find the net price.

Problems

	1.	2.	3.	4.	5.	6.	7.
List Price	\$150.00	\$210.00	\$42.30	\$642.60	\$12.98	\$71.46	\$213.74
Trade-discount Rate	20%	15%	30%	25%	15%	12%	8%
Trade Discount	\$30.00	?	?	?	?	?	?
Net Price	?	?	?	?	?	?	?

8. $30 list price for blender.
15% trade discount.
What is the net price?

9. $150 list price for record changer.
24% trade discount.
What is the net price?

10. A Time Saver calculator carries a catalogue price of $275. The Time Saver Company offers a 17% trade discount. What is the net price?

11. Best Buys discount stores purchases hand-held hairdryers from Calvin Wholesalers. Find the trade discount and net price for each hairdryer.

Item	List	Trade-discount Rate
a. Typhoon 900W	$24.99	20%
b. Hurricane 800W	22.99	25%
c. Stormy 850W	19.95	30%

12. Calco, Inc., manufactures automobile parts and accessories. A portion of their "Parts and Accessories Price List" for auto dealers is shown. Calco offers the trade discounts listed to their dealers. Find the dealer price for each item.

CLASS	PART NUMBER	SUGGESTED LIST PRICE	TRADE DISCOUNT
A	D2UZ 8200-B	$89.60	40%
HX	D27Z 82-A	57.20	35%
CQ	COYT 8A178	2.90	42%
D	C7ZZ 8B90	7.80	25%

13. Precision Paper Company offers a 75% discount to commercial users of their products. Green Glass, Inc., placed this order. What is the net price?

QUANTITY	UNITS	DESCRIPTION	PRICE PER UNIT
25	SHEETS	H1163 32 × 40 14 PLY #33 MAT Cut to 150 PCS. 11 × 14 125 PCS. 11 × 12	$144.40 TOTAL ?

14. Parkins Sales Agency receives a 27% trade discount on farm equipment from Rose Equipment, Inc. What is the net price for each order?

Quantity	Stock No.	Description	Unit Price
2	1-1	Heavy nuts	$0.35
2	1-2	Starlock washer	$0.15
2	61 HOB	Blades	$4.75

Maintaining Your Skills

Look up the skills in parentheses if you need help or more practice.

Find the percentage. Round answers to the nearest cent. (Skill 30)

15. 10% × $19.20
16. 18% × $50
17. 20% × $97.80
18. 35% × $9.80
19. 50% × $416.70
20. 42% × $989.46

Subtract. (Skill 6)

21. $19.20 − $1.92
22. $50 − $9
23. $9.18 − $1.84

14-2 Trade Discount—Complement Method

Objective Compute the net price using the complement method.

The **complement method** is another method of finding the net price. When you receive a trade discount, you can subtract the discount rate from 100%. This gives you the complement of the trade discount rate, or the percent you actually pay. Multiply the complement by the list price to find the net price.

▶ **Net Price = Complement of Trade-discount Rate × List Price**

Example

Skills 30, 6 **Application** L **Term** Complement method

Southland Hi-Fi is also purchasing Karan turntables, Model 3620, which have a list price of $340. Southland receives a trade discount of 30%. What is the net price of each turntable?

Solution

1. Find the **complement of trade-discount rate.**
   100% − 30% = 70%
2. Find the **net price.**
   Complement of Trade-discount Rate × List Price
   70% × $340.00 = $238.00 net price

A. 75%
B. $315

▶ **Self-Check** Cover the answers and complete the problems.

A. What is the complement of 25%?

B. List price: $420.
Trade-discount rate: 25%.
Find the net price.

Problems

Use the complement method to solve each problem.

	1.	2.	3.	4.	5.	6.	7.
List Price	$150.00	$94.00	$12.00	$65.45	$298.79	$2.80	$76.53
Trade-discount Rate	20%	10%	15%	60%	32%	12%	27%
Complement	80%	?	?	?	?	?	?
Net Price	?	?	?	?	?	?	?

8. $124.70 list price for lawnmower.
   40% trade discount.
   What is the net price?

9. $363.79 list price for desk.
   20% trade discount.
   What is the net price?

10. Parkins Sales Agency receives a 25% trade discount on these items. What will Parkins Sales pay for each order?

Quantity	Stock No.	Description	Unit Price	Total List
1	84-D	DAXCO cutter-complete	$282.00	$282.00
2	84-06	Blades	4.00	8.00

11. All Things Automotive, Inc., receives a 37% trade discount on these items. What is the net price for each?

Number	Description	Quantity	List	Total
ED-1	Water pump	1	$32.52	$32.52
400	Headlight	1	3.96	3.96
10S	Shock absorber	1	14.95	14.95
PF-91	Sound module	1	47.75	47.75

12. The Clothes Closet receives a 48% trade discount from Kadet Clothes Company. Find the net price for each total listed.

ORDER NO.	DESCRIP.	STYLE NO.	NO. OF UNITS	UNIT PRICE	TOTAL
11260467	BLAZER	87024	2	$19.76	$39.52
11260467	JACKET	87026	5	19.24	96.20
11260467	SHIRT	87233	9	7.28	65.52
11260467	SHIRT	87237	5	7.80	39.00
11260467	SHIRT	87533	6	6.24	37.44

13. The Village Cycle Shop purchases a bicycle from Gendron Wheel Company for a list price of $149.95. Village receives a 36% trade discount.
   a. What net price does the Village Cycle Shop pay for the bicycle?
   b. What is the dollar value of the discount?

14. Everything Electrical receives a 13% discount from Farber's Electric Supply. Find the net price for each amount on the invoice.

FIXTURE LOCATION	DESCRIPTION	DEPT	QTY	PRICE	AMOUNT
NA3	Fastway	4	1	7.10	7.10
EQ25	ITE	4	3	5.67	17.01
QPI-BR15	Rite Corp.	4	6	2.65	15.90
762OC	EFCO	4	2	.88	1.76

Maintaining Your Skills

Look up the skills in parentheses if you need help or more practice.

Find the complement. (Application L)

15. 25% 16. 33% 17. 15% 18. 10% 19. 35% 20. 44% 21. 65% 22. 72%

Find the percentage. (Skill 30)

23. 75% × $72.00 24. 67% × $124.00 25. 85% × $24.80

26. 65% × $248.60 27. 53% × $916.43 28. 34% × $1461.83

14-3 Trade-discount Rate

Objective

Compute the trade-discount rate.

Some wholesalers and manufacturers publish only the list prices and net prices in their catalogues. If you know both of these prices, you can calculate the **trade-discount rate.**

▶ **Trade Discount = List Price − Net Price**

▶ **Trade-discount Rate** $= \dfrac{\textbf{Trade Discount}}{\textbf{List Price}}$

Example

Skills 6, 31 **Application** A **Term** Trade-discount rate

The Siesta Swimsuit Manufacturing Company shows a $35 list price in its catalogue for a style A3091 swimsuit. The net price to the retailer is $21. What is the trade-discount rate?

Solution

1. Find the **trade discount.**
   List Price − Net Price
   $35.00 − $21.00 = $14.00
2. Find the **trade-discount rate.**
   Trade Discount ÷ List Price
   $14.00 ÷ $35.00 = 40% trade-discount rate

A. $19
B. 25%

▶ **Self-Check** Cover the answers and complete the problems.

A. List price: $76.
Net price: $57.
Find the trade discount.

B. Trade discount: $19.
List price: $76.
Find the trade-discount rate.

Problems

Round the trade-discount rate to the nearest tenth of a percent.

	1.	2.	3.	4.	5.	6.	7.
List Price	$4.00	$42.50	$2.79	$86.74	$224.37	$2015.20	$847.32
Net Price	$3.00	$32.30	$2.09	$54.21	$147.91	$1491.73	$567.14
Trade Discount	$1.00	?	?	?	?	?	?
Trade-discount Rate	?	?	?	?	?	?	?

8. List price is $236.48.
Net price is $177.36.
What is the trade-discount rate?

9. List price is $124.62.
Net price is $83.08.
What is the trade-discount rate?

10. Decade Wholesalers, Inc., offers discounts on most items they sell. What is the trade-discount rate for each of the items listed?

Item	List	Net
Complexion brushes	$13.99	$ 9.43
Lighted make-up mirror	32.99	23.74
Cleansing brushes	12.98	8.74
4-way make-up mirror	24.95	17.97

11. Master Plumbers Supply, Inc., varies the rate of discount with the item purchased. This invoice was received by Lawson Home Plumbing Service. What is the trade-discount rate for each gross amount?

USE INVOICE NUMBER ABOVE ON ALL CORRESPONDENCE

ALL ITEMS NET UNLESS OTHERWISE SPECIFIED

QUANTITY SHIPPED	QUANTITY ORDERED	DESCRIPTION	UNIT PRICE	GROSS AMOUNT	DISC.	NET AMOUNT
4	4	0001-WH-TOUCH-UP	2.20	8.80	2.20	6.60
2	2	0001-BL-TOUCH-UP	2.20	4.40	0.00	4.40
2	2	0001-AVO-TOUCH-UP	2.20	4.40	1.54	2.86
2	2	0001-GLD-TOUCH-UP	2.20	4.40	1.76	2.64

12. Find the trade-discount rate for each gross amount listed on this invoice.

QUANTITY	PRODUCT CODE	PRICE PER UNIT	GROSS AMOUNT	NET AMOUNT
140	00010	$ 7.99	$1118.60	$ 894.88
160	00015	19.67	3147.20	2360.40
110	00018	24.84	2732.40	1912.68
100	00022	36.71	3671.00	3010.22
200	00024	14.35	2870.00	2598.75
24	00500	8.43	202.32	131.51
48	00502	6.20	297.60	214.27

Maintaining Your Skills

Look up the skills in parentheses if you need help or more practice.

Subtract. (Skill 6)

13. $78.00 – $62.40
14. $16.40 – $12.30
15. $162.70 – $113.89
16. $45.84 – $33.92
17. $875.91 – $788.32
18. $2471.83 – $2101.06

Divide. Round answers to the nearest tenth. (Skill 11)

19. $15.60 ÷ $78.00
20. $4.10 ÷ $16.40
21. $39.57 ÷ $197.84
22. $202.50 ÷ $843.75
23. $488.36 ÷ $1479.89
24. $2714.53 ÷ $7143.49

14-4 Chain Discounts

Objective

Compute the final net price after a chain of discounts.

A supplier may offer **chain discounts** to sell out a discontinued item or to encourage you to place a larger order. A chain discount is a series of trade discounts. For example:

35% less 20% less 15%

- This does not mean 70% off
- Often written as 35/20/15
- 35% discount is deducted from list price to yield first net price.
- 20% of first net price is deducted from first net price to yield second net price.
- 15% of second net price is deducted from second net price to yield final net price.

Example

Skills 30, 2, 6 **Application** A **Term** Chain discount

The Goshen Corporation produces a 3-horsepower pump which is priced at $520. A new model is ready for production, so the 3-horsepower model is to be discontinued. Goshen is offering a chain discount of 30% less 20%. What is the final net price of the pump?

Solution

1. Find the **first discount.**
   Discount Rate × List Price
   30% × $520.00 = $156.00
2. Find the **first net price.**
   List Price − Discount
   $520.00 − $156.00 = $364.00
3. Find the **second discount.**
   20% × $364.00 = $72.80
4. Find the **final net price.**
   $364.00 − $72.80 = $291.20 final net price

A. $120
B. $96

▶ **Self-Check** Cover the answers and complete the problems.

List price is $200. Chain discount is 40% less 20%.

A. Find the first net price.

B. Find the final net price.

Problems

	1.	2.	3.	4.	5.
List Price	$450	$840	$950	$1460	$2575
Chain Discount	20% less 10%	25% less 20%	20% less 15%	30% less 10%	40% less 20%
First Discount	$90	$210	?	?	?
First Net Price	$360	$630	?	?	?
Second Discount	?	?	?	?	?
Final Net Price	?	?	?	?	?

6. $84.50 list price for world globe. 20/15 chain discount. What is the net price?

7. $64.50 list price for toaster-oven. 30/12 chain discount. What is the net price?

8. Beta Lumber Yard offers Zink Construction chain discounts of 20/15. What is the net price for each item?

Item	List Price
Storm door	$74.60
Storm window	48.00
Foil-backed insulation	11.50/roll

9. The Reality Corporation offers trade discounts and additional discounts to encourage large orders. What is the net price per item if each invoice total is high enough to obtain the additional discount?

Item	List Price	Trade Discount	Additional Discount
Masonry drill	$84.20	50%	5% (invoice total over $250)
Raw-tip drill	76.40	20%	10% (invoice total over $500)
Hole saw	56.20	30%	20% (invoice total over $1000)

10. The Co-Op offers chain discounts of 20% less 15% less 10% for membership, early payment, and early pickup of fertilizer. Find the final net price for an order of fertilizer listing at $1460.

11. What is the final net price for each of these overstocked items from the Continental Auto Parts price list?

Part Number	Suggested List Price	Chain Discount
B7S	$112.30	40% less 25% less 15%
C37X	74.60	42% less 20% less 10%
173A	31.85	35% less 15% less 5%
B62Y	65.16	25% less 10% less 7%

Maintaining Your Skills

Look up the skills in parentheses if you need help or more practice.

Round to the place value indicated. (Skill 2)

12. $14.87496 (nearest cent)
13. $78.61 (nearest whole number)
14. 42.89% (nearest tenth of a percent)
15. $14,146.8074 (nearest cent)
16. 29.89% (nearest percent)
17. 149.499 (nearest whole number)

Subtract. (Skill 6)

18. $74.89 − $4.49
19. $17.75 − $3.55
20. $861.73 − $258.52
21. $226.25 − $56.56
22. $1498.14 − $524.35
23. $16,917.84 − $845.89

Find the percentage. (Skill 30)

24. 15% × $860.00
25. 20% × $415.80
26. 30% x $971.85
27. 10% × $1718.90
28. 5% × $9456.79
29. 25% × $24,478.82

14-5 Chain Discounts—Complement Method

Objective Compute the final net price using the complement method.

The complement method applied to trade discounts can also be applied to chain discounts. First subtract each discount from 100% to find the complements. Then multiply the complements to find the percent that you actually pay. This percent is called the **net-price rate.** To find the net price, multiply the net-price rate by the list price.

You can also use the complement method to find the discount. Find the complement of the net-price rate by subtracting it from 100%. That percent is called the **single equivalent discount (SED).** Multiply the SED by the list price to find the discount.

- **Net-price Rate = Product of Complements of Chain-discount Rates**
- **Net Price = Net-price Rate × List Price**
- **SED = Complement of Net-price Rate**
- **Discount = SED × List Price**

Example

Skills 30, 2, 6 **Application** A **Term** Chain discount

The Goshen Corporation produces a 5-horsepower pump, which is priced at $780. A new model is ready for production, so the 5-horsepower model is to be discontinued. Goshen is offering a chain discount of 30% less 20% less 10%. What is the net price of the pump? What is the SED and the discount?

Solution

1. Find the **net-price rate.**
   Product of Complements of Chain-discount Rates
   70% × 80% × 90% = 50.4%
2. Find the **net price.**
   Net-price Rate × List Price
   50.4% × $780 = $393.12 net price
3. Find the **SED.**
   Complement of the Net-price Rate
   100% − 50.4% = 49.6% SED
4. Find the **discount.**
   SED × List Price
   49.6% × $780 = $386.88 discount

A. 63.75%; $255
B. 36.25%
$145

Self-Check Cover the answers and complete the problems.

List price is $400. Chain discount is 25% less 15%.

A. Find the net price rate and the net price.

B. Find the SED and the discount.

Problems

	1.	2.	3.	4.	5.	6.
List Price	$500	$800	$480	$1470	$600	$1540
Chain Discounts	20% less 10%	30% less 15%	25% less 10%	30% less 10%	30% less 20% less 10%	20% less 10% less 5%
Net Price	$360	$476	?	?	?	?
SED	?	?	?	?	?	?
Discount	?	?	?	?	?	?

7. $112.50 list price for desk set.
20/10 chain discount.
What is the net price?
What is the discount?

8. $49.40 list price for food mixer.
30/12 chain discount.
What is the net price?
What is the discount?

9. Beta Lumber Yard offers Rose Construction chain discounts of 20/10. What is the net price for each item? What is the SED?

Item	List Price
Storm door	$74.60
Storm window	48.00
Foil-backed insulation	11.50/roll

10. The Reality Corporation offers trade discounts and additional discounts to encourage large orders. What is the net price per item if each invoice total is high enough to obtain the additional discount?

Item	List Price	Trade Discount	Additional Discount
Masonry drill	$84.20	50%	10% (invoice over $500)
Raw-tip drill	76.40	20%	20% (invoice over $1000)
Hole saw	56.20	30%	5% (invoice over $250)

11. What is the net price for each of these overstocked items from the Continental Auto Parts price list?

Part Number	Suggested List Price	Chain Discount
B75	$112.30	40% less 25% less 15%
X24T	98.60	30% less 20% less 10%
416V	74.20	25% less 15% less 5%
C32U	46.35	20% less 10% less 5%

Maintaining Your Skills

Look up the skills in parentheses if you need help or more practice.

Round to the nearest cent. (Skill 2)

12. $47.6251 13. $96.7463 14. $516.0817 15. $1978.4891 16. $816.7951

Find the complement. (Application L)

17. 25% 18. 30% 19. 5% 20. 35% 21. 27% 22. 42% 23. 83%

Find the percentage. (Skill 31)

24. 72% × $418.00 25. 60% × $117.40 26. 50.4% × $1347.80

14-6 Cash Discounts—Ordinary Dating

Objective Compute the cash price when discount is based on ordinary dating.

You receive an **invoice** for each purchase you make from a supplier. The invoice lists the quantities and costs of the items purchased. To encourage prompt payment, the supplier may offer a **cash discount** if the bill is paid within a certain number of days. The exact terms of the discount are stated on the invoice. Many suppliers use **ordinary dating.** For example:

2/10, net 30 or **2/10, n/30**

2/10: 2% cash discount if paid within 10 days of date of invoice

net 30: Net price must be paid within 30 days of date of invoice

▶ **Cash Discount = Cash-discount Rate × Net Price**

▶ **Cash Price = Net Price − Cash Discount**

Example

Skills 30, 6 **Application** A **Term** Ordinary dating

This is part of the invoice that Stacey's Department Store received for a shipment of lamps. What is the cash price of the lamps if the bill is paid within 10 days?

Oak Hill Lighting, Inc.

SHIP TO: STACEY'S DEPT. STORE
INVOICE NO: 4-3467

DATE	ORDER NO.	TERMS	ACCT. NO.
6-5 --	LH 3019	3/10, NET 30	712E 4

STYLE	COLOR	QUANTITY	PRICE	AMOUNT
A9407	BGE	6	$37.25	$223.50
A9841	GRN	5	42.15	210.75
J2113	NVY	8	26.40	211.20
J2114	GRN	8	26.40	211.20
			TOTAL	$856.65

Solution

1. Find the **cash discount.**
   Cash-discount Rate × Net Price
   3% of $856.65 = $25.6995 = $25.70

2. Find the **cash price.**
   Net Price − Cash Discount
   $856.65 − $25.70 = $830.95 cash price

A. $42
B. $798

▶ **Self-Check** Cover the answers and complete the problems.

Terms are 5/10, n/30. Net price is $840.

A. Find the cash discount.   B. Find the cash price.

Problems

	Invoice Date	Terms	Date Paid	Net Price	Cash Discount	Cash Price
1.	9/15	2/10, n/30	9/24	$ 850.00	$17.00	?
2.	11/6	3/10, n/30	11/10	$ 463.20	?	?
3.	2/6	5/10, n/30	2/19	$ 216.75	?	?
4.	8/24	1/10, n/30	9/2	$2461.73	?	?

5. $1645.70 net price for dishware. Terms are 3/15, n/30. Date of invoice is October 15. Date paid is October 29. What is the cash price?

6. $424.50 net price for vinyl boots. Terms are 2/10, net 30. Date of invoice is December 6. Invoice is paid on December 15. What is the cash price?

7. The net price of goods from Precision Paper Company to the photo lab of Green Glass, Inc., was $36.80. The invoice was paid on March 10. According to the terms on the invoice, what was the cash price?

DATE SHIPPED	VIA		SALES PERSON	DATE OF INVOICE
2/24/--	OUR DELIVERY	TERMS: 2%–20 DAYS	65	2/23/--

8. The net price of goods from Hinges, Inc., to Abbotts' Hardware Store is $182. If the invoice is paid on December 11, what is the cash price?

CUSTOMER NUMBER -1238 SHIPPING DATE 11-22-~ INVOICE DATE 12-05-~

CUST. ORDER NO.	CUST. ORDER DATE	TERMS	TYPE	SHPG. ORDER NO.
211	11-15-~	2/10, n/30	02	847710

9. What is the cash price for these materials from Allon Founders, Inc., to Strange Electric Corporation? The invoice is paid on November 20.

CUSTOMER'S COPY
INVOICE NO. NO. 4559 DATE NOV. 7 TERMS 1% – 10 days Net 30 days

YOUR ORDER NO.	PIECES	PATTERN NO.	WEIGHT	PRICE	AMOUNT
7552	240	641-7		$0.61	$146.40
1319	157	320-5		0.53	83.21
					$229.61

Maintaining Your Skills

Look up the skills in parentheses if you need help or more practice.

Subtract. (Skill 6)

10. $470.00 – $9.40
11. $117.80 – $11.78
12. $862.00 – $25.86
13. $19.80 – $0.99
14. $1691.75 – $67.67
15. $216.84 – $4.34

Find the percentage. (Skill 30)

16. 2% × $350.00
17. 3% × $916.00
18. 4% × $85.80
19. 5% × $161.40

14-7 Cash Discounts—EOM Dating

Objective

Compute the cash price when discount is based on end-of-month dating.

Many wholesalers use **end-of-month dating** when granting cash discounts. With this method of dating, your business receives a cash discount if you pay for your merchandise within a certain number of days after the end of the month. For example:

6/15 EOM ← A 6% cash discount if bill is paid within 15 days after the end of the month in which the invoice is issued.

- **Cash Discount = Cash-discount Rate × Net Price**
- **Cash Price = Net Price − Cash Discount**

Example

Skills 30, 6 Application A Term End-of-month dating

Stacey's Department Store received this invoice from Furniture Wholesalers. What is the last day that Stacey's can take advantage of the 4% cash discount? What is the cash price?

DATE	INVOICE NO.	ACCT. NO.	STORE NO.	TERMS	VENDOR NO.
JUL 12	213788-6	208 712	34	4/10 EOM	12/7

SALESPERSON	VIA	ORDER REG. NO.
38L	OV DIRECT TRUCK	0547-6

CUSTOMER ORDER NO.	STYLE	QUANTITY	PRICE	AMOUNT
L3075	02673	8	$ 63.00	$ 504.00
L3076	02677	8	71.00	568 00
L3077	02681	12	84.00	1008.00
			TOTAL	$ 2080.00

Solution

1. Find the last day a discount can be taken.
   Date of invoice: July 12
   Terms: 4/10 EOM
   Discount can be taken until August 10, 10 days after the end of July.
2. Find the **cash discount.**
   Cash-discount Rate × Net Price
   4% × $2080.00 = $83.20
3. Find the **cash price.**
   Net Price − Cash Discount
   $2080.00 − $83.20 = $1996.80 cash price

A. Oct. 15
B. $8148

▸ **Self-Check** Cover the answers and complete the problems.

An invoice dated September 12 has terms of 3/15 EOM. Net price is $8400. The invoice is paid October 13.

A. Find the last day a discount can be taken.

B. Find the cash price.

Problems

	Invoice Date	Terms	Date Paid	Net Price	Cash Discount	Cash Price
1.	March 20	2/10 EOM	Feb. 9	$ 450.00	$9.00	?
2.	Oct. 3	4/15 EOM	Nov. 12	$ 1240.00	?	?
3.	May 21	3/10 EOM	June 21	$ 4854.80	?	?
4.	Dec. 19	5/10 EOM	Jan. 10	$14,684.20	?	?

5. Net price is $612.70.
Terms are 2/10 EOM.
Invoice date is September 20.
Invoice is paid on October 9.
What is the cash price?

6. Net price is $2461.50.
Terms are 3/10 EOM, n/30 EOM.
Invoice date is June 15.
Invoice is paid on July 7.
What is the cash price?

7. An invoice from Plumbers Supply, Inc., to Lawson Drilling showed a net price of $16.50. The terms are 2/10 EOM, n/30 EOM. The invoice was dated March 7. The invoice was paid April 15. What is the cash price?

8. An invoice from Kalo Foods Corporation to Pajella Wholesalers carried a net price of $14,221.50. The terms are 8/10 EOM. The invoice is dated July 8. The invoice is paid August 7. What is the cash price?

9. An invoice from Kadet Clothes to the Clothes Factory has a list price of $425.88. After a trade discount of 18%, the net price is $349.22. A portion of the invoice is shown. What is the cash price if the invoice is paid on January 5?

INVOICE # K85616

CUSTOMER NO.	DATE	PACKING SLIP NO.
626121	DEC 03	11280467

TERMS: 8/10/EOM FOB SFC

Maintaining Your Skills

Look up the skills in parentheses if you need help or more practice.

Subtract. (Skill 6)

10. $716.00 − $14.32
11. $427.80 − $21.39
12. $1971.25 − $78.85

Find the percentage. (Skill 30)

13. 2% × $716.00
14. 4% × $3215.50
15. 3% × $124.80
16. 98% × $716.00
17. 97% × $5155.40
18. 95% × $116.94

Reviewing the Basics

Skills

Round to the nearest cent.

(Skill 1)

1. $1375.728
2. $753.4565
3. $171.4545
4. $191.8042

Solve. Round to the nearest cent.

(Skill 6)

5. $99.53 − $89.73
6. $1061.99 − $153.64
7. $417.80 − $83.50
8. $9814.90 − $3435.22
9. $21,476.86 − $2147.64
10. $418 − $117.40

(Skill 30)

11. 15% of $705.95
12. 27% of $178.56
13. 2% of $861.80
14. 8% of $3451.90
15. 50% of $1347.80
16. 72% of $418

Round to the nearest tenth of a percent.

(Skill 31)

17. $1.92 is what percent of $2.50?
18. $35.50 is what percent of $85.55?
19. $171.35 is what percent of $685.40?
20. $1478 is what percent of $14,780?

Applications

Find the complement.

(Application L)

21. 15%
22. 56%
23. 53.55%
24. 2%
25. 62%

Terms

Use each term in one of the sentences.

26. Trade discount
27. Chain discount
28. Complement method
29. Ordinary dating
30. End-of-month dating
31. Trade-discount rate

a. *3/5 net 30* is an example of __?__ .

b. The __?__ is a way to find the net price when a trade-discount rate is given.

c. A series of discounts on an item is called a __?__ .

d. *9/5 EOM* is an example of __?__ .

e. If you know both the trade discount and the list price, you can calculate the __?__ .

f. A discount from the catalogue price is called a __?__ .

g. The catalogue price is the __?__ .

Refer to your reference files at the back of the book if you need help.

Problem Solving Working Backwards

Word problems that involve a sequence of events or actions can sometimes be solved most efficiently by working backwards. If the final result of the problem is given, start your solution at that point and work backwards through the steps to arrive at the beginning conditions of the problem.

Example 1 Peter starts at point *A* and enters a fun house. He pays $1.00 to get in and loses half of the money in his possession while he is in the fun house. He then pays $1.00 when he exits at *B*. He goes to the next entrance (*C*), pays $1.00 to get in, loses half of his money, and pays $1.00 to exit at *D*. This is repeated until he exits at *Z* and gives his last $1.00 to get out. How much money did he start with?

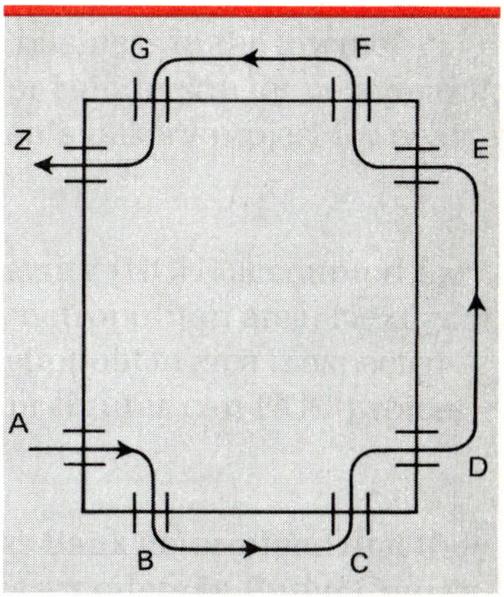

Solution Work backwards, starting at Z.
Inside at *Z* he had $1.00 left, which he paid to get out. Double this to $2.00 (since it was halved). Add $1.00 to make it $3.00 at *F*. Add $1.00 to make it $4.00 inside of *E*. Double it to $8.00. Add $1.00 to make it $9.00 at *E*. Add $1.00 to make it $10 inside at *D*. Double it to $20. Add $1.00 to make it $21 outside at *C*. Add $1.00 to make it $22 inside at *B*. Double it to $44. Add $1 to make it $45 at *A*. Therefore, Peter must have started with $45.

Example 2 Lisa baked some cookies. She put one-half of them away for the next day. Then she divided the remaining cookies evenly among her 3 sisters so that each received 4. How many cookies did Lisa bake?

Solution Each of her 3 sisters received 4 cookies. Therefore, Lisa divided 12 cookies. The 12 cookies represent one-half of what she baked; therefore, she must have baked 24 cookies.

Problems

1. Two barrels, A and B, contain unspecified amounts of cider, A containing more than B. From A, pour into B as much cider as B already contains. Then from B, pour into A as much cider as A now contains. Finally, pour from A into B as much cider as B presently has. Both barrels now contain 64 liters of cider. How many liters of cider were in each barrel at the start of the process?

2. A water lily doubles itself in size each day. From the time the original plant was placed in a pond until the surface was completely covered took 30 days. How long did it take for the pond to be half-covered?

Unit Test

Lesson 14-1

1. Slumber Rest purchases its Elite mattress from a factory outlet. The list price of each mattress is $359.44. Slumber Rest receives a 25% trade discount. What is the net price for each mattress?

Lesson 14-2

2. Roller-Rama purchases Smooth-Glide roller skates that have a list price of $89.23 a pair. Roller-Rama receives a trade discount of 18%. What is the net price that Roller-Rama pays for each pair of skates? Use the complement method to solve.

Lesson 14-3

3. Stroll Co., a manufacturer of light bulbs, lists the price of its Low-Glare bulb as $3.55. The net price to the retailer is $2.84. What is the trade-discount rate?

Lesson 14-4

4. Cameroon Aircraft Company manufactures a four-place helicopter that is priced at $754,400. When it introduced its new Lightning model, Cameroon offered a chain discount of 25% less 20% on the old model. What is the final net price of the helicopter?

Lesson 14-5

5. Sports, Inc., produces a golf shoe that is priced at $74.80. When the new model was introduced, the $74.80 shoe was discounted at 25% less 15%. What is the net price of the old model using the complement method?

Lesson 14-6

6. Complete Data received an invoice dated September 13 from Integrated Computer Supply. The invoice carried terms of 3/15, net 30. The net price of the invoice was $3350.78. What is the cash price of the invoice if it is paid on September 25?

Lesson 14-7

7. Complete Data received this invoice from High Technology. What is the last day that Complete Data can make use of the 5% cash discount? What is the cash price?

DATE	INVOICE NO.	ACCT. NO.	STORE NO.	TERMS	VENDOR NO.
April 16	181818	75239	07	5/10 EOM	16
CUSTOMER ORDER NO.		TYPE	QUANTITY	PRICE	AMOUNT
F78906		08765	13	$10.56	$137.78
F90875		98765	09	$ 6.98	62.82
L56329		09453	24	$23.74	569.76
				TOTAL	$770.36

Unit 15 Sales

▸ **After your business has purchased goods from suppliers, you must determine the selling price, or *retail price,* of the product. You want to sell the product at a higher price than its *cost,* or the amount you paid for it. The difference between the cost and the selling price is your *markup,* or *gross profit.* When your markup is larger than your overhead or operating expenses, you make a *net profit.***

Supply and demand affects what a consumer is willing to pay for goods.

15-1 Markup

Objective Compute the markup in dollars.

The **cost** of a product is the amount that your business pays for it. The cost includes expenses such as freight charges. Your business sells the product at a **selling price,** or **retail price,** which is higher than the cost. The **markup** is the difference between the selling price and the cost. The markup may also be called the **gross profit** or **profit margin.**

▸ **Markup = Selling Price − Cost**

Example

Skill 6 Application A Term Markup

Tarko Associates purchased Royston 40-channel CB radios for $87.20 each. Tarko sells the radios for $129.95 each. What is the markup on each radio?

Solution

Find the **markup.**
Selling Price − Cost
$129.95 − $87.20 = $42.75 markup

A. $18.20
B. $587.99

▸ **Self-Check** Cover the answers and complete the problems.

Find the markup.

A. Selling price: $47.99. Cost: $29.79.

B. Selling price: $979.99. Cost: $392.

Problems

	Item	Selling Price	−	Cost	=	Markup
1.	Ruler	$ 0.29	−	$ 0.05	=	?
2.	Suitcase	$ 149.50	−	$ 101.40	=	?
3.	Necklace	$ 2.98	−	$ 0.97	=	?
4.	Auto engine cover	$ 3.87	−	$ 2.32	=	?
5.	CB radio	$ 129.95	−	$ 87.10	=	?
6.	Roll of wallpaper	$ 6.15	−	$ 4.07	=	?
7.	Sailboat	$2147.00	−	$1653.19	=	?
8.	Wall clock	$ 87.80	−	$ 30.73	=	?

9. Sweater.
Cost is $25.75.
Selling price is $49.99.
What is the markup?

10. Dozen bran muffins.
Cost $0.60 to prepare.
Selling price is $1.99.
What is the markup?

11. Felt-tip pen.
Cost is $0.27.
Selling price is $0.69.
What is the markup?

12. Hairdryer.
Cost is $14.72.
Selling price is $24.95.
What is the markup?

13. Pfiefer's Hardware purchases paper towel holders at a cost of $6.00. Pfiefer's sells the paper towel holders for $9.98 each. What is the markup?

14. House Furniture, Inc., purchased a living room suite from Warehouse Furniture Company for $420.50. House Furniture, Inc., sold the living room suite for $624.88. What was the markup on the living room suite?

15. Quality Jewelers advertised this digital watch in yesterday's paper. The cost of the watch to Quality Jewelers is $17.97. What is the markup on each watch?

16. Barbara Morse operates a bicycle repair shop. Recently she bought an older bicycle for $14 at a local auction. She repaired the bicycle for a cost of $7.45 in parts. She painted and polished the bicycle for a cost of $4.54. When the bicycle was finished, Barbara sold it for $40. What was the markup on the bicycle?

17. Action Stamping and Manufacturing Company sells finished automobile engine covers to wholesale auto parts companies for $33.48 a dozen. It costs Action Stamping $17.78 a dozen to produce the covers.
a. What is the markup on a dozen engine covers?
b. What is the markup on one engine cover?

18. Hibbards' Family Market buys lettuce by the crate. Each crate contains 24 heads of lettuce. The cost of each crate to Hibbards' is $12.50. Hibbards' sells the lettuce for $0.69 per head.
a. What is the markup per head of lettuce?
b. What is the markup per crate of lettuce?

Maintaining Your Skills

Look up the skills in parentheses if you need help or more practice.

Subtract. (Skill 6)

19. $24.29 − $14.30
20. $9.79 − $3.92
21. $116.79 − $93.43
22. $249.50 − $157.19
23. $74.99 − $52.49
24. $416.95 − $333.56
25. $55.45 − $34.38
26. $916.95 − $644.87
27. $1979 − $1088.45

15-2 Markup Rate

Objective Compute the markup as a percent of the selling price.

The **markup rate** is markup expressed as a percent. Businesses usually express the markup as a percent of the selling price.

▶ **Markup Rate** $= \dfrac{\textbf{Markup}}{\textbf{Selling Price}}$

Example

Skills 6, 31 **Application** A **Term** Markup rate

The Flower Pot buys roses for $7.77 a dozen. It sells them for $12.95 a dozen. What is the markup rate based on the selling price?

Solution

1. Find the **markup.**
   Selling Price − Cost
   $12.95 − $7.77 = $5.18
2. Find the **markup rate** based on selling price.
   Markup ÷ Selling Price
   $5.18 ÷ $12.95 = 0.40 = 40% markup rate

A. $29.34, 30%
B. $17.48, 53%

▶ **Self-Check** Cover the answers and complete the problems.

Find the markup and the markup rate.

A. Selling price: $97.79.
   Cost: $68.45.

B. Selling price: $32.95.
   Cost: $15.47.

Problems

Round answers to the nearest tenth of a percent.

1. Wallet.
   Cost is $4.22.
   Selling price is $9.45.
   What is the markup rate based on the selling price?
2. Dishwasher.
   Cost is $149.74.
   Selling price is $227.
   What is the markup rate based on the selling price?
3. Electric can opener.
   Cost is $6.88.
   Selling price is $13.59.
   What is the markup rate based on the selling price?
4. Sound movie system.
   Cost is $381.54.
   Selling price is $564.12.
   What is the markup rate based on the selling price?

5. John Cortez is a buyer for Lamps, Inc. He purchases Artform lamps for $12.37 each. Lamps, Inc., sells them for $29.99 each. What is the markup for each lamp? What is the markup rate based on the selling price?

6. The Ultimate, Inc., purchases cotton shirts for $8.99 each. The selling price of each shirt is $17.99. What is the markup rate based on the selling price of each shirt?

7. Automobile seat covers cost Odesky's Auto Glass $58.75. Odesky's sells the seat covers for $99.99. What is the markup rate based on the selling price?

8. Convenient Carryout purchases corn for 64¢ a dozen. The selling price of the corn is 99¢ per dozen. What is the markup rate based on the selling price?

9. Mirabella's Marina pays $2475 each, including freight, for Delmar sailboats. Mirabella's sells each sailboat for $3417.55. What is the markup rate based on the selling price?

10. Jenkins Hardware sells a door lock for $17.95. The cost to Jenkins is $8.20 per lock. What is the markup rate based on the selling price?

11. A series L gas water heater sells for $207.95 at City Sales. Each water heater costs City Sales $143.88. What is the markup rate based on the selling price?

12. Hopewell's Furniture held a special sale on wicker magazine racks, planters, and end tables. The selling price of each item was $39.95. What was the markup rate based on the selling price for each item?

ITEM NUMBER	DESCRIPTION	COST
C3455	Magazine rack	$18.55
C3456	Planter	23.86
C3457	End table	27.49

Maintaining Your Skills

Look up the skills in parentheses if you need help or more practice.

Subtract. (Skill 6)

13. $14.89 − $11.91
14. $49.97 − $37.48
15. $97.49 − $68.24
16. $145.59 − $87.35
17. $249.99 − $149.99
18. $799.79 − $439.88

Find the rate. Round answers to the nearest tenth of a percent. (Skill 31)

19. $2.98 ÷ $14.89
20. $12.49 ÷ $49.97
21. $42.95 ÷ $95.45
22. $47.90 ÷ $119.97
23. $476.80 ÷ $780.00
24. $1243.50 ÷ $1749.90

15-3 Net Profit

Objective Compute the net profit in dollars.

The markup on the products that you sell must cover your **overhead** or **operating expenses.** These expenses include wages and salaries of employees, rent, utility charges, and taxes.

The exact overhead expense on each item sold is difficult to determine accurately. You may approximate the overhead expense of each item. For example, if your total overhead expenses are 40% of total sales, you may estimate the overhead expense of each item sold to be 40% of its selling price. When the markup of an item is greater than its overhead expenses, you make a **net profit** on the item.

- **Overhead = Overhead Percent × Selling Price**
- **Net Profit = Markup − Overhead**

Example

Skills 30, 2 **Application** A **Term** Net profit

Posadny's Appliance Center purchases small vacuum sweepers for $41 each. Posadny's sells the sweepers for $87.50 each. John Posadny estimates his overhead expenses to be 40% of the selling price of his merchandise. What is the net profit on each sweeper?

Solution

1. Find the **markup.**
   $87.50 − $41.00 = $46.50
2. Find the **overhead.**
   Overhead Percent × Selling Price
   40% × $87.50 = $35.00
3. Find the **net profit.**
   Markup − Overhead
   $46.50 − $35.00 = $11.50 net profit

A. $140, $72, $68
B. $49, $30.80, $18.20

Self-Check Cover the answers and complete the problems.

Find the markup, overhead, and net profit.

A. Selling price: $240.
Cost: $100.
Overhead: 30% of selling price.

B. Selling price: $88.
Cost: $39.
Overhead: 35% of selling price.

Problems

	Markup	−	(Overhead Percent	×	Selling Price)	=	Net Profit
1.	$50.00	−	(20%	×	$80.00)	=	?
2.	$15.00	−	(30%	×	$30.00)	=	?
3.	$18.00	−	(40%	×	$45.00)	=	?

Estimate the net profit based on the estimated overhead.

4. Power lawn mower costs $104.15.
Selling price is $149.90.
Overhead is $23.86.
What is the net profit?

5. Tape dispenser costs $4.15.
Selling price is $6.64.
Overhead is $1.18.
What is the net profit?

6. Pair of jeans costs $10.47.
Selling price is $17.95.
Overhead is 26% of selling price.
What is the net profit?

7. Tropical plant costs $13.65.
Selling price is $23.88.
Overhead is 42% of selling price.
What is the net profit?

8. Universal Carpet Company purchases carpet padding for $0.72 per square yard. Universal charges customers $2.50 per square yard to have the carpet padding installed. The labor and other overhead total $1.25 per square yard. What is the net profit per square yard?

9. Sterns Auto Glass purchases windshields at a cost of $58.75. The selling price of an installed windshield is $99.99. The overhead for the windshield is estimated to be 24% of the selling price. What is the net profit?

10. Gloria Jackson is a buyer for Country Clothes. She purchased some shirts for a cost of $6.66 each. Her store sells them for $11.68 each. Operating expenses are estimated to be 25% of the selling price. What is the net profit per shirt?

11. DP Computer Associates, assemblers of specialized central processing units, delivered a system to Ajax Trucking for a selling price of $970,000. The cost of the unit to DP Computer Associates was $300,000. The system design, programming, and other overhead expenses were estimated to be 33.5% of the selling price. What net profit did DP Computer Associates make on the sale?

12. Allison Marjoram manages Fabric Fair. She buys Trim-Ease scissors at $7.56 a dozen. She sells them at $1.49 a pair. The overhead is estimated to be 21% of the selling price. What is the net profit per pair?

Maintaining Your Skills

Look up the skills in parentheses if you need help or more practice.

Round to the nearest cent. (Skill 2)

13. $4.667 14. $1.0296 15. $.3129 16. $1.9435 17. $9.7986 18. $18.3985

Subtract. (Skill 6)

19. $45.75 − $23.86 20. $10.23 − $10.03 21. $41.24 − $23.99

Find the percentage. Round answers to the nearest cent. (Skill 30)

22. 35% × $140 23. 26% × $17.95 24. 42% × $23.88

25. 21% × $7.56 26. 42.5% × $817.50 27. 17.3% × $12,749.49

15-4 Net-profit Rate

Objective

Compute the net profit as a percent of the selling price.

You may want to know the **net-profit rate** of an item your business sells. The net-profit rate is net profit expressed as a percent of the selling price of the item.

▶ **Net-profit Rate** = $\frac{\textbf{Net Profit}}{\textbf{Selling Price}}$

Example

Skills 6, 30, 31 **Application** A **Term** Net-profit rate

Corbette's sells a vinyl handbag for $10.95. The cost of the handbag to Corbette's is $4.38. Corbette's estimates the overhead expenses on the handbag to be 40% of its selling price. What is the net-profit rate based on the selling price of the handbag?

Solution

1. Find the **net profit.**

   Markup − Overhead
   ($10.95 − $4.38) − (40% of $10.95)
   $6.57 − $4.38 = $2.19

2. Find the **net-profit rate.**

   Net Profit ÷ Selling Price
   $2.19 ÷ $10.95 = 0.20 = 20% net-profit rate

A. $7.96; $5.97; $1.99; 10%
B. $29.95; $22.48; $7.47; 17%

▶ **Self-Check** Cover the answers and complete the problems.

Find the markup, overhead, net profit, and net-profit rate.

A. Selling price: $19.90.
Cost: $11.94.
Overhead expense: 30% of selling price.

B. Selling price: $44.95.
Cost: $15.00.
Overhead expense: 50% of selling price.

Problems

Round to the nearest percent.

	Markup	−	(Overhead Percent	×	Selling Price)	=	Net Profit	Net-profit Rate
1.	$30.00	−	(25%	×	$ 50.00)	=	?	?
2.	$31.00	−	(10%	×	$ 70.00)	=	?	?
3.	$15.99	−	(40%	×	$ 24.60)	=	?	?
4.	$67.41	−	(30%	×	$149.80)	=	?	?
5.	$ 9.95	−	(20%	×	$ 19.90)	=	?	?

Round answers to the nearest tenth of a percent.

6. Selling price of $12.50.
Markup is $9.16.
Overhead is $4.16.
What is the net-profit rate?

7. Selling price of $437.30.
Markup is $262.83.
Overhead is $131.64.
What is the net-profit rate?

8. Pair of boots costs $10.47.
Selling price is $17.95.
Overhead is 26% of selling price.
What is the net-profit rate?

9. Floor lamp costs $31.75.
Selling price is $63.49.
Overhead is 11% of selling price.
What is the net-profit rate?

10. Spud's Hardware sells these workshop vacuum cleaners. The cost to Spud's Hardware is $27 per machine. Overhead is estimated to be 20% of the selling price. What is the net-profit rate?

Clean up with savings!
Big Buy Shop Vacs $46.15
at Spud's Hardware

11. Saver's Discount Shops sell Luande slacks for $15.38 per pair. The markup on each pair is $7.38. Overhead is estimated to be 40.3% of the selling price. What is the net-profit rate?

12. Kitchen Cupboard purchases boned turkey for $0.58 per box. They sell the turkey for $0.79 per box. Kitchen Cupboard estimates the overhead at 15% of the selling price. What is the net-profit rate on each box?

13. Fred Johnson manufactures fishing lures. He pays $0.42 for the parts for each of his famous "wigglers." Fred assembles the lures and sells them to discount stores for $0.69. He estimates the overhead to be 8.5% of the selling price of the lures. What is the net-profit rate per lure?

14. Klips Cotter Pins purchases steel bars for $2.50 per bar. An automatic machine cuts and forms a bar into 30 cotter pins. The 30 pins are packaged in 1 box and marked up $1.40. Overhead expenses are estimated to be 4% of the selling price per box. What is the net-profit rate per box?

15. Channel One Stereo buys tape cassettes for $29.60 a carton. A carton contains 24 cassettes. Channel One sells the cassettes for $2.30 each or $52.80 per carton. Overhead is estimated to be 20% of selling price.
a. What is the net-profit rate on the sale of 1 cassette?
b. What is the net-profit rate when a carton of cassettes is sold?

Maintaining Your Skills

Look up the skills in parentheses if you need help or more practice.

Find the percentage. Round answers to the nearest cent. (Skill 30)

16. 15% × $74.80
17. 22% × $116.75
18. 9% × $41.80
19. 30% × $199.99
20. 25% × $789.89
21. 18% × $1416.74

Find the rate. Round answers to the nearest tenth of a percent. (Skill 31)

22. $5.60 ÷ $22.40
23. $19.76 ÷ $98.80
24. $11.60 ÷ $33.33

15-5 Determining Selling Price—Markup Based on Selling Price

Objective

Compute the selling price of an item based on cost and markup rate.

You can use records of past sales and expenses to plan the markup rate needed to cover overhead expenses and yield a profit. You can use the cost of an item and the desired markup rate based on selling price to figure the best selling price.

$$\text{Selling Price} = \frac{\text{Cost}}{\text{Complement of Markup Rate}}$$

Example

Skills 28, 11 **Application** L **Term** Markup rate

The Walker Sport Shop knows from past expense records that it must aim for a markup that is 35% of the selling price of its merchandise. The shop received a shipment of tennis rackets at a cost of $26 each. What is the minimum selling price that the shop should charge?

Sales for Month:	$27,000	% of Total Sales
Cost of goods sold:	$17,550	65%
Overhead expenses:	5,400	20
Profit:	4,050	15
	$27,000	100%

Solution

1. Find the **complement of markup rate.**
   100% − 35% = 65%
2. Find the **selling price.**
   Cost ÷ Complement of Markup Rate
   $26.00 ÷ 65% = $40.00 selling price

A. $600
B. $6.00

Self-Check Cover the answers and complete the problems.

Find the selling price.

A. Cost: $420.
Markup: 30% of selling price.

B. Cost: $4.50.
Markup: 25% of selling price.

Problems

	1.	2.	3.	4.	5.	6.
Cost	$5.00	$96.50	$11.00	$74.40	$235.43	$45.40
Markup Rate (selling price base)	60%	50%	45%	38%	76%	29%
Complement of Markup Rate	40%	?	?	?	?	?
Selling Price (nearest cent)	?	?	?	?	?	?

7. Round campaign button.
Costs $0.40.
Markup is 60% of selling price.
What is the selling price?

8. Calculator.
Costs $4.75.
Markup is 75% of selling price.
What is the selling price?

9. Cortell's Newsstand marks up paperbacks 25% of their selling price. A paperback costs Cortell's $1.60. What is the selling price of the book?

10. Color Haven purchases paint from the manufacturer and marks it up 60% of its selling price. Color Haven pays $6.34 per gallon for white latex paint. What is the selling price for the paint?

11. Mark Harris prices items at Custom Draperies, Inc. Custom Draperies has a storewide policy to mark up each item 56% of the selling price. What selling price does Mark stamp on these items?

ITEM	COST
Adjustable rod (white)	$0.22
Sash rod	0.26
Adjustable brass rod	0.56

12. Mary Woo manages the stationery department of Discount City. What selling price will Mary charge for each of these items?

LINE	ITEM CODE	DESCRIPTION	UNIT	STORE COST	MARKUP RATE SELLING PRICE BASE
14	063614	TAPE MASKING 300 IN. ROLL	RL	0.545	44.8%
15	071358	PROTRACTOR W/RULER PKG	EA	0.105	44.7
16	031615	TWINE COTTON 4 PLY 430 FT.	EA	0.302	45.1
17	062824	CHALK WHITE 12 STICK PKG.	PK	0.149	44.8
18	120057	MODEL CEMENT PLASTIC 1 OZ	EA	0.218	44.1

Maintaining Your Skills

Look up the skills in parentheses if you need help or more practice.

Find the complement. (Application L)

13. 32% 14. 44% 15. 26% 16. 17% 17. 58.7% 18. 35.4%

Divide. Round answers to the nearest cent. (Skill 11)

19. $18.00 ÷ 45%
20. $94.00 ÷ 25%
21. $459.00 ÷ 60%
22. $167.60 ÷ 40%
23. $1.13 ÷ 50%
24. $2571.83 ÷ 50%
25. $967.80 ÷ 42%
26. $71.48 ÷ 56%
27. $0.418 ÷ 52.3%

15-6 Markup Rate Based on Cost

Objective

Compute the markup rate based on cost.

Your business may use the cost of a product as the base for markup rate. The markup rate in supermarkets is relatively low. Milk may be marked up 5% of the cost. Other dairy products may be marked up 20% to 30% of the cost. The markup of clothing may be from 80% to 140% of the cost. The markup of jewelry may be 100% or more of the cost.

▶ $\text{Markup Rate} = \dfrac{\text{Markup}}{\text{Cost}}$

Example

Skills 6, 31 **Application** A **Term** Markup

Kiello Furniture purchases a 3-piece furniture grouping for $370 from the manufacturer. Kiello sells the 3-piece set for $610.50. What is the markup rate based on cost?

Solution

1. Find the **markup.**
   Selling Price − Cost
   $610.50 − $370.00 = $240.50

2. Find the **markup rate** based on cost.
   Markup ÷ Cost
   $240.50 ÷ $370.00 = 0.65 = 65% markup rate

A. 294%
B. 50%

▶ **Self-Check** Cover the answers and complete the problems.

Find the markup rate based on cost.

A. Selling price: $98.50.
Cost: $25.

B. Selling price: $90.
Cost: $60.

Problems

Round answers to the nearest tenth of a percent.

	(Selling Price	−	Cost	=	Markup)	÷	Cost	=	Markup Rate
1.	($ 1.50	−	$ 1.25	=	?)	÷	$ 1.25	=	?
2.	($ 35.00	−	$ 20.00	=	?)	÷	$ 20.00	=	?
3.	($963.99	−	$321.33	=	?)	÷	$321.33	=	?
4.	($ 19.95	−	$ 15.96	=	?)	÷	$ 15.96	=	?
5.	($342.50	−	$163.50	=	?)	÷	$163.50	=	?

6. Selling price is $47.95.
Cost is $19.18.
What is the markup rate based on cost?

7. Selling price is $2.79 each.
Cost is $1.39 each.
What is the markup rate based on cost?

8. Kiddie Clothes buys children's jeans at $4.18 per pair. They mark up each pair $3.50. What is the markup rate based on cost?

9. Chuck Beech, a salesperson at Cranshaw's Department Store, sold a pair of golf shoes for $46.15. The shoes cost Cranshaw's $24 a pair. What is the markup rate based on cost?

10. Marie Hernandez is a buyer for the Smart Tops Shoppe. She purchased an assortment of knit vests for $7.49 each. They were marked to sell for $18.95 each. What is the markup rate based on cost?

11. Andy O'Toole works at his parents' outdoor market. He sells tomatoes at $0.79 a pound. The tomatoes cost $0.24 a pound. What is the markup rate based on cost?

12. Janet Rinehart is in charge of newspapers and magazines at Corner News. What is the markup rate based on cost for each item?

Item	Cost	Selling Price
Local Sunday Paper	$0.20	$0.25
Map of City	$0.25	$0.35
Out-of-State Sunday Paper	$0.38	$0.50
Antique Magazine	$1.60	$2.00

13. Sound-Off, Inc., sells quality stereo components. The following listing indicates cost and selling price of various components. Find the markup rate based on cost for each.

ITEM	COST	SELLING PRICE
P636 Receiver	$147.50	$355.00
P552D Turntable	147.50	329.00
P9191 Cassette Deck	232.00	458.00
P88A Speakers (pair)	210.00	522.38

Maintaining Your Skills

Look up the skill in parentheses if you need help or more practice.

Subtract. (Skill 6)

14. $78.50 − $62.80
15. $149.50 − $89.70
16. $7.47 − $3.73
17. $97.80 − $73.35
18. $1496.96 − $478.89
19. $32,416.50 − $21,597.83

Find the rate. Round answers to the nearest tenth of a percent. (Skill 31)

20. $15.70 ÷ $62.80
21. $59.80 ÷ $89.70
22. $7.40 ÷ $14.80

15-7 Determining Selling Price—Markup Based on Cost

Objective

Compute the selling price based on cost and markup rate.

You can use the cost of an item and desired markup rate based on cost to figure the selling price of the item.

▶ **Markup = Markup Rate × Cost**

▶ **Selling Price = Cost + Markup**

Example

Skills 30, 5 **Application** A **Term** Markup rate

An importer sells Sinbano electronic watches to jewelry stores for \$79.95 each. The importer calculates the suggested retail price and attaches it to each watch. The retail price is computed by marking up the cost to the jewelry store by 120%. What is the suggested retail selling price?

Solution

1. Find the **markup.**
   Markup Rate × Cost
   120% × \$79.95 = \$95.94

2. Find the **selling price.**
   Cost + Markup
   \$79.95 + \$95.94 = \$175.89 selling price

A. \$32.00; \$72.00
B. \$137.25; \$320.25

▶ **Self-Check** Cover the answers and complete the problems.

Find the markup and the selling price.

A. Cost: \$40.
Markup: 80% of cost.

B. Cost: \$183.
Markup: 75% of cost.

Problems

	1.	2.	3.	4.	5.	6.
Cost	\$120.00	\$70.00	\$430.00	\$45.00	\$72.60	\$216.40
Markup Rate	50%	140%	80%	60%	125%	215%
Markup	\$60.00	\$98.00	?	?	?	?
Selling Price	?	?	?	?	?	?

7. Hammers.
Cost \$5.20 each.
Markup is 40% of cost.
What is the selling price?

8. Binoculars.
Cost \$14.65.
Markup is 75% of cost.
What is the selling price?

9. The Domst Wholesale Company purchases spice racks directly from the manufacturer for $6.00 each. The spice racks are marked up 50% of cost and sold to retail department stores, drugstores, and discount stores. What is the selling price of each spice rack?

10. Columbus Manufacturing produces truck dock seals. Their 9′ × 10′ seal costs $250 to produce. Markup is 40% of cost. What is the selling price?

11. King Appliance buys Laundry Mate washers from a distributor for $199.74 each. The markup is 51.6% of cost. What is the selling price?

12. Open-Hearth Furnaces, Inc., calculates the cost of manufacturing a particular furnace to be $1373.80. Open-Hearth marks up each furnace 35% based on cost. What is the selling price?

13. Pet Pal, Inc., buys dog collars from a manufacturer for $0.29 each. Pet Pal marks up each collar 613.8% of cost. What is the selling price?

14. The Olde England Company manufactures candles. Cost per gross (144 candles) is $79.20. The candles are sold at a markup of 80% based on cost.
    a. What is the selling price for a dozen candles?
    b. What is the selling price for a single candle?

15. Slumber City carries bedspreads which cost $11.37 each. Slumber City operates on a markup of 186% of cost.
    a. What is the selling price?
    b. What is the markup as a percent of the selling price?

16. Arden Exporters buys hinges from a U.S. manufacturer for $7.56 a dozen. Arden marks up the hinges 136.5% based on cost. The overhead is estimated to be 32% of the selling price.
    a. What is the selling price per hinge?
    b. What is the markup as a percent of the selling price?
    c. What is the net profit?
    d. What is the net profit as a percent of the selling price?

Maintaining Your Skills

Look up the skills in parentheses if you need help or more practice.

Add. (Skill 5)

17. $75 + $30 18. $60.00 + $72.00 19. $249.60 + $312.00

Find the percentage. Round answers to the nearest cent. (Skill 30)

20. 40% × $75.00 21. 120% × $60.00 22. 125% × $249.60

23. 115% × $44.00 24. 225% × $1464.40 25. 75% × $1.32

26. 157% × $213.89 27. 94% × $71.63 28. 215% × $0.44

15-8 Markdown

Objective

Compute the markdown in dollars and as a percent of the regular selling price.

Your business may sell some merchandise at sale prices to attract customers or to make room for new merchandise. The **markdown** or **discount** is the difference between the regular selling price of an item and its sale price. The **markdown rate** is markdown expressed as a percent of the regular selling price of the item.

▶ **Markdown = Regular Selling Price − Sale Price**

▶ **Markdown Rate = $\frac{\text{Markdown}}{\text{Regular Selling Price}}$**

Example

Skills 31, 6 **Application** A **Term** Markdown rate

Hanson's Department Store sells a typewriter at a regular selling price of $139.99. For one week only, Hanson's has marked down the price to $104.99. What is the markdown rate?

Solution

1. Find the **markdown.**
   Regular Selling Price − Sale Price
   $139.99 − $104.99 = $35.00

2. Find the **markdown rate.**
   Markdown ÷ Regular Selling Price
   $35.00 ÷ $139.99 = 0.2500 = 25% markdown rate

A. $10.00; 20%
B. $40.00; 50%

▶ **Self-Check** Cover the answers and complete the problems.

Find the markdown and markdown rate.

A. Regular selling price: $50.
   Sale price: $40.

B. Regular selling price: $79.95.
   Sale price: $39.95.

Problems

Round answers to the nearest tenth of a percent.

	(Regular Price	−	Sale Price	=	Markdown)	÷	Regular Price	=	Markdown Rate
1.	($ 12.00	−	$ 9.00	=	?)	÷	$ 12.00	=	?
2.	($ 25.00	−	$ 16.00	=	?)	÷	$ 25.00	=	?
3.	($ 79.98	−	$ 39.99	=	?)	÷	$ 79.98	=	?
4.	($1495.00	−	$1395.00	=	?)	÷	$1495.00	=	?

5. Regular price is $7.47.
Sale price is $4.66.
What is the markdown rate?

6. Regular price is $999.95.
Sale price is $799.88.
What is the markdown rate?

7. Regular price is $450.
Sale price is $300.
What is the markdown rate?

8. Regular price is $64.
Sale price is $56.
What is the markdown rate?

9. A plastic drop cloth can be purchased for 19¢ with a coupon, 29¢ without. What is the markdown rate if the coupon is used?

10. Tiger Stores is participating in a national promotion of Clean Sweep vacuum cleaners. The lowest-priced vacuum cleaner is marked down $4.96 to a sale price of $59.99. Their middle-priced vacuum cleaner is marked down from $114.95 to $69.99. Their top-priced vacuum cleaner is marked down $24.96 from the regular price of $124.95. What is the markdown rate for each?

11. Slate-look blocks are marked down.
   a. Find the markdown rate for each.
   b. Find the best buy, based on dollar markdown.
   c. Find the best buy, based on percent markdown.

SLATE-LOOK BLOCKS

Reg.	2.17	Outside Radius,	1.77
Reg.	1.67	Border Block,	1.27
Reg.	2.17	Inside Radius,	1.77
Reg.	3.57	24X24" Block,	2.77
Reg.	1.77	16X16" Block,	1.44

12. Miller Furniture has marked down this group of furniture.
   a. What is the markdown rate on each item?
   b. Based on markdown rate, which is the best buy?
   c. Based on dollar markdown, which is the best buy?

	Reg.	Sale
Sofa..........	$319	**$229**
Love Seat...	$269	**$188**
Chair..........	$189	**$129**
Tables ea....	$ 68	**$ 49**

Maintaining Your Skills

Look up the skill in parentheses if you need help or more practice.

Subtract. (Skill 6)

13. $97.50 − $57.50
14. $7.90 − $4.80
15. $14.79 − $9.99

Find the rate. (Skill 31)

16. $30.00 ÷ $50.00
17. $3.10 ÷ $15.50
18. $4.99 ÷ $19.96
19. $12.50 ÷ $125.00
20. $17.50 ÷ $35.00
21. $4.80 ÷ $96.00
22. $22.02 ÷ $146.80
23. $87.43 ÷ $249.80
24. $20.00 ÷ $97.50
25. $70.00 ÷ $499.50
26. $51.88 ÷ $197.98
27. $461.84 ÷ $13,471.80

Reviewing the Basics

Skills

Round to the nearest tenth of a percent.

(Skill 2) 1. 7.04% 2. 141.454% 3. 34.851% 4. 71.99%

Solve.

(Skill 5) 5. \$136.50 + \$201.49 6. \$419.40 + \$83.88

7. \$821.93 + \$37.52 8. \$1.47 + \$2.94 + \$201.33

(Skill 6) 9. \$48.30 − \$18.09 10. \$44.40 − \$8.88 11. \$278.55 − \$109.87

Round to the nearest cent.

(Skill 11) 12. \$18.00 ÷ 0.27 13. \$175.00 ÷ 0.25 14. \$202.46 ÷ 0.63

Write as a decimal.

(Skill 28) 15. 29% 16. 35% 17. 47.4% 18. 48.7% 19. 171%

Round to the nearest cent.

(Skill 30) 20. 76% of \$82.51 21. 145% of \$153.00 22. 14.7% of \$417.89

Round to the nearest tenth of a percent.

(Skill 31) 23. \$4.83 is what percent of \$21.95? 24. \$118.40 is what percent of \$572.00?

Applications

Find the complement.

(Application L) 25. 39% 26. 73% 27. 14.6% 28. 68.1%

Terms

Match each term with its definition on the right.

29. Markup
30. Markup rate
31. Net profit
32. Net-profit rate
33. Markdown rate

a. the difference between the cost and the selling price, expressed as a percent
b. net profit expressed as a percent
c. the difference between the markup and the overhead when the markup is greater than the overhead
d. the amount that your business pays for a product
e. the discount expressed as a percent of the selling price
f. the difference between the selling price and the cost of a product

Refer to the reference files at the back of the book if you need help.

Inside Economics

The Gross National Product (GNP) is a measure of a nation's economic performance. The GNP is the estimated total value of all goods and services produced by a nation during a year. Only goods and services which add to the national income are included; do-it-yourself activities are not.

Who uses the GNP and why?

- Consumers may use the GNP to compare their current standard of living with that of prior years.
- Businesses may watch the GNP to plan increases or decreases in production to meet changes in consumer demands.
- The government uses the GNP to monitor the country's economic "health."

Answer the questions about GNP.

1. Which of the following would be included in the GNP?
   - **a.** Purchasing a new automobile
   - **b.** Washing your own car
   - **c.** Doing your homework
   - **d.** Taking a bus to work
   - **e.** Buying groceries
   - **f.** Working for a housekeeping service

2. Inflation may make a nation's GNP appear to be growing faster than it really is. For this reason, economists calculate the real GNP, or the adjusted GNP, to compare economic growth from year to year.

$$\text{Real GNP} = \text{GNP} - (\text{Inflation} \times \text{GNP})$$

Calculate the real GNP.

	Inflation Rate	Approximate GNP in Billions
a.	3%	\$ 976
b.	8%	\$1516
c.	13%	\$1192

3. A nation's per-capita GNP can provide an indication of the nation's standard of living.

$$\text{Per-Capita GNP} = \frac{\text{GNP}}{\text{Population}}$$

Nation	GNP in Millions	Population in Millions
South Korea	\$ 62,200	41.3
United States	\$3,080,700	234.2
Japan	\$1,128,000	119.2

The table shows GNP's and populations for a recent year. Find the per-capita GNP for each country.

Unit Test

Lesson 15-1

1. Hi-Resistance Industrial Electrical Supply buys an electric motor for $150.75. The company sells the motor for $185.90. What is the markup?

Lesson 15-2

2. Chem-Supply buys test tubes for $1.88 a dozen. They sell them for $2.35 a dozen. What is the markup rate based on the selling price?

Lesson 15-3

3. Universal Drill, an oil equipment wholesaler, purchases drilling bits for $2922.50 each. Universal sells the bits for $3500.75. Universal estimates the overhead expenses for each bit to be 15% of the selling price. What is the estimated net profit on each bit?

Lesson 15-4

4. Water-Tite sells a waterproofing compound for $52.95. It costs Water-Tite $40.55 to purchase the compound. The overhead is estimated to be 15% of the selling price of the compound. What is the estimated net-profit rate based on the selling price?

Lesson 15-5

5. Ocean-Born Salt Water Fish Shop operates on a markup of 28% of the selling price of its merchandise. During December, the shop bought a tank of the fish listed on the invoice. What selling price will Ocean-Born charge per fish?

SHIP TO: Ocean-Born Salt Water Fish	
ITEM DESCRIPTION	COST
007-31	$6.48 ea.

Lesson 15-6

6. Bey's, a shoe boutique, sells Italian-made shoes for $132.95. Bey's purchases the shoes for $75.55. To the nearest tenth of a percent, what is the markup rate based on cost?

Lesson 15-7

7. During the winter, Sun Grown buys oranges from overseas growers. Sun Grown determines the retail selling price of the oranges by marking up the cost by 76%. The fruit company bought its first shipment of winter oranges for $0.60 per dozen. What is the retail selling price per dozen of this shipment?

Lesson 15-8

8. Al Summer's, an automobile dealership, buys its cars from the Clarion Motor Coach Company. Because of a downturn in sales, Clarion decided to reduce the price of its new StraightAway model. The regular selling price is $7000.95. The StraightAway's new selling price is $6700.95. What is the markdown rate to the nearest tenth of a percent?

Unit 16 Marketing

▸ The Marketing Department of your business estimates the number, the price, and the amount of sales you can expect to make on your product. This *market share* can be based on *opinion surveys* of people who buy your product and *sales projections,* or estimates of future sales. The Marketing Department also handles the advertising costs for your product.

Sales projections have an impact on a company's plans for growth.

16-1 Opinion Surveys

Objective Compute the rate of a particular response in an opinion survey.

When you develop a new product, you will want to know how well it is likely to sell. You may conduct a **product test** by asking a group of people to try the product. You may hire an **opinion research firm** that specializes in product testing to conduct the test. Panels of volunteers try the product and respond to questions in an **opinion survey.** The opinion-research firm tabulates the answers to the questions and submits tables of results to you.

Percent of Particular Response = Number of Times Particular Response Occurs ÷ Total Number of Responses

$$\text{Percent of Particular Response} = \frac{\text{Number of Times Particular Response Occurs}}{\text{Total Number of Responses}}$$

Example

Skill 31 **Application** C **Term** Opinion survey

Countryside Cereal Company conducted an opinion survey of 2600 people for its new Good Morning Cereal. This table shows the responses of the 2600 people in the survey.

		Age Group		
Opinion	**Under 18 Years**	**18 to 40 Years**	**40 or Over**	**Total**
Excellent	200	430	1450	2080
Good	25	108	215	348
Fair	10	12	99	121
Dislike	5	10	36	51
Total	240	560	1800	2600

What percent of the total responses were for "excellent"?

Solution

Find the **percent of particular response.**

Number of Times Particular Response Occurs ÷ Total Number of Responses

2080 ÷ 2600 = 0.80 = 80% percent of "excellent" responses

A. 90%
B. 84%

Self-Check Cover the answers and complete the problems.

A. 1800 out of 2000 people surveyed own computers. What percent of the total surveyed own computers?

B. 420 out of 500 people surveyed like Mintie toothpaste. What percent of the total surveyed like it?

Problems

	Product in Survey	Number of "Good" Responses	÷	Total Number of Responses	=	Percent of "Good" Responses
1.	Ice cream	20	÷	40	=	?
2.	Soup	80	÷	200	=	?
3.	Cleanser	400	÷	1000	=	?
4.	Movie	250	÷	1250	=	?
5.	Calculator	1050	÷	2000	=	?

6. Survey for orange juice.
358 responses.
196 "fair" responses.
What is the percent of "fair" responses?

7. Survey for silicone glue.
505 responses.
86 negative responses.
What is the percent of negative responses?

8. Caravel Motors conducted a survey on their dealer service. The people surveyed were asked to choose one answer for this question: "If you DO NOT usually go to the dealer from whom you bought your car, why?" The choices and number of responses received for each were:

27	Moved away from vicinity	112	Crowded service department
7	Disliked quality of service	11	Location not convenient
95	Service charges too high	41	Some other reason

What is the percent of each response?

9. Before beginning production of its new vacuum cleaner, the Devaney Company conducted an opinion survey of 1740 people. When asked if they would purchase the product, the people gave these responses:

Response	Under 25	25–34	35–49	50 or over
Definitely	43	138	116	74
Probably	75	194	213	130
Possibly	97	145	136	109
No	91	47	48	84

a. Based on 1740 responses, what is the percent of each response?
b. What is the percent of "definitely" responses in each age group?
c. Which age group is most likely to purchase the item?

Maintaining Your Skills

Look up the skills in parentheses if you need help or more practice.

Find the percentage. Round to the nearest tenth. (Skill 30)

10. 5% of 70 11. 60% of 80 12. 40% of 500 13. 35% of 1200

Use the table at the right to answer the following questions. (Application C)

14. What is the income of Energy, Inc.?

15. What is the dividend per share of Ohren Corp.?

16. What is the total income of the three companies?

Company	Income (in millions)	Dividends per Share
Irgo	$ 26	$1.53
Energy, Inc.	7	None
Ohren Corp.	140	0.77

16-2 Sales Potential

Objective

Compute the annual sales potential of a new product.

Before your business mass produces a new product, you may try to determine the product's **sales potential.** The sales potential is an estimate of the sales volume of a product during a specified period of time. You may manufacture a small number of the product for a selected group of people to try. This group is called a **sample.** The sales potential of your new product is based on:

(1) the percent of the people in the sample who would purchase your product,
(2) an estimate of the size of the **market,** and
(3) the average number of times that an individual might purchase this type of product during a specified period of time.

The market is the total number of people who might purchase the type of product that you make.

▶ **Annual Sales Potential** = **Percent of Potential Purchasers** × **Estimated Market Size** × **Individual Rate of Purchase**

Example

Skills 31, 30 **Application** A **Term** Sales potential

Lipscomb Corporation has developed a new denture cream called Hold Tight. Lipscomb chose a sample of denture wearers to try the new product. Of the 2000 people in the sample, 800 said they would purchase Hold Tight. Lipscomb Corporation estimates that there are 1,000,000 people who wear dentures. Lipscomb's surveys indicate that each denture wearer purchases about 5 tubes of denture cream per year. What is the sales potential for Hold Tight for 1 year?

Solution

1. Find the **percent of potential purchasers.**
   800 ÷ 2000 = 40%

2. Find the **annual sales potential.**

Percent of Potential Purchasers		Estimated Market Size		Individual Rate of Purchase	
40%	of	1,000,000	×	5	= 2,000,000 annual potential sales

3,000,000

▶ **Self-Check** Cover the answer and complete the problem.

Out of 1500 secretaries in a sample, 900 preferred a new typewriter ribbon. The estimated market size is 100,000. Each secretary uses about 50 ribbons a year. What is the annual sales potential?

Problems

	1.	2.	3.	4.	5.
Number in Sample	100	200	500	1000	2000
Number of Potential Purchasers	17	6	15	240	240
Percent of Potential Purchasers	17%	?	?	?	?
Estimated Market Size	800,000	2,800,000	4,700,000	25,000,000	32,600,000
Individual Rate of Purchase	1 can	once	once	8 boxes	12 issues
Annual Sales Potential	?	?	?	?	?

6. Long-lasting razor.
350 people in sample.
50 would purchase razor.
Market size about 20,000,000.
Each person buys about 5 per year.
What is the annual sales potential?

7. Notebook/organizer.
175 people in sample.
70 would purchase product.
Market size about 15,000,000.
Each person buys about 3 per year.
What is the annual sales potential?

8. Fonfball is a new game. Out of 1000 people surveyed, 57 said they would buy it. The estimated market size is 2,500,000. The company estimates that each person would buy the game once per year. What is the annual sales potential?

9. National Optometrics Company is marketing a new, softer, more pliable contact lens. Out of a sample of 200 users, 38 preferred the new lenses. There is an estimated total market of 565,000 contact lens users. The average contact lens wearer purchases 2 lenses per year. What is the sales potential for the new lenses for 1 year?

10. The Automotive Institute has developed a new spark plug for automobiles using unleaded gasoline. The Marketing Department had a sample of 2000 automobile owners use the new spark plug. Sixty owners responded favorably and stated that they would buy the spark plug. There are approximately 5,000,000 automobiles using unleaded gasoline. An average number of 6 spark plugs would be purchased per automobile per year. What is the sales potential for the new spark plug for 1 year?

Maintaining Your Skills

Look up the skill in parentheses if you need help or more practice.

Find the percentage. Round answers to the nearest hundredth. (Skill 30)

11. $4\frac{3}{4}$% of 33 12. $\frac{7}{8}$% of 83 13. 125% of $2140 14. 16.5% of 0.175

16-3 Market Share

Objective Compute the market share of a new product.

To find out how well your product is selling in the marketplace, you may want to find your product's **market share.** Market share is that percent of the total market that purchases your product instead of a competitor's. You can calculate your percent of the total market sales by using either the number of units sold or the dollar value of sales.

▶ **Market Share** = $\frac{\textbf{Total Product Sales}}{\textbf{Total Market Sales}}$

Example

Skill 31 Application A Term Market share

Luxura, an air conditioner manufacturer, sold 1,800,000 air conditioners during the year. During the same period, a total of 8,000,000 air conditioners were purchased in the entire U.S. market. What was Luxura's market share for the year?

Solution

Find the **market share.**

Total Product Sales ÷ Total Market Sales

1,800,000 ÷ 8,000,000 = 0.225 = 22.5% market share

A. 25%
B. 5%

▶ **Self-Check** Cover the answers and complete the problems.

Find the market share.

A. Total product sales: 4,000,000. Total market sales: 16,000,000.

B. Total product sales: $8,000,000. Total market sales: $160,000,000.

Problems

	Company	Total Product Sales	÷	Total Market Sales	=	Market Share
1.	Diwan	$4,000,000	÷	$20,000,000	=	?
2.	Merrill	$ 800,000	÷	$32,000,000	=	?
3.	Hersh	$1,200,000	÷	$ 5,000,000	=	?
4.	TBC	$1,000,000	÷	$42,000,000	=	?

5. Typewriters.
Product sales total 1,800,000 units.
Market sales total 20,000,000 units.
What is the market share?

6. Lawn mowers.
Product sales total 940,000 units.
Market sales total 5,000,000 units.
What is the market share?

7. Light bulbs.
Product sales total $5,000,000.
Market sales total $80,000,000.
What is the market share?

8. Carpets.
Product sales total $40,060,000.
Market sales total $250,000,000.
What is the market share?

9. Tread-Evers, Inc., sells approximately 8,000,000 automobile tires per year. There are approximately 32,000,000 automobile tires sold per year in the entire market. What is Tread-Ever's market share?

10. Affordable Paperbacks Company sells approximately 1,000,000 paperback books per year. Sales for the entire paperback book market total approximately 12,000,000 books per year. What is Affordable Paperback Company's market share of paperbacks?

11. Easy-Light Grille Company sells about $975,000 in outdoor grilles annually. The total annual sales of outdoor grilles in the entire market is about $2,500,000. What is Easy-Light's market share for outdoor grilles?

12. Quality Linen Sheet Company has sales totaling approximately $820,000 in fitted sheets. Total sales of fitted sheets by all companies is approximately $3,450,000. What is Quality Linen's market share of fitted sheets?

13. Sky Hi Shoes had sales totaling approximately $435,000 in basketball shoes last year. Last year, sales of all basketball shoes totaled approximately $950,000. What market share of basketball shoes did Sky Hi Shoes have?

14. Last year there were total sales of approximately $4,500,000 in dining-room furniture. Bridgeport Chair Company's sales of dining-room furniture totaled approximately $987,500 last year. What was Bridgeport Chair Company's market share of dining-room furniture?

15. Graham Brothers, a company that manufactures games, sells an electronic game called Match. Last year, sales for Match totaled approximately $432,000. Total market sales for electronic games were approximately $2,350,000. What market share did Match have?

16. Bayshore Nursery provides a landscaping service in Markham County. Bayshore's landscaping business totaled approximately $89,500 for the year. The landscaping business for Markham County totaled about $157,000 for the year. What market share did Bayshore have?

Maintaining Your Skills

Look up the skills in parentheses if you need help or more practice.

Find the rate. Round to the nearest tenth of a percent. (Skill 31)

17. $n\%$ of 64 = 24 18. $n\%$ of 96 = 68 19. $n\%$ of 25 = 50 20. $n\%$ of 36 = 80

Divide. Round to the nearest tenth. (Skill 10)

21. 920 ÷ 30 22. 1125 ÷ 85 23. 968 ÷ 324 24. 821 ÷ 11 25. 574 ÷ 42

16-4 Sales Projections

Objective Use a graph to compute projected sales.

A **sales projection** is an estimate of the dollar volume or unit sales that might occur during a future time period. A sales projection is usually based on past sales. You may use a sales projection to plan for production or purchasing. New products or changing economic conditions may result in sales figures that differ from the figures you projected.

You may use a graph to project a rough estimate of future sales.

(1) Construct a graph of past sales.

(2) Draw a straight line from the first year of data, approximately through the "middle" of the data, to the year for which the projection is being made.

(3) Read the number or dollar value, the sales projection.

Example

Skills ____ **Application** N **Term** Sales projection

The Marketing Department of Stanton Stores wishes to project sales for 1990. The sales history of Stanton Stores is:

Year	1979	1980	1981	1982	1983	1984	1985
Sales (in millions)	\$2.1	\$2.0	\$2.8	\$4.2	\$3.0	\$5.0	\$4.1

Using a graph, what sales projection might the Marketing Department make for 1990?

Solution

1. Graph the sales from 1979 through 1985.
2. Starting from sales in 1979, draw a straight line, approximately through the "middle" of the data, to 1990.
3. Read the sales projection for 1990.

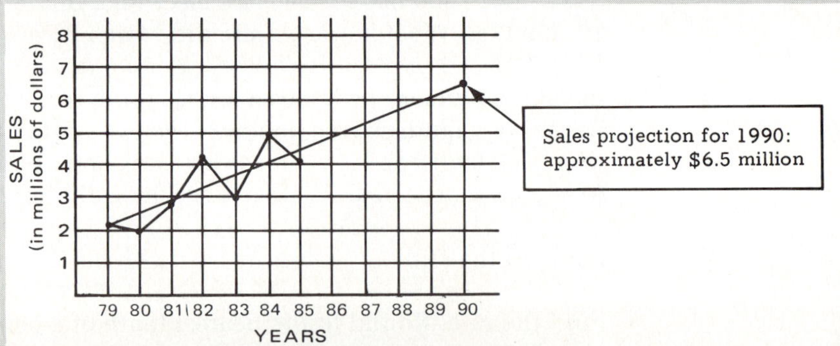

\$5.3 million
\$5.7 million
\$6.1 million

▸ **Self-Check** Cover the answers and complete the problems.

Refer to the above graph. Read the approximate sales projections for 1987, 1988, and 1989.

Problems

Use the graph to answer exercises 1–6.

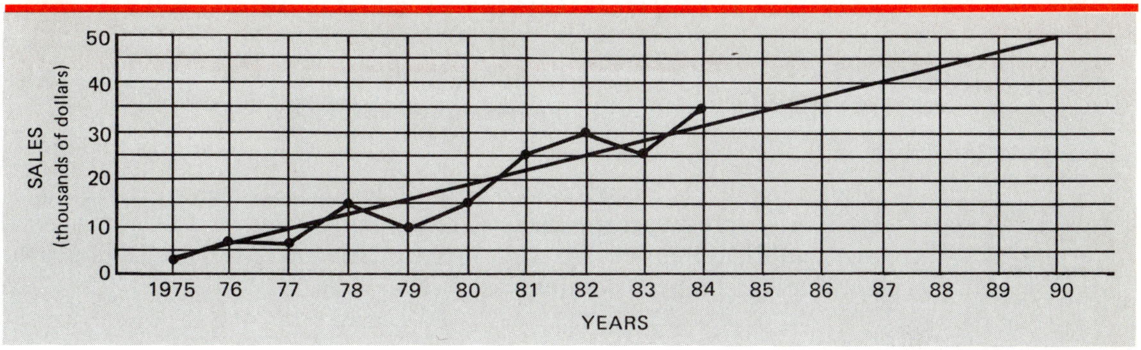

	1.	2.	3.	4.	5.	6.
Year	1985	1986	1987	1988	1989	1990
Sales Projection	?	?	?	?	?	?

Construct a graph for each exercise and draw a straight line through the "middle" of the data to project sales.

7. Sales records for Target Sporting Goods show this data:

Year	1975	1976	1977	1978	1979	1980	1981	1982	1983	1984
Sales (in millions)	$4.3	$4.7	$5.0	$4.9	$5.1	$5.3	$5.0	$5.3	$5.4	$5.5

What sales might be projected for 1989?

8. The sales history of Bi-Low Markets shows:

Year	1965	1970	1975	1980
Sales (in millions)	$12	$14	$10	$12

Project sales for 1985, 1990, and 1995.

9. Digico manufactures computers. Production records for the first half of the year show this information for the Model 4330 computer:

Month	Jan.	Feb.	Mar.	Apr.	May	June
Production (thousands of units)	17	12	15	10	12	9

Project production quantities for July, August, and September.

Maintaining Your Skills

Look up the skills in parentheses if you need help or more practice.

Multiply. Round to the nearest thousandth. (Skill 8)

10. $342.86 × 100

11. $1240.03 × 1000

12. 0.014 × 400

13. 0.00161 × 16.42

14. 0.525 × 243

Write each number in words. (Skill 1)

15. 107 16. 32 17. 0.7 18. 5104 19. 15,746

16-5 Sales Projections—Factor Method

Objective

Use the factor method to compute projected sales.

The **factor method** is another way to project sales. The **factor** is your company's present market share. Your company may use federal government publications or other sources to find the total sales projected for the entire market for the coming year.

▶ **Projected Sales = Market-share Factor × Projected Market Sales**

Example

Skill 30 Application A Term Factor method

SaveAway Food Stores have a 3% share of the market for food sales in the Chicago Heights area. Food sales in the Chicago Heights area for next year are estimated to be $27,300,000. What is the projected sales figure for SaveAway Food Stores in the Chicago Heights area for next year?

Solution

Find the **projected sales.**

Market-share Factor × Projected Market Sales

3% of $27,300,000 = $819,000 projected sales

A. $441,000
B. $1,435,000

▶ **Self-Check** Cover the answers and complete the problems.

Find the projected sales.

A. Market-share factor: 7%.
Projected market sales: $6,300,000.

B. Market-share factor: 14%.
Projected market sales: $10,250,000.

Problems

	Company	Market Share	×	Projected Market Sales	=	Projected Sales
1.	GDS	5%	×	$20,000,000	=	?
2.	Corbin	7%	×	$12,400,000	=	?
3.	Haskins	3.5%	×	$60,000,000	=	?
4.	Ahren	8.2%	×	$90,000,000	=	?

5. Long-Life Luggage Distributors.
15% share of market.
$1,600,000 estimated market total for next year.
What is Long-Life's sales projection?

6. Quick Copy Machines.
12% share of market.
$4,100,000 estimated market total for next year.
What is Quick Copy's sales projection?

7. National Mini-Carts, Inc., now has 12% of the market for electric golf carts. The total estimated sales of new electric carts for next year is $8,000,000. What sales volume should National Mini-Carts project for next year?

8. Three Oaks Bowling Lanes has traditionally had 7% of the bowling business in the East Northport area. The total estimated bowling business in this area for next year is $720,000. What bowling business can Three Oaks project for next year?

9. Mrs. Turlap's Bakery does 22% of the bakery business in the town of Henderson. The estimated bakery business for Henderson for next year is $142,500. What sales volume can Mrs. Turlap's Bakery project for next year?

10. This year enrollment at Urban Central University (UCU) accounted for 85% of all university students in the greater metropolitan area. Next year the total university enrollment in the greater metropolitan area is expected to be about 48,000 students. What enrollment figures can UCU project for next year?

11. Based on past records, Third National Savings Bank determined that they have a market share of 8.5% of all the savings accounts in Butler County. Estimated savings accounts in Butler County for next year are 270,000. Project the number of savings accounts that Third National Savings Bank will have next year.

*12. **A Brief Case** Many retail businesses increase their market share by opening additional stores.

a. Prince Dairy Outlets now do 8% of the dairy business in Leland County with their 4 outlets. Estimated dairy products sales in Leland County for next year are $1,416,000. Prince is adding 1 more outlet next year and hopes to increase its market share by 25%. Project Prince Dairy's sales for next year.

b. Cobbler's 3 stores now do 14% of the shoe repair business in North County. Estimated shoe repair business in the North County market area for next year is $720,000. Cobbler is adding 2 stores and expects to increase its share of the market by 40%. Project Cobbler's shoe repair business for next year.

Maintaining Your Skills

Look up the skill in parentheses if you need help or more practice.

Find the percentage. Round answers to the nearest hundredth. (Skill 30)

13. 6.8% of 310
14. 8.81% of $22,500
15. 3.1% of $120,000
16. 125% of 90
17. $\frac{3}{8}$% of $400
18. 0.7% of $40,000
19. 1000% of 60
20. 1% of $35,000
21. 14% of 92

16-6 Newspaper Advertising Costs

Objective

Compute the cost of advertising in a newspaper.

You may advertise your products or services in a newspaper to increase your sales. The cost of the advertisement depends on whether it is in color or black and white and on the amount of space it uses. The amount of space is figured in lines of a given width. Generally, newspapers charge a certain rate for each line. Some newspapers offer a reduced rate if you contract a year in advance for a specified amount of space.

▶ **Advertisement Cost = Space in Number of Lines × Rate per Line**

Example

Skills 8, 2 **Application** A **Term** ______

Northshore Real Estate, Inc., contracted with Hamilton News for 28,000 lines of advertising for the year. Northshore plans to advertise a new subdivision in the evening newspaper. The advertisement is the equivalent of 560 lines long. What is the cost of the advertisement each time it appears?

HAMILTON NEWS — ADVERTISING RATES

	Evening Per Line	Sunday Per Line
No Contract Rates:	$0.88	$1.04
Annual Contract Rates:		
700 lines	0.735	0.865
1400 lines	0.730	0.860
4200 lines	0.725	0.855
7000 lines	0.720	0.850
14,000 lines	0.715	0.845
28,000 lines	0.710	0.840
48,000 lines	0.705	0.835
70,000 lines	0.700	0.830
84,000 lines	0.695	0.825
100,000 lines	0.690	0.820
200,000 lines	0.685	0.815

Solution

Find the **advertisement cost.**

Space in Number of Lines × Rate per Line

560 × $0.710 = $397.60 advertisement cost

A. $648.00
B. $527.00

▶ **Self-Check** Cover the answers and complete the problems.

Use the above table to find the advertisement costs for Carpet Store.

Carpet Store has contracted for 7000 lines annually.

A. An advertisement in the evening paper is 900 lines long.

B. An advertisement in the Sunday paper is 620 lines long.

Problems

Use the table on page 370 to find the advertisement cost per run.

	Contract	Edition	Space in Lines	×	Rate per Line	=	Advertisement Cost
1.	200,000	Evening	945	×	$0.685	=	?
2.	48,000	Evening	560	×	$0.705	=	?
3.	7000	Sunday	140	×	?	=	?
4.	No contract	Evening	1400	×	?	=	?
5.	84,000	Evening	1890	×	?	=	?

6. Homer's Sporting Goods.
Contracts for 4200 lines annually.
42-line ad in Sunday paper.
What is the cost of the advertisement?

7. Miyaura's Department Store.
Contracts for 100,000 lines annually.
2835-line ad in evening paper.
What is the cost of the advertisement?

8. Big Buy Supermarket has an annual contract for 200,000 lines of advertising. In the Wednesday paper, Big Buy has a 2-page advertisement equivalent to 5670 lines. What is the cost of the advertisement?

9. The Laurel Ski Lodge has a contract for 14,000 lines of advertising for the year. Sunday's paper will carry an advertisement for the lodge equivalent to 708 lines. How much does the advertisement cost?

*10. **A Brief Case** Jo-Del Corporation placed this advertisement in the Monday Hamilton News. The advertisement is 2 columns wide and 2 inches deep. Each column inch is equivalent to 14 lines.

a. For how many lines will Jo-Del be charged?

Columns Wide × Inches Deep × 14 Lines

b. Jo-Del does not have a contract. What will the advertisement cost?

Maintaining Your Skills

Look up the skill in parentheses if you need help or more practice.

Multiply. Round answers to the nearest thousandth. (Skill 8)

11. 2.9 × 1.8
12. 0.96 × 0.11
13. 0.008 × 1000
14. 62.05 × 0.005
15. 0.932 × 0.16
16. 0.12 × 200
17. 56.00 × 0.75
18. 0.3194 × 25

16-7 Television Advertising Costs

Objective

Compute the cost of advertising on television.

You may advertise your products or services on television. The cost of a television advertisement depends on the time of day, program ratings, and the length of the advertisement. Television commercials are generally 10, 30, or 60 seconds long. The cost of a 10-second advertisement is usually one half the cost of a 30-second advertisement. The cost of a 60-second advertisement is usually twice the cost of a 30-second advertisement.

▶ **Cost of 10-Second Ad = $\frac{1}{2}$ × Cost of 30-Second Ad**

▶ **Cost of 60-Second Ad = 2 × Cost of 30-Second Ad**

Example

Skill 20 Application A Term ________

The Marketing Department of the McMahon Company plans to advertise several new hairdryers on network television. The advertising campaign calls for these commercials:

Number	Length	Time
3	30-second	Daytime
1	10-second	Daytime
2	30-second	Prime Time
1	60-second	Prime Time

The rates are $6200 for a 30-second daytime commercial and $22,450 for a 30-second prime-time commercial. What is the total cost for this advertising campaign?

Solution

1. Find the **cost of 30-second ads.**
   a. Daytime: $6200 × 3 = $18,600
   b. Prime Time: $22,450 × 2 = $44,900

2. Find the **cost of 10-second ads.**
   $\frac{1}{2}$ × Cost of 30-Second Ad
   Daytime: ($\frac{1}{2}$ × $6200) × 1 = $3100

3. Find the **cost of 60-second ads.**
   2 × Cost of 30-Second Ad
   Prime Time: (2 × $22,450) × 1 = $44,900

4. Find the **sum of the costs.**
   $18,600 + $44,900 + $3100 + $44,900 = $111,500 total cost

A. \$143,000
B. \$202,500

▸ **Self-Check** Cover the answers and complete the problems.

A. A 30-second ad costs \$13,000. Find the total costs for three 30-second ads and four 60-second ads.

B. A 30-second ad costs \$45,000. Find the total costs for four 30-second ads and one 10-second ad.

Problems

	Rate per 30-Second Ad	Number of 10-Second Ads	Number of 30-Second Ads	Number of 60-Second Ads	Total Costs
1.	\$ 3000	0	2	1	?
2.	\$ 26,000	1	3	0	?
3.	\$100,000	0	2	2	?
4.	\$ 50,000	0	4	3	?

5. Daytime television.
60-second ad for dog food.
30-second ad costs \$8400.
What is the cost of the 60-second ad?

6. Prime-time television.
10-second ad for radio station.
30-second ad costs \$21,400.
What is the cost of the 10-second ad?

7. The manufacturers of Brighten Toothpaste have planned an advertising campaign that includes these commercials:

Number	Length	Time
2	30-second	Daytime
3	10-second	Daytime
2	30-second	Prime Time
2	60-second	Prime Time

The rates are \$5800 for a 30-second daytime commercial and \$23,500 for a 30-second prime-time commercial. What is the cost of the advertising campaign?

8. Stadium New Car Sales is interested in sponsoring the sports coverage on the local television station. The rate per 30-second commercial is \$275. Stadium New Car Sales contracts for ten 30-second advertisements, five 10-second advertisements, and five 60-second advertisements. What is the total cost for these advertisements?

9. Crown Amusement Park has planned a special television campaign. They will use a daytime show for twenty 10-second ads and five 30-second ads. They will use a prime time show for ten 30-second ads and five 60-second ads. The rates are \$6250 per 30-second daytime ad and \$18,500 per 30-second prime time ad. What is the cost of the television campaign?

Maintaining Your Skills

Look up the skill in parentheses if you need help or more practice.

Multiply. Express the answers in lowest terms. (Skill 20)

10. $\frac{1}{2} \times \frac{2}{5}$ 11. $\frac{3}{10} \times \frac{5}{8}$ 12. $6 \times 8\frac{2}{7}$ 13. $2\frac{2}{7} \times 1\frac{1}{8}$ 14. $4 \times 16\frac{1}{2}$

16-8 Pricing

Objective Compute the selling price that will result in the highest possible net profit.

You must set the selling prices of your products high enough to cover expenses and to make a profit. You may determine the selling price of an item by estimating the net profit for each of several possible selling prices and choosing the selling price that will result in the highest profit. In manufacturing, the cost per item varies with the number manufactured and ultimately sold. Often the higher the selling price of an item, the fewer items will be sold.

▸ **Possible Net Profit** = (**Selling Price per Unit** − **Total Cost per Unit**) × **Estimated Unit Sales**

Example

Skills 6, 8 **Application** A **Term** Net profit

Gamma Electronics manufactures preprinted circuit boards. Gamma has a fixed overhead of $60,000. The variable costs to produce the circuit boards are $0.70 per unit. To determine the best selling price, Gamma estimates the number of units that could be sold at various selling prices.

Fixed Costs per Unit = Fixed Costs ÷ Estimated Unit Sales

Total Cost per Unit = Fixed Costs Per Unit + Variable Costs Per Unit

Selling Price per Unit	Estimated Unit Sales	Fixed Costs	Fixed Costs per Unit	Variable Costs per Unit	Total Cost per Unit
$ 4.00	20,000	$60,000	$ 3.00	$0.70	$ 3.70
5.50	15,000	60,000	4.00	0.70	4.70
8.00	10,000	60,000	6.00	0.70	6.70
15.00	5000	60,000	12.00	0.70	12.70

What selling price will maximize Gamma's profits?

Solution

1. Find the **possible net profit** for each selling price.

(Selling Price per Unit − Total Cost per Unit) × Estimated Unit Sales

a.	($ 4.00	−	$ 3.70)	×	20,000 = $ 6000.00
b.	($ 5.50	−	$ 4.70)	×	15,000 = $12,000.00
c.	($ 8.00	−	$ 6.70)	×	10,000 = $13,000.00
d.	($15.00	−	$12.70)	×	5000 = $11,500.00

2. Find the **greatest possible profit** $13,000.00

3. Find the **selling price** $8.00

A. $290,000
$195,000
B. $2.50

▸ **Self-Check** Cover the answers and complete the problems.

Selling Price per Unit	Estimated Unit Sales	Fixed Costs	Fixed Costs per Unit	Variable Costs per Unit	Total Cost per Unit
$2.50	200,000	$200,000	$1.00	$0.05	$1.05
4.00	100,000	200,000	2.00	0.05	2.05

A. Find the possible net profit for each price.
B. Which selling price yields the greatest possible profit?

Problems

Round answers to the nearest cent.

	Selling Price per Unit	Estimated Unit Sales	Total Fixed Costs	Fixed Costs per Unit	Variable Costs per Unit	Total Cost per Unit	Possible Net Profit
1.	$ 42.99	10,000	$110,000.00	$11.00	$18.98	$29.98	?
2.	$ 64.79	7500	$110,000.00	$14.67	$18.98	?	?
3.	$ 81.99	6150	$110,000.00	?	$18.98	?	?
4.	$124.95	2400	$110,000.00	?	$18.98	?	?

5. Complete the table for the Bancroft Company. Which selling price yields the greatest possible profit? How many units should be produced?

Selling Price per Unit	Estimated Unit Sales	Total Fixed Costs	Fixed Costs per Unit	Variable Costs per Unit	Total Cost per Unit	Possible Net Profit
$ 2.75	2,000,000	$2,000,000	?	$0.06	?	?
4.95	1,000,000	2,000,000	?	0.06	?	?
9.10	500,000	2,000,000	?	0.06	?	?
21.25	100,000	2,000,000	?	0.06	?	?

6. Fernald Games developed these cost and price figures for their new Space Explorer game. Which combination should Fernald choose to handle the smallest number of items and to exceed $200,000 in possible profit?

Selling Price per Unit	Estimated Unit Sales	Total Fixed Costs	Fixed Costs per Unit	Variable Costs per Unit	Total Cost per Unit	Possible Net Profit
$7.95	50,000	$220,000	?	$0.85	?	?
6.95	90,000	220,000	?	0.85	?	?
5.95	120,000	220,000	?	0.85	?	?
4.49	150,000	220,000	?	0.85	?	?
3.09	200,000	220,000	?	0.85	?	?

Maintaining Your Skills

Look up the skill in parentheses if you need help or more practice.

Subtract. (Skill 6)

7. 541.96 − 82.78 8. 16.9 − 1.017 9. 41 − 27.4 10. 46.21 − 42.98

Reviewing the Basics

Skills

Solve.

(Skill 6) **1.** \$7.00 − \$3.25 **2.** \$89.30 − \$24.17 **3.** \$2174.60 − \$356

(Skill 8) **4.** 342 × \$0.835 **5.** \$6.74 × 11,000 **6.** 62.1 × 96.8

(Skill 6) **7.** 475.25 − 45.4 **8.** 26.2 − 0.138 **9.** 7.5 − 1.95

(Skill 20) **10.** $\frac{1}{2}$ × \$871 **11.** $\frac{1}{2}$ × \$11,483 **12.** $\frac{1}{4}$ × \$16,260

(Skill 30) **13.** 57% of 19,840 **14.** 17% of \$16,820,000

Round to the nearest tenth of a percent.

(Skill 31) **15.** 892 is what percent of 3670? **16.** \$28,640 is what percent of \$83,580?

Applications

(Application C) **17.** What is the total of "very interested" responses in this survey?

Opinion	City	Suburbs	Rural	Total
Very interested	200	3019	2515	5734
Interested	515	2272	1475	4262
Slightly interested	1150	1400	1640	4190
Not interested	347	217	250	814
Total	2212	6908	5880	15,000

(Application N) **18.** What production level is projected for Lime Rock Mineral Company for 1989?

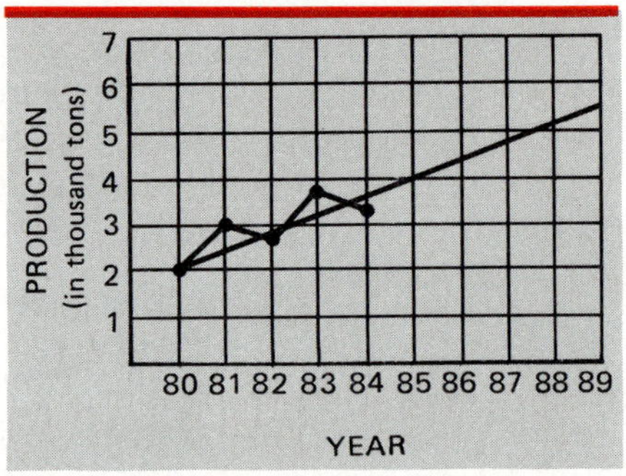

Terms

Write your own definition for each term.

19. Opinion survey **20.** Sales potential **21.** Market share

22. Sales projection **23.** Factor method **24.** Net profit

Refer to your reference files at the back of the book if you need help.

A Microcomputer Activity
Sales Analysis

To complete this microcomputer application, you will need the diskette *Microcomputer Applications for Business Mathematics,* which accompanies this textbook.

Select option 6, Sales Analysis, from the menu. Input the information in the following problems to find the break-even point, gross sales, and gross profit.

1. Use the FIXED COSTS and VARIABLE COSTS given in the program. Change the SELLING PRICE and EXPECTED SALES to the amounts indicated.

	10—SELLING PRICE	11—EXPECTED SALES	BREAK-EVEN POINT	GROSS SALES	GROSS PROFIT
a.	$10.95	17,000	?	?	?
b.	11.45	15,500	?	?	?
c.	11.95	15,000	?	?	?
d.	12.45	14,000	?	?	?
e.	12.95	13,583	?	?	?
f.	13.45	11,500	?	?	?

2. Sobo Electronic Calculator Company developed these cost and price figures for their new Model 101 calculator.

```
FIXED COSTS:
1-RENT            2-TAXES         3-MARKETING       4-MANUFACTURING
  10,200            62,400          75,000            81,500

VARIABLE COSTS:
5-SUPPLIES        6-LABOR     7-UTILITIES     8-PACKAGING     9-SHIPPING
  1.50              3.75        .85             .25             .25
```

Complete the table below to determine the best selling price and greatest profit.

	10—SELLING PRICE	11—EXPECTED SALES	BREAK-EVEN POINT	GROSS SALES	GROSS PROFIT
a.	$6.95	900,000	?	?	?
b.	7.25	800,000	?	?	?
c.	7.45	750,000	?	?	?
d.	8.45	400,000	?	?	?
e.	8.95	200,000	?	?	?
f.	9.45	50,000	?	?	?

Unit Test

Lesson 16-1

1. Oakshire High School conducted an opinion survey of the student council's performance. Of the 300 students surveyed, 57 gave an "average" response. What percent of the total gave "average" responses?

Lesson 16-2

2. Of a sample of 400 swimming pool owners, 63 said they would buy Leisure Living's new chemical for swimming pools. There is a market of about 9,000,000 potential users. Each user would purchase the chemical once per year. What is the sales potential for the chemical for 1 year?

Lesson 16-3

3. Last year Gendron Wheel Works sold about 516,000 wagons. Approximately 1,230,000 wagons were sold in the entire market last year. What was Gendron's market share?

Lesson 16-4

4. Use a graph to project Suntyme's sales for 1989 based on this sales history.

Year	1981	1982	1983	1984
Sales in Millions	$2.0	$6.0	$3.1	$4.9

Lesson 16-5

5. Finder's Clothing has 6.7% of the Willow Run metropolitan clothing market. Next year's clothing sales in the area are estimated to be $17,000,000. Use the factor method to project Finder's sales for next year.

Lesson 16-6

6. Landa's Music Studio has a contract for 20,000 lines of advertising in the Herald this year. What is the cost of a 500-line advertisement?

Annual Contract Rates	Per Line
16,000 lines	$0.815
20,000 lines	0.810
30,000 lines	0.805

Lesson 16-7

7. Countryside Cereal plans to advertise on television. Costs for 30-second commercials are $8000 for daytime and $28,645 for prime time. The cost of a 10-second commercial is $\frac{1}{2}$ the cost of a 30-second commercial. The cost of a 60-second commercial is 2 times the cost of a 30-second commercial. What is the total advertising cost?

Number	Length	Time
5	10-second	Daytime
4	30-second	Prime Time
2	60-second	Prime Time

Lesson 16-8

8. Complete the chart. Which selling price will maximize profits?

Selling Price per Unit	Estimated Unit Sales	Total Fixed Costs	Fixed Costs per Unit	Variable Costs per Unit	Total Cost per Unit	Possible Net Profit
$30,000	16	$200,000	$12,500	$3,500	?	?
37,500	14	200,000	?	3,500	?	?
40,500	12	200,000	?	3,500	?	?

Cumulative Review (Skills/Applications) Units 12–16

Skills

Round to the nearest cent.

(Skill 2) **1.** \$10.422 **2.** 15.6¢ **3.** \$836.5283 **4.** 96.48¢

Round to the nearest tenth of a percent.

5. 6.32% **6.** 14.53% **7.** 42.956% **8.** 97.751%

Solve.

(Skill 5) **9.** \$175.90 + \$351.96 **10.** \$457.54 + \$68.63

11. \$61.76 + \$90.68 + \$55.92 **12.** \$216.55 + \$44.47 + \$7.29

13. \$916.54 + \$792.85 **14.** \$670.89 + \$462.50 + \$9.89

(Skill 6) **15.** \$55.86 − \$23.63 **16.** \$17.83 − \$9.74 **17.** \$64.16 − \$17.60

18. \$140.85 − \$81.61 **19.** \$6036.48 − \$5432.83

20. \$971.83 − \$74.97 **21.** \$8264.67 − \$3476.81

(Skill 8) **22.** \$1.98 × 5 **23.** \$96.50 × 3 **24.** \$431 x 0.83

25. \$29.35 × 1.7 **26.** \$158.70 × 2.4 **27.** \$0.43 × 0.06

(Skill 11) **28.** \$97.34 ÷ 62 **29.** \$14.40 ÷ 12 **30.** \$275.74 ÷ 6.8

31. \$56,376 ÷ 1.16 **32.** \$17,850 ÷ 2.50 **33.** \$54.93 ÷ 0.03

(Skill 16) **34.** $\frac{1}{2} + \frac{3}{8}$ **35.** $\frac{3}{4} + \frac{5}{8}$ **36.** $\frac{2}{3} + \frac{1}{4}$

37. $3\frac{1}{8} + 2\frac{2}{5}$ **38.** $4\frac{1}{2} + 2\frac{1}{3}$ **39.** $\frac{7}{8} + 2\frac{1}{3} + 1\frac{1}{2}$

(Skill 20) **40.** $\frac{3}{4} \times \frac{2}{3}$ **41.** $\frac{3}{4} \times \frac{1}{3}$ **42.** $4 \times \frac{3}{8}$

43. $\frac{1}{2} \times \$14$ **44.** $\frac{1}{4} \times \$28$ **45.** $\frac{1}{3} \times \$42$

Write as a decimal.

(Skill 28) **46.** 25% **47.** 37.5% **48.** 46.8% **49.** 61.87%

50. 3% **51.** 2.4% **52.** 0.1% **53.** 0.47%

Solve. Round to the nearest cent or tenth of a percent.

(Skill 30) **54.** 15% of \$380 **55.** 20% of \$195 **56.** 25% of \$12.56

57. 60% of \$543.96 **58.** 10% of \$1678.50 **59.** 1% of \$186,243

(Skill 31) **60.** ?% of \$4856 = \$582.72 **61.** ?% of 56 = 8.4

62. ?% of \$46.52 = \$16.80 **63.** ?% of \$37.86 = \$1.98

Applications

Use the table to answer the following.

(Application C) 64. What is the regular price of a package of 50 mL beakers?

65. What is the sale price of a package of 100 mL beakers?

66. For 250 mL beakers, how much less is the sale price than the regular price?

Package of 12 Beakers		
Size (mL)	Regular Price	Sale Price
50	$ 5.25	$ 4.20
100	8.05	7.08
250	13.00	11.40

67. For 50 mL beakers, the sale price is what percent of the regular price?

Find the complement.

(Application L) 68. 50% 69. 16.3% 70. 42% 71. 3.75%

Use the graph to answer the following.

(Application N) 72. How many units were sold in 1978?

73. In what year were the most units sold?

74. How many more units were sold in 1980 than in 1976?

75. What is the sales projection for 1985?

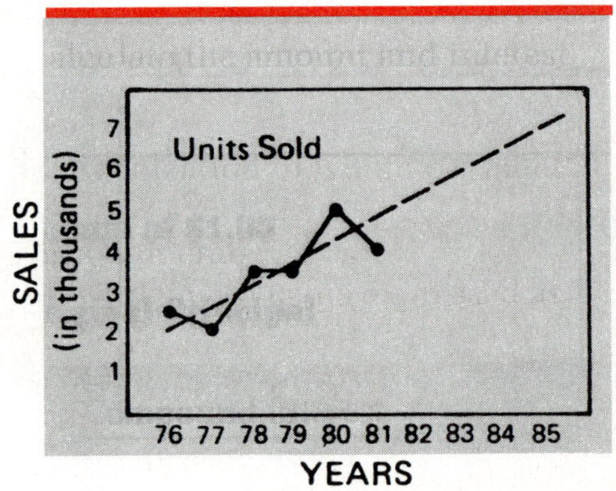

Find the mean.

(Application Q) 76. 7, 15, 2 77. 2, 7, 5, 4, 3

78. 72, 96, 26, 67 79. 3.2, 4.0, 9.1, 6.3

80. 51.8, 54.6, 66.0 81. 62.22, 55.92, 49.78

Terms

Write your own definition for each term.

82. Employee benefits 83. Release time 84. Salary scale
85. Travel expenses 86. Prime cost 87. Break-even point
88. Quality control 89. Time-study 90. Trade discount
91. Chain discount 92. Trade-discount rate 93. End-of-month dating
94. Markup 95. Markup rate 96. Net profit
97. Markdown rate 98. Opinion survey 99. Sales potential
100. Market share 101. Factor method 102. Sales projection

Cumulative Review Test Units 12–16

Lesson 12-2

1. Bev Houlson is a statistician, level 3, for the Fidelity Insurance Corporation. She presently earns $16,960 annually. Bev will receive a 6.8% cost-of-living increase and a 5.4% salary increase. What will Bev's new salary be?

Lesson 12-3

2. Rudy Maher works for the Hollis Security Corporation. He earns $18,720 per year. Hollis provides these benefits for Rudy. To the nearest percent, what is the rate of benefits?

Vacation: 2 weeks
Holidays: 10 days
Compensation insurance: 4.2%
Unemployment insurance: 3.8%
Social security: 7.15%

Lesson 12-4

3. Jean Fargo is a computer consultant. Her travel expenses this month include: air fare $950, taxi fares $65.50, meals $342.20, hotels $695.30, and miscellaneous $42.30. What is Jean's total travel expense?

Lesson 12-5

4. The Personnel Department of the Cabinet Company is sending 5 of its employees to a local 2-day training program. The 5 employees' combined wages total $420 per day. The company paid the registration fee of $80 for each person. The company also paid for the employees' lunches for the 2 days at a total cost of $71.40. What was the total cost for the Cabinet Company to send their employees to the training program?

Lesson 13-1

5. Cable Steel produces cable from scrap steel that it buys for $205 per ton. The company produces 4 rolls of cable from each ton. The direct labor charge is $31.80 an hour. The employees produce 2 rolls of cable per hour. To the nearest tenth of a cent, what is the prime cost of manufacturing 1 roll of cable?

Lesson 13-4

6. City Center Hospital made a time-study of its outpatient pharmacy. The hospital recorded the times that Jill Muirfield spent unpacking cartons of bandages.

	Observations in Seconds				
Task	**#1**	**#2**	**#3**	**#4**	**#5**
Pick up carton	5.0	5.5	4.0	5.5	4.5
Open carton	4.0	4.5	5.5	4.0	5.0
Remove boxes	13.0	14.0	12.5	13.5	14.5
Discard carton	5.0	5.5	5.0	5.0	5.5

If Jill takes a 10-minute break each hour, how many cartons can she unpack in 1 hour?

Lesson 13-6

7. Al-Can ships its industrial containers in wooden boxes. The containers are arranged in one layer. Each box is made of $\frac{1}{8}$-inch thick wood with $\frac{1}{8}$-inch thick wood spacers. Each container is 8 inches high and $4\frac{1}{8}$ inches in diameter. What are the dimensions of each wooden box?

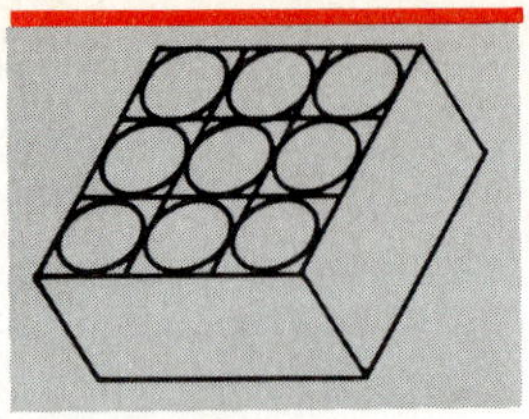

Lesson 14-3

8. Luminate, Inc., a manufacturer of light bulbs, lists the price of its Low-Glare bulb as $2.98. The net price to the retailer is $2.09. What is the trade-discount rate?

Lesson 14-5

9. The Links Co. produces a golf shoe that is priced at $112.40. When the new model was introduced, the $112.40 shoe was discounted at 25% less 15%. What is the net price of the old model using the complement method?

Lesson 14-6

10. Local Software, Inc., received an invoice from Integrated Computer Supply dated September 13. The invoice carried terms of 2/10, net 30. The net price of the invoice was $4681.60. What is the cash price of the invoice if it is paid on September 22?

Lesson 15-3

11. Field Company, an oil equipment wholesaler, purchases drilling bits for $3147.50 each. Field sells the bits for $5036. Field estimates the overhead expenses for each bit to be 20% of the selling price. What is the estimated net profit on each bit?

Lesson 15-7

12. During the winter, Tree Grown buys oranges from overseas growers. Tree Grown determines the retail selling price of the oranges by marking up the cost by 120%. The fruit company bought its first shipment of winter oranges for $0.72 per dozen. What is the retail selling price per dozen of this shipment?

Lesson 16-2

13. Of a sample of 500 swimming pool owners, 106 said they would buy Litehouse Living's new chemical for swimming pools. There is a market of about 9,000,000 potential users. Each user would purchase the chemical once per year. What is the sales potential for the chemical for one year?

Lesson 16-6

14. Main Street Meat Market has a contract for 20,000 lines of advertising at $0.810 per line in the Herald this year. What is the cost of a 700-line advertisement?

Lesson 16-8

15. Sailaway builds sailboats. Complete the chart. Which selling price will maximize Sailaway's profits?

Selling Price per Unit	Estimated Unit Sales	Total Fixed Costs	Fixed Costs per Unit	Variable Costs per Unit	Total Cost per Unit	Possible Net Profit
$35,000	16	$200,000	$12,500	$3750	?	?
$42,500	14	$200,000	?	$3750	?	?
$47,500	12	$200,000	?	$3750	?	?

Unit 17

Warehousing and Distribution

▸ **The Warehousing and Distribution Department keeps a record of all merchandise, or *inventory,* stored in your warehouse. *Inventory cards* show the number of each item in stock, the number of incoming items (*receipts*), and the number of outgoing items (*issues*). Your inventory must be large enough to meet your sales needs. The size and value of your inventory will affect your insurance, taxes, and worth of your business. The Warehousing and Distribution Department is also responsible for shipping your product.**

Retail stores periodically conduct an inventory of their stock on hand.

17-1 Storage Space

Objective Compute the total storage space.

Your business needs **warehouse** or **storage space** in which to keep raw materials or products until you are ready to use them. You may need space in a large warehouse or only a small stockroom, depending on the size of the items you are storing.

▶ **Storage Space = Volume per Item × Number of Items**

Example

Skill 8 **Application** Z **Term** Storage space

Mahr Corporation manufactures portable television sets. Each set is packaged in a carton measuring 2 feet long, 1.5 feet wide, and 1.5 feet high. How many cubic feet of space does Mahr need to store 1200 television sets?

Solution

1. Find the **volume per item.**
   2 ft × 1.5 ft × 1.5 ft = 4.5 ft³ volume per item

2. Find the **storage space.**
   Volume per Item × Number of Items
   4.5 ft³ × 1200 items = 5400 ft³ storage space

A. 12,000 in³
B. 2400 cm³

▶ **Self-Check** Cover the answers and complete the problems.

Find the storage space needed.

A. Dimensions of item: 2″ × 3″ × 4″. 500 items.

B. Dimensions of item: 4 cm × 3 cm × 1 cm. 200 items.

Problems

	Carton Dimensions (Length	× Width	× Height	= Volume)	×	Number of Items	=	Storage Space
1.	(1.7 ft	× 1.3 ft	× 0.5 ft	= ?)	×	700	=	?
2.	(1 ft	× 1.17 ft	× 0.42 ft	= ?)	×	160	=	?
3.	(0.7 m	× 0.2 m	× 0.35 m	= ?)	×	1200	=	?
4.	(0.15 m	× 0.15 m	× 2 m	= ?)	×	450	=	?
5.	(2.5 m	× 0.8 m	× 0.6 m	= ?)	×	250	=	?
6.	(1.5 ft	× 0.85 ft	× 0.85 ft	= ?)	×	325	=	?

7. Artificial fireplace logs.
Carton is 1.25 ft by 1 ft by 0.7 ft.
100 cartons to be stored.
How much storage space is required?

8. Black Forest cuckoo clocks.
Carton is 0.5 m by 0.3 m by 0.3 m.
500 cartons to be stored.
How much storage space is required?

9. Harrison bedroom lamps are packed 2 in a carton. The dimensions of the carton are 2.5 feet by 1.2 feet by 2 feet. The Bette-Lighting Store wants to order 300 pairs of lamps. How much storage space will Bette-Lighting need to store the lamps?

10. Wholesale Drugs, Inc., has ordered 200 cartons of rubber cement. Each carton of bottles measures 1 foot by 0.75 feet by 0.375 feet. How much storage space is required?

11. Major Electronics produces console television sets. The sets are packaged in cartons measuring 4 feet long, 2.5 feet wide, and 3 feet high. How much storage space is required for 2500 television sets?

12. Olde English Furniture Company packs its desks in cartons measuring 2 meters long, 1 meter wide, and 0.75 meters high. Great Britain Importing has ordered 600 desks. How much storage space is needed?

13. The Yishawa electric lawn mower is housed in a carton measuring 0.85 meters by 0.85 meters by 0.45 meters. General Western Company ordered 1755 Yishawa lawn mowers. How much storage space is needed?

14. Bananas are delivered to grocery stores in crates measuring 3 feet by 18 inches by 12 inches. How many cubic feet of storage space are required for 20 crates of bananas?

15. Patio torches are delivered in boxes that are 0.6 meters by 30 centimeters by 15 centimeters. How many cubic meters of storage space are required to store 1000 patio torches?

16. Thirty-six clock radios are packed in a carton measuring 1 meter long, 1 meter wide, and 0.5 meters high. How much storage space is required for 250 cartons of 36 clock radios each? How much space is required for 7200 clock radios packaged in cartons of 36 clock radios each?

Maintaining Your Skills

Look up the skill in parentheses if you need help or more practice.

Multiply. (Skill 8)

17. 4 ft x 5 ft × 2 ft

18. 3 m × 2 m × 1 m

19. 6 in × 5 in × 3 in

20. 8 mm × 6 mm × 5 mm

21. 4.2 cm × 3.6 cm × 3.5 cm

22. 1.5 yd × 0.8 yd × 0.2 yd

23. 40 ft^3 × 250

24. 8 cubic m × 750

25. 125 in^3 × 620

17-2 Taking an Inventory

Objective Compute total inventory.

An **inventory** shows the number of each item that your business has in stock. You may use an **inventory card** to keep a record of the items on hand. The number of incoming items, called **receipts,** are added to the previous inventory. The number of outgoing items, called **issues,** are subtracted from the previous inventory.

▶ **Inventory = Previous Inventory + Receipts − Issues**

Example

Skills 3, 4 **Application** A **Term** Inventory

The inventory card for Save-More Paint Supply shows that 100 gallons of white latex paint, stock number WL-9, were in inventory on February 1. Each week, the inventory record is updated. How many gallons were on hand March 1?

Solution

ITEM: *Latex paint, white*
STOCK NUMBER: *WL-9*

Inventory at beginning of week → Opening Balance

Week of	Opening Balance	Receipts (Quantity IN)	Issues (Quantity OUT)	Inventory at End of Week
Feb 1	100	50	78	72
Feb 8	72	0	40	32
Feb 16	32	200	103	129
Feb 22	129	50	156	23
Mar 1	23			

Number on hand March 1 → 23

A. 220
B. 35

▶ **Self-Check** Cover the answers and complete the problems.

Find the inventory.

A. Previous inventory, 200; receipts, 70; issues, 50.
B. Previous inventory, 60; receipts, 50; issues, 75.

Problems

Complete the inventory record.

	Month	Opening Balance	+	Receipts (Quantity IN)	−	Issues (Quantity OUT)	=	Number on Hand
1.	January	250	+	75	−	100	=	?
2.	February	225	+	125	−	50	=	?
3.	March	?	+	50	−	75	=	?
4.	April	?	+	0	−	40	=	?

5. Inventory of wooden pallets.
125 on hand at beginning of week.
Received 50 during the week.
Issued 32 on Friday.
How many on hand at end of week?

6. Inventory of Pure-Red Tomato Sauce.
Started the week with 73 cases.
Received 20 cases during the week.
Shipped out 47 cases during the week.
How many on hand at end of week?

7. The Corner Service Station takes an inventory on a monthly basis. What is the balance on hand for each item?

Item Number	Description	Beginning Balance	Quantity IN	Quantity OUT	Balance on Hand
X 23	10 W-30 Oil	173 qt	100 qt	112 qt	?
W 15	Washer solvent	41 gal	50 gal	33 gal	?
C 5	5-gallon can 30 W oil	16 cans	10 cans	3 cans	?

8. How many of each item does Office Supply Specialists have on hand at the end of this month?

Stock No.	Description	Opening Balance	+ Receipts	− Issues	= On Hand
11398	STPLR-RS153	210			?
11402	BOX STPLS-RT11	1205	500		?
11610	BOX PA CLPS-L3	10,341		1423	?
11682	BOX PA CLPS-L7	433	3500	1255	?

9. How many clamps did A-Z Hardware have in stock on each date?

Last Action Date	Receipts	Issues	On Hand
9/3	150		150
9/7		35	?
9/10	70		?
9/15		75	?
9/20	70		?
9/25		87	?

10. The Stock Status Summary for all of the stores in the Lamely's Department Stores chain is printed monthly. How many of each item are in stock?

DATE 8-28 STOCK STATUS SUMMARY

MATER'L CLASS	STOCK NUMBER	DESCRIPTION	OPENING BALANCE	TRANSACTIONS RECEIVED	ISSUED	ON HAND
174	146301	MARKER-FP	49		16	?
174	146334	MARKER-W/2	7			?
175	121	NOTEBOOK	2780	3000	3675	?
175	6841	NOTEBOOK	975	1200	1208	?

(Continue on next page.)

11. Whitewater Wholesalers Supply Company carries this inventory card on a claw hammer. Find the balance for each date.

Part No.: CH-13	Date	Receipts	Issues	Balance
Description	6/1	50		50
Claw hammer	6/12		15	?
Location: 24	6/18	25		?
	6/26		30	?

12. Complete this plumbing supply house inventory record for a faucet set:

Date	Receipts	Shipments	Balance
9/1			75
9/9	50		?
10/12		60	?
10/30		45	?
11/15	50		?
12/11		50	?

13. An inventory record for Sunrise Wholesale Electronics shows receipt of 100 Model C-TI hand calculators on 4/1. Another shipment of 80 arrived 4/15. One hundred ten calculators were shipped out on 5/10. Fifty arrived on 6/2 and 25 were shipped out on 6/17. Find the balance as of 6/17.

14. On January 1, Ceramics Wholesalers started with an inventory of 96 ceramic vases. Shipments of 23, 87, and 32 were sent out on 1/13, 2/18, and 3/11, respectively. Receipts of 25, 50, and 40 were received on 1/20, 2/25, and 3/15, respectively. Find the balance as of 3/15.

15. The Farm Store started on August 1 with a balance of 121 cases of oil. On August 6, 40 cases were shipped out. On August 16, 50 cases were received. On August 27, 35 cases were shipped out, and on August 31, 60 cases were received. Complete an inventory control card showing the balance after each transaction.

16. Boatwick-Brahams Wholesale Hardware Supply Company began July 1 with a balance of forty-three 14-inch gas chain saws. It received shipments of 20, 35, and 15 chain saws on July 20, August 20, and September 20, respectively. It made shipments of 17, 38, and 33 chain saws on July 17, August 22, and September 30, respectively. Complete an inventory control card showing the balance after each transaction.

Maintaining Your Skills

Look up the skills in parentheses if you need help or more practice.

Add. (Skill 3)

17. 100 + 25 + 65   18. 50 + 40 + 25   19. 37 + 45 + 96

Subtract. (Skill 4)

20. 223 − 146   21. 643 − 75   22. 8752 − 125   23. 6421 − 899

24. 21,465 − 3022   25. 18,347 − 12,399   26. 63,194 − 49,999

17-3 Valuing an Inventory

Objective Use the average-cost method to compute the inventory value.

You must calculate the value of your inventory when you purchase insurance, pay taxes, or compute the worth of your business. The **average-cost method** is one way of calculating the value of an inventory. Because the cost of incoming items may change, you calculate the value of the inventory based on the average cost of the goods you received.

▶ **Inventory Value = Average Cost per Unit × Number on Hand**

Example

Skills 8, 11 Application Q Term Average-cost method

Save-More Paint Supply is valuing its inventory of white latex paint.

ITEM: Latex paint, white
STOCK NUMBER: WL-9

Number on hand at beginning of week → 100 (Feb 1 Receipts)
Average cost in January → $1.90 (Feb 1 Unit Cost)

Week of:		Receipts	Unit Cost	Total Cost
Feb 1	Opening Balance:	100	$1.90	$190.00
		50	2.00	100.00
Feb 8		0	--	--
Feb 15		200	2.05	410.00
Feb 22		50	2.40	120.00
	Total	400		$820.00

On March 1, Save-More had 23 gallons of white latex paint on hand. What is the value of the inventory on March 1?

Solution

1. Find the **average cost per unit.**
   Total Cost of Units ÷ Number Received
   $820.00 ÷ 400 = $2.05

2. Find the **inventory value.**
   Average Cost per Unit × Number on Hand
   $2.05 × 23 = $47.15 inventory value

A. $3.40
B. $238

▶ **Self-Check** Cover the answers and complete the problems.

100 items purchased at $3.00 each. 50 items purchased at $4.20 each.

A. Find the average cost per unit.

B. Find the inventory value if 70 items were on hand.

Problems

1. Merchandise Furniture Mart is valuing its inventory of lawn chairs. On June 1, Merchandise Furniture Mart had 28 lawn chairs on hand.
   a. What is the average cost per lawn chair?
   b. What is the value of the inventory of lawn chairs on June 1?

Week of:	Receipts	Unit Cost	Total Cost
May 1	50	$2.60	$130.00
May 8	20	2.65	53.00
May 15	—	—	—
May 22	20	2.75	55.00
May 29	30	2.70	81.00
Total	120		$319.00

2. General Pen, Inc., is valuing its inventory of the Diplomat pen. On April 1, General Pen had 46 Diplomat pens on hand.
   a. What is the average cost per pen?
   b. What is the value of the inventory of Diplomat pens on April 1?

Date	Receipts	Unit Cost	Total Cost
January 12	40	$5.00	$200.00
February 11	35	4.80	168.00
March 21	60	4.40	264.00
Total	135		$632.00

3. a.

Date	Receipts	Unit Cost	Total Cost
3/10	1200	$0.15	?
4/22	850	$0.12	?
Total	?		?

   b. Average cost per unit: ?
   Number on hand on 4/30: 1000
   c. Value of inventory: ?

4. a.

Date	Receipts	Unit Cost	Total Cost
12/18	700	$5.30	?
1/10	600	$4.85	?
2/6	750	$5.50	?
Total	?		?

   b. Average cost per unit: ?
   Number on hand on 2/8: 650
   c. Value of inventory: ?

5. Records for Williams Market show this opening balance and these receipts for Homestock soup in November. At the end of the month, 24 cans were on hand. What is the value of the inventory?

Date		Receipts	Unit Cost
11/1	Opening bal.	120	$0.24
11/10		144	0.24
11/17		96	0.30
11/24		120	0.22

6. TBA, Inc., inventory records for boxes of spark plugs show this information. At the end of the month, 110 boxes were on hand. Find the value of the inventory.

Date	Receipts	Unit Cost
5/1	130	$6.95
5/19	200	8.00
5/31	100	7.50

7. City Furniture's inventory records for Forstan end tables cover a 6-month period.

Date	Unit Cost	Receipts	Sales	Balance
12/31	$22.50	24		24
1/24			15	9
2/15	$27.50	20		29
4/23			18	11
5/2	$30.50	20		31
6/30			10	21

a. What is the average cost per unit as of June 30?

b. What is the value of the inventory on June 30?

8. Aromatiques, Ltd., inventory records for LaNice Perfume are shown.

a. What is the balance after December 31?

b. What is the average cost per unit as of December 31?

c. What is the value of the inventory as of December 31?

Date	Unit Cost	In	Out	Balance
12/1	$5.00	200		200
12/8			75	125
12/12			67	58
12/20	6.50	150		208
12/22			73	135
12/31	7.00	100		?

*9. **A Brief Case** Sun Time, a swimming pool supply company, uses the FIFO (first in, first out) method of valuing inventory. The FIFO method assumes that the first items received are the first items sold.

a. Complete Sun Time's records for receipts of plastic stacking chairs.

Date	Receipts	Unit Cost	Total Cost
5/1	50	$5.50	?
5/8	100	$7.00	?
5/19	50	$7.50	?
Total	?		?

As of May 20, Sun Time had sold 130 of the chairs.

b. Use the FIFO method to calculate the cost of the 130 chairs sold.

50 at $5.50 = ?
80 at $7.00 = ? ← Of the 100 received on 5/8
130 Total: ?

c. Use the FIFO method to calculate the value of the 70 chairs in stock.

→ 20 at $7.00 = ?
50 at $7.50 = ?
70 Total: ?

Maintaining Your Skills

Look up the skills in parentheses if you need help or more practice.

Multiply. (Skill 8)

10. 40 × $7.50 11. 100 × $4.58 12. 75 × $12.40 13. 125 × $2.60

14. 72 × $4.63 15. 27 × $11.47 16. 118 × $6.91 17. 193 × $14.28

Divide. Round answers to the nearest cent. (Skill 11)

18. $935.00 ÷ 130 19. $515 ÷ 70 20. $207.85 ÷ 45 21. $318.92 ÷ 42

22. $3077.22 ÷ 245 23. $79.73 ÷ 63 24. $346.22 ÷ 38 25. $3824.63 ÷ 179

17-4 Carrying an Inventory

Objective

Compute the annual cost of carrying an inventory.

Your business must keep a sufficient inventory of goods to meet production or sales needs. Based on past records, you may estimate the annual cost of maintaining, or carrying, the inventory at a certain level for the coming year. The annual cost of carrying the inventory includes taxes, insurance, storage fees, handling charges, and so on. The annual cost is often expressed as a percent of the value of the inventory.

▸ **Annual Cost of Carrying Inventory = Percent × Inventory Value**

Example

Skill <u>30</u> **Application** <u>A</u> **Term** ________

Metal Fabricators, Inc., produces washers, nuts, bolts, and similar items. The company maintains an inventory of metal sheets and bars to be used in production. The company also maintains an inventory of finished products to fill customers' orders. Metal Fabricators estimates the annual cost of maintaining the inventory to be 30% of the value of the inventory. This year the company plans to maintain an inventory totaling $500,000 in value. What amount should be set as the estimated cost of carrying the inventory for the year?

Solution

Find the **annual cost of carrying inventory.**

Percent × Inventory Value

30% × $500,000 = $150,000 annual cost of carrying inventory

A. $140,000
B. $610,000

▸ **Self-Check** Cover the answers and complete the problems.

Find the annual cost of carrying an inventory.

A. Inventory value is $700,000. Estimated annual cost is 20% of value.

B. Inventory value is $2,440,000. Estimated annual cost is 25% of value.

Problems

	1.	2.	3.	4.	5.	6.
Value of Inventory	$750	$70,000	$8244	$132.50	$4794.80	$1476.70
Percent	10%	15%	25%	38%	12%	27.5%
Annual Cost of Carrying Inventory	?	?	?	?	?	?

7. D.J. Music Company.
Maintains $48,000 inventory.
Costs about 25% of value to maintain.
What is the approximate annual cost of carrying the inventory?

8. Bender Stamping Company.
Maintains $120,000 inventory.
Costs about 32% of value to maintain.
What is the approximate annual cost of carrying the inventory?

9. General Clothing Store estimates the cost of carrying its inventory of clothing to be 18% of the value of the merchandise. About how much does it cost General Clothing annually to carry a $70,000 inventory?

10. Ben Iron and Steel, Inc., carries a $90,000 inventory. The company estimates the annual cost of carrying the inventory at 27% of the value of the inventory. What is the approximate annual cost of carrying the inventory?

11. Mancuso Warehouse Food Market estimates the annual cost of carrying its inventory at 26% of the value of the inventory. The 26% is broken down as follows:

Type of Expense	Percent
Spoilage and physical deterioration	7.0%
Interest	11.0%
Handling	4.0%
Storage facilities	2.0%
Transportation	1.0%
Taxes	0.5%
Insurance	0.5%
Total	26.0%

a. Mancuso generally carries an inventory valued at $170,000. What is the approximate annual cost for each expense of carrying the inventory?
b. Mancuso was able to reduce its inventory to one half of its value and still meet customer orders. About how much did Mancuso save on the total cost of carrying its inventory?

12. In Exercise 11, if Mancuso carried an inventory valued at $247,560, what would be the approximate annual cost for each expense of carrying the inventory?

Maintaining Your Skills

Look up the skill in parentheses if you need help or more practice.

Multiply. (Skill 30)

13. 25% × $720,000
14. 30% × $480,000
15. 20% × $916,000
16. 40% × $114,000
17. 15% × $3,472,500
18. 37% × $216,400
19. 24% × $72,600
20. 42% × $18,430
21. 5% × $1,643,000
22. 7% × $916,480
23. 0.5% × $93,484
24. 0.75% × $216,724

17-5 Transportation by Air

Objective

Compute the total shipping cost by air freight.

Your business may ship goods to its customers by **air freight.** Either you or the customer pays the shipping charges. The total cost of shipping goods by air freight includes pickup and delivery to the airport, air freight charges to the city of destination, local delivery from the airport to the final destination, and federal tax where it applies.

▶ **Total Shipping Cost** = **Pickup Charge** + **Air Freight Charge** + **Delivery Charge** + **Federal Tax**

Example

Skills 2, 30 **Application** A **Term** Air freight

Avenue Video Corporation is shipping television equipment by air freight. The packaged equipment weighs 650 pounds. Reliable Pickup Service charges $2.80 for each 100 pounds or fractional part to transport the equipment to the airport. The air freight rate is $16.86 per 100 pounds, charged on the actual number of pounds shipped. Speedy Delivery Service charges $3.20 per 100 pounds, in increments of 50 pounds, to deliver the equipment. In addition, there is a 5% federal tax on the air freight charge because it involves interstate transportation. What is the total cost of shipping the equipment?

Solution

1. Find the **pickup charge.**
   $2.80 × (700 ÷ 100) = $ 19.60
2. Find the **air freight charge.**
   $16.86 × (650 ÷ 100) = 109.59
3. Find the **delivery charge.**
   $3.20 × (650 ÷ 100) = 20.80
4. Find the **federal tax.**
   5% of $109.59 = $5.4795 = 5.48
   Total Shipping Cost $155.47

A. $5.73
B. $132.43

▶ **Self-Check** Cover the answers and complete the problems.

A. Federal tax is 6% of the air freight charge.
Air freight charge is $95.50.
Find the federal tax.

B. Pickup charge is $14.75.
Air freight charge is $95.50.
Delivery charge is $16.45.
Federal tax is $5.73.
Find the total shipping cost.

Problems

	Goods	Cost of Pickup	+	Cost of Air Freight	+	Cost of Delivery	+	Federal Tax	=	Total Cost
1.	Auto parts	\$ 11.00	+	\$ 32.33	+	\$ 8.22	+	\$ 1.62	=	?
2.	Arc welders	\$ 34.16	+	\$ 84.23	+	\$ 28.79	+	\$ 5.92	=	?
3.	Cameras	\$108.75	+	\$358.88	+	\$114.38	+	\$23.33	=	?
4.	Books	\$173.28	+	\$424.08	+	\$148.20	+	\$21.20	=	?

For Exercises 5 and 6, charges are per 100 pounds or fractional part.

5. 400 pounds of battery chargers.
   \$3.75 per 100 pounds for pickup.
   \$13.95 per 100 pounds for air freight.
   \$2.80 per 100 pounds for delivery.
   $5\frac{1}{2}$% federal tax on air freight charge.
   What is the total shipping cost?

6. 335 pounds of 10-speed bicycles.
   \$3.80 per 100 pounds for pickup.
   \$11.82 per 100 pounds for air freight.
   \$3.25 per 100 pounds for delivery.
   $6\frac{1}{2}$% federal tax on air freight charge.
   What is the total shipping cost?

7. A crate of musical instruments that weighs 470 pounds is shipped by air freight. The costs per 100 pounds or fractional part are: \$3.25 for pickup, \$13.60 for air freight, and \$2.90 for delivery. A $7\frac{1}{4}$% federal tax on the air freight charge is added. What is the total shipping cost?

8. Luxura is shipping 1460 pounds of prototype auto parts by air freight. Pickup is \$4.25 per 100 pounds or fractional part. Delivery is \$4.10 per 100 pounds or fractional part. The air freight is \$12.40 per 100 pounds, charged on the actual number of pounds shipped. There is a 4% tax on the air freight and delivery charges. What is the total shipping cost?

9. Consumers' League is shipping 73 pounds of computer tapes by air freight from Falmouth to Lowell. Express Service charges \$3.85 per 100 pounds, in increments of 50 pounds, to transport the tapes to the airport. Northway Air Freight charges \$0.20 per pound, with a minimum charge of \$16. Central Carrier charges \$4.15 per 100 pounds or fractional part to deliver the tapes. There is no federal tax. What is the total shipping cost?

Maintaining Your Skills

Look up the skill in parentheses if you need help or more practice.

Find the percentage. Round answers to the nearest cent. (Skill 30)

10. 5% × \$95.60 11. 6% × \$114.95 12. $5\frac{1}{2}$% × \$64.95 13. $7\frac{1}{2}$% × \$81.40

Add. (Skill 2)

14. \$17.50 + \$104.70 + \$19.80 + \$5.24

15. \$26.90 + \$123.80 + \$24.75 + \$7.43

16. \$21.90 + \$99.80 + \$20.85 + \$7.49 17. \$15.45 + \$102.85 + \$16.70 + \$0

17-6 Transportation by Truck

Objective Compute the total shipping cost by truck.

Your business may ship its goods by truck. The **basic rate** for shipping by truck depends on the classification of the goods that are shipped. The classification is determined by the type and value of the goods. The basic rates differ according to the weight of the goods and the distance that they are being shipped, and they may also vary from carrier to carrier. The basic rates are often given per 100 pounds. Most companies charge only for the actual weight of the goods being shipped.

▶ **Total Shipping Cost = Weight × Basic Rate**

Example

Skill <u>**8**</u> **Application** <u>**C**</u> **Term** <u>**Basic rate**</u>

Lebret Auto is shipping 4470 pounds of spark plugs (Class 100) via Eldridge Express. Eldridge will pick up the spark plugs at the manufacturing plant and deliver them to an automobile parts distributor. The distance is 415 miles. What is the total shipping cost?

ELDRIDGE EXPRESS . . . BASIC RATES PER 100 POUNDS . . . CLASS 100 ITEMS

Weight Group (in pounds)	Distance (in miles) 1-100	101-200	201-300	301-400	401-500	501-600	601-700	701-800	801-900
0-500	$5.49	$6.45	$7.33	$8.18	$9.00	$9.86	$10.71	$11.67	$12.37
501-1000	4.78	5.71	6.70	7.60	8.47	9.27	10.11	11.00	11.66
1001-2000	4.22	5.06	5.97	6.80	7.54	8.24	9.02	9.74	10.34
2001-5000	3.46	4.28	5.10	5.88	6.64	7.28	7.99	8.74	9.28
5001-10,000	2.81	3.43	4.17	4.79	5.38	5.91	6.50	7.11	7.56

Solution

1. Find the **basic rate.** (Refer to above table.)
   Weight group: 2001-5000 pounds
   Distance: 401-500 miles . . . . $6.64 basic rate
2. Find the **weight** in hundreds of pounds.
   $4470 \div 100 = 44.7$
3. Find the **total shipping cost.**
   Weight × Basic Rate
   $44.7 \times \$6.64 = \$296.808 = \$296.81$ total shipping cost

A. $8.24
B. $140.08

▶ **Self-Check** Cover the answers and complete the problems.

A. Find the basic rate to ship 1700 pounds 535 miles.
B. Find the total shipping cost.

Problems

Use the table on page 396 to find the basic rates.

	1.	2.	3.	4.	5.	6.
Weight (pounds)	450	630	4980	7495	9500	4575
Distance (miles)	315	885	93	743	161	517
Basic Rate	$8.18	$11.66	?	?	?	?
Total Shipping Cost	?	?	?	?	?	?

7. 1800 pounds of typewriters.
Transported 447 miles.
What is the total shipping cost?

8. 865 pounds of hiking boots.
Transported 345 miles.
What is the total shipping cost?

9. A shipment of wool jackets weighing 485 pounds is transported by truck from the North Country Clothing plant to its Highland warehouse. The distance is 806 miles. What is the shipping cost?

10. A shipment of frozen butter is sent by truck from Green Bay, Wisconsin to St. Louis, Missouri. The distance is 445 miles. The frozen butter weighs 985 pounds. What is the shipping cost?

11. An 863-pound shipment of lamps and lamp shades is transported from The Lamp Factory, Inc., to the national distribution center in Chicago. The distance is 174 miles. What is the shipping charge?

12. A 300-pound shipment of darkroom equipment and chemicals is taken from Rochester, New York to Louisville, Kentucky. The distance is 591 miles. What is the shipping charge?

13. A shipment of Alvadore condensed coffee is being sent from Alvadore's central warehouse to a local distribution center in Vancouver, British Columbia. The distance is 725 miles. The coffee weighs a total of 283 pounds. What is the shipping cost?

Maintaining Your Skills

Look up the skill in parentheses if you need help or more practice.

Multiply. (Skill 8)

14. 54.20 × $7.56
15. 7.86 × $4.78
16. 14.15 × $5.06
17. 32.66 × $5.10
18. 0.87 × $8.18
19. 4.10 × $9.00
20. 6.74 × $9.27
21. 18.32 × $9.02
22. 12.11 × $9.74
23. 47.85 × $9.28
24. 20.12 × $3.46
25. 28.41 × $5.88
26. 8.17 × $11.66
27. 63.47 × $2.81
28. 98.61 × $7.11
29. 5.01 × $10.11
30. 97.5 × $14.00
31. 14.5 × $12.70

Reviewing the Basics

Skills

Round to the nearest hundred.

(Skill 2) **1.** 546 **2.** 2156 **3.** 1873 **4.** 3717

Solve.

(Skill 3) **5.** 72 + 38 **6.** 536 + 278 **7.** 478 + 394

(Skill 4) **8.** 65 − 42 **9.** 721 − 355 **10.** 273 − 195

(Skill 8) **11.** \$53.25 × 2.8 **12.** \$99.99 × 87 **13.** \$121.63 × 92

Round answers to the nearest cent.

(Skill 11) **14.** \$145.10 ÷ 17 **15.** \$2005 ÷ 36 **16.** \$1975.17 ÷ 24

(Skill 30) **17.** 9.7% of \$218.30 **18.** 19% of \$400,000 **19.** 4.8% of \$347.85

Applications

(Application C) Find the shipping charge.

20. Weight: 4718 pounds
Distance: 619 miles

21. Weight: 7149 pounds
Distance: 1335 miles

Basic Rates Per 100 Pounds		
Weight Group (in pounds)	Distance (in miles)	
	500–1000	1001–1500
2001–5000	\$36.19	\$48.95
5001–10,000	\$30.72	\$40.33

(Application Q) Find the average cost to the nearest cent.

22. 100 units at \$2.40 each and 50 units at \$2.45 each

23. 120 units at \$0.35 each and 115 units at \$0.42 each

(Application Z) Find the volume.

24. Box is 8 ft by 2 ft by 10.5 ft

25. Box is 23 m by 12 m by 33 m

Terms

Match each term with its definition on the right.

26. Inventory

27. Average-cost method

28. Air freight

29. Basic rate

30. Storage space

a. a way of calculating an inventory's value

b. the number of each item you have in stock

c. the charge per pound to ship goods by truck

d. a way of shipping goods to customers

e. a place in which to keep products until they are needed

f. includes taxes, storage fees, and handling costs

Refer to your reference files at the back of the book if you need help.

Problem Solving
Drawing a Picture or Diagram

Some word problems, particularly those that involve lengths, widths, and dimensions, can be simplified if you make an illustration. Pictures and diagrams can also help you keep track of information in multi-step problems.

Example 1 A 45-foot piece of fencing is cut into 3 pieces. The second piece is 3 feet longer than the first piece and the third piece is 9 feet longer than the first piece. Find the lengths of the 3 pieces.

Solution

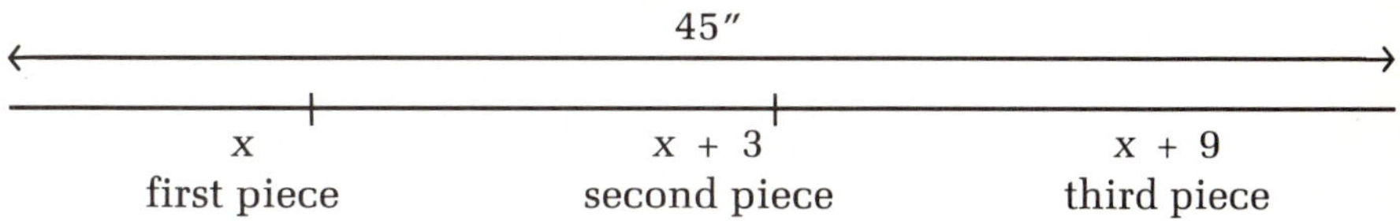

$$\begin{aligned} x + (x + 3) + (x + 9) &= 45 \\ 3x + 12 &= 45 \\ 3x &= 33 \\ x &= 11'' \quad \text{length of first piece} \\ x + 3 &= 14'' \quad \text{length of second piece} \\ x + 9 &= \underline{20''} \quad \text{length of third piece} \\ & 45'' \quad \text{total length of fencing} \end{aligned}$$

Example 2 A two-volume set of classics is bound in $\frac{1}{8}$-inch covers. The text in each volume is 2 inches thick. The two volumes are side by side on the shelf. A bookworm travels from inside the front cover of Volume I to the inside back cover of Volume II. How far does the bookworm travel?

Solution

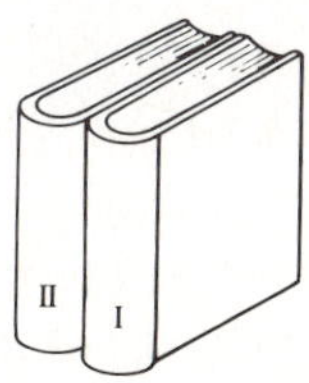

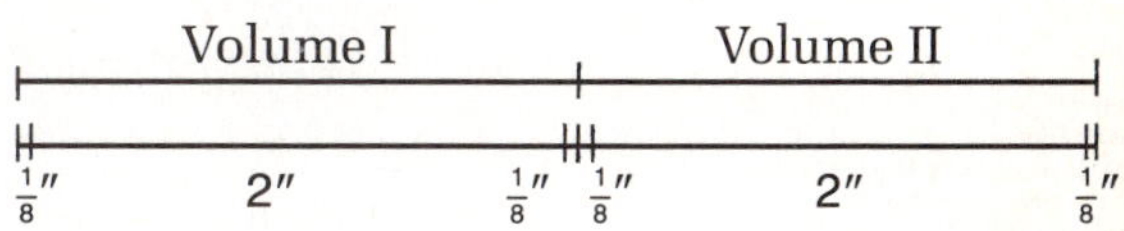

Bookworm travels: $2'' + \frac{1}{8}'' + \frac{1}{8}'' + 2'' = 4\frac{2}{8}'' = 4\frac{1}{4}''$

The bookworm travels $4\frac{1}{4}$ inches from the front cover of Volume I to the back cover of Volume II.

Problems Make a drawing, figure, or graph to help you solve these problems.

1. A rectangle has a perimeter of 72 inches. The length of the rectangle is 8 inches larger than the width. What is the width of the rectangle?

2. A barn has dimensions 40 feet by 60 feet. A cow is tethered to one corner of the barn with a 50-foot rope. The cow always stays outside the barn. How many square feet of grazing land can the cow reach?

Unit Test

Lesson 17-1

1. Office Equipment Company manufactures file cabinets. Each cabinet is stored in a box measuring 4 feet high, 1.5 feet wide, and 2.5 feet long. How many cubic feet of space does Office Equipment need to store 1700 file cabinets?

Lesson 17-2

2. How many StraightAway models does Clarion Motor Coach have on its lot on October 1?

Item: StraightAway
Stock Number: JR 2201 F

Month of:	Opening Balance	Receipts	Issues	Inventory at End of Month
August	65	5	20	?
September	?	6	22	?
October	?			

Lesson 17-3

3. Down-Home Overalls took inventory of its heavy cotton overalls. On March 1, Down-Home had 40 pairs of heavy cotton overalls in stock. What is the value of the inventory on March 1?

Item: Heavy Cotton Overalls
Stock Number: 3251

Date		Receipts	Unit Cost	Total Cost
Feb 1	Opening bal.	30	$15.00	$450.00
Feb 18		50	15.50	775.00
Feb 27		40	15.75	630.00
	Total	?		?

Lesson 17-4

4. No-Phosphates maintains a soap inventory valued at $60,000. The cost of maintaining the inventory is approximately 15% of the value of the inventory. What is the approximate annual cost of carrying this inventory?

Lesson 17-5

5. Cloud Publishing is shipping 150 pounds of books by air freight. The costs per 100 pounds or fractional part are: $3.10 for pickup, $10.75 for air freight, and $2.65 for delivery. A 5% federal tax is added to the air freight charge. What is the total cost to deliver the books?

Lesson 17-6

6. Hyland Company is shipping 376 pounds of photographic equipment. There-Now Trucking will handle shipment. The distance is 275 miles. What is the total shipping cost?

THERE-NOW TRUCKING Class 100 Items
Basic Rates Per 100 Pounds

Weight Group (in pounds)	Distance (in miles)		
	1–100	101–200	201–300
0–500	$4.23	$5.20	$6.18
501–1000	$3.65	$4.43	$5.10

Unit 18 Services

▶ Rent for your office space, maintenance costs of the building, and costs of utilities are all part of business *services.* Many businesses hire professional services for building maintenance and telephone installation and repair. In addition, your business may hire professional *consultants* to advise you on particular problems.

Companies may have service contracts for equipment that is used frequently, such as copier machines.

18-1 Building Rental

Objective Compute the monthly rental charge.

Your business may **rent** or **lease** a building or a portion of a building, usually on an annual basis. The building owner may charge a certain rate per square foot per year. Your total monthly rental charge depends on the number of square feet that your business occupies.

$$\text{Monthly Rental Charge} = \frac{\text{Annual Rate} \times \text{Number of Square Feet}}{12}$$

Example

Skills 10, 2 **Application** X **Term** Rent

Renie Machine Company rents a portion of a building owned by R & J Associates. The floor space of Renie's portion of the building measures 100 feet by 50 feet. R & J charges an annual rate of $3.00 per square foot. To the nearest dollar, what is Renie's monthly rental charge?

Solution

1. Find the **number of square feet.**
   Length × Width
   100 ft × 50 ft = 5000 ft²

2. Find the **monthly rental charge.**
   (Annual Rate × Number of Square Feet) ÷ 12
   ($3.00 × 5000) ÷ 12
   $15,000 ÷ 12 = $1250 monthly rental charge

A. 4000, $2000
B. 4375, $911.46

Self-Check Cover the answers and complete the problems.

Find the number of square feet and the monthly rental charge.

A. 100 ft × 40 ft at $6.00 per square foot annually
B. 125 ft × 35 ft at $2.50 per square foot annually

Problems

Find the monthly rental charge to the nearest dollar.

	Dimensions	=	(Number of Square Feet	×	Annual Rate per Square Foot)	÷ 12	=	Monthly Rental Charge
1.	20 ft by 20 ft	=	(?	×	$6.00)	÷ 12	=	?
2.	30 ft by 20 ft	=	(?	×	$6.20)	÷ 12	=	?
3.	130 ft by 100 ft	=	(?	×	$2.30)	÷ 12	=	?

4. Daily Tribune warehouse.
Dimensions are 35 feet by 40 feet.
Annual rental charge is
$2.40 per square foot.
What is the monthly rental charge?

5. Precious Gems branch store.
Dimensions are 30 feet by 50 feet.
Annual rental charge is
$9.50 per square foot.
What is the monthly rental charge?

6. Judson Manufacturing Company is considering the rental of additional manufacturing space at $2.60 per square foot per year. The space Judson wants to rent measures 130 feet by 110 feet. What monthly rent will Judson pay for the additional space?

7. The Blue Demon has rented additional mall space to expand its arcade operation. The space measures 40 feet by 50 feet and rents for $11.75 per square foot per year. What monthly rent does the Blue Demon pay for the additional space?

8. Diane's Fine Clothes is opening a store at the Nottingham Mall. The rent is $3.00 per square foot per year plus 5% of the store's gross sales. The area of the store is 3000 square feet. If Diane's had $270,000 in gross sales the first year, what monthly rent will it pay?

9. Now Sound rents a 20 foot by 40 foot area at the Nottingham Mall. Now Sound pays $2.75 per square foot per year plus 5% of gross sales. Last year Now Sound had $165,000 in gross sales. What was its monthly rent?

10. The local office of Community Dental Services, Inc., is planning to open a branch office in the Heatherdowns area. The office in the advertisement suits its needs. What is the cost per square foot per year for this office?

> **HEATHERDOWNS AREA**
> OFFICE SPACE
> 1400 sq. ft., carpeted and paneled, gas heat and air-conditioned. $450 per mo.
> 482-6619

*11. **A Brief Case** Some businesses rent office space for 2 or 3 years at a time. Often the lease for the office space includes a **rent escalation clause.** The rent escalation clause states that the rent may increase by a certain percent after a certain period of time.

Makin Legal Clinic rents a suburban office space for $10.00 per square foot per year. The dimensions of the office are 35 feet by 40 feet. Makin signed a 2-year lease with a 5% rent escalation clause. The rent will increase by 5% for the second year of the lease. What is the monthly rental charge for each year of the lease?

Maintaining Your Skills

Look up the skill in parentheses if you need help or more practice.

Divide. Round answers to the nearest dollar. (Skill 10)

12. $15,000 ÷ 12
13. $18,000 ÷ 12
14. $14,425 ÷ 12
15. $3750 ÷ 12
16. $20,135 ÷ 12
17. $321,462 ÷ 12

18-2 Maintenance and Improvement

Objective

Compute the total building maintenance charge.

If your business owns a building, you will need people to clean and maintain the building. You may have your own Maintenance Department or you may hire a service firm. The total cost of a particular maintenance job generally includes a **labor charge** and a **materials charge.** The labor charge is the cost of paying the people who do the job and is calculated on an hourly basis for each service person.

▸ **Total Charge = Labor Charge + Materials Charge**

Example

Skill 8 **Application** A **Term** Labor charge

Mid-Continent Insurance hired Perfect Painting Service to paint its offices. Four painters each worked 23 hours to complete the job. The regular hourly rate for each painter is $14. All work was done on weekends, so the painters were paid time and a half. The painters used 37 gallons of paint, for which they charged $10.75 per gallon. What was the total charge for painting the offices?

Solution

1. Find the **labor charge.**
   ($14.00 × $1\frac{1}{2}$) × 23 hr × 4 = $1932.00
2. Find the **materials charge.**
   37 × $10.75 = $397.75
3. Find the **total charge.**
   Labor Charge + Materials Charge
   $1932.00 + $397.75 = $2329.75 total charge

A. $580
B. $304.50

▸ **Self-Check** Cover the answers and complete the problems.

Find the total charge.

A. 4 people worked 10 hours each at $4.00 per hour.
Materials cost $420.

B. 3 people worked 8 hours each at $8.00 per hour.
9 gallons of paint at $12.50/gal.

Problems

	(Time Required	×	Number of Employees	×	Hourly Rate	=	Labor Charge)	+	Materials Charge	=	Total Charge
1.	(12 hours	×	2	×	$10.50	=	?)	+	$370.00	=	?
2.	(3 hours	×	12	×	$ 4.10	=	?)	+	0	=	?
3.	($4\frac{1}{2}$ hours	×	3	×	$ 9.80	=	?)	+	$825.00	=	?

4. 3 people replace fluorescent lights. Hourly rate of $4.75 each. Job takes 3 hours. What is the labor charge?

5. 4 people clean office area. Hourly rate of $3.85 each. Job takes 16 hours. What is the labor charge?

6. Independent Insurance Agency is moving from its present office to another office in the same building. It takes 9 people $3\frac{3}{4}$ hours to complete the move. The hourly rate per person is $7.45. What is the total charge?

7. Moncrief Cleaners hired 3 carpenters to remodel its store. The carpenters each earn $15.10 per hour. Each carpenter worked $21\frac{1}{2}$ hours. The materials charge was $1447.50. Find the total charge.

8. All-State Trucking Company used 5 loads of crushed stone on a terminal lot at $75.70 a load. Ten people were each paid $4.20 an hour to spread the stone. It took 3 hours. What was the total charge?

9. Main Supermarket hired 5 people to refinish its floor. The regular hourly rate for each person was $5.40. Because the floor was refinished on Sunday, each worker was paid double time. The job took $6\frac{1}{4}$ hours to complete. The materials charge was $247.58. What was the total charge?

10. Mazzeo Hardware Company had a minor fire in its store. Mazzeo received these estimates for repair work.

	Hourly Rate	Time Required	Materials Charge
Electrician	$15.85	13 hours	$146.70
Carpenter	$13.60	$12\frac{1}{2}$ hours	$872.50
Painter	$11.15	$12\frac{1}{4}$ hours	$131.65

Mazzeo must decide if it should hire these 3 people or have its own employees repair the damage. Materials would cost Mazzeo a total of $938.26. It would take 5 Mazzeo employees 13 hours each to do the job. Three of the employees each earn $8.90 an hour. The other 2 employees each earn $7.15 an hour. If the price is the only consideration, what should Mazzeo Hardware do?

*11. **A Brief Case** Clean Sweep Janitorial Service cleans the Morrison Bank daily from Monday through Saturday. One Clean Sweep employee works at the bank for 3 hours each day. The employee is paid $4.35 per hour on weekdays and time and a half on Saturdays. When calculating Morrison's total charge, Clean Sweep adds 70% of the cost of labor to cover its overhead. How much does Clean Sweep charge the Morrison Bank per week?

Maintaining Your Skills

Look up the skill in parentheses if you need help or more practice.

Multiply. (Skill 8)

12. 15 × 6 × 10

13. 15.25 × 6.5 × 4

14. 12 × 1.5 × 7

15. 14.50 × 8 × 1.5

16. 4 × 12 × 4.1

17. 4.5 × 9.8 × 3

18-3 Equipment Rental

Objective

Compute the total equipment rental cost.

Your business may find that it is more economical to rent than to buy certain equipment or furniture. Generally, the total cost of renting items is determined by the length of time for which you rent them. Some states charge a use tax on items that are rented.

▸ **Total Rental Cost = (Rental Charge × Time) + Tax**

Example

Skill 30 Application A Term Rent

Davison Accounting is renting new furniture for a small, temporary office. Office Rental Company charges 10% of the list price of new furniture per month. The list prices of the pieces that Davison is renting are:

Item	List Price
Desk, 60 in × 30 in	$209.95
File cabinet, 4-drawer	100.00
Swivel arm chair	122.90
2 guest chairs ($108.35 ea)	216.70
Total	$649.55

In addition to the rental charge, there is a $4\frac{1}{2}\%$ tax. What is the total cost of renting the furniture for 4 months?

Solution

1. Find the **rental charge** per month.
   10% of list price total
   10% of $649.55 = $64.955 = $64.96
2. Find the **tax.**
   $4\frac{1}{2}\%$ of ($64.96 × 4)
   $4\frac{1}{2}\%$ of $259.84 = $11.6928 = $11.69
3. Find the **total rental cost.**
   (Rental Charge × Time) + Tax
   ($64.96 × 4) + $11.69
   $259.84 + $11.69 = $271.53 total rental cost

A. $231.00
B. $282.15

▸ **Self-Check** Cover the answers and complete the problems.

Find the total rental cost.

A. Rental charge: 10% of $550.
Tax rate: 5%.
Time: 4 months.

B. Rental charge: 8% of $675.
Tax rate: 4.5%.
Time: 5 months.

Problems

	Item	List Price	Monthly Charge	Monthly Charge	Time	Tax	Rental Cost
1.	Table	$ 103.50	10% of list	?	4 months	3%	?
2.	Book shelves	$ 114.25	11% of list	?	12 months	6%	?
3.	Copy machine	$1170.00	9% of list	?	8 months	$6\frac{1}{4}$%	?

4. Bulldozer rented for 3 weeks.
Rental charge is $1250 per week.
Tax is 5%.
What is the total rental cost?

5. Jackhammer rented for 7 days.
Rental charge is $147.95 per day.
No tax.
What is the total rental cost?

6. The list price of a word processor is $5695. The monthly rental charge is $9\frac{1}{2}$% of the list price. There is a $3\frac{1}{4}$% tax per month. What is the total rental charge for 1 year?

7. Strong Cement Company plans to rent a backhoe for 2 days. The rental charge is $185.50 per day plus a flat fee of $35 for delivery and pick-up. There is no tax. What is the total rental cost?

8. Law-For-All, Inc., is renting additional furniture for 6 months. The monthly rental charge is 10.5% of the list price. There is an 8% use tax. What is the total rental cost of the following:

Item	List Price
3 desks with chairs	$279.95 each
10 guest chairs	79.95 each
1 typewriter	647.50 each

9. Joseph Laine and Sons, Plumbers, rent a ditcher as they need it. The daily rental charge is $162.75. There is a $4\frac{1}{2}$% tax. What is the total cost of renting the ditcher for 11 days?

*10. **A Brief Case** Adams Travel needs an additional 4-drawer file cabinet for 5 months. Adams received cost information from 2 rental agencies and 1 used-equipment store. If price is the only consideration, which company should Adams contact?

CA Furniture Rental	Direct Office Rental	Economy Office Furniture
Per month:	Per month:	• $74.50 purchase price
• $14 rental charge	• $12 rental charge	• 5% sales tax
• 5% insurance charge	• No insurance charge	• $7 delivery charge
• 5% use tax	• 5% use tax	

Maintaining Your Skills

Look up the skill in parentheses if you need help or more practice.

Find the percentage. Round answers to the nearest cent. (Skill 30)

11. 5% of $464 12. 8% of $27.70 13. 6% of $174.50 14. 11% of $324.42

18-4 Utilities Costs—Telephone

Objective

Compute the monthly telephone cost.

To operate your business, you will need several **utilities**. Utilities are public services such as telephone, electricity, water, and gas. Each utility uses a different cost structure for charging its customers. For example, the **basic monthly charge** for your telephone service depends on the number of incoming lines, the type of equipment, and the type of service that you have. If you make more calls than the number included in your basic monthly charge, you must pay an additional amount. A federal excise tax is also added to your telephone charge each month.

▶ **Total Cost for Month** = **Basic Monthly Charge** + **Cost of Additional Calls** + **Federal Excise Tax**

Example

Skills 5, 30 **Application** A **Term** Basic monthly charge

Fast Printing has one incoming telephone line. The basic monthly charge for this line is \$14.85. The basic monthly charge includes 80 outgoing local calls per month. The cost of additional outgoing local calls is \$0.09 each. Fast Printing made 105 local calls last month. A 3% federal excise tax is added to the bill. What is the total cost of Fast Printing's telephone service for the month?

Solution

1. Find the **cost of additional calls.**
   (105 − 80) × \$0.09
   25 × \$0.09 = \$2.25
2. Find the **federal excise tax.**
   3% of (\$14.85 + \$2.25)
   3% of \$17.10 = \$0.513 = \$0.51
3. Find the **total cost for month.**
   Basic Monthly Charge + Cost of Additional Calls + Federal Excise Tax
   \$14.85 + \$2.25 + \$0.51 = \$17.61 total cost

A. \$17.51
B. \$26.88

▶ **Self-Check** Cover the answers and complete the problems.

Find the total cost for the month.

A. Base charge of \$15.85.
10 additional calls at \$0.115.
3% federal tax.

B. Base charge of \$22.50.
30 additional calls at \$0.12 each.
3% federal tax.

Problems

	Basic Monthly Charge	Additional Local Calls	Cost per Additional Call	Cost of All Additional Calls	3% Federal Excise Tax	Total Cost for Month
1.	$15.90	15	$0.08	$1.20	$0.51	?
2.	$28.75	30	$0.09	?	?	?
3.	$42.35	110	$0.10	?	?	?
4.	$34.75	0	$0.08	?	?	?

Add a 3% federal excise tax to find the total cost for the month.

5. Bounty Sales has 1 telephone.
Basic monthly charge is $19.30.
Includes 90 outgoing local calls.
Each additional call costs $0.10.
Made 115 outgoing local calls.
What is the total cost for the month?

6. Newtown Pizza has 1 telephone.
Basic monthly charge is $16.30.
Includes 80 outgoing local calls.
Each additional call costs $0.08.
Made 42 outgoing local calls.
What is the total cost for the month?

7. Donegal Insurance Agency has 2 telephones. The total basic monthly charge for both telephones is $29.35. The charge includes 160 outgoing local calls. The cost of additional local calls is $0.09 each. Donegal made 223 outgoing local calls this month. What is the total cost of telephone service, including tax, for the month?

8. Brooks, Sanchez, and Johnson, brokers for stocks and bonds, have 5 telephones for a total basic monthly charge of $88.60. The charge includes 500 local outgoing calls. The cost of additional local calls is $0.08 each. Last month, 863 outgoing local calls were made from the office. In addition, long distance calls totaling $488.32 were made. The federal excise tax is charged on the basic charge, extra local calls, and long distance calls. What is the total cost for the month?

9. Citywide Real Estate has 2 incoming lines. Each has a rotary telephone with an illuminated dial. Citywide also has 2 extension telephones. The monthly charge includes 160 outgoing local calls. Each additional local call costs $0.15. Citywide made 180 outgoing local calls this month. What is the total cost of telephone service, including tax, for the month?

Monthly Charge Per Phone	
Rotary Telephone	$17.15
Extension Telephone	3.00
Illuminated Dial	3.30

Maintaining Your Skills

Look up the skill in parentheses if you need help or more practice.

Add. (Skill 5)

10. $15.25 + $11.20 + $0.79

11. $24.80 + $38.17 + $1.86

12. $83.60 + $6.29 + $162.48

13. $744 + $86.90 + $3.19 + $0.98 + $1.32

18-5 Utilities Costs—Electricity

Objective Compute the monthly costs for electricity.

The monthly cost of electricity for your business depends on the **demand charge** and the **energy charge.** The demand charge is based on the **peak load** during the month. The peak load is the greatest number of **kilowatts** (kW) that your business uses at one time during the month. The energy charge is based on the total number of **kilowatt-hours** (kW•h) that your business uses during the month. Meters installed at your business record the kilowatt demand and the number of kilowatt-hours. The electric company may add a **fuel adjustment charge** to your monthly bill to help cover increases in the cost of fuel needed to produce your electricity.

▶ **Total Cost for Month** = **Demand Charge** + **Energy Charge** + **Fuel Adjustment Charge**

Example

Skill 8 **Application** A **Term** Fuel adjustment charge

Sherwood Tool and Die had a peak load of 100 kilowatts of electricity during April. The demand charge is $6.60 per kilowatt. Sherwood used a total of 30,000 kilowatt-hours of electricity during the month. The energy charge for the first 1000 kilowatt-hours is $0.063 per kilowatt-hour. The cost of the remaining kilowatt-hours is $0.046 per kilowatt-hour. The fuel adjustment charge for April is $0.0135 per kilowatt-hour. What is the total cost of the electricity that Sherwood used in April?

Solution

1. Find the **demand charge.**
   100 kW × $6.60 = $660.00
2. Find the **energy charge.**
   a. 1000 kW•h × $0.063 = $63.00
   b. (30,000 − 1000) kW•h × $0.046 = $1334.00
   $63.00 + $1334.00 = $1397.00
3. Find the **fuel adjustment charge.**
   30,000 kW•h × $0.0135 = $405.00
4. Find the **total cost for month.**
   Demand Charge + Energy Charge + Fuel Adjustment Charge
   $660.00 + $1397.00 + $405.00 = $2462.00 total cost

$1649.00

▶ **Self-Check** Cover the answer and complete the problem.

Find the total cost for month.

Demand Charge Energy Charge Fuel Adjustment Charge

100 × $6.79 + [(1000 × $0.065) + (9000 × $0.045)] + 10,000 × $0.05

Problems

1. Bayes Seasonings Company.
Used 22,000 kW•h in May.
Peak load of 110 kW.
Demand charge: $5.00 per kW.
Energy charge: $0.05 per kW•h.
Fuel adjustment: $0.02 per kW•h.
What is the total cost for May?

2. Magnus Department Store.
Used 12,500 kW•h in April.
Peak load of 80 kW.
Demand charge: $3.10 per kW.
Energy charge: $0.07 per kW•h.
Fuel adjustment: $0.03 per kW•h.
What is the total cost for April?

3. ABC Rentals used 15,000 kilowatt-hours of electricity with a peak load of 100 kilowatts in June. The demand charge is $2.95 per kilowatt. The energy charge is $0.05 per kilowatt-hour for the first 1000 kilowatt-hours and $0.04 per kilowatt-hour for all kilowatt-hours over 1000. The fuel adjustment charge is $0.03 per kilowatt-hour. What is the total cost of electricity for ABC Rentals for June?

4. Willocroft Towers Restaurant used 23,500 kilowatt-hours of electricity last month. The peak load for the month was 150 kilowatts. The demand charge is $3.07 per kilowatt. The energy charge per kilowatt-hour is $0.07 for the first 10,000 kilowatt-hours and $0.05 for all kilowatt-hours over 10,000. The fuel adjustment charge is $0.03 per kilowatt-hour. What is the total cost of electricity for Willocroft Towers for last month?

5. Ellison Printing used 24,161 kilowatt-hours of electricity in August. The peak load during the month was 93 kilowatts. The demand charge is $3.05 per kilowatt. The energy charge per kilowatt-hour is $0.06 for the first 10,000 kilowatt-hours and $0.04 for all kilowatt-hours over 10,000. What is the total cost of electricity for Ellison Printing in August?

Refer to the chart for Exercises 6–8.

NORTHWEST POWER	
Demand Charge	
First 50 kW	$3.80/kW
Over 50 kW	$3.40/kW
Energy Charge	
First 250 kW·h	$0.085/kW·h
Next 750 kW·h	$0.069/kW·h
Next 2000 kW·h	$0.05/kW·h
Next 2000 kW·h	$0.039/kW·h
Next 5000 kW·h	$0.035/kW·h
Over 10,000 kW·h	$0.03/kW·h
Fuel Adjustment	$0.0215/kW·h

6. Markson, Inc., used 12,200 kilowatt-hours of electricity this month. The peak load during the month was 75 kilowatts. What is the total cost of electricity for the month?

7. Chesterton Temporary Services used 8400 kilowatt-hours of electricity for the month. Chesterton's peak load during the month was 80 kilowatts. What is the total cost of electricity for this month?

8. J and J Industries, Inc., used a total of 32,116 kilowatt-hours of electricity for the month. The peak load during the month was 112 kilowatts. What is the total cost of electricity for the month?

Maintaining Your Skills

Look up the skill in parentheses if you need help or more practice.

Multiply. (Skill 8)

9. 100 × 5.3
10. 1000 × 0.15
11. 9000 x 0.015
12. 150 × 8.91

18-6 Professional Services

Objective

Compute the total cost of professional services.

You may hire professional **consultants** to advise your business on a particular problem. The method of determining each **consultant's fee** may vary. Some consultants charge a flat fee, some charge a percent of the cost of the project, and some charge by the hour.

▶ **Total Cost = Sum of Consultants' Fees**

Example

Skills 5, 8, 30 **Application** A **Term** Consultant's fee

Art Lighting Manufacturing Company plans to expand its use of computers to include billing, payroll, and inventory control. The company plans to construct a new building at a cost of $250,000 to house the new computer installation. Art Lighting hires an architect, a systems analyst, and a computer programmer. The architect charges 5% of the total cost of the building. The systems analyst charges a flat fee of $3000. The computer programmer charges $25 an hour and works 175 hours. What is the total cost of the professional services?

Solution

Find the **sum of consultants' fees.**

a. Architect: 5% of $250,000 . . . . $12,500
b. Systems analyst: Flat fee 3000
c. Computer programmer: $25 × 175 4375
Total Cost $19,875

A. $16,320
B. $2000

▶ **Self-Check** Cover the answers and complete the problems.

Find the total cost.

A. Architect: 6% of $200,000.
Lawyer: 42 hours at $60 per hour.
Surveyor: $1800 flat fee.

B. Engineer: 30 hours at $40 per hour.
Computer programmer: $500 flat fee.
Computer time: $300 flat fee.

Problems

Professional Service	Fee Structure	Project Information	Total Fee
1. Attorney	$47.50 per hour	Worked 15 hours	?
2. Industrial nurse	$8.50 per hour	Worked 40 hours	?
3. Work-flow specialist	$3750 flat fee	Worked 27 hours	?
4. Project engineer	9.5% of project cost	$850,000 project	?

5. Stock underwriter.
Paid 8% of total stocks issued.
$10,000,000 issued.
What is the total fee?

6. Real estate broker.
Paid $6\frac{1}{2}$% of total sale.
Sale totals $462,500.
What is the total fee?

7. Keller Industries hired an in-service training specialist to conduct a workshop for 92 of its employees. The specialist was paid $75 per person. What was the total cost of the specialist's services?

8. Franchot-Highland Hospital hired an industrial engineering firm to conduct a work sampling on the average nurse's day. IE, Inc., did the work sampling and charged $42.50 per hour. It took IE 32 hours to complete the task. What did the work sampling cost Franchot-Highland Hospital?

9. Northwest Feed and Grain Company plans to issue $17,500,000 in bonds to pay for an extensive expansion. One bond broker will sell the bonds for a fee of 9.5% of the $17,500,000 face value of the bonds. What will the broker's services cost Northwest?

10. Watson's Department Store plans to buy a large piece of commercial property. The selling price of the property is $4,250,000. Sam Reston will handle the entire transaction for 7.4% of the selling price. What will it cost Watson's Department Store to have Reston handle the transaction?

11. Security Loan Company plans to build a new branch office. Security hired Alice Bayona, an architectural consultant. Alice's fee is 6.5% of the cost of the project. Security hired Tom Woods as project engineer for the new branch. Tom's fee is 4.5% of the cost of the project. Security hired Sara North to take care of the legal aspects of the project. Sara charges $45 per hour. It took her 8.7 hours to prepare the legal documents. The cost of the branch office is estimated to be $3,100,000. What amount should Security plan to pay for the 3 professional services?

12. Inkton Stores has installed automated checkout equipment in its 6 stores. Inkton hired Anthony Stone to consult on its installation. Stone charged 7.5% of the installation cost. Inkton hired Justina Ming to instruct its employees on the use of the equipment. Justina conducted six 4-hour sessions at $70 per session. The cost of the equipment and installation totaled $475,830. What was the total cost for professional services?

Maintaining Your Skills

Look up the skills in parentheses if you need help or more practice.

Add. (Skill 5)

13. 328.83 + 400 + 501.17

14. 9000 + 58.5 + 629.5

15. 5722 + 826.1 + 1.598

16. 543.34 + 553.51 + 7.70

Multiply. (Skill 8)

17. 0.05 × 14,400

18. 24.5 × 800

19. 0.15 × 3210

Reviewing the Basics

Skills

Round to the nearest dollar.

(Skill 2) 1. \$378.09 2. \$12.931 3. \$2489.51 4. \$108.509

Solve. Round to the nearest hundredth.

(Skill 3) 5. \$645 + \$195 + \$86 6. \$802 + \$9551 + \$679 7. \$57,120 + \$8934

(Skill 4) 8. 672 − 498 9. 72,000 − 14,300 10. 92,000 − 9221

(Skill 5) 11. 1420.84 + 1043.89 12. 928.44 + 213.904 13. 51.71 + 4.223

(Skill 6) 14. 6543.21 − 341.45 15. 3585.65 − 743.93 16. 367.22 − 154.77

(Skill 7) 17. 120 × 60 18. 24 × 68 19. 45 × 126

(Skill 8) 20. 43.11 × 564 21. 3198.53 × 0.563 22. 43,528 × 0.57

(Skill 10) 23. 2587 ÷ 12 24. 14,361 ÷ 12 25. 45,323 ÷ 24

(Skill 30) 26. 32% × 582,000 27. 11.5% × 358,400 28. 6.1% × 844,700

Applications

Find the area of each rental space.

(Application X) 29. 40 ft by 24 ft 30. 45 ft by 27 ft 31. 113 ft by 86 ft

32. 7 ft by 12 ft 33. 12.5 ft by 64.5 ft 34. 124.5 ft by 35.75 ft

Terms

Use each term in one of the sentences.

35. Rent
36. Labor charge
37. Basic monthly charge
38. Fuel adjustment charge
39. Consultant's fee

a. Your __?__ for telephone service may include a certain number of outgoing local calls.

b. You will pay a __?__ when you hire a professional person to advise your business on a particular project.

c. A __?__ may be added to your electricity bill.

d. Your business may __?__ rather than buy office space or equipment.

e. The __?__ is the greatest number of kilowatts that your business uses at one time during the month.

f. The cost of having a maintenance firm do a particular job may include a __?__ and a charge for materials.

Refer to your reference files at the back of the book if you need help.

Inside Economics

Inflation is an economic condition during which there are price increases in the cost of goods and services.

INDICATORS	
Consumers see increases in:	**Businesses see increases in:**
Cost of food and clothing	Wholesale prices
Cost of housing	Utilities costs
Cost of transportation	Cost of production and shipping
Cost of medical services	Demands for scarcer materials

Some of the causes of inflation are (1) heavy spending, resulting in high demand, (2) increased production costs while producers try to maintain profit levels, and (3) lack of competition. Inflation is expressed as a percent increase over a specified time period. You can find the inflation rate of a commodity or service by dividing the price increase over a specific time period by the original price of the commodity or service.

$$\text{Inflation Rate} = \frac{\text{Price Increase}}{\text{Original Price}}$$

For example, you find the total cost of leasing an automobile in your home town has increased from about \$24 a day to about \$28 a day in a one-year period. What is the inflation rate?

$$\text{Inflation Rate} = \frac{\$4}{\$24} = 16.7\% \text{ per year}$$

Find the inflation rate to the nearest tenth of a percent.

1. A house was sold for \$65,000. A year later, the same house was sold for \$80,000.

2. Utilities costs to run Joseph's Department Store have increased from \$890 per month to \$1050 per month for basically the same amount of usage.

3. The price of fuel oil in January is \$1.044 per gallon. In March, fuel oil costs \$1.124 per gallon.

4. The regular selling price of a refrigerator was \$459.95. This year, the same model refrigerator sells for \$479.95.

5. In June, an apartment rented for \$525 a month. The following January, the same apartment rents for \$545 a month.

Unit Test

Lesson 18-1

1. Tedrow Farm Equipment rents warehouse space at an annual rate of $3.45 per square foot. Tedrow's warehouse measures 185 feet by 150 feet. What is the monthly rental charge?

Lesson 18-2

2. Motel Nine hired 3 high school students to clean its swimming pool. Each student worked for 2.5 hours at an hourly rate of $3.90. The materials charge was $18.54. What was the total charge for this service?

Lesson 18-3

3. Clyde Bailey is opening a branch tax office for 4 months. He plans to rent this furniture at a monthly charge of 9% of the list price plus a 4% tax.

Item	List Price
2 desks	$247.95 each
2 desk chairs	47.95 each
4 guest chairs	42.45 each
1 file cabinet	74.95
1 bookcase	97.50

What is the total rental cost for the furniture?

Lesson 18-4

4. Fonda Imports, Inc., pays a basic monthly charge of $34.95 for telephone service. The charge includes 110 outgoing local calls. Each additional outgoing local call costs $0.09. Last month, a total of 184 outgoing local calls were made from Fonda Imports. A 3% federal tax was added to the bill for the month. What was the total cost of telephone service for the month?

Lesson 18-5

5. The Batson Corporation used 18,500 kilowatt-hours of electricity with a peak load of 120 kilowatts in April. The demand charge is $5.12 per kilowatt. The energy charge per kilowatt-hour is $0.08 for the first 10,000 kilowatt-hours and $0.06 for all kilowatt-hours over 10,000. The fuel adjustment charge is $0.04 per kilowatt-hour. What is the total cost of electricity for the Batson Corporation for April?

Lesson 18-6

6. National Industries, Inc., hired a consultant from the Beisecker Engineering Firm to help develop a new engine. The consultant's fee was 11.5% of the cost of the project. The project cost $975,000. What was the total cost of the consultant's services?

Unit 19 Accounting

▶ **The Accounting Department prepares the company *payroll register,* which is a record of all your employees' income. The department also keeps records of all your business expenses and the value of your equipment. The value of your equipment goes down, or *depreciates,* yearly because of age or wear and tear. There are four ways to determine depreciation: *straight-line method, sum-of-the-years' digit method, double-declining-balance method,* and *accelerated cost recovery system.***

Instead of writing payroll checks, many companies now use computers to have wages automatically deposited to employees' bank accounts.

19-1 Payroll Register

Objective

Complete a payroll register.

A **payroll register** is a record of the gross income, deductions, and net income of your company's employees. You may use a computer to prepare your payroll register. If you prepare the register by hand, you will probably refer to tables to determine the amount of income tax to withhold from each employee's pay.

Example

Skills 5, 6, 8 **Application** C **Term** Payroll register

Fallon Grain pays its employees weekly. Rita Jones prepares the payroll register for Fallon's 5 employees from this information.

Name	Regular Pay	Overtime Pay	Income Tax Information	Health Insurance Coverage
J. D. Grimes	$10.00/h	Time and a half	Married, 3 allow.	Family
H. M. Fox	$5.60/h	Time and a half	Single, 2 allow.	—
R. A. Jones	$300/week	—	Single, 1 allow.	Single
J. B. Crow	$410/week	—	Married, 3 allow.	Family
C. M. Able	$260/week plus 8% commission	—	Married, 0 allow.	Family

Solution

Rita has prepared the payroll register for the week of March 2. What is the net pay for the week?

$260 plus 8% of $2240 in sales

From federal withholding tables

7.15% of gross pay

Payroll register for week of: *March 2*

Name	Hours Worked		Hourly Rate	Gross Pay	Deductions				Net Pay
	Reg	OT			FIT	FICA	Hosp Ins.	Total	
J. D. Grimes	40	2	$10.00	$430.00	$49.80	$30.75	$9.80	$90.35	$339.65
H. M. Fox	40	—	5.60	224.00	22.50	16.02	—	38.52	185.48
R. A. Jones	—	—	—	300.00	41.90	21.45	3.92	67.27	232.73
J. B. Crow	—	—	—	410.00	46.00	29.32	9.80	85.12	324.88
C. M. Able	—	—	—	439.20	62.50	31.40	9.80	103.70	335.50
Total				$1803.20	$222.70	$128.94	$33.32	$384.96	$1418.24

$346.14

▸ **Self-Check** Cover the answer and complete the problem.

Use this information to find the net pay: gross weekly pay, $440; FIT, $52.00; FICA, $31.46; health insurance, $10.40.

Problems

	Name	Gross Pay	FIT	FICA	Total Deductions	Net Pay
1.	Baxter, L.	$460	$ 77.30	$32.89	$110.19	?
2.	Dunlop, J.	$215	$ 24.30	$15.37	?	?
3.	Harris, R.	$830	$203.60	$59.35	?	?
4.	Jones, S.	$430	$ 62.50	$30.75	?	?
5.	Renaldo, T.	$270	$ 12.60	$19.31	?	?
6.	Welsh, D.	$450	$ 66.90	$32.18	?	?

7. Thalmann Manufacturing Company.
Payroll for week.
Total gross pay: $8765.50.
Total deductions: $1873.79.
What is the total net pay?

8. Aiello's Department Store.
Payroll for week.
Total gross pay: $2160.
Total deductions: $461.74.
What is the total net pay?

Use the tables on pages 566–567, for federal withholding tax (FIT). Use the social security (FICA) tax rate of 7.15%.

9. Southern Paving Company employs students in the summer. Southern pays a standard hourly rate of $7.00. The only deductions are federal withholding, social security, and city income tax (CIT). Complete the payroll register for the week.

Payroll Register for Southern Paving Company Date: June 8, 19—

Employee	Income Tax Information	Hours Worked	Gross Pay	FIT	FICA	CIT	Total Deductions	Net Pay
Cole, Dean	Single, 1 allow	32	?	?	?	$2.04	?	?
Drake, Ann	Single, 0 allow	36	?	?	?	2.30	?	?
Lusetti, Marie	Single, 1 allow	33	?	?	?	2.10	?	?
Pappas, Mike	Single, 0 allow	34	?	?	?	2.17	?	?
Smith, Luellen	Single, 1 allow	32	?	?	?	2.04	?	?
Trotter, Robert	Single, 1 allow	35	?	?	?	2.23	?	?
Young, Ann	Single, 0 allow	38	?	?	?	2.42	?	?
Total			?	?	?	?	?	?

(Continue on next page.)

10. Collaborative, Inc., consulting engineers, pays its employees weekly salaries. Deductions include state and city income taxes. Complete the payroll register for the week.

Name	FIT Information	Weekly Salary	Deductions						Net Pay
			FIT	FICA	SIT	CIT	Hosp. Ins.	Total Deductions	
David	M, 5 allow.	$650.00	?	?	$14.63	$ 9.75	$12.50	?	?
Eberly	M, 6 allow.	610.00	?	?	13.73	9.15	15.00	?	?
Houlles	S, 1 allow.	585.00	?	?	13.16	8.78	10.00	?	?
Isetta	M, 0 allow.	595.00	?	?	13.39	8.93	10.00	?	?
Saba	S, 3 allow.	615.00	?	?	13.84	9.23	10.00	?	?
North	M, 8 allow.	785.00	113.20	?	17.66	11.78	20.00	?	?
O'Hara	M, 2 allow.	349.75	?	?	7.87	5.25	10.00	?	?
TOTAL		?	?	?	?	?	?	?	?

11. Super Subscription Service hires high school and college students to sell subscriptions. The students earn $3.00 an hour plus 17% commission on all sales. Deductions include $5.00 per week for health insurance. Complete the payroll register.

Name	FIT Information	Hours Worked	Total Sales	Gross Pay	Deductions				Net Pay
					FIT	FICA	Insurance	Total	
Adams	S, 1 allow.	40	$640	?	?	?	$5.00	?	?
Broski	S, 0 allow.	40	585	?	?	?	5.00	?	?
Frank	S, 2 allow.	39	735	?	?	?	5.00	?	?
Hunt	S, 1 allow.	40	554	?	?	?	5.00	?	?
Miller	S, 0 allow.	35	630	?	?	?	5.00	?	?
Vasco	S, 3 allow.	39	700	?	?	?	5.00	?	?
Zook	S, 1 allow.	36	815	?	?	?	5.00	?	?
Total			?	?	?	?	?	?	?

***12. A Brief Case** Tops-n-Bottoms Boutique pays its employees weekly. Deductions include federal withholding, state income tax, social security, and 1% of gross pay for unemployment compensation. Use this information to prepare the payroll for the week of November 30.

Name	FIT Information	Position	Pay Plan	Total Sales or Hours Worked
Cross	S, 2 allow.	Sales	7% straight commission	$6460
Miller	S, 1 allow.	Sales	7% straight commission	3425
Nystrand	M, 3 allow.	Sales	7% straight commission	4850
Petros	S, 1 allow.	Sales Trainee	5% straight commission	4350
Baker	S, 1 allow.	Maintenance	$5.50 per hour	40 hours

TOPS-N-BOTTOMS BOUTIQUE Payroll Register Week of: November 30, 19 —

Name	Gross Pay	FIT	SIT	FICA	Tax for Unemployment Compensation	Total Deductions	Net Pay
Cross	?	?	$7.91	?	?	?	?
Miller	?	?	4.20	?	?	?	?
Nystrand	?	?	5.94	?	?	?	?
Petros	?	?	3.81	?	?	?	?
Baker	?	?	3.85	?	?	?	?
Total	?	?	?	?	?	?	?

Maintaining Your Skills

Look up the skills in parentheses if you need help or more practice.

Add. (Skill 5)

13. $22.54 + $181.62 **14.** 427.6 + 8.143 **15.** 1.002 + 0.076

Subtract. (Skill 6)

16. $45.04 − $43.29 **17.** $183.88 − $89.54 **18.** $1023.45 − $456.02

19-2 Business Expenses

Objective

Compute the percent a particular business expense is of the total expenses.

Your business must keep accurate records of all its expenses. You will use the information when you prepare income tax forms and when you calculate your company's profits. You may determine your total expenses monthly, quarterly, or annually. To plan for future spending, you may calculate the percent each expense is of the total.

▶ **Percent of Total** $= \dfrac{\textbf{Particular Expense}}{\textbf{Total Expenses}}$

Example

Skills 5, 31 **Application** A **Term** ______

The Earthenware Company manufactures clay mugs, planters, and other products. Records of Earthenware's expenses for the first quarter of the year show:

Amount	Expense
$ 62,942.17	Payroll
3,000.00	Advertising
125,981.06	Raw materials
4,000.00	Factory and showroom rental
329.42	Office supplies
721.19	Insurance
3,026.16	Utilities
$200,000.00	Total

What percent of total expenses did Earthenware spend on advertising during the quarter?

Solution

Find the **percent of total.**

Particular Expense ÷ Total Expenses

$3,000.00 ÷ $200,000.00 = 0.015 = 1.5% of total was spent on advertising

A. 6%
B. 2%

▶ **Self-Check** Cover the answers and complete the problems.

A. Particular expenses: typing, $800; drafting, $1500; payroll, $22,700. Drafting is what percent of the total?

B. Particular expenses: utilities, $1100; security, $1200; raw materials, $28,000; payroll, $29,700. Security is what percent of the total?

Problems

	Expenses for Supplies	÷	Total Expenses	=	Percent of Total
1.	$ 5000	÷	$ 40,000	=	?
2.	$ 1200	÷	$ 15,000	=	?
3.	$ 360	÷	$ 5000	=	?
4.	$150,000	÷	$6,000,000	=	?

5. Holness Marketing Consultants.
Expenses total $80,000 for quarter.
Payroll for quarter is $46,800.
Payroll expense is what percent of the total?

6. Bright's Restaurant.
Annual expenses total $120,000.
Annual building rental is $7500.
Building rental is what percent of the total?

7. Acme Tree Service had the following business expenses last year:

Payroll	$178,800.00
Advertising	1,200.00
Equipment	48,000.00
Office/Garage Rental	2,400.00
Supplies	1,500.00
Insurance	6,000.00
Utilities	100.00

What is the total of the expenses? Payroll is what percent of the total?

8. Violet's Clothes Closet had the following business expenses last month:

Payroll	$ 2,450.00
Advertising	900.00
Cost of Goods Sold	15,230.00
Store Rental	1,000.00
Supplies	125.00
Insurance	100.00
Utilities	195.00

Cost of goods sold is what percent of the total business expenses for the month?

Maintaining Your Skills

Look up the skills in parentheses if you need help or more practice.

Add. (Skill 5)

9. $\begin{array}{r} \$178,000 \\ 24,300 \\ +\ 67,700 \\ \hline \end{array}$

10. $\begin{array}{r} \$2240.00 \\ 768.15 \\ +\ 326.50 \\ \hline \end{array}$

11. $\begin{array}{r} 127.6 \\ 0.642 \\ +\ 16.9021 \\ \hline \end{array}$

12. $\begin{array}{r} 4582.51 \\ 438.90 \\ +\ 111.24 \\ \hline \end{array}$

Find the rate. Round answers to the nearest tenth of a percent. (Skill 31).

13. n% of $80 = $50

14. n% of 80 = 200

15. n% of 72 = 36

16. n% of $120,000 = $60,000

17. n% of $2000 = $180

19-3 Apportioning Expenses

Objective

Compute a department's share of the total business expense.

Your business may **apportion**, or distribute, certain expenses among its departments. Each department is charged a certain amount of the particular expense. Often, the amount that each department is charged depends on the space that it occupies.

▶ $$\text{Amount Paid} = \frac{\text{Square Feet Occupied}}{\text{Total Square Footage}} \times \text{Total Expense}$$

Example

Skills 10, 8 **Application** X **Term** Apportion

The DePietro Corporation occupies a building that contains 500,000 square feet. The Engineering Department occupies an area that measures 100 feet by 350 feet. The annual cost of security for the entire building is $41,650. DePietro apportions the cost based on square feet occupied. What annual amount for security does DePietro charge to its Engineering Department?

Solution

1. Find the **square feet occupied.**
   100 ft × 350 ft = 35,000 square feet

2. Find the **amount paid.**
   (Square Feet Occupied ÷ Total Square Footage) × Total Expense
   (35,000 ÷ 500,000) × $41,650
   0.07 × $41,650 = $2915.50 amount paid

A. $300
B. $200

▶ **Self-Check** Cover the answers and complete the problems.

A. Dental Associates occupies 40 feet by 50 feet in a building with 80,000 square feet. Annual cost of air conditioning is $12,000. What does Dental Associates pay?

B. Antiques, Ltd., occupies 20 feet by 40 feet in a mall with 40,000 square feet. Monthly mall expenses are $10,000. What does Antiques, Ltd., pay?

Problems

	(Square Feet Occupied	÷	Total Square Footage)	×	Total Expense	=	Amount Paid
1.	(57,000	÷	950,000)	×	$50,000	=	?
2.	(23,750	÷	950,000)	×	$50,000	=	?
3.	(13,300	÷	950,000)	×	$50,000	=	?

Round to the nearest thousandth and to the nearest cent.

4. Building area: 2,000,000 sq. ft. Annual maintenance cost: $252,000. Auditing Dept. is 75 ft by 60 ft. What does the Auditing Department pay annually for maintenance?

5. Building area: 150,000 sq. ft. Annual insurance charge: $1270. Marketing Dept. is 35 ft by 45 ft. What does the Marketing Department pay annually for insurance?

6. Ling Company apportions the annual cost of utilities among its departments. The total cost for the year was $122,700. The total area of the building is 615,000 square feet. The company's library occupies an area that is 18 feet by 54 feet. What did the library pay for utilities for the year?

7. Botteron Manufacturing pays a total of $724,500 per year to rent its building. The total area of the building is 230,000 square feet. Botteron apportions the rental charge among its departments. How much does each of these departments pay for rent?
   a. Industrial Design, 60 ft by 75 ft   **b.** Personnel, 36 ft by 45 ft

8. Aries Corporation owns and maintains a 6-store shopping mall. Aries paid $4000 for advertising last year. It apportions the cost based on square footage. How much does Aries charge each store?

Shop	Dimensions
Bali-Hi Cleaners	45 ft by 50 ft
R & B Ice Cream	30 ft by 50 ft
Electro-Shack	45 ft by 100 ft
Wheeler's Bicycle Shop	60 ft by 100 ft
Terry's Travel	30 ft by 100 ft
Nicholson's Gifts	70 ft by 50 ft

*9. **A Brief Case** Some businesses apportion costs among their departments on the basis of gross sales. The gross sales for Discount City totaled $2,400,000 last year. It distributed these annual expenses:

Maintenance	Utilities	Security
$4800	$21,000	$3360

   a. The Children's Wear Department had $650,000 in gross sales last year. How much did it pay for each of the annual expenses?
   b. The Jewelry Department had $800,000 in gross sales last year. How much did it pay for each of the annual expenses?
   c. The Automotive Service Department had $300,000 in gross sales last year. How much did it pay for each of the annual expenses?

Maintaining Your Skills

Look up the skills in parentheses if you need help or more practice.

Divide. Round answers to the nearest tenth. (Skill 10)

10. 86 ÷ 8   11. 426 ÷ 20   12. 980 ÷ 42   13. 4210 ÷ 400

Multiply. (Skill 8)

14. 2.6 × 1.7   15. 5.55 × 0.03   16. 26.1 × 1000   17. 7.21 × 130

19-4 Depreciation—Straight-Line Method

Objective Use the straight-line method to compute the annual depreciation of an item.

For tax purposes, the Internal Revenue Service allows you to recognize the **depreciation** of many of the items that your business owns. Depreciation is a decrease in the value of an item because of its age or condition. The **straight-line method** is one way of determining the annual depreciation of an item. This method assumes that the depreciation is the same from year to year. To calculate the depreciation, you must know the original cost, the **estimated life**, and the **resale value** of the item. The estimated life of an item is the length of time, usually in years, that it is expected to last. The resale value is the estimated trade-in, salvage, or scrap value at the end of the item's expected life.

▸ **Annual Depreciation** = $\frac{\textbf{Original Cost} - \textbf{Resale Value}}{\textbf{Estimated Life}}$

Example

Skills 4, 10 **Application** A **Term** Straight-line method

The law firm of Mayo and Morgan purchased a new typewriter that cost $975. The life of the typewriter is estimated to be 5 years. The total resale value after 5 years of use is estimated to be $400. Using the straight-line method, find the annual depreciation of the typewriter.

Solution

Find the **annual depreciation.**

(Original Cost − Resale Value) ÷ Estimated Life
($975 − $400) ÷ 5
$575 ÷ 5 = $115 annual depreciation

A. $250
B. $1800

▸ **Self-Check** Cover the answers and complete the problems.

Find the annual depreciation.

A. $3000 original cost; 10 years estimated life; $500 resale value.

B. $10,000 original cost; 5 years estimated life; $1000 resale value.

Problems

	(Original Cost	−	Resale Value)	÷	Estimated Life	=	Annual Depreciation
1.	($ 87,500	−	$ 7500)	÷	8 years	=	?
2.	($ 745	−	$ 200)	÷	5 years	=	?
3.	($146,500	−	$10,000)	÷	10 years	=	?
4.	($ 5250	−	$ 500)	÷	7 years	=	?

5. New fork lift.
Cost is $25,000.
Estimated life of 5 years.
Salvage value estimated at $5000.
What is the annual depreciation?

6. New delivery truck.
Cost is $9600.
Estimated life is 6 years.
Salvage value estimated at $600.
What is the annual depreciation?

7. Westside Bowling Lanes recently purchased new pinsetters for a total cost of $120,000. The estimated life of the pinsetters is 10 years. The trade-in value of the setters after 10 years is estimated to be $7000. What is the annual depreciation?

8. Sue Wong is a certified public accountant. She purchased equipment for her office for $4800. The trade-in value of the equipment is estimated to be $250 after 7 years of use. What is the annual depreciation?

9. Acme Stamping Company purchased a punch press for $145,000. After 12 years of useful life, the estimated salvage value is $1000. What is the annual depreciation?

10. John Baxter is a representative for a carpet manufacturer. John uses his car entirely for business. He trades in his car every 2 years. John paid $6700 for his present car. He expects to receive $4000 for the car when he trades it in after 2 years. What is the annual depreciation?

11. The Montville Company recently purchased 3 new heat-sealing machines. Each machine cost $17,500. The estimated life of each machine is 4 years. The salvage value of each machine is expected to be $900 at the end of the 4 years. What is the annual depreciation for all 3 machines?

12. The Newton Cleaning Center recently purchased 7 new washing machines at a cost of $290 each and 5 new clothes dryers at a cost of $245 each. The salvage value of the washers is expected to be $50 each after 4 years of use. The salvage value of the dryers is expected to be $75 each after 5 years of use. What is the total annual depreciation?

Maintaining Your Skills

Look up the skills in parentheses if you need help or more practice.

Subtract. (Skill 4)

13. $4582 − 3291

14. $100,450 − 96,722

15. $419 − 29

16. $10,046 − 853

Divide. Round answers to the nearest tenth. (Skill 10)

17. 281 ÷ 26

18. 1218 ÷ 106

19. 86 ÷ 18

20. 9416 ÷ 182

Multiply. (Skill 8)

21. $260.00 × 0.07

22. $2345.70 × 0.05

23. $43.50 × 0.06

24. $123.00 × 0.075

19-5 Book Value

Objective Use the straight-line method to compute the book value of an item.

Book value is the approximate value of an item after you have owned it for a period of time. The book value is the original cost minus the **accumulated depreciation.** The accumulated depreciation is the total depreciation to date. At the end of an item's life, its book value and resale value should be equal.

▶ **Book Value = Original Cost – Accumulated Depreciation**

Example

Skill 7 **Application** A **Term** Book Value

Mayo and Morgan purchased a new typewriter for $975. The total resale value after 5 years is estimated to be $400. Using the straight-line method, Mayo and Morgan determined that the typewriter will depreciate $115 per year. What will the book value be at the end of 4 years?

Solution

1. Find the **accumulated depreciation** for fourth year.
   $115 per year × 4 years = $460
2. Find the **book value.**
   Original Cost – Accumulated Depreciation
   $975 – $460 = $515 book value

$1600
$1000

▶ **Self-Check** Cover the answers and complete the problems.

A computer cost $3400. Its estimated life is 4 years. The annual depreciation is $600. What will the book value be at the end of 3 years? At the end of 4 years?

Problems

	Annual Depreciation	× Number of Years	= Accumulated Depreciation	Original Cost	– Accumulated Depreciation	= Book Value
1.	$ 1500	× 7 years	= ?	$16,000	– ?	= ?
2.	$12,000	× 4 years	= ?	$70,000	– ?	= ?
3.	$ 920	× 3 years	= ?	$ 5000	– ?	= ?

Use the straight-line method of depreciation. Round to the nearest dollar.

4. $7200 for refrigeration equipment. Depreciates $700 per year. What is the book value after 2 years?
5. $5800 for sprinkler system. Depreciates $600 per year. What is the book value after 3 years?

6. New food-service equipment.
Cost is $28,000.
Estimated life is 5 years.
Resale value of $1200.
What is the book value after 4 years?

7. New copy machine.
Cost is $2375.
Estimated life of 6 years.
Resale value of $300.
What is the book value after 5 years?

8. Sue Wong purchased new office equipment for $4800. The trade-in value of the equipment is estimated to be $1550 after 5 years of use. Find the book value after each year of use.

Year	Annual Depreciation	Accumulated Depreciation	Book Value
1	$650	$ 650	?
2	650	1300	?
3	650	?	?
4	650	?	?
5	650	?	?

9. Chris Oulette purchased a new tractor for $22,500. The tractor is expected to have a trade-in value of $5000 after 4 years. What is the book value at the end of each of the 4 years?

Year	Annual Depreciation	Accumulated Depreciation	Book Value
1	?	?	?
2	?	?	?
3	?	?	?
4	?	?	?

*10. **A Brief Case** The units-of-production method is another method of computing depreciation. This method bases depreciation on the number of units produced rather than the years of estimated life.

$$\text{Depreciation} = \frac{\text{Number of Units Produced}}{\text{Estimated Units of Production}} \times \left(\begin{matrix}\text{Original}\\ \text{Cost}\end{matrix} - \begin{matrix}\text{Salvage}\\ \text{Value}\end{matrix}\right)$$

a. Skyler Manufacturing purchased a new automatic lathe for $140,000. The lathe is expected to be worth $10,000 after 200,000 units are produced. Use the units-of-production method to calculate the book value of the lathe after 20,000 units are produced.

b. Harlan Plastics purchased a new folding machine for $37,288. The salvage value of the machine is estimated to be $5000 after 500,000 units are produced. Use the units-of-production method to calculate the book value after 27,000 units are produced.

Maintaining Your Skills

Look up the skill in parentheses if you need help or more practice.

Multiply. (Skill 7)

11. 920 × 10 12. 7500 × 45 13. 62 × 27 14. 19,250 × 70 15. 324 × 40

16. 12,000 × 64 17. 1262 × 5200 18. 960 × 610 19. 815 × 230

19-6 Sum-of-the-Years'-Digits Method

Objective Use the sum-of-the-years'-digits method to compute the yearly depreciation.

The **sum-of-the-years'-digits method** is another way to determine the annual depreciation of an item. This method assigns a fraction of the total depreciation to each year of the item's life. More depreciation is allowed in the early years of the item's life than in the later years. For example, if the life of a machine is estimated to be 4 years, the years' digits are 4, 3, 2, and 1. The sum of the years' digits is: 4 + 3 + 2 + 1 = 10. The fractions of the total depreciation are: $\frac{4}{10}$ for the first year, $\frac{3}{10}$ for the second year, $\frac{2}{10}$ for the third year, and $\frac{1}{10}$ for the fourth year. To use the sum-of-the-years'-digits method, you must know the original cost, the estimated life, and the resale value.

▶ **Depreciation for Year** = **Fraction of Total Depreciation** × (**Original Cost** − **Resale Value**)

Example

Skills 20, 2 **Application** A **Term** Sum of the years' digits

Durgin Manufacturing purchased a new automatic drill press for $50,000. The expected life of the drill press is 5 years. At the end of the 5 years, the resale value of the machine is expected to be $5000. Durgin uses the sum-of-the-years'-digits method to calculate depreciation. To the nearest dollar, what is the depreciation for the first year of the life of the drill press?

Solution

1. Find the **fraction of total depreciation.**

$$\frac{\text{Assigned Digit}}{\text{Sum of the Years' Digits}} = \frac{5}{5 + 4 + 3 + 2 + 1} = \frac{5}{15}$$

2. Find the **depreciation for year.**

Fraction of Total Depreciation × (Original Cost − Resale Value)

$\frac{5}{15}$ × ($50,000 − $5000)

$\frac{5}{15}$ × $45,000 = $15,000 depreciation for first year

A. $12,000
B. $9000

▶ **Self-Check** Cover the answers and complete the problems.

A. $\frac{4}{15}$ × $45,000 = ? depreciation for second year.

B. $\frac{3}{15}$ × $45,000 = ? depreciation for third year.

Problems

Use the sum-of-the-years'-digits method to calculate depreciation. Round to the nearest dollar.

	Original Cost	Resale Value	Estimated Life	Depreciation 1st Year Fraction	1st Year Amount	2nd Year Fraction	2nd Year Amount
1.	\$ 6000	\$1200	3 years	$\frac{3}{6}$	\$2400	$\frac{2}{6}$	?
2.	\$ 2500	\$ 400	6 years	$\frac{6}{21}$	\$ 600	?	?
3.	\$24,000	\$6000	2 years	?	?	?	?
4.	\$ 950	\$ 250	7 years	?	?	?	?

5. Sander.
Estimated life of 2 years.
Original cost is \$4500.
Resale value is \$900.
What is the depreciation for the first year?

6. High-speed motor.
Estimated life of 3 years.
Original cost is \$3700.
Resale value is \$1200.
What is the depreciation for the first year?

7. The original cost of a small industrial drill is \$5200. After 4 years, the resale value is estimated to be \$700. What is the depreciation for each of the 4 years?

8. The Word Pro Typing Service recently purchased 2 new typewriters for a total cost of \$1950. The typewriters have an estimated life of 5 years. After 5 years, the resale value of each typewriter is expected to be \$400. What is the depreciation for each typewriter for each of the 5 years?

*9. **A Brief Case** The ALCO Company recently purchased a plastic-molding machine for \$50,000. The estimated life of the machine is 4 years. The resale value after 4 years is expected to be \$2500. What is the book value of the machine after each year of use?

Year	Depreciation for Year	Accumulated Depreciation	Book Value
First	?	?	?
Second	?	?	?
Third	?	?	?
Fourth	?	?	?

Maintaining Your Skills

Look up the skill in parentheses if you need help or more practice.

Multiply. Express the answers in lowest terms. (Skill 20)

10. $\frac{1}{2} \times \frac{2}{5}$
11. $\frac{3}{8} \times 1\frac{1}{2}$
12. $\frac{3}{4} \times 14$
13. $2\frac{1}{8} \times 1\frac{3}{4}$
14. $\frac{1}{3} \times \frac{2}{9}$
15. $5\frac{1}{6} \times 10\frac{1}{8}$
16. $14\frac{1}{2} \times 2\frac{2}{5}$
17. $\frac{11}{12} \times \frac{2}{33}$

19-7 Double-Declining-Balance Method

Objective Use the double-declining-balance method to compute the annual depreciation and book value.

Your business may use the **double-declining-balance method** to find the annual depreciation of an item. This method allows the highest possible depreciation in the first year of the item's life. The depreciation declines as the item approaches the end of its life. To use this method, you must know the item's original cost and estimated life.

▶ **Annual Depreciation Rate** $= 2 \times \dfrac{100\%}{\text{Estimated Life}}$

▶ **Depreciation for Year** = **Annual Depreciation Rate** × **Previous Declining Balance**

Example

Skill 30 **Application** A **Term** Double-declining balance

Delona Studios purchased television equipment for $50,000. Delona estimates that the equipment will become obsolete in 5 years. What are the depreciation and book value for each of the 5 years?

Solution

1. Find the **annual depreciation rate.**
   2 × (100% ÷ Estimated Life)
   2 × (100% ÷ 5) = 40%
2. Find the **depreciation for year.**
   Annual Depreciation Rate × Previous Declining Balance

40% of ORIGINAL COST

PREVIOUS DECLINING BALANCE − DEPRECIATION FOR YEAR

Year	Depreciation for Year	Accumulated Depreciation	Declining Balance (Book Value)
First	$20,000	$20,000	$30,000
Second	12,000	32,000	18,000
Third	7,200	39,200	10,800
Fourth	4,320	43,520	6,480
Fifth	2,592	46,112	3,888 ← Resale value

40% of PREVIOUS DECLINING BALANCE OF $30,000

A. $20,000
$80,000
B. $16,000
$64,000

▶ **Self-Check** Cover the answers and complete the problems.

A machine that cost $100,000 will be obsolete in 10 years.

A. Find the depreciation and book value for the first year.
B. Find the depreciation and book value for the second year.

Problems

	Estimated Life	Annual Depreciation Rate	×	Previous Declining Balance	=	Depreciation for Year
1.	5 years	40%	×	$ 5200	=	?
2.	4 years	?	×	$ 900	=	?
3.	6 years	?	×	$ 90,000	=	?
4.	8 years	?	×	$120,000	=	?

5. New delivery van cost $9600. Estimated life of 6 years. What is the depreciation for the first year? The book value?

6. New generator cost $27,000. Estimated life of 4 years. What is the depreciation for the first year? The book value?

7. Mart Company purchased new check-out equipment for $60,000. The life of the equipment is estimated to be 4 years. The resale value is expected to be $3750. Complete the depreciation table for each of the 4 years.

Year	Depreciation for Year	Accumulated Depreciation	New Declining Balance
First	$30,000	?	?
Second	$15,000	?	?
Third	?	?	?
Fourth	?	?	?

*8. **A Brief Case** When the double-declining-balance method is used to calculate depreciation, the declining balance in the last year of an item's life might not equal the expected resale value. When this happens, the depreciation for the last year is adjusted so that the final declining balance and the resale value are the same.

Laker's purchased office equipment for $5200. The resale value is expected to be $350 after a 5-year life. Complete the depreciation table.

Year	Depreciation for Year	Accumulated Depreciation	New Declining Balance
1	$2080	?	?
2	?	?	?
3	?	?	?
4	?	?	?
5	?	?	$350

This amount is computed to be $270. It must be increased so that the book value is $350.

Maintaining Your Skills

Look up the skill in parentheses if you need help or more practice.

Find the percentage. Round answers to the nearest hundredth. (Skill 30)

9. 70% of $1350
10. 6.92% of 50
11. $\frac{4}{5}$% of 45
12. 130% of 93.6
13. $16\frac{1}{8}$% of 120
14. 1200% of 90
15. $3\frac{1}{2}$% of 0.18
16. 32.25% of 60

19-8 Accelerated Cost Recovery System (ACRS)

Objective Use the Accelerated Cost Recovery System method to compute annual depreciation.

The **Accelerated Cost Recovery System** (ACRS) is another method of computing depreciation. Introduced by the Economic Recovery Tax Act of 1981, ACRS allows businesses to calculate depreciation over a short period of time for greater tax deductions. This method encourages businesses to replace equipment earlier than they would if they used other depreciation methods. Under ACRS, property can be fully depreciated over a period of 3, 5, 10, or 15 years according to fixed percents.

▸ **Annual Depreciation = Fixed Percent × Original Cost**

▸ **Book Value = Original Cost − Accumulated Depreciation**

Example

Skills 30, 3 **Term** Accelerated cost recovery system

Ruskin Drugs purchased a new automobile for $6000 to use in delivering prescriptions. ACRS allows automobiles to be fully depreciated in 3 years according to 3 fixed percents: 25% the first year, 38% the second year, and 37% the third year. What are the annual depreciation and book value for each of the 3 years?

Solution

1. Find the **annual depreciation.**

Fixed Percent	×	Original Cost
25%	×	$6000 = $1500
38%	×	$6000 = $2280
37%	×	$6000 = $2220

2. Find the **book value.**

Original Cost	−	Accumulated Depreciation	
$6000	−	$1500	= $4500
$6000	−	$3780	= $2220
$6000	−	$6000	= $0

$1500 + $2280 = $3780

A. $2500, $3800, $3700
B. $2125, $3230, $3145

▸ **Self-Check** Cover the answers and complete the problems.

ACRS depreciates light trucks in 3 years according to the same fixed percents used for automobiles.

A. Find the annual depreciation for each of the 3 years for a light truck that cost $10,000.

B. Find the annual depreciation for each of the 3 years for a light truck that cost $8500.

Problems

Use the Accelerated Cost Recovery System (ACRS) to find the annual depreciation. Round answers to the nearest dollar.

	Item	Original Cost	Fully Depreciated	Depreciation 1st Year	2nd Year	3rd Year
1.	Tractor unit	$ 36,000	3 years	$9000	$13,680	?
2.	Taxi	$ 12,000	3 years	?	?	?
3.	Research equipment	$180,000	3 years	?	?	?
4.	Breeder hogs	$ 5200	3 years	?	?	?
5.	Race horse	$ 29,000	3 years	?	?	?

6. Experimental equipment.
Original cost is $475,000.
Fully depreciated in 3 years.
Fixed percents are 25%, 38%, and 37%.
What is the depreciation for each of the 3 years?

7. Limousine.
Original cost is $38,500.
Fully depreciated in 3 years.
Fixed percents are 25%, 38%, and 37%.
What is the depreciation for each of the 3 years?

8. Stein's Travel Agency purchased a van for $32,400 to transport clients to and from the airport. The van is fully depreciated in 3 years. Fixed percents are 25%, 38%, and 37%. What are the depreciation and book value for each of the 3 years?

Aircraft, computers, and copiers are fully depreciated in 5 years according to 5 fixed percents: 15% the first year, 22% the second year, 21% the third year, 21% the fourth year, and 21% the fifth year. Use these percents to solve problems 9 and 10.

9. Darien Plastics purchased a computer for $47,900. It will be fully depreciated in 5 years. Find the depreciation for each of the 5 years.

10. Hobbs Manufacturing purchased a copier for $17,600. It will be fully depreciated in 5 years. Find the depreciation for each of the 5 years.

11. Railroad tank cars are fully depreciated in 10 years. Western Railroad purchased a tank car for $325,000. Fixed percents are:

Year	Percent	Year	Percent	Year	Percent	Year	Percent	Year	Percent
1	8%	3	12%	5	10%	7	9%	9	9%
2	14%	4	10%	6	10%	8	9%	10	9%

Find the annual depreciation and book value for the first 4 years.

Maintaining Your Skills

Look up the skill in parentheses if you need help or more practice.

Find the percentage. (Skill 30)

12. 12% of $62,500 13. 37.5% of $1240 14. 175% of $90 15. $\frac{3}{4}$% of $60

Reviewing the Basics

Skills

Round to the nearest dollar.

(Skill 2) **1.** \$8472.536 **2.** \$10,439.80 **3.** \$21,964.76

Solve.

(Skill 4) **4.** \$16,245 − \$2371 **5.** \$135,000 − \$12,480 **6.** \$327 − \$219

(Skill 5) **7.** \$22.45 + \$18.93 + \$5.40 **8.** \$1578.54 + \$267.33 **9.** \$1.99 + \$3.21

(Skill 6) **10.** \$2481.39 − \$130.52 **11.** \$11,192.86 − \$6481.75

(Skill 7) **12.** \$254 × 7 **13.** \$680 × 90 **14.** \$1320 × 4

(Skill 8) **15.** 0.03 × \$21,489 **16.** 0.10 × 300 **17.** 0.12 × \$1650

(Skill 20) **18.** $\frac{3}{10}$ × \$18,456 **19.** $\frac{1}{4}$ × \$1260 **20.** $\frac{5}{21}$ × \$26,397

Round to the nearest thousandth.

(Skill 10) **21.** 2180 ÷ 23,000 **22.** 16,000 ÷ 655,000 **23.** 37 ÷ 164

Find the percent. Round to the nearest cent.

(Skill 30) **24.** 50% of \$27,500 **25.** 12.5% of \$2180 **26.** 5.1% of \$784

27. 125% of \$14,055 **28.** 10% of \$255.75 **29.** 3% of \$264

Round to the nearest tenth of a percent.

(Skill 31) **30.** \$327 is what percent of \$20,465? **31.** \$2371 is what percent of \$17,488?

32. \$475 is what percent of \$28,900? **33.** \$580 is what percent of \$18,642?

Application

Find the area.

(Application X) **34.** 120 ft by 37 ft **35.** 245 ft by 85 ft

Terms

Write your own definition for each term.

36. Payroll register
37. Straight-line method
38. Apportion
39. Sum-of-the-years'-digits method
40. Book value
41. Double-declining-balance method
42. Accelerated Cost Recovery System

Refer to your reference files at the back of the book if you need help.

A Microcomputer Application
Payroll Deductions

To complete this microcomputer application, you will need the diskette *Microcomputer Applications for Business Mathematics,* which accompanies this textbook.

Select option 7, Payroll Deductions, from the menu. The Kreo Ice Cream Store employs high school students after school and in the summer. Kreo pays a standard hourly rate of $4.25. Deductions are taken for federal withholding (FIT), social security (FICA), and city income tax (CIT). Input the information in the following problems to complete the payroll register.

1. Week of: June 15

Employee	Employee Number	Hours Worked	Income Tax Information
a. Cole, Dean	1001	34	Single, 1 allowance
b. Drake, Ann	1002	35	Single, 0 allowance
c. Lusetti, Marie	1003	27	Single, 1 allowance
d. Pappas, Mike	1004	32	Single, 0 allowance
e. Smith, Luellen	1005	30	Single, 1 allowance
f. Trotter, Robert	1006	40	Single, 1 allowance
g. Young, Ann	1007	39	Single, 0 allowance

2. Week of: June 22

Employee	Employee Number	Hours Worked	Income Tax Information
a. Cole, Dean	1001	36	Single, 1 allowance
b. Drake, Ann	1002	39	Single, 0 allowance
c. Lusetti, Marie	1003	25	Single, 1 allowance
d. Pappas, Mike	1004	34	Single, 0 allowance
e. Smith, Luellen	1005	22	Single, 1 allowance
f. Trotter, Robert	1006	26	Single, 1 allowance
g. Young, Ann	1007	36	Single, 0 allowance

3. Week of: June 29

Employee	Employee Number	Hours Worked	Income Tax Information
a. Cole, Dean	1001	38	Single, 1 allowance
b. Drake, Ann	1002	40	Single, 0 allowance
c. Lusetti, Marie	1003	37	Single, 1 allowance
d. Pappas, Mike	1004	36	Single, 0 allowance
e. Smith, Luellen	1005	27	Single, 1 allowance
f. Trotter, Robert	1006	25	Single, 1 allowance
g. Young, Ann	1007	39	Single, 0 allowance

Unit Test

Lesson 19-1

1. Complete the payroll register. Use the FICA tax rate of 7.15%.

Name	Gross Income	FIT	FICA	SIT	Total Deductions	Net Pay
Bisset	$415.00	$53.90	?	$8.32	?	?
Heller	442.00	64.70	?	8.84	?	?
Molnar	455.00	54.20	?	9.10	?	?
Stein	438.00	54.00	?	8.76	?	?
Total	?	?	?	?	?	?

Lesson 19-2

2. Carson's Insurance had these expenses during the last quarter:

Payroll	$15,670.00
Advertising	845.00
Office Rental	2,360.00
Office Supplies	755.00
Utilities	370.00

What is the total of the expenses? Payroll is what percent of the total?

Lesson 19-3

3. Bynarowicz Manufacturing Company pays $125,000 per year for security. The company apportions the cost among its departments based on space. The Research Department occupies an area that measures 50 feet by 70 feet. The building contains 1,500,000 square feet. How much does the Research Department pay for security?

Lesson 19-4

4. Central Copy purchased a new copy machine for $6480. The machine has an estimated life of 4 years. The resale value is expected to be $600. Use the straight-line method to find the annual depreciation.

Lesson 19-5

5. Find the book value after each year of use for the copy machine in Exercise 4.

Lesson 19-6

6. Oberly's Gifts purchased new store fixtures for a new branch store. The cost of the fixtures was $8270. The salvage value after 5 years is estimated to be $1040. Use the sum-of-the-years'-digits method to find the depreciation for each of the 5 years.

Lesson 19-7

7. The original price of a slitting and creasing machine is $33,000. The machine is expected to have a life of 3 years. Use the double-declining-balance method to find the annual depreciation and the book value for the first 2 years. Round answers to the nearest dollar.

Lesson 19-8

8. Ace Construction purchased a new light truck for $9000. Ace will use the ACRS method to fully depreciate the truck over 3 years. The fixed percents for the 3 years are 25%, 38%, and 37%. Use the ACRS method to find the annual depreciation for each of the 3 years.

Unit 20 Accounting Records

▸ The Accounting Department also keeps other records that show the income, expenses, and value of your business. A *balance sheet* shows your cash, or *assets,* and money you owe others, or your *liabilities.* An *income statement* details your income and operating expenses.

Assets must equal liabilities plus equity in order for a balance sheet to balance.

20-1 Assets, Liabilities, and Equity

Objective Compute the total assets, liabilities, and owner's equity.

When you start a business, you will need to buy either merchandise to sell or materials with which to make your products. You may need to purchase office supplies, equipment, buildings, or land. In addition, you must have cash to make change, to pay bills, and to meet other expenses. **Assets** are the total of your cash, the items that you have purchased, and any money that your customers owe you.

You may borrow money to start your business or you may purchase merchandise on credit. **Liabilities** are the total amount of money that you owe to creditors. **Owner's equity, net worth,** or **capital** is the total value of assets that you own, after you subtract liabilities.

▸ **Owner's Equity = Assets − Liabilities**

Example

Skills 5, 6 **Application** A **Term** Owner's equity

Kevin and Sue Halbrock recently opened an auto parts store. They used $50,000 of their own money and a bank loan of $20,000. From the $70,000, Kevin and Sue paid $21,000 for merchandise and $1000 for supplies. This left a cash balance of $48,000. They received another shipment of $5000 worth of merchandise. They did not pay for this merchandise immediately. What was their owner's equity?

Solution

1. Find the **assets.**

a. Cash: $70,000 − ($21,000 + $1000)	$48,000
b. Merchandise: $21,000 + $5000	26,000
c. Supplies	1000
	$75,000

2. Find the **liabilities.**

a. Bank loan	$20,000
b. Unpaid merchandise	5000
	$25,000

3. Find the **owner's equity.**

   Assets − Liabilities
   $75,000 − $25,000 = $50,000 owner's equity

A. $43,000
B. $45,000

▸ **Self-Check** Cover the answers and complete the problems.

Find the owner's equity.

A. Assets: $125,000.
Liabilities: $82,000.

B. Assets: $18,000 cash, $49,000 merchandise.
Liabilities: $22,000.

Problems

	Assets	–	Liabilities	=	Owner's Equity
1.	$ 47,559.37	–	$ 5690.11	=	?
2.	$ 54,094.21	–	$22,569.24	=	?
3.	$ 66,494.12	–	$19,786.35	=	?
4.	$198,357.65	–	$99,664.59	=	?

5. $14,640.00 in cash.
$27,671.50 in merchandise.
$1461.90 in supplies.
$16,750.00 owed to bank.
$671.85 owed in taxes.
What are the total assets?
What are the total liabilities?
What is the owner's equity?

6. $26,800.00 in cash.
$12,530.00 in supplies.
$2163.85 in equipment.
$3250.00 owed for supplies.
$5000.00 owed to bank.
What are the total assets?
What are the total liabilities?
What is the owner's equity?

7. The Speedie-Quick Corner Grocery has these assets and liabilities.

Cash: $1261.73	Supplies: $315.79	Unpaid merchandise: $9874.70
Inventory: $14,124.85	Building: $75,000.00	Taxes owed: $715.80
Equipment: $24,716.50	Land: $15,000.00	Real estate loan: $63,400.00

What are the total assets? What are the total liabilities? What is the owner's equity?

8. Village Clothiers, Ltd., has these assets and liabilities.

Cash on hand: $2156.79	Store fixtures: $945.00	Unpaid merchandise: $1145.90
Customers owe: $4193.48	Building: $49,500.00	Taxes owed: $516.74
Inventory: $10,471.72	Land: $10,000.00	Wages owed: $145.93
Supplies: $247.93	Bank loan: $9845.50	Mortgage loan: $44,900.00

What are the total assets? What are the total liabilities? What is the owner's equity?

Maintaining Your Skills

Look up the skills in parentheses if you need help or more practice.

Add. (Skill 5)

9. 42.0 + 1.67 + 19.852

10. 2.50 + 0.67 + 118.92

11. 114.2 + 8.197

12. 0.006 + 4.1 + 9.14

Subtract. (Skill 6)

13. 92.624 – 1.87

14. 17,482.86 – 6,141.00

15. 194.77 – 18.621

16. 96.24 – 88.17

20-2 Balance Sheet

Objective Complete a balance sheet.

A **balance sheet** shows the financial position of your company on a certain date. You may prepare a balance sheet monthly, quarterly, or annually. The balance sheet shows your total assets, total liabilities, and owner's equity. The balance sheet is designed so that the assets appear on the left. The liabilities and owner's equity appear on the right. The sum of the assets must equal the sum of the liabilities and owner's equity.

Example

Skill 3 **Application** ______ **Term** Balance sheet

The Halbrocks used $50,000 of their own money plus a $20,000 loan to start their auto parts business. They received two shipments of merchandise. They paid cash for the $21,000 shipment. They did not pay immediately for the $5000 shipment. They bought supplies for $1000. The Halbrocks have $48,000 left in cash. What does their balance sheet show?

Solution

Halbrock Auto Parts
Balance Sheet
May 21, 19—

Assets			Liabilities		
Cash	$48,000		Notes payable	$20,000	
Merchandise inventory	26,000		Accounts payable	5,000	
Supplies	1,000		Total liabilities		$25,000
			Owner's Equity		
			Capital		50,000
Total assets		$75,000	Total liabilities and owner's equity		$75,000

Owed to bank

Owed for Merchandise

Shipments valued at $21,000 and $5000

ASSETS — LIABILITIES

Must be equal. Note double lines under totals.

A. $700,000
B. $700,000

▶ **Self-Check** Cover the answers and complete the problems.

Assets		Liabilities	
Cash	$180,000	Notes payable	$100,000
Inventory	$500,000	Accounts payable	$ 40,000
Supplies	$ 20,000	Owner's equity	$560,000

A. Total assets . . . . ?

B. Total liabilities and owner's equity . . . . . . ?

Problems

	1.	2.	3.	4.
Cash	$22,000	$ 82,000	$12,000	$180,000
Inventory	2000	160,000	87,100	596,000
Supplies	1500	4000	16,900	42,800
Total Assets	?	?	?	?
Bank Loan	8000	0	62,000	210,700
Taxes Owed	1400	3200	1000	8900
Total Liabilities	?	?	?	?
Owner's Equity	?	?	?	?

5. $20,000 in cash.
$32,000 in merchandise.
$22,000 owed to bank.
What are the total assets?
What are the total liabilities?
What is the owner's equity?

6. $52,200 in cash.
$72,340 in equipment.
$15,000 owed to bank.
What are the total assets?
What are the total liabilities?
What is the owner's equity?

7. Complete the balance sheet for Village Clothiers, Ltd.

Village Clothiers, Ltd.
Balance Sheet
September 30, 19—

Assets		Liabilities	
Cash on hand	2156 79	Bank loan	9845 50
Accts. receivable	4193 48	Accts. payable	1145 90
Inventory	10471 72	Taxes owed	516 74
Supplies	247 93	Wages owed	145 93
Store fixtures	945 00	Mortgage loan	44900 00
Building	49500 00		?
Land	10000 00	Owner's Equity	
		Capital	?
		Total liabilities	
Total Assets	?	and owner's equity	?

(Continue on next page.)

8. Complete the balance sheet for Suburban TV and Radio Sales.

Suburban TV and Radio Sales
Balance Sheet
March 31, 19--

Assets		Liabilities	
Cash on hand	3600 70	Bank loan	18940 10
Accts. receivable	8250 80	Accts. payable	3290 40
Inventory	22160 10	Taxes owed	560 30
Supplies	580 30	Wages owed	240 80
Store fixtures	1820 10	Mortgage loan	80190 40
Building	92840 80	Total Liabilities	?
Land	21850 00	Owner's Equity	
		Capital	?
		Total Liabilities	
Total Assets	?	and Owner's Equity	?

Prepare a balance sheet for each business.

9. Nationwide Foods, Inc., had these assets and liabilities on December 31:

Assets		Liabilities	
Cash	$ 7,700,000	Accounts payable	$96,520,000
Accounts receivable	48,310,000	Notes payable	46,360,000
Inventory	46,540,000	Income taxes	1,750,000
Property	64,320,000	Other liabilities	11,270,000
Investments	280,000		
Other assets	29,120,000		

10. Clean-Metals, Inc., shows these assets and liabilities as of June 30:

Assets (in millions of dollars)		Liabilities (in millions of dollars)	
Cash	$ 1.7	Notes payable	$13.8
Accounts receivable	10.1	Accounts payable	3.8
Inventories	32.9	Income taxes	2.6
Property	21.9	Stock	7.1
Foreign investments	3.1	Other liabilities	10.5
Other assets	7.9		

Maintaining Your Skills

Look up the skill in parentheses if you have difficulty with the problems.

Add. (Skill 3)

11. 17,459 + 1,910 **12.** 132,964 + 86,790 **13.** 524,769 + 82,347

14. 842,379 + 862,592 + 1,263 **15.** 34,126 + 329 + 81,231

20-3 Income Statement

Objective

Complete an income statement.

An **income statement,** or **profit-and-loss statement,** shows in detail your income and operating expenses. If your gross profit is greater than your total operating expenses, your income statement will show a **net income,** or **net profit.**

- **Gross Profit = Net Sales − Cost of Goods Sold**
- **Net Income = Gross Profit − Total Operating Expenses**

Example

Skills 3, 4 Application A Term Income statement

Three months after opening their store, Kevin and Sue Halbrock prepare an income statement. Sales for the first three months totaled \$50,500. Merchandise totaling \$500 was returned to them. The Halbrock's inventory records show that the goods they sold cost them \$30,000. Records show that the cost of their operating expenses totaled \$12,280. What is the net income?

Solution

HALBROCK AUTO PARTS
Income Statement
For the Quarter Ended March 31, 19—

Income:		
Sales	\$50,500	
Less: Sales returns and allowances	500	
Net sales		\$50,000
Cost of goods sold		30,000
Gross profit on sales		\$20,000
Operating expenses:		
Salaries and wages	\$8980	
Delivery expenses	1000	
Rent	600	
Advertising	250	
Utilities	483	
Supplies	107	
Depreciation	500	
Insurance	146	
Miscellaneous	214	
Total operating expenses		\$12,280
Net Income		\$7720

TOTAL SALES − RETURNS (→ \$50,000)

NET SALES − COST OF GOODS SOLD (→ \$20,000)

GROSS PROFIT − TOTAL OPERATING EXPENSES (→ \$7720)

A. $15,000
B. $167,000; $7000

▸ **Self-Check** Cover the answers and complete the problems.

A. Sales: $120,000.
Returned merchandise: $5000.
Cost of goods sold: $40,000.
Total operating expenses: $60,000.
Find the net income.

B. Sales: $250,000.
Returned merchandise: $3000.
Cost of goods sold: $80,000.
Total operating expenses: $160,000.
Find the gross profit on sales.
Find the net income.

Problems

	Total Sales	Returns	Net Sales	Cost of Goods Sold	Gross Profit	Operating Expenses	Net Income
1.	$ 2500	$ 150	$2350	$ 1645	$705	$ 205	?
2.	$ 13,765	$2775	?	$ 6476	?	$ 1465	?
3.	$ 92,147	$3918	?	$ 46,074	?	$ 8173	?
4.	$ 23,300	$1000	?	$ 10,500	?	$ 10,000	?
5.	$700,000	0	?	$400,000	?	$280,000	?

Complete the income statements.

6. Income:

Sales	$900,000	
Less: Sales returns and allowances	42,000	
Net sales		?
Cost of goods sold		$475,000
Gross profit on sales		?
Operating expenses:		
Salaries and wages	$300,000	
Taxes	10,000	
Utilities	5,000	
Depreciation	15,000	
Total operating expenses		?
Net Income		?

7. Income:

Sales	$620,000	
Less: Sales returns and allowances	210,000	
Net sales		?
Cost of goods sold		$300,000
Gross profit on sales		?
Operating expenses:		
Total operating expenses		$190,000
Net Loss		?

Prepare an income statement for each business in Exercises 8–11.

8. Last month, Speedie-Quick Corner Grocery Store had total sales of $9745. Merchandise totaling $250 was returned. The goods that were sold cost Speedie-Quick $7147. Operating expenses for the month were $573.

9. Lopez Housecleaning Referral Service prepares an income statement monthly. For this past month, they collected $2850 from homeowners for services provided (sales). They had $75 in allowances for jobs which had to be redone. Salaries and wages totaled $1710, supplies cost $279, and other operating expenses totaled $165.

10. During the past quarter, Village Clothiers, Ltd., had total sales of $14,760 and returns of $125. The cost of goods sold amounted to $4918. Operating expenses for the quarter included: salaries and wages of $1495, real-estate loan payment of $1050, advertising at $150, utilities and supplies of $175, bank loan payment of $330, and other operating expenses of $360.

11. Clean-Metals, Inc., prepares an annual income statement for distribution to its stockholders. This past year, net sales totaled $77.8 million. Cost of goods sold totaled $45.6 million. Operating expenses included: wages and salaries of $16.0 million, depreciation and amortization of $2.1 million, general taxes of $1.0 million, interest paid totaling $1.0 million, income taxes of $6.3 million, and miscellaneous operating expenses of $1.1 million.

*12. **A Brief Case** If your total operating expenses are greater than your gross profit, your income statement will show a net loss.

Net Loss = Total Operating Expenses − Gross Profit

Find the net loss for this quarter.

Gross profit		1923 80
Operating expenses:		
Wages	1675 65	
Supplies	483 00	
Miscellaneous	107 00	
Total		?
Net Loss		?

Maintaining Your Skills

Look up the skills in parentheses if you need help or more practice.

Add. (Skill 3)

13. $410.22 + 67.86 + 100.03

14. $ 12,260.00 + 8,861.76 + 176,259.18

15. $5629.86 + 4941.07 + 1000.00

16. $ 86.21 + 7.67 + 112.81

Subtract. (Skill 4)

17. $120,242 − 87,951

18. $812.40 − 744.81

19. $1200 − 1600

20. $122,840 − 366,982

20-4 Income Statement Analysis

Objective

Compute the percent that an item on the income statement is of the net sales.

Your business may analyze the dollar amounts for the items on its income statement by converting each amount to a percent of net sales.

▶ **Percent of Net Sales** = $\dfrac{\textbf{Amount for Item}}{\textbf{Net Sales}}$

Example

Skill 31 **Application** A **Term** Net sales

Kevin and Sue Halbrock analyze their income statement for the quarter. The statement shows:

Income:		
Sales	$50,500	
Less: Sales returns and allowances	500	
Net sales		$50,000
Cost of goods sold		30,000
Gross profit on sales		$20,000

What is the gross profit as a percent of net sales?

Solution

Find the **percent of net sales.**

Amount for Item ÷ Net Sales

$20,000 ÷ $50,000 = 0.4 = 40% of net sales

A. 1.2%
B. 80%

▶ **Self-Check** Cover the answers and complete the problems.

A. Sales returns: $1200.
Net sales: $100,000.
What are the sales returns as a percent of net sales?

B. Cost of goods sold: $240,000.
Net sales: $300,000.
What is the cost of goods sold as a percent of net sales?

Problems

Round answers to the nearest tenth of a percent.

Item	Amount for Item	÷	Net Sales	=	Percent of Net Sales
1. Cost of goods sold	$30,000	÷	$50,000	=	?
2. Operating expenses	$12,280	÷	$50,000	=	?
3. Net income	$ 7720	÷	$50,000	=	?

4. Speedie-Quick's income statement for one month showed these figures.

	Net sales	$9495
	Cost of goods sold	7147
a.	Gross profit on sales	?

b. What is the gross profit as a percent of net sales?

5. The income statement for one quarter for Village Clothiers, Ltd., showed these figures.

	Net sales	$14,635
	Cost of goods sold	4,918
a.	Gross profit on sales	?
	Total operating expenses	$3,560
b.	Net Income	?

c. What is the cost of goods sold as a percent of net sales?
d. What is the gross profit on sales as a percent of net sales?
e. What are the operating expenses as a percent of net sales?
f. What is the net income as a percent of net sales?

6. The income statement for Nationwide Foods, Inc., for the past year showed these figures.

	Net sales	$400,389,092
	Cost of goods sold	260,847,797
a.	Gross profit on sales	?
	Total operating expenses	$128,955,411
b.	Net Income	?

c. What is the cost of goods sold as a percent of net sales?
d. What is the gross profit on sales as a percent of net sales?
e. What are the operating expenses as a percent of net sales?
f. What is the net income as a percent of net sales?

Maintaining Your Skills

Look up the skill in parentheses if you need help or more practice.

Find the rate. Round answers to the nearest tenth of a percent. (Skill 31)

7. n% of 90 = 45
8. n% of 120 = 80
9. n% of 21 = 63
10. n% of 75 = 25
11. n% of 200 = 50
12. n% of 280 = 7

20-5 Comparing Two Income Statements

Objective Compare two income statements and compute the percent of change.

You can compare your income statements for different periods by computing **percent changes** from one income statement to another. When computing percent change, the dollar amount on the earlier statement is the **base figure.** The **amount of change** is the difference between the base figure and the figure on the current statement. If the amount for an item decreases from one income statement to the next, both the amount of change and the percent change are negative.

▸ **Percent Change** = $\frac{\text{Amount of Change}}{\text{Base Figure}}$

Example

Skill 31 Application A Term Percent change

Kevin and Sue Halbrock prepare the second income statement and compare it to the one for the previous quarter. To the nearest tenth of a percent, what is the percent change for each item?

Solution

CURRENT FIGURE − BASE FIGURE

	Last Quarter (Base)	This Quarter	Amount of Change	Percent Change
Net sales	$50,000	$64,000	$14,000	28%
Cost of goods sold	30,000	40,000	10,000	33.3
Gross profit on sales	$20,000	$24,000	$ 4000	20
Operating expenses	12,280	10,650	−1630	−13.3
Net income	$ 7720	$13,350	$ 5630	72.9

Operating expenses decreased. Change is negative.

The Halbrocks' net income increased by 72.9%.

A. 50%
B. − 10%

▸ **Self-Check** Cover the answers and complete the problems.

Find the percent change from last quarter to this quarter.

A. Last quarter: $50,000. This quarter: $75,000.

B. Last quarter: $200,000. This quarter: $180,000.

Problems

Round to the nearest tenth of a percent.

	Last Year (Base)	This Year	Amount of Change	Percent Change
1.	$565,000	$670,000	$105,000	?
2.	$135,000	$100,000	− $ 35,000	?
3.	$ 47,376	$ 58,971	?	?

4. Last year net sales were $230,000. This year net sales are $210,000. What is the amount of change? What is the percent change?

5. Last week cost of goods sold was $1000. This week cost of goods sold is $1500. What is the amount of change? What is the percent change?

6. Last year net sales were $465,000. This year net sales are $600,000. What is the amount of change? What is the percent change?

7. Last month net income was $215,649. This month net income is $186,375. What is the amount of change? What is the percent change?

8. Income statements for Speedie-Quick Corner Grocery Store showed these figures for March and April.

	March	April	Amount of Change	Percent Change
a. Net sales	$9495	$11,046	?	?
b. Cost of goods sold	7147	8314	?	?
c. Gross profit on sales	2348	2732	?	?
d. Operating expenses	573	646	?	?
e. Net income	1775	2086	?	?

9. Income statements for Nationwide Foods, Inc., showed the following figures.

	Last Year	This Year	Amount of Change	Percent Change
a. Net sales	$400,389,092	$402,777,546	?	?
b. Cost of goods sold	260,847,797	288,680,629	?	?
c. Gross profit on sales	139,541,295	114,096,917	?	?
d. Operating expenses	128,955,411	114,091,463	?	?
e. Net income	10,585,884	5,454	?	?

10. Two annual income statements for Clean-Metals, Inc., showed these figures.

	1980 (in millions)	1981 (in millions)	Amount of Change	Percent Change
a. Net sales	$77.8	$81.7	?	?
b. Cost of goods sold	45.6	45.6	?	?
c. Gross profit on sales	32.2	36.1	?	?
d. Operating expenses	27.5	25.6	?	?
e. Net income	4.7	10.5	?	?

Maintaining Your Skills

Look up the skill in parentheses if you need help or more practice.

Find the rate. Round answers to the nearest tenth of a percent. (Skill 31)

11. $n\%$ of 80 = 56

12. $n\%$ of 90 = 70

13. $n\%$ of 1800 = 1400

14. $n\%$ of $2000 = $4000

15. $n\%$ of 15,000 = 2500

Reviewing the Basics

Skills

Solve.

(Skill 3)

1. $675,442 + $936,716 + $275,481 + $63,275

2. $52,658,913 + $18,429,073 + $9,465,032 + $11,629,745

(Skill 4)

3. $18,500 − $7613
4. $19,562 − $8798
5. $29,506 − $17,992
6. $38,421 − $29,682

(Skill 5)

7. $1571.30 + $230.56 + $48.16 + $2571.56

8. $2358.42 + $16,391.77 + $615.08 + $22,598.40

(Skill 6)

9. $8473.51 − $652.09
10. $26,471.58 − $1191.79
11. $63,209.80 − $1987.65
12. $297,463.25 − $263,745.76

Round to the nearest tenth of a percent.

(Skill 31)

13. $16,000 is what percent of $82,000?
14. $7654 is what percent of $26,050?
15. n% of 80 = 24
16. n% of 42 = 178

Applications

Use the formula Gross Profit = Net Sales − Cost of Goods Sold to solve.

(Application A)

17. Net sales are $24,000.
Cost of goods sold is $15,932.
What is the gross profit?
18. Cost of goods sold is $32,544.
Net sales are $57,350.
What is the gross profit?
19. Net sales are $2,700,000.
Cost of goods sold is $2,650,000.
What is the gross profit?
20. Net sales are $5600.
Cost of goods sold is $5490.
What is the gross profit?

Terms

Match each term with its definition on the right.

21. Owner's equity
22. Balance sheet
23. Income statement
24. Percent change

a. the total amount of money that you owe to creditors
b. shows your sales, operating expenses, and net profit
c. the difference between your assets and liabilities
d. is negative if the amount decreases from one income statement to the next
e. shows your assets, liabilities, and owner's equity

Refer to your reference files at the back of the book if you need help.

Career Wise

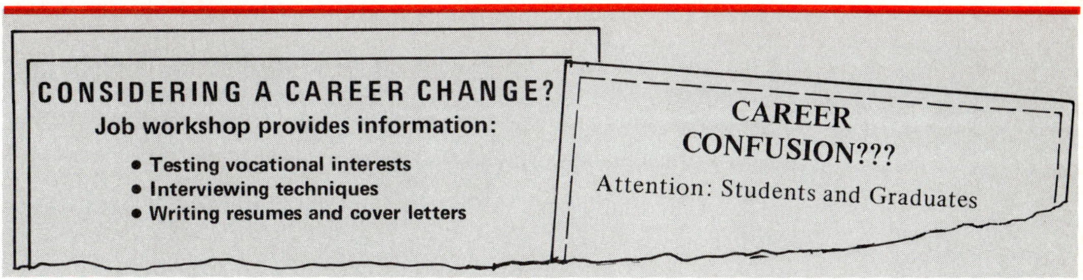

For what type of job are you best suited? Your past activities can help you decide what type of work you might like. Several tests are available to help you decide what type of work you might like. Several tests are available to help you identify your abilities and interests. Your counselor may help you find information about tests. Once you have identified your talents and interests, you can choose a job that requires characteristics like your own. Publications such as the *Occupational Outlook Handbook* and the *Dictionary of Occupational Titles* may help you identify the skills and education needed for various careers.

John Milton and Noreen Domenica took work-related activities tests at school. Their test results are listed on charts. The information at the left helps John and Noreen see how their scores compare to the scores of other students who took the test.

NAME John Milton		TEST DATE 3-24		
compared with other students	PEOPLE Score	DATA Score	THINGS Score	IDEAS Score
Upper ¼	20		17	
Middle ½		7		
Lower ¼				3

NAME Noreen Domenica		TEST DATE 3-24		
compared with other students	PEOPLE Score	DATA Score	THINGS Score	IDEAS Score
Upper ¼		18		17
Middle ½	8		10	
Lower ¼				

1. In which areas did John and Noreen receive their highest scores?

2. In each area, how do John's and Noreen's scores compare to the scores of other students?

3. Based on his scores on this test, would John be best suited to work as a computer programmer, an appliance repairperson, or a host in a restaurant?

4. Based on her scores on this test, would Noreen be best suited to work as a salesperson, a computer programmer, or a mechanic?

Unit Test

Lesson 20-1

1. The Gift Horse had these assets and liabilities on June 30.

   Cash: $2300.
   Inventory: $18,000.
   Supplies: $237.
   Store furnishings: $8455.
   Unpaid merchandise: $7500.
   Taxes owed: $432.
   Bank loan: $2500.

   What are the total assets? What are the total liabilities? What is the owner's equity?

Lesson 20-2

2. Prepare a balance sheet for The Gift Horse. Use the information in Exercise 1.

Lesson 20-3

3. Stellar Manufacturing, Inc., prepares an annual income statement. This past year, Stellar had net sales of $400,389. The cost of goods sold totaled $260,847. Operating expenses included: wages and salaries totaling $105,761; depreciation and amortization totaling $6449; interest paid totaling $8017; product recall totaling $3596; and miscellaneous operating expenses of $2463. Prepare an income statement for Stellar Manufacturing.

Lesson 20-4

4. The income statement for Reliable Appliances showed these figures.

   a.

Net sales	$38,000
Cost of goods sold	25,000
Gross profit on sales	?

   b. To the nearest tenth of a percent, what is the gross profit as a percent of net sales?

Lesson 20-5

5. Income statements for Village Clothiers, Ltd., showed the following figures for two quarters. Round to the nearest tenth of a percent.

	First Quarter	Second Quarter	Amount of Change	Percent Change
a. Net sales	$15,000	$9000	?	?
b. Cost of goods sold	5000	4300	?	?
c. Gross profit on sales	10,000	6000	?	?
d. Operating expenses	3500	3500	?	?
e. Net income	6000	1500	?	?

Unit 21 Financial

▶ Financial management of your business includes your investments, loans, and taxes. You may borrow money by taking a *commercial loan,* or business loan, from a bank. If you want to raise money, you can sell shares of your business, or stocks and bonds. Money that you don't need for day-to-day expenses can be invested to earn additional interest. Your business may invest its surplus cash in *United States Treasury Bills* or in other businesses with a high credit rating.

Today, corporations own more than 90 percent of all stocks and bonds traded on a stock exchange.

21-1 Corporate Income Taxes

Objective

Compute the taxable income and the corporate income tax.

Your business must pay federal income taxes. The tax rates vary depending on the size and type of your business. **Corporations,** businesses owned by stockholders, are subject to federal tax rates ranging from 17% to 46% of **taxable income.** Taxable income is the portion of your company's gross income that remains after normal business expenses are deducted. Normal business expenses include wages, rent, utilities charges, interest paid on loans, property taxes, depreciation, and so on. The structure for federal **corporate income taxes** is graduated.

If TAXABLE INCOME is:		TAX is:	
Over—	But not over—		Of the amount over—
0	$25,000	17%	0
$25,000	50,000	$4,250 + 20%	$25,000
50,000	75,000	9,250 + 30%	50,000
75,000	100,000	16,750 + 40%	75,000
100,000	—	26,750 + 46%	100,000

▶ **Taxable Income = Annual Gross Income − Deductions**

Example

Skill 30 Application C Term Corporate income taxes

The Atlee Corporation had a gross income of $852,261 for the year. Atlee had these business expenses during the year:

Wages	$635,820	Utilities	$17,325	Depreciation	$27,358
Rent	$38,950	Interest	$8,432	Other deductions	$14,126

What federal corporate income tax must the Atlee Corporation pay?

Solution

1. Find the **deductions.**
$635,820 + $38,950 + $17,325 +
$8432 + $27,358 + $14,126 = $742,011

2. Find the **taxable income.**
Annual Gross Income − Deductions
$852,261 − $742,011 = $110,250

3. Find the **federal corporate income tax.** (Refer to the above table.)
$26,750 + 46% of ($110,250 − $100,000)
$26,750 + 46% of $10,250 = $31,465 federal corporate income tax

A. $18,750
B. $2550

▶ **Self-Check** Cover the answers and complete the problems.

Use the tax table to find the federal corporate income tax.

A. Gross income: $200,000. Deductions: $120,000.

B. Gross income: $90,000. Deductions: $75,000.

Problems

Refer to the table on page 456 for federal corporate income taxes.

	Annual Gross Income	–	Deductions	=	Taxable Income	Total Tax
1.	$ 51,378	–	$ 37,898	=	$13,480	?
2.	$170,400	–	$127,800	=	?	?
3.	$379,440	–	$286,530	=	?	?
4.	$380,009	–	$271,435	=	?	?
5.	$400,500	–	$352,225	=	?	?

6. McCabe Food Corporation. Annual gross income is $261,885. Deductions total $194,735. What is the federal corporate income tax?

7. D and J Manufacturing, Inc. Annual gross income is $808,395. Deductions total $615,920. What is the federal corporate income tax?

8. Finlandia Corporation had these business expenses for the year:

Wages	$974,785.00	Property taxes	$196,417.80
Rent	$52,750.00	Depreciation	$34,167.00
Utilities	$12,615.00	Other deductions	$10,871.00
Interest	$5,162.50		

Finlandia Corporation had a gross income of $1,420,000 for the year.
a. What is the total of Finlandia's business expenses?
b. What is Finlandia's taxable income?
c. What is Finlandia's federal corporate income tax for the year?

9. Dr. Robin Cooper formed a corporation. The corporation had these business expenses for the year:

Wages	$28,750.00	Property taxes	$9173.00
Utilities	$14,435.00	Depreciation	$4191.70
Interest	$3654.85	Other deductions	$2816.50

Dr. Cooper's gross income for the year was $187,200. What is Dr. Cooper's federal corporate income tax for the year?

Maintaining Your Skills

Look up the skill in parentheses if you need help or more practice.

Find the percentage. Round answers to the nearest hundredth. (Skill 30)

10. 6% of 98 11. 8.56% of 60 12. $\frac{7}{8}$% of 140 13. 130% of 500

14. 25% of 3500 15. 9.7% of 200 16. $16\frac{1}{2}$% of 690 17. 2.64% of 1200

Subtract. (Skill 6)

18. 56,200 – 12.44 19. 124,344 – 345.78 20. 92.215 – 18.831

21-2 Issuing Stocks and Bonds

Objective Compute the selling expenses and the net proceeds from an issue of stocks and bonds.

Your business may raise money by issuing **stocks** or **bonds.** When you issue stocks, the buyer becomes a part owner of your business. When you issue bonds, the buyer is lending money to your business.

When you issue stocks or bonds, you must pay certain expenses. One expense is an **underwriting commission,** a commission to the investment banker who helps you distribute the stocks or bonds. Other expenses include accounting costs, legal fees, and printing costs. The amount that your business actually receives from the sale of the stocks or bonds after paying these expenses is the **net proceeds.**

▶ **Net Proceeds = Value of Issue − Total Selling Expenses**

Example

Skill 30 **Application** A **Term** Net proceeds

Glanzman Company is planning a major expansion program. To finance the program, Glanzman plans to sell an issue of 400,000 shares of stock at $38.50 per share. The underwriting commission will be 6.5% of the value of the stocks. Accounting fees, legal fees, printing costs, and other expenses are estimated to be 0.8% of the value of the stocks. If all the shares of stock are sold, what net proceeds will Glanzman Company receive?

Solution

1. Find the **value of issue.**
   $38.50 × 400,000 = $15,400,000
2. Find the **total selling expenses.**
   a. Underwriting commission
   6.5% of $15,400,000 . . . . $1,001,000
   b. Other expenses
   0.8% of $15,400,000 . . . . $ 123,200
   $1,124,200
3. Find the **net proceeds.**
   Value of Issue − Total Selling Expenses
   $15,400,000 − $1,124,200 = $14,275,800 net proceeds

A. $4,975,000
B. $19,930,000

▶ **Self-Check** Cover the answers and complete the problems.

Find the net proceeds.

A. Value of stock: $5,000,000. Total selling expenses: 0.5% of value of stocks.

B. Value of stock: $20,000,000. Underwriting commission: $30,000. Total selling expenses: 0.2% of value of stock.

Problems

	Underwriting Commission: Value of Issue	× Percent =	Dollar Amount	+ Other Expenses	= Total Selling Expenses	Net Proceeds
1.	($ 8,000,000 ×	7%	= $560,000)	+ $107,000 =	?	?
2.	($ 675,000 ×	10%	= ?)	+ $ 37,425 =	?	?
3.	($14,000,000 ×	6.5%	= ?)	+ $112,500 =	?	?

4. Value of stocks is $890,000. Underwriting commission is 11%. Other expenses total $45,390. What are the net proceeds?

5. Value of bonds is $4,730,000. Underwriting commission is 7.7%. Other expenses total $89,870. What are the net proceeds?

6. Allen Manufacturing Corporation sold 1,500,000 shares of stock at $31.90 per share. The investment banker's commission was 5.2% of the value of the stock. The other expenses were 1% of the value of the stock. What net proceeds did Allen Manufacturing Corporation receive?

7. Central Energy Company issued 10,000 shares of stock at $50 per share. Find the net proceeds after these selling expenses are deducted.

Underwriting Expenses	
Commissions	$50,000
Legal fees	$8,000
Advertising	$5,000
Miscellaneous	$4,000

Other Expenses	
Printing costs	$15,000
Legal fees	$30,000
Accounting fees	$25,000
Miscellaneous	$9,100

Use the table to solve Exercises 8 and 9.

Size of Issue (in millions)	Underwriting Commission	Other Expenses
Under $0.5	7.7%	8.2%
$0.5—$0.9	7.5%	3.7%
$1.0—$1.9	7.3%	3.6%
$2.0—$4.9	4.5%	1.4%

8. The Dutton-Graber Company sold 10,000 bonds at $45 per bond. What are the net proceeds?

9. Bonfiglio Business Machines Corporation sold 150,000 bonds for $22.70 per bond. What are the net proceeds?

*10. **A Brief Case** The Hawthorne City Council needs to raise $1.3 million dollars to continue required services for the remainder of the year. The City Council plans to sell bonds at $24.30 each to raise the money. Selling expenses are expected to be 7.4% of the value of the bonds. How many bonds must the City Council sell to obtain net proceeds of $1.3 million?

Maintaining Your Skills

Look up the skill in parentheses if you need help or more practice.

Find the percentage. Round answers to the nearest hundredth. (Skill 30)

11. 10.2% of $30,000 12. $6\frac{1}{4}$% of $8200 13. 0.72% of 9600 14. 240% of 60

21-3 Borrowing

Objective Compute the maturity value of a commercial loan.

Your business may borrow money to buy raw materials, products, or equipment. **Commercial loans,** or business loans, are similar to personal loans. The interest rates and amounts borrowed are generally much greater for commercial loans than for personal loans. The **prime rate** is the lowest rate of interest available to commercial customers at a given time.

The **maturity value** of your loan is the total amount you repay. The maturity value includes both the principal borrowed and the interest owed on the loan. Commercial loans usually charge ordinary interest at exact time. That is, the length of time of the loan is calculated by dividing the exact number of days of the loan by 360 days.

▶ **Maturity Value = Principal + Interest Owed**

Example

Skills 28, 10 **Application** A **Term** Maturity value

Armstrong Hardware borrowed $20,000 from Hilltop Commercial Bank to pay for a shipment of lawn tractors and garden supplies. The bank loaned the money at 16.75% ordinary interest for 60 days. What is the maturity value of the loan?

Solution

1. Find the **interest owed.**

   Principal × Rate × Time

   $\$20{,}000 \times 16.75\% \times \frac{60}{360}$

   $(\$20{,}000 \times 0.1675 \times 60) \div 360$

   $\$201{,}000 \div 360 = \$558.333 = \$558.33$

2. Find the **maturity value.**

   Principal + Interest Owed

   $20,000 + $558.33 = $20,558.33 maturity value

A. $103,562.50
B. $159,187.50

▶ **Self-Check** Cover the answers and complete the problems.

Find the maturity value of the loan.

A. $100,000 borrowed at 14.25% ordinary interest for 90 days.

B. $150,000 borrowed at 15.75% for 140 days.

Problems

Use ordinary interest at exact time to solve.

	Principal	×	Rate	×	Time	=	Interest	Maturity Value
1.	$20,000	×	16%	×	90/360	=	?	?
2.	$30,000	×	12%	×	100/360	=	?	?
3.	$25,500	×	16.5%	×	120/360	=	?	?

4. $12,000 is borrowed for 60 days. Interest rate is 15%. What is the interest owed? What is the maturity value?

5. $26,500 is borrowed for 180 days. Interest rate is 18%. What is the interest owed? What is the maturity value?

6. First Federal Savings loaned $90,000 to Martinez Home Builders, Inc. The term of the loan was 270 days. The interest rate was 14.75%. What was the maturity value of the loan?

7. To take advantage of a toy manufacturer's closeout special on children's games, Stoll Toy and Novelty borrowed $6500 from the Union Trust Company. Union Trust charged 17% interest on the loan. The term of the loan was 175 days. What was the maturity value of the loan?

8. Osovich Construction Company was granted a 1-year construction loan of $250,000 to finance the construction of an apartment complex. Osovich borrowed the money from Banco, Inc., at an interest rate of 15.75%. What is the maturity value of the loan?

*9. **A Brief Case** Scott Manufacturing Company needs $260,000 for 270 days to help finance the production of an experimental solar hot water heater. The financial manager arranged financing from 3 sources. Each loan charges ordinary interest at exact time.

Swancreek Trust Company
$120,000 loan for 270 days
Interest rate: 16.75%

George S. Hefner Investment Company
$90,000 loan for 270 days
Interest rate: 17.75%

Investment Bankers, Inc.
$50,000 loan for 270 days
Interest rate: 17.25% for first 90 days
17.5% for next 180 days

a. What is the total interest for the 3 loans?
b. What is the total maturity value?

Maintaining Your Skills

Look up the skill in parentheses if you need help or more practice.

Write the percents as decimals. (Skill 28)

10. 18.6% 11. 200% 12. 0.7% 13. $\frac{3}{10}$% 14. 0.06%

21-4 Investments

Objective Compute the cost of a Treasury Bill.

Your business may invest surplus cash that is not needed for day-to-day operations. One way to invest your money is to purchase **United States Treasury Bills.** You may purchase the bills through a bank. Some banks charge a service fee for the paperwork involved in obtaining the bills.

When you purchase a Treasury Bill, you are actually lending money to the government. In return, you receive interest at the rate that is in effect at the time you purchase the bill. The interest is ordinary interest at exact time. Treasury Bills are issued on a **discount basis.** That is, the interest is computed and then subtracted from the **face value** of the bill to determine the cost of the bill. The face value of the Treasury Bill is the amount of money you will receive on the **maturity date** of the bill. Maturity dates for Treasury Bills range from 91 days to a year.

▶ **Cost of Treasury Bill = (Face Value of Bill − Interest) + Service Fee**

Example

Skills 8, 10 **Application** A **Term** U.S. Treasury Bill

The financial manager of Seaboard Manufacturing Company has decided to invest the company's surplus cash in a $30,000 United States Treasury Bill for 120 days. The interest rate is 12.5%. The bank charges a service fee of $25 to obtain the Treasury Bill. What is the cost of the Treasury Bill?

Solution

1. Find the **interest.**

Principal × Rate × Time

$30,000 × 12.5% × $\frac{120}{360}$

($30,000 × 0.125 × 120) ÷ 360

$450,000 ÷ 360 = $1250

2. Find the **cost of Treasury Bill.**

(Face Value of Bill − Interest) + Service Fee

($30,000 − $1250) + $25

$28,750 + $25 = $28,775 cost of Treasury Bill

$96,763.89

▶ **Self-Check** Cover the answers and complete the problems.

Find the cost of the Treasury Bill.

$100,000 Treasury Bill purchased at 13% interest for 91 days. Service fee is $50.

Problems

Use ordinary interest at exact time to solve.

	Face Value of Treasury Bill	Interest Rate	Time in Days	Interest	Bank Service Fee	Cost of Treasury Bill
1.	$ 40,000	13%	100	?	$30	?
2.	$ 50,000	12%	120	?	$25	?
3.	$ 70,000	11.5%	180	?	$34	?
4.	$110,000	12.25%	91	?	No fee	?

5. Face value of bill is $30,000.
Interest rate is 11%.
Bill matures in 120 days.
Bank service fee is $30.
What is the interest?
What is the cost of the bill?

6. Face value of bill is $80,000.
Interest rate is 12.5%.
Bill matures in 100 days.
Bank service fee is $25.
What is the interest?
What is the cost of the bill?

7. Williams Family Electric Company purchased a $60,000 Treasury Bill at 10% ordinary interest for 120 days. The bank charged a service fee of $20. What is the interest? What is the cost of the Treasury Bill?

8. Zarik Investment Company purchased a $40,000 United States Treasury Bill at 11.75% interest. The Treasury Bill matures in 91 days. The bank service fee is $35. What is the cost of the Treasury Bill?

9. Marker Printing Company had $20,000 in surplus cash. Sue Marker decided to invest in United States Treasury Bills at an interest rate of 10.45%. She purchased a 110-day bill with a face value of $20,000. The bank charged a $15 service fee. What was the cost of the Treasury Bill?

10. Samuel Hobinski sold his apartment house for $240,000. After paying various expenses, he had $210,000 to invest. He purchased a $100,000 United States Treasury Bill for 120 days at an interest rate of 11.25%. He purchased a second Treasury Bill with a face value of $110,000. The second bill was for 91 days at an interest rate of 12%. Samuel's bank does not charge a service fee on Treasury Bills over $100,000. What is the total cost of the 2 bills?

Maintaining Your Skills

Look up the skills in parentheses if you need help or more practice.

Multiply. (Skill 8)

11. 15.22 × 9 12. 5.125 × 6.7 13. 0.61 × 0.32 14. 15.86 × 0.108

Divide. (Skill 10)

15. 86 ÷ 8 16. 6.29 ÷ 1000 17. 706 ÷ 50 18. 500 ÷ 24

21-5 Investments—Net Interest

Objective

Compute the net interest from commercial paper.

Your business may invest extra cash in **commercial paper.** Commercial paper is a 30-day to 270-day **promissory note** issued by a company with a very high credit rating. When your business invests in commercial paper you are actually lending money to another company. Commercial paper earns ordinary interest at exact time. Usually your business obtains commercial paper through a bank. The bank may charge a service fee. Your **net interest** is the total interest that you earn after any bank service fees have been deducted.

▶ **Net Interest = Total Interest Earned – Service Fee**

Example

Skills 8, 10 **Application** A **Term** Commercial paper

Boat Works has a $90,000 cash surplus. The financial manager used the cash to make the following investments in commercial paper:

G.L.A.C. Corporation	$20,000 at 11.9% for 30 days
A.B.C. Financial Corporation	$40,000 at 11.75% for 60 days
R&R Telephone Company	$30,000 at 11.625% for 120 days

The bank charges a service fee of $35. What net interest will Boat Works earn?

Solution

1. Find the **total interest earned.**
   a. G.L.A.C. Corporation
   $\$20,000 \times 11.9\% \times \frac{30}{360} = \$\ 198.33$
   b. A.B.C. Financial Corporation
   $\$40,000 \times 11.75\% \times \frac{60}{360} = \$\ 783.33$
   c. R&R Telephone Company
   $\$30,000 \times 11.625\% \times \frac{120}{360} = \underline{\$1162.50}$
   $2144.16

2. Find the **net interest.**
   Total Interest Earned – Service Fee
   $2144.16 – $35.00 = $2109.16 net interest

$1208.33

▶ **Self-Check** Cover the answer and complete the problem.

$100,000 invested at 14.5% interest for 30 days. No service fee. Find the net interest.

Problems

Use ordinary interest at exact time to solve.

	Borrower	Value of Note	Interest Rate	Time in Days	Interest	Bank Service Fee	Net Interest
1.	J and J Corp.	$ 40,000	11.5%	30	?	$20	?
2.	Worldwide Sales	$100,000	11.75%	90	?	$18	?
3.	Hudson Aircraft	$400,000	11.375%	180	?	No fee	?

4. $90,000 note for 60 days.
Purchased from G.M.D. Corporation.
Interest rate is 11.5%.
Bank service fee is $20.
What is the net interest?

5. $80,000 note for 40 days.
Purchased from Highland Railroad.
Interest rate is 11.75%.
Bank service fee is $25.
What is the net interest?

6. Wilksborough Tomato Canning Factory has a $40,000 cash surplus. The financial manager made the following investments in commercial paper:

Rural Farm Electric	$25,000 at 11.5% for 270 days
Farmers' Grain Supply	$15,000 at 11.625% for 180 days

The bank charges a service fee of $15. What net interest will Wilksborough Tomato Canning Factory receive?

7. Carla Winslow, the financial manager of The Winslow Sporting Goods Factory, made the following investments in commercial paper:

Western Copper Mines, Ltd.	$60,000 at 10.75% for 270 days
Georgia Lumber Mills Corp.	$80,000 at 11% for 180 days
Transview, Inc.	$40,000 at 11.5% for 30 days

The bank charges a service fee of $8.00 per promissory note. What net interest will The Winslow Sporting Goods receive?

*8. **A Brief Case** On June 1, Clinton Land Development Company received $240,000 for the sale of an industrial park. Joseph DeAngelo, the financial manager for the company, invested the money in the following commercial paper:

Acme Corporation	$60,000 at 11.95% until June 16
Rural Telephone	$80,000 at 11.5% until September 25
Downtown Electric	$50,000 at 11.25% until October 20
Automotive Wholesalers	$50,000 at 11.45% until November 4

The bank service fee is $25 for each note under $51,000 and $15 for each note over $51,000. What net interest will Clinton Land Development Company receive from these investments?

Maintaining Your Skills

Look up the skill in parentheses if you need help or more practice.

Multiply. (Skill 8)

9. $41.16 \times 10{,}000$ 10. 0.929×0.52 11. 7140×0.003 12. 6.143×0.085

21-6 Growth Expenses

Objective Compute the total cost of expanding a business.

You may expand your business in several different ways. You may purchase or build an addition to your present building. You may purchase another business to become part of your business. Your business may **merge,** or combine, with another business to form a new business. **Growth expenses** for your business may include construction fees, consultation fees, legal fees, and so on.

▶ **Total Cost of Expansion = Sum of Individual Costs**

Example

Skill 5 **Application** A **Term** Growth expenses

The owner of Miller's Party Shop plans to expand the business by opening a new store. Growth expenses include:

Marketing survey $2500 plus $3.75 per person for 150 people interviewed
Land $60,000
Building construction $95,000
Architect's fee 5% of the cost of construction
Surfacing parking lot $0.80 per square foot for 6000 square feet
Legal fees $740
Equipment and fixtures $22,000
Additional stock $13,000
Miscellaneous expenses $4000

What is the total cost for the expansion of Miller's Party Shop?

Solution

Find the **sum of individual costs.**

a. Marketing survey: $2500 + ($3.75 × 150)	$ 3062.50
b. Land	60,000.00
c. Building construction	95,000.00
d. Architect's fee: 5% of $95,000	4750.00
e. Surfacing parking lot: $0.80 × 6000	4800.00
f. Legal fees	740.00
g. Equipment and fixtures	22,000.00
h. Additional stock	13,000.00
i. Miscellaneous expenses	4000.00
Total Cost of Expansion	$207,352.50

A. $31,700
B. $77,860

▶ **Self-Check** Cover the answers and complete the problems.

Find the total cost of expansion.

A. Expansion costs for Gordon Jewelry store: legal fees, $1000; display cases, $18,000; safe, $9500; miscellaneous, $3200.

B. Expansion for Connie's Fashions: new construction, $27,000; display racks, $8600; clothing stock, $42,000; advertisements, $260.

Problems

1. Expand newspaper route.
   Purchase new route for $45.
   Purchase new bicycle for $87.50.
   What is the total cost of expansion?

2. Expand lawn-care business.
   Purchase new truck for $8200.
   Purchase supplies for $737.
   What is the total cost of expansion?

3. Hillside Farms is opening a new cheese outlet in the Northside Mall. Hillside paid rent in advance for 3 months. The rental charge is $964 per month. Hillside made a 20% down payment on refrigeration equipment that costs a total of $11,360. Hillside also purchased additional supplies for $4200. What is the total cost of expansion?

4. Shop Mart is adding a new department that will specialize in fishing equipment. Shop Mart pays $675 for redecoration of an area of the store. Shop Mart also makes a 45% down payment on new stock that costs a total of $14,600. What is the total of these growth expenses?

5. Cahill's Shoes is expanding by adding a new department specializing in children's shoes. Cahill is converting a storage area of 1800 square feet to a selling area. The costs of the expansion are:

   Construction permit $80
   Removal of two walls $945
   New lighting fixtures and installation $2420
   Carpet and installation $14.50 per square yard
   Down payment on merchandise 25% of $3617
   Equipment and fixtures $2740

   What is the total cost for Cahill's Shoes to expand?

6. Cope Financial Services plans to open a new branch office. Cope purchased property for $22,500. Construction costs for a new building totaled $87,500. In addition, Cope paid an architect's fee of 3.1% of the cost of construction. Legal fees for the expansion totaled $3570. New equipment and fixtures cost $18,430. Other expenses came to $3425. What is the total cost of Cope's expansion?

7. Norton Enterprises is searching for a new business to buy. Search, Inc., a company that specializes in locating firms for sale, has located a small stamping plant. If Norton Enterprises purchases the firm, it must pay Search a finder's fee of 3.25% of the total worth of the stamping plant. To acquire the plant, Norton must pay the plant its total worth of $1.75 million. Norton must pay legal fees amounting to 0.2% of the total worth of the stamping plant. In addition, Norton must pay the debts of the stamping plant. The debts amount to $76,846.25 plus 18% interest for 1 year. What is the total cost of the expansion?

Maintaining Your Skills

Look up the skill in parentheses if you need help or more practice.

Add. (Skill 5)

8. 92.05 + 156.714

9. 486.2 + 16.004

10. 986.149 + 107.63

11. 2.091 + 0.717

12. 3513.93 + 2657.49

Reviewing the Basics

Skills

Solve.

(Skill 5)

1. $5384.20 + $67,920.57 + $671.09

2. $95,000.00 + $8432.65 + $30,198.40 + $16,840.00

(Skill 8)

3. $25,000 × 0.135 × 90

4. $40,000 × 0.119 × 120

5. $14,000 × 0.12 × 120

6. $25,000 × 16.2 × 120

Round to the nearest cent.

(Skill 10)

7. $350,000 ÷ 360

8. $680,000 ÷ 360

9. $960,000 ÷ 360

Write as a decimal.

(Skill 28)

10. 12.3%

11. $7\frac{1}{4}$%

12. 10.75%

13. 3400%

Solve.

(Skill 30)

14. 17% of $23,500

15. 46% of $12,430

16. 0.8% of 70

17. 130% of 120

18. $3\frac{1}{2}$% of 4000

19. $\frac{1}{4}$% of 50

Application

Find the tax bracket.

(Application C)

20. Taxable income is $38,400.

21. Taxable income is $87,530.

If TAXABLE INCOME is:		Your TAX is:	
Over—	But not over—		Of the amount over—
0	$25,000	17%	0
$25,000	50,000	$4,250 + 20%	$25,000
50,000	75,000	9,250 + 30%	50,000
75,000	100,000	16,750 + 40%	75,000
100,000	—	26,750 + 46%	100,000

Terms

Write your own definition for each term.

22. Corporate income taxes

23. Net proceeds

24. Maturity value

25. U.S. Treasury Bill

26. Commercial paper

27. Growth expenses

Refer to your reference files at the back of the book if you need help.

Consumer Note

An alternative to investing in financial assets, such as certificates of deposit and savings bonds, is a wide category called **tangible assets.** Tangible assets that may bring in profits include stamps, coins, rare books, antiques, and paintings. They cannot be converted to cash as readily as financial assets, so they should be considered long-term investments.

Stamps and coins have been a traditional hedge against inflation. To assemble a valuable collection, specialize and buy high-quality items. The value of a coin depends on rarity, condition, demand, date, and authenticity. If you want to start a stamp collection, specialize in one type, such as airmail stamps or stamps from one country. You can learn about collecting from collectors, dealers, or books and magazines. Join a club to meet people who share your interests.

Limited editions and nostalgia items should be viewed with caution. Limited editions such as medallions, plates, and spoons are marketed extensively. Nostalgia items include old postcards, old toys, and political campaign buttons. A safe rule is to buy only items that you will enjoy owning even if their values do not increase.

Investing in tangible items can start in your grandparents' attic or at a flea market or auction. Remember that the value of an item depends on its quality, condition, rarity, and historical importance. The best rules to follow are to specialize in one type of item, to avoid fads, and to buy only what you like.

Complete the following exercises concerning tangible assets.

1. Some collectors pay experts for advice on purchases. Suppose an expert charges 8% of the total price. How much would a collector pay the expert on the purchase of a coin for $150?

2. If a book was purchased for $25 and is sold for $400, what is the percent of profit?

Unit Test

Lesson 21-1

1. Harbor View Corporation had an annual gross income of $748,995 for the year. Business deductions totaled $652,847. What federal corporate income tax must Harbor View pay for the year?

If TAXABLE INCOME is:		TAX is:	
Over—	But not over—		Of the amount over—
0	$25,000	17%	0
$25,000	50,000	$4,250 + 20%	$25,000
50,000	75,000	9,250 + 30%	50,000
75,000	100,000	16,750 + 40%	75,000
100,000	—	26,750 + 46%	100,000

Lesson 21-2

2. Central Energy, Inc., is issuing 600,000 shares of stock. Each share will be sold at $37.80. The underwriting commission is 3% of the value of the stock. The other expenses are expected to be 0.6% of the value of the stock. If all the shares of stock are sold, what net proceeds will Central Energy receive?

Lesson 21-3

3. Johnson Fruit Farms borrowed $23,500 for 150 days at the prime rate of interest. When Johnson borrowed the money, the prime rate was 17.5%. The bank charged ordinary interest at exact time. What was the maturity value of the loan?

Lesson 21-4

4. Carlton Manufacturing had $50,000 in surplus cash. The financial manager decided to invest in a United States Treasury Bill with a face value of $50,000. The bill matured in 126 days. The interest rate was 11.9% ordinary interest at exact time. The bank service fee was $20. What was the cost of the Treasury Bill?

Lesson 21-5

5. South Hickory Furniture Company sold its entire stock of kitchen cupboards for installation in an apartment complex. South Hickory invested the $90,000 income from the sale in the following commercial paper:

National Telephone, Inc.	$60,000 for 30 days at 12%
International Steel, Inc.	$30,000 for 60 days at 11.75%

Both investments earn ordinary interest at exact time. The bank service fee is $30. What net interest will South Hickory Furniture Company receive?

Lesson 21-6

6. Henderson Clothing opened a new branch store. Henderson is renting the store for $1100 per month. Henderson paid the rent in advance for 2 months. Henderson purchased new furniture and store fixtures for $21,500. Henderson also purchased additional merchandise for $18,470. Miscellaneous expenses totaled $3260. What was the total cost of expansion?

Unit 22 Information Management

▸ Your Data Processing Department may use computers for *information management,* the organizing and analyzing of your business data. Information management includes four basic activities: *posting,* or the recording of data; *classifying* data into areas, or *files; storing* files; and *processing* all business transactions on a daily basis. Processing data on a computer is also called *real-time processing.* Computers can analyze data faster than a person can manually.

Most businesses today use computers to process or store data.

The Production Department of The Piano Works receives daily costs and inventory reports from the Data Processing Department. The information in the reports below and on page 473 will be referred to and used to answer the problems throughout this unit.

TO: PRODUCTION SUPERVISOR Date: January 13
FROM: DATA PROCESSING DEPARTMENT

PRODUCTION COSTS: FOR DAY OF January 9

Date	Project	Dept.	Part #	Amt/Used
01/09/86	188N2	11	2683	3
01/09/86	188N2	12	3102	2
01/09/86	182R3	11	2693	2
01/09/86	188N2	12	2070	2
01/09/86	188N2	10	201	2
01/09/86	185N4	12	3155	3
01/09/86	185N4	12	848	2
01/09/86	182R3	11	2683	12
01/09/86	172R9	19	128	1
01/09/86	188N2	12	6023	1
01/09/86	188N2	12	2080	3
01/09/86	188N2	12	2315	2
01/09/86	188N2	12	2310	1
01/09/86	185N4	12	2080	3
01/09/86	185N4	12	2310	1
01/09/86	185N4	12	2315	2
01/09/86	188N2	11	845	4
01/09/86	185N4	14	2420	1
01/09/86	185N4	14	2418	2
01/09/86	183N5	13	5816	21
01/09/86	183N5	13	5817	22
01/09/86	183N5	13	5818	11
01/09/86	183N5	13	5819	11
01/09/86	183N5	13	5907	11
01/09/86	183N5	13	2070	1
01/09/86	182R3	11	1373	1
01/09/86	189R1	15	2060	1
01/09/86	189R1	15	2010	2
01/09/86	188N2	12	8641	1
01/09/86	183N5	13	311	8
01/09/86	182R3	11	527	8
01/09/86	185N4	12	843	8
01/09/86	188N2	12	413	8
01/09/86	182R3	12	418	8
01/09/86	183N5	15	501	8
01/09/86	183N5	15	2503	6
01/09/86	189R1	12	412	8
01/09/86	189R1	12	845	4
01/09/86	189R1	12	848	6
01/09/86	185N4	14	310	8
01/09/86	189R1	15	501	2

TO: PRODUCTION SUPERVISOR   Date: January 13
FROM: DATA PROCESSING DEPARTMENT

STATUS OF INVENTORY: WEEK ENDING January 9

Part#	Description	Stock	Cost	Retail	Restck	Order/On
1003	STOP RAIL	2	26.00	35.00	09/08/84	1
1004	KEYFRAME	1	83.00	89.00	08/12/84	1
1006	BAL. RAIL	3	24.00	30.00	11/18/85	2
1007	KEYBED	2	200.00	289.00	03/14/84	0
1011	HOPPER	3	8.90	9.99	04/16/83	1
1018	DEHUMID	2	18.42	23.50	06/23/84	1
1373	POLISH	12	2.90	3.50	08/12/84	6
1374	POLISH	10	22.60	28.00	08/12/84	6
1423	FLY	2	1.60	2.40	09/11/84	1
1428	PIVOT	0	7.60	8.65	10/08/85	12
1701	PEDALPAD	2	7.00	8.50	11/18/84	1
1850	TUNFORK	5	7.90	9.95	09/11/84	2
2009	BOLTS	20	4.38	4.98	02/18/84	15
2010	TACKS	40	1.27	1.98	09/12/84	50
2011	SCREWS	12	2.40	2.98	02/18/84	6
2012	SCREWS	2	2.58	3.90	02/18/84	1
2050	TOGGLES	33	1.20	1.49	09/11/84	20
2060	BLOCKS	3	78.00	82.00	06/23/84	2
2070	SHIMS	23	1.00	1.39	07/18/84	20
2080	BUTTONS	18	0.60	0.85	07/18/84	20
2310	BUSHINGS	23	2.50	3.20	07/20/84	15
2315	AGRAFFES	60	0.90	1.20	06/23/84	20
2381	PINS	22	1.70	2.00	02/18/84	20
2418	WIRE	31	3.00	3.75	08/12/84	15
2419	WIRE	23	5.00	6.00	08/12/84	15
2420	WIRESET	6	17.40	19.10	11/18/84	2
2501	STRINGS	36	1.80	2.98	11/18/84	25
2502	STRINGS	40	1.84	2.98	11/18/84	25
2503	STRINGS	30	1.81	2.98	11/18/84	25
2504	STRINGS	20	5.14	7.98	11/18/84	10
2683	HAMMER	56	4.60	5.49	10/09/85	20
2693	VARNISH	18	12.24	13.59	01/04/85	15
3101	GROMMETS	18	1.84	2.10	02/18/84	20
3102	CLIPS	17	1.40	1.65	06/23/84	20
3153	FLANGES	12	2.50	3.25	02/18/84	12
3155	PLATES	18	2.00	3.75	02/18/84	12
5816	WHITE KEY	75	0.59	0.89	04/18/85	60
5817	BLACK KEY	60	0.69	0.89	04/18/85	50
5818	KEYTOPS	72	0.29	0.69	08/23/84	60
5819	KEYTOPS	78	0.29	0.65	04/18/84	45
5891	DAMP HEAD	21	2.19	2.79	12/20/84	20
5893	DAMP FELT	39	0.43	0.98	12/20/84	20
5902	SHARPS	12	16.50	22.00	04/18/84	10
5907	WEDGES	80	1.60	2.25	06/23/84	30
6011	PINS	73	1.20	1.50	02/18/84	20
6023	LEADS	12	3.00	4.20	01/28/84	12
6274	SPRINGS	16	2.50	3.29	02/18/84	20
8641	PEDAL L	8	8.18	10.50	10/18/84	5
8642	PEDAL R	8	8.18	10.50	10/18/84	5
8643	PEDAL C	10	8.18	10.50	10/18/84	5
8720	WIPPENS	6	2.20	3.19	09/18/84	2
8731	SHANKS	18	0.90	1.23	10/28/83	12

22-1 Posting

Objective Use data field keys to update and create new records.

Data, pieces of information related to the activities of your business, must be gathered and recorded. You can then analyze the data to plan and control the business. Data may be the names of your clients, sales figures, production costs, inventory, and so on.

Posting, the recording of data, may be performed manually on ledger books or electronically using computers. Whatever the medium, you must first organize the data into **records.** Each line in the reports on pages 472 and 473 represents one record. A record may contain many **fields,** each containing a fact about that one aspect of the business. The data fields at the right deal with production costs.

DATA FIELD KEY for Production Costs	
Field	**Description**
1	Date of Work
2	Project Number
3	Department Number
4	Part/Employee Number
5	Amount Used/Hours Worked

Example

Skills _____ **Application** C **Term** Posting

Jill Donnelly, a production assistant at The Piano Works, received this notice to update the list of production costs for January 10. She uses the key above to organize the data into fields to create the new records. What will each record look like after she has entered the data into the computer?

PRODUCTION COSTS UPDATE

Dept. # 12 Date 1/10/86

Project # 189R1

Part #	PART	Used
2501	Strings	2
2381	Pins	4

Solution

1. Use the data field key above to organize each record.

DATE	PROJECT	DEPT.	PART	AMOUNT USED

2. Organize the data in the production costs update using the key.

```
01/10/86   189R1   12   2501   2
01/10/86   189R1   12   2381   4
```

2/01/86 223K
12 2009 8

▶ **Self-Check** Cover the answers and complete the problems.

Organize the data using the above data field key.

Dept. #12 Date: 2/1/86 Project #223K Part #2009 Part: bolts Used 8

Problems

Use the reports on pages 472 and 473 to answer the following.

	Part Number	Part Description	Number in Stock	Unit Cost	Retail Price	Date Part was Last Restocked
1.	2315	Agraffes	60	?	?	June 23, 1984
2.	5902	?	?	$16.50	?	?
3.	2060	?	?	?	$82.00	?
4.	3153	?	?	?	?	?

5. Melvin Stockwell is a production assistant at The Piano Works. He received this notice to update the list of production costs for January 12. What will each record look like after he has entered the data into the computer?

PRODUCTION COSTS UPDATE

Dept. # 15 Date 1/12/86

Project # 172R9

Part #	PART	Used
2420	Wireset	1
2010	Tacks	3
6023	Leads	1
8642	Rt Pedal	1
8643	Center Ped	1

Approved: BZK

6. Frankie Tolljian wants to know how many man-hours were spent on project 189R1 on January 9. The fourth data field of each production costs record is either a 3- or a 4-digit number. Employees at The Piano Works are identified by 3-digit numbers. Which records of production costs on January 9 represent man-hours worked on project 189R1?

7. Hollis Landau is a project supervisor at The Piano Works. He keeps a record of all production costs for his project. His project number is 188N2. List those records from the production costs report that are related to Hollis's project on January 9.

8. Nolan Lidell wants to know what progress has been made on project 183N5 on January 9. He uses the production costs report to locate the part numbers of the parts used that day on the project. Which records from the production costs report involve parts used on project 183N5?

9. The inventory report lists the unit cost of each part in the fourth data field. What is the unit cost of each part used on January 9 on project 183N5?

Maintaining Your Skills

Look up the skill in parentheses if you need help or more practice.

Put the numbers in order from smallest to largest. (Skill 1)

2768 2670 2671 2714 2709 2693 2683 2783

22-2 Classification

Objective Classify information into files.

After data are gathered, they must be sorted and classified into **files.** A file groups related material. A file may contain the names of all employees in one department, all transactions for one day, the current inventory for one department, or other related information. The production costs report on page 472 is contained in one file. The collection of files describing activities of your business is called a **data base.**

Example

Skills 1, 4 **Application** C **Term** Files

Mark Roberts must enter a new part listing into the inventory file. Between which 2 records will the new part listing be found?

NOTIFICATION TO UPDATE INVENTORY
THE PIANO WORKS, INC.

Inventory Update [ ]
New Part [✓]
Part Obsolete [ ]

Part #	Description	Stock	Cost	Retail	Restocked	Order/On
2695	Varnish C	15	$17.45	$20.95	1/10/86	5

Solution

The inventory file on page 473 is arranged by part number in ascending order.

1. Find the part number that is the next smaller than the new part number. **2693**
2. Find the part number that is the next larger than the new part number. **3101**
3. Write the data from the report as a record using the data field structure shown on page 473.

```
2695  VARNISH C  15  17.45  20.95  01/10/86   5
```

4. Enter the new part listing between the 2 records.

```
2693  VARNISH    18  12.24  13.59  01/04/85  15
2695  VARNISH C  15  17.45  20.95  01/10/86   5
3101  GROMMETS   18   1.84   2.10  02/18/84  20
```

▶ **Self-Check** Cover the answers and complete the problem.

Enter the new part listing between 2 records. Refer to page 473.

```
3109  rings    15  1.75  1.95  05/15/86   5
```

Answer:

```
3102  clips    17  1.40  1.65  06/23/84  20
3109  rings    15  1.75  1.95  05/15/86   5
3153  flanges  12  2.50  3.25  02/18/84  12
```

Problems

Use the reports on pages 472 and 473 to answer the following.

1. Felix Byronoski.
   Received this data.
   Part number 6011.
   Unit cost is now $1.35.
   Which file or files are affected?

2. Martha La Cour.
   Received this data.
   Project number 189R1.
   Used part number 2060.
   Which file or files are affected?

3. Larry Miller.
   Received this data.
   Project number 185N4.
   Employee number 848.
   Worked 2 hours.
   Which file or files are affected?

4. Olivia Young.
   Received this data.
   Part number 1428 obsolete.
   Effective January 15.
   Approved by Jerry Wise.
   Which file or files are affected?

5. Chris Vidallo received this notice to update the inventory file. What will the record for the new part listing look like after Chris enters the data into the computer? Between which 2 records will the new part listing be located in the inventory file?

NOTIFICATION TO UPDATE INVENTORY THE PIANO WORKS, INC.				Inventory Update [ ] New Part [✓] Part Obsolete [ ]		
Part #	Description	Stock	Cost	Retail	Restocked	Order/On
2008	T-Bolt	11	$3.18	$3.75	1/15/86	2

6. Arlene Hoi works in the Finishing Department. She has ordered 2 cans of varnish, part number 2693, for a repair project. Which file contains the information on the number of cans in stock? Which record is affected by this order?

7. Marianne Fegley has ordered 3 bolts needed to repair a piano. The part number of the bolt is 2009. Which file contains the information on the number of bolts in stock? Which record is affected?

8. Mark Roberts received 2 reports showing that projects 188N2 and 185N4 have ordered 12 and 9 agraffes, respectively. Are there enough parts in stock to fill this order? What will the record look like after the computer updates the inventory file?

Maintaining Your Skills

Look up the skill in parentheses if you need help or more practice.

Subtract. (Skill 6)

9. $8.65 − $7.60
10. $9.95 − $7.90
11. $2.98 − $1.87
12. $5.49 − $4.60
13. $35.00 − $26.50
14. $19.10 − $17.40
15. $365.42 − $216.78

22-3 Storage

Objective Use information that has been stored in a data base.

Files may be stored **on-line** in the computer's internal memory for immediate access. Files that contain information for future data processing may be stored **off-line** on magnetic disks or magnetic tapes. In off-line storage, a data processing operator mounts the storage device onto the computer system to access the information.

Example

Skills ____ **Application** C **Term** On-line or off-line storage

Derek Morris mounts the magnetic disk that contains the payroll file for The Piano Works onto the computer system. He loads the file into the computer's memory to update the employee records with the number of hours worked on January 9. After the data have been processed, how many hours will the record show that Elton Wagner has worked to date?

EMPLOYEE NUMBER	NAME	HOURS THRU 1/8
845	Mary Wright	40
311	Tomas Villa	40
413	Clara Miller	42
527	Elton Wagner	40
843	Roger Tory	38

Solution

1. Find the employee number. 527
2. Find the number of hours worked through January 8. 40
3. Find the number of hours worked on January 9. 8

Total hours worked to date 48

A. 40 + 8 = 48
B. 40 + 8 = 48

▶ **Self-Check** Cover the answers and complete the problems.

Update the files for employee numbers 311 and 845.

A. 311: Hours through 1/8 = ?
Hours worked on 1/9 = ?
Total hours worked = ?

B. 845: Hours through 1/8 = ?
Hours worked on 1/9 = ?
Total hours worked = ?

Problems

Use the reports on pages 472 and 473 to answer the following.

1. Clara Miller.
42 hours through January 8.
How many hours on January 9?
How many hours in all?

2. Roger Tory.
38 hours through January 8.
How many hours on January 9?
How many hours in all?

3. Tomas Villa.
How many hours through January 8?
How many hours on January 9?
How many hours in all?

4. Mary Wright.
How many hours through January 8?
How many hours on January 9?
How many hours in all?

5. Nolan Lidell wants to know what progress has been made on project 183N5. By determining which parts were used, he can estimate what progress has been made. List the records of production costs involving project 183N5 on January 9. What are the names of the parts used?

6. Patty Wake is the supervisor on project 182R3. The project was completed on January 9. She wants to calculate the total cost of the parts used to repair the piano. The Data Processing Department retrieved the following information. What is the total parts cost for the 3 days of work?

PRODUCTION COSTS FOR PROJECT **** 182R3 ****

DATE	PROJECT	DEPT.	PART #	PARTS/USED
01/07/86	182R3	11	5816	15
01/07/86	182R3	11	527	6
01/07/86	182R3	12	418	8
01/07/86	182R3	11	5891	7
01/08/86	182R3	11	5817	3
01/08/86	182R3	12	418	8
01/08/86	182R3	11	527	4
01/08/86	182R3	11	5893	4

7. Chester Burke is in charge of maintaining the inventory. He regularly checks the inventory file to determine which parts are in short supply. He places orders for those parts which are on or below the order-on quantity. He enters a command to the computer to place a check mark beside those part numbers which are on or below the order-on quantity. Which part numbers will the computer place a check mark by?

8. The personnel file at The Piano Works is stored on magnetic tape. Harriet Porter, personnel manager, asked Gino Lascios to find the names of employees who have worked 10 or more years for the company. Gino sent the report shown to Harriet.

NAME	EMPL #	DEPT. #	SERVICE (YEARS)
CARL APPLE	724	50	13
ARLENE HOI	848	12	17
NOLAN LIDELL	412	20	33
HARRIET PORTER	571	36	26
OLIVIA YOUNG	418	12	24

How many employees have worked for the company for 10 or more years? What is the average number of years served by these employees?

Maintaining Your Skills

Look up the skill in parentheses if you need help or more practice.

Add. (Skill 3)

9. 112 + 9 + 35
10. 13 + 892 + 24
11. 51 + 28 + 116

22-4 Real-time Processing

Objective Update a data base where real-time processing is involved.

Planning and controlling a business requires current information. When your computer processes data and immediately updates the data base to keep up with the day's events, the computer is used in **real-time processing.** Applications of real-time processing include computer-based patient files in hospitals, business inventories, fast-food checkouts, and electronic funds transfer systems offered by banks.

Example

Skills 3, 4 **Application** C **Term** Real-time processing

Ellen Curtain is a sales clerk in the Retail Department. To process a customer's purchase, she enters the part number and the quantity purchased. The computer produces the sales slip and updates the inventory file. What will each record for the parts purchased look like after the computer updates the inventory file?

```
            THE PIANO WORKS

QUAN  ITEM     STOCK #   PRICE   TOTAL
 1    WIRE      2418     $3.75   $3.75
 2    WEDGES    5907     $2.25   $4.50
 2    FLY       1423     $2.40   $4.80
 1    HAMMER    2683     $5.49   $5.49
                      SUBTOTAL  $18.54
                      TAX (3%)    0.55
               >>>>>>>>>>>TOTAL $19.09

  THANK YOU - PLEASE CALL AGAIN
```

Solution

1. Find the records of the parts purchased in the inventory file.

```
2418   WIRE     31   3.00   3.75   08/12/84   15
5907   WEDGES   80   1.60   2.25   06/23/84   30
1423   FLY       2   1.60   2.40   09/11/84    1
2683   HAMMER   56   4.60   5.49   10/09/85   20
```

2. Find the part number and the quantity purchased from the sales receipt.

Part Number:	2418	5907	1423	2683
Quantity Purchased:	1	2	2	1

3. Update the inventory records for these parts.

```
2418   WIRE     30   3.00   3.75   08/12/84   15
5907   WEDGES   78   1.60   2.25   06/23/84   30
1423   FLY       0   1.60   2.40   09/11/84    1
2683   HAMMER   55   4.60   5.49   10/09/85   20
```

▸ **Self-Check** Cover the answer and complete the problems.

A customer purchased 8 keytops, part #5818. Update the file.

Answer:
```
5818   KEYTOPS 64   0.29   0.69   08/23/84   60
```

Problems

Use the reports on pages 472 and 473 to answer the following.

	Part Number	Quantity	Description	Retail Price	Total
1.	2050	2	?	$1.49	$2.98
2.	1011	1	Hopper	?	?
3.	2683	3	?	?	?
4.	5893	8	?	?	?

5. Sally Daggett is a sales clerk in the Retail Department. She processes a customer's purchase by entering the part numbers and the quantities purchased into the store's terminal. The terminal produces this sales slip. What will each of the records for the parts purchased look like after the computer updates the inventory file?

```
           THE PIANO WORKS
QUAN  ITEM     STOCK #  PRICE    TOTAL
 2    WIRE      2419     $6.00  $12.00
 4    DP FEL    5893     $0.98  $ 3.92
 1    POLISH    1373     $3.50  $ 3.50
                      SUBTOTAL  $19.42
                      TAX (3%)    0.58
               >>>>>>>>>>>TOTAL $20.00
   THANK YOU - PLEASE CALL AGAIN
```

6. Sherri O'Neal wants to know how many projects are currently in production at The Piano Works. She programs the computer to display a list of those projects that were in production on January 9. What project numbers will be displayed on the terminal?

7. Chester Burke received this invoice for replacement parts. What will the records for these parts look like after he updates the inventory file?

INVOICE ORDER	INVOICE DATE	PURCHASE ORDER	ACCOUNT NO.
X 60349	Jan. 15, 1986	A44385	493-PW

QUANTITY	DESCRIPTION	UNIT PRICE	TOTAL
30	Pivots, stock #1428	7 60	$228 00
60	Grommets, stock #3101	1 84	110 40
48	Clips, stock #3102	1 40	67 20
24	Springs, stock #6274	2 50	60 00
	AMOUNT DUE:		$465 60
	TERMS: Net, 30 days		

Maintaining Your Skills

Look up the skill in parentheses if you need help or more practice.

Multiply. (Skill 8)

8. $9.40 × 27 9. $6.98 × 15 10. $7.21 × 11 11. $8.64 × 48

12. $256.49 × 12 13. $91.89 × 20 14. $220.22 × 18.5 15. $116.32 × 1000

22-5 Data Analysis

Objective

Retrieve and analyze data.

Electronic data processing has made it possible for businesses to process vast amounts of data. The techniques that you may use to analyze data vary, depending on the applications involved. **Data analysis** can be as simple as finding totals and averages. It can be as complex as computer modeling and simulation, used to predict a company's growth when factors affecting the company are expected.

Example

Skills ______ **Application** C **Term** **Data analysis**

Carl Apple, a production supervisor, wants to know which carpenter worked on project 189R1 on January 9. Carpenters at The Piano Works are listed under Department 15. Who was the carpenter working on project 189R1 on January 9?

Employee	Employee #
Bill Udall	409
Tina Wells	501
Folkert Reese	511
May Setti	630

Solution

1. List the records of production costs for January 9 that involve project 189R1 and Department 15.

Date	Project	Dept.	Part	Amt/Used
01/09/86	189R1	15	2060	1
01/09/86	189R1	15	2010	2
01/09/86	189R1	15	501	2

2. Employees at The Piano Works have 3-digit identification numbers. Piano parts have 4-digit identification numbers. List the records involving employees.

```
01/09/86    189R1    15    501    2
```

3. Use the chart above to find the name of the carpenter who worked on project 189R1 on January 9.

   Employee number 501: Tina Wells

▶ **Self-Check** Cover the answers and complete the problems.

List the records involving employees who worked on project 182R3.

Answer:

```
01/09/86    182R3    11    527    8
01/09/86    182R3    12    418    8
```

Problems

Use the reports on pages 472 and 473 to answer the following.

1. Patty Wake, production supervisor for project 182R3, requested the Data Processing Department to prepare the report shown below.

TO: PRODUCTION SUPERVISOR DATE: 1/13/86
FROM: DATA PROCESSING DEPARTMENT

PRODUCTION COSTS FOR PROJECT NUMBER 182R3

DATE	PROJECT	DEPT.	PART #	PARTS/USED
01/09/86	182R3	11	2693	2
01/09/86	182R3	11	2683	12
01/09/86	182R3	11	1373	1
01/09/86	182R3	11	527	8
01/09/86	182R3	12	418	8

a. How many man-hours were spent on the project on January 9?

b. If the employees working on the project earn $7.85 an hour, how much in wages went toward production costs?

c. What was the total parts cost on the project on January 9?

d. What was the total production cost on January 9?

2. Carl Apple is the production supervisor for project 189R1. He needs to keep track of various costs while the project is being completed. Carl requested the Data Processing Department to prepare the report shown below.

TO: PRODUCTION SUPERVISOR DATE: 1/13/86
FROM: DATA PROCESSING DEPARTMENT

PRODUCTION COSTS FOR PROJECT NUMBER 189R1

DATE	PROJECT	DEPT.	PART #	PARTS/USED
01/09/86	189R1	15	2060	1
01/09/86	189R1	15	2010	2
01/09/86	189R1	12	412	8
01/09/86	189R1	12	845	4
01/09/86	189R1	12	848	6
01/09/86	189R1	15	501	2

a. If Nolan Lidell, employee number 412, earns $10.35 an hour, and the other employees working on the project earn $8.15 an hour, how much in wages went toward production costs?

b. What was the total parts cost for the day?

c. What was the total production cost for the day?

(Continue on next page.)

3. Chester Burke uses a program that searches the inventory file and lists those parts that are on or below the order-on quantity. He places orders for each of these parts. The numbers of each part ordered must bring the number in stock up to 4 times the order-on quantity. How many of each part should he order?

STATUS OF INVENTORY: WEEK ENDING January 9

	Part #	Part	Stock	Cost	Retail	RESTCK	Order/On
->	1004	KEYFRAME	1	83.00	89.00	081284	1
->	1428	PIVOT	0	7.60	8.65	100885	12
->	2010	TACKS	40	1.27	1.98	091284	50
->	2080	BUTTONS	18	0.60	0.85	071884	20
->	3101	GROMMETS	18	1.84	2.10	021884	20
->	3102	CLIPS	17	1.40	1.65	062384	20
->	3153	FLANGES	12	2.50	3.25	021884	12
->	6023	LEADS	12	3.00	4.20	012884	12
->	6274	SPRINGS	16	2.50	3.29	021884	20

4. Patty Wake, the project supervisor for project 183N5, turned in the time card information for the employees who worked on the project during the week ending January 9. The personnel file contains the information shown below for these employees.

WEEK ENDING JANUARY 9		
#501	WELLS	40 HOURS
#848	HOI	38 HOURS
#311	VILLA	48 HOURS

EMPLOYEE	Employee Number	Date of Employment	Marital Status	Exemptions Claimed	Hourly Rate
WELLS, TINA	501	03/15/77	M	1	$7.90
HOI , ARLENE	848	11/29/82	S	0	$8.15
VILLA, TOMAS	311	06/01/82	M	3	$7.45

a. The overtime rate at The Piano Works is time-and-a-half for employees who work over 40 hours per week. What were the gross earnings for the week for each employee?

b. How much was withheld for federal income tax from each employee's paycheck? Use the tables on pages 566 and 567.

c. The state tax rate is 5.375% of gross earnings. How much was withheld from each employee's paycheck for state income tax?

d. The social security tax rate is 7.15% of the first $42,600 earned. How much was withheld from each employee's paycheck for social security taxes?

e. If these are the only deductions, what was each employee's net pay for the week?

Use the personnel file on page 484 to answer the following.

5. Employees at The Piano Works earn paid vacation according to the schedule shown. In the table, 5 working days is equivalent to a week. How long has Tina Wells, Arlene Hoi, and Tomas Villa each worked for the company? According to the schedule, how many weeks of paid vacation will each receive this year?

VACATION POLICY as of 1/15/86	
Employment	Vacation
6 months or less	None
1 year or less	5 days
1 to 3 years	10 days
4 to 8 years	15 days
9 to 15 years	20 days
16 years or more	25 days

6. Carl Apple is the project supervisor for project 172R9. Since the project was recently completed, he must determine the total cost of repairing the piano. He requested the Data Processing Department to prepare a complete listing of production costs on the project. He received the report shown below.

PRODUCTION COSTS Project 172R9

Date	Project	Dept.	Part	Amt/Used
01/05/86	172R9	13	311	3
01/06/86	172R9	13	311	8
01/06/86	172R9	13	2683	3
01/07/86	172R9	13	2503	4
01/08/86	172R9	19	128	6
01/09/86	172R9	19	128	1

a. What is the total retail value of the parts used to repair the piano?

b. The table at the right shows the hourly rates for employees working on the project. What is the total labor cost for the project?

Department Number	Hourly Rate
13	$7.90
19	$12.85

c. What is the total cost of parts and labor?

d. Carl uses the formula below to calculate the charge to the customer for work performed. How much is the customer charged for having the piano repaired?

Billing Amount for Repairs = 90% of Total Retail Value of Parts + Labor Cost

Maintaining Your Skills

Look up the skills in parentheses if you have difficulty with the problems.

Write as decimals. (Skill 28)

7. 90% 8. $66\frac{2}{3}$% 9. 11.9% 10. 16.3% 11. $7\frac{1}{2}$%

Multiply. (Skill 8)

12. 7.21 × 0.66 13. 8.00 × 1.19 14. 0.32 × 21.4 15. 18.6 × 0.53

22-6 Reporting

Objective

Analyze the information given in a report.

The reports that a data processing department prepares must meet many requirements. They must be accurate. They must be timely. They must be sent to and received by a person who can use the information in decision-making. They must be planned so that they are easy to read and use. The **format** is the physical arrangement of the information in the report. The person who needs the report will usually make the decision as to what will be included in the report.

Example

Skills ________ **Application** _ **Term** **Format**

Emily Conti is the manager of the Retail Department for Piano Works. She has requested the Data Processing Department to prepare a report showing the unit cost, retail price, and markup for piano accessories. She received the report shown. She wants to calculate the average markup. What is the average markup, based on the retail price?

PART #	PART	COST	RETAIL	MARKUP	PERCENT
9011	BENCH	$74.00	$85.00	$11.00	12.9%
9015	BENCH	$83.00	$89.00	$ 6.00	6.7%
9110	CUSHION	$24.00	$30.00	$ 6.00	20.0%
9113	CUSHION	$18.00	$21.00	$ 3.00	14.3%
9260	METRONOME	$41.00	$53.00	$12.00	22.6%
9265	METRONOME	$24.00	$32.50	$ 8.50	26.2%
9308	COVER	$60.00	$72.00	$12.00	16.7%
9310	COVER	$76.60	$84.50	$ 7.90	9.3%
9410	LAMP	$28.80	$36.95	$ 8.15	22.1%

Solution

1. Find the **sum** of the markups.
   $11.00 + $6.00 + $6.00 + $3.00 + $12.00 + $8.50 + $12.00 + $7.90 + $8.15 = $74.55

2. Find the **sum** of the retail prices.
   $85.00 + $89.00 + $30.00 + $21.00 + $53.00 + $32.50 + $72.00 + $84.50 + $36.95 = $503.95

3. Find the **average markup.**
   Sum of Markups ÷ Sum of Retail Prices
   $74.55 ÷ $503.95 = 0.14793 = 14.8% average markup

Self-Check Cover the answers and complete the problem.

Prepare a report that shows the number in stock and the number on order for parts 8641, 8642, and 8643.

Answer:

Part No.	Part	Stock	Restock	Order/On
8641	Pedal L	8	10/18/84	5
8642	Pedal R	8	10/18/84	5
8643	Pedal C	10	10/18/84	5

Problems

Use the report on page 486 to answer the following.

Emily Conti has put together a "starter package" to offer customers purchasing a new piano. The package includes a bench, stock number 9011; a cushion, stock number 9113; a cover, stock number 9308; and a lamp, stock number 9410.

1. What is the total cost of the "starter package"?

2. What is the total retail price of the package?

3. What is the average markup of the package?

4. Emily wants to price the "starter package" at a 10% markup based on selling price. What will be the sale price of the "starter package"?

SALES:	September Units	September Sales	Year-to-Date Units	Year-to-Date Sales
Model E32	3	$5,987.85	35	$69,858.25
Model E72	1	2,095.95	12	25,151.40
Model E92	1	2,495.95	9	22,463.55
Model 10	4	6,359.80	43	68,367.85
Model 12	2	3,379.90	15	25,349.25
Model 22	2	3,579.90	13	23,269.35
Model H-12C	1	2,989.95	13	38,869.35
Model H-22C	0	0.00	8	26,319.60
Model H-32C	1	3,589.95	5	17,949.75

SERVICE:	September	Year-to-Date
Total Service Contracts:	$19,354.99	$124,504.04
TOTAL SALES:	$49,834.24	$442,102.39

Ken Best, sales manager, received this report for the month of September. The report shows the sales figures for pianos manufactured for the month and year-to-date.

Use the report to answer the following.

5. How many different models of pianos does The Piano Works manufacture?

6. For September, which model produced the greatest amount of sales?

7. In September, which model produced the least amount of sales?

8. For the year-to-date, which model produced the least amount of sales?

9. For September, what percent of gross sales were service contracts?

10. What percent of total sales are service contracts, on the average?

***11. A Brief Case** Companies are required every 3 months to pay a share of their **estimated income taxes** for the year. The accumulated profits for the preceding quarter or quarters are used to estimate the tax due for the year. At the end of the year, an income tax return is filed using figures for all 12 months.

QUARTER	GROSS INCOME	BUSINESS EXPENSES	PROFIT OR <LOSS>	QUARTERLY TAX
JANUARY FEBRUARY MARCH	41 326 38 462 38 062	30 604 26 400 38 992	10 722 12 062 -930	5 429.10
TOTAL TO DATE	117 850	95 996	21 854	5 429.10
APRIL MAY JUNE	45 625 51 009 59 823	29 904 28 432 31 644	15 721 22 577 28 179	25 578.16
TOTAL TO DATE	274 307	185 976	88 331	31 007.26
JULY AUGUST SEPTEMBER	62 842 55 119 49 834	43 405 29 680 31 762	19 437 25 439 18 072	24 143.58
TOTAL TO DATE	442 102	290 823	151 279	55 150.84
OCTOBER NOVEMBER DECEMBER	42 824 37 477 39 093	27 498 26 821 29 014	15 326 10 656 10 079	?
TOTAL TO DATE	561 496	374 156	187 340	?

Timothy Blake, an accountant at The Piano Works, received from the Data Processing Department a summary of gross incomes, business expenses, and taxable incomes for the last 12 months. The report shows that the company had paid estimated taxes of $5429.10, $25,578.16, and $24,143.58 for the first 3 quarters of the year. He is now completing the company's income tax return for the year.

a. What were the company's total profits for the year?

b. He uses the tax table at the right to compute the tax. What is the total corporate income tax for the year?

c. How much had been paid for the first 3 quarters?

If TAXABLE INCOME is:		TAX is:	
Over—	But not over—		Of the amount over—
0	$25,000	17%	0
$25,000	50,000	$4,250 + 20%	$25,000
50,000	75,000	9,250 + 30%	50,000
75,000	100,000	16,750 + 40%	75,000
100,000	—	26,750 + 46%	100,000

d. What amount of the tax due remains to be paid for the year?

***12. A Brief Case** Henry Wordsworth, president of The Piano Works, reviews the report below prior to its release to stockholders. The 5-year **financial summary** shows the company's earnings, operating costs, and so on for each of the past 5 years.

SUMMARY OF OPERATIONS	THE PIANO WORKS (Year Ending December 31)				
(in Thousands) excluding shares	1985	1984	1983	1982	1981
Sales	$561.5	$482.9	$412.8	$390.6	$253.8
Operating Costs	374.2	342.0	328.5	230.9	192.4
Taxes on Income	66.9	45.6	20.5	53.9	12.7
Net Income	120.4	95.3	63.8	105.8	48.7
Percent of Sales	21.4%	19.7%	15.5%	27.0%	19.2%
Earnings per share	$0.60	$0.47	$0.32	$0.53	$0.24
Dividends Paid per Share	$0.25	$0.15	0	$0.25	0
Average Number of Shares Outstanding	200,005	201,132	201,650	198,512	200,000

a. Which year in the company's 5-year history showed the greatest increase in sales?

b. Although the company may make a profit, it may decide not to pay dividends to its stockholders. Instead, the money is reinvested in the company for growth. In which years did the company not pay dividends to its stockholders?

c. In 1985, what was the dividend paid per share to stockholders?

d. In 1985, what total amount did the company pay its stockholders in dividends?

e. In the company's 5-year history, which year showed the greatest profit margin?

f. In the last 5 years, how much has the company grossed?

g. In the last 5 years, how much has the company netted?

h. Can the number of stockholders be determined by examining this financial report?

Maintaining Your Skills

Look up the skills in parentheses if you need help or more practice.

Find the percentage. (Skill 30)

13. $7\frac{1}{2}\% \times 70$ **14.** $374.2 \times 10\%$ **15.** $19.7\% \times 95.3$ **16.** $208.62 \times 100\%$

Subtract. (Skill 4)

17. 19,437 − 930 **18.** 151,279 − 88,331 **19.** 225,482 − 163,931

Reviewing the Basics

Applications

Use the report to answer the following.

(Application C)

1. What were the company's gross sales in 1984?

2. What were the company's gross sales in 1985?

3. By what percent did gross sales increase from 1984 to 1985?

SUMMARY OF OPERATIONS	THE PIANO WORKS	
(in Thousands) excluding shares	1985	1984
Sales	$561.5	$482.9
Operating Costs	374.2	342.0
Taxes on Income	66.9	45.6
Net Income	120.4	95.3
Percent of Sales	21.4%	19.7%
Earnings per share	$0.60	$0.47
Dividends Paid per Share	$0.25	$0.15
Average Number of Shares Outstanding	200,005	201,132

4. A stockholder owns 50 shares of The Piano Works stock. What were the annual dividends paid to this stockholder in 1984? What were the annual dividends paid to the stockholder in 1985?

Terms

Use each term in one of the sentences.

5. Posting
6. File
7. On-line
8. Off-line
9. Real-time processing
10. Data analysis
11. Data
12. Format
13. Data base

a. Computers have made it possible for businesses to process vast amounts of __?__ .

b. The act of recording data is called __?__ .

c. __?__ occurs when the work that a computer does keeps up with the day's events.

d. The collection of files describing the activities of the business is called a __?__ .

e. The physical arrangement of the information in a report is called its __?__ .

f. A group of records, all containing related material, is called a __?__ .

g. When information is stored __?__ , a data processing operator must mount the storage device onto the computer system in order for the computer to access the information.

h. __?__ can be as simple as finding totals and averages.

i. When information is stored __?__ , the information is stored in the computer's internal memory for immediate access.

Refer to your reference files at the back of the book if you need help.

The Changing Face of Business

The Universal Product Code (UPC) is a series of lines, bars, and numbers that can be seen on the packages of most products that consumers purchase. The UPC is used with an electronic cash register, an optical scanning device, and a computer to prepare a sales slip for the customer.

0 38000 40421

Code number of the manufacturer. | Code number of the product.

The scanning device "reads" the UPC and signals the computer to search its memory for the price. In many instances, the computer returns not only the price but also the name and size of the product. An electronic cash register prints this information on a sales slip. As a sale is made, the computer keeps a running total of each product sold so that the store manager can be advised at all times of the status of the inventory.

Answer the following questions concerning the UPC.

1. In the UPC shown above, which number identifies the manufacturer?

2. Which number identifies the product?

3. Which cash register tape shown indicates that the UPC system was used?

```
01.00GR A :
01.89MT A :
01.99GR A :
00.37PR A :
00.37PR A :
00.00TX A :
05.62 : A SBTL
05.62 : A TL
THANK YOU :
COME AGAIN
  -62-
```

```
CLNWSH DETGNT     2.31
FR ORNGE JCE      0.99
BT APPLE JCE      0.75
SU LGHT TUNA      0.98
QT MILK           0.59
TAX DUE           0.12
TOTAL             5.66
CASH TEND        10.00
CHG DUE           4.34
 1/24 19:05 0232/14
```

4. Many businesses believe that if the UPC system is used, the practice of pricing each individual product should be eliminated. How would this practice affect the business manager? How would this practice affect the consumer?

Unit Test

Use the reports on pages 472 and 473 to answer the following.

Lesson 22-1

1. Earl Nix ordered 2 bushings, part number 2310. What does the current record for this part look like in the inventory file?

Lesson 22-2

2. Wilma Holland received this notice to add a new part to the inventory file. Insert the new part listing in the inventory file and show the 2 records between which the new listing will be found.

NOTIFICATION TO UPDATE INVENTORY
THE PIANO WORKS, INC.

Inventory Update [ ]
New Part [✓]
Part Obsolete [ ]

Part #	Description	Stock	Cost	Retail	Restocked	Order/On
4220	JACK	36	$2.40	$3.25	1/29/85	12

Lesson 22-3

3. The monthly sales figures for The Piano Works are stored on a magnetic tape. Each month, Derek Morris mounts the tape onto the computer system and prepares a report showing the company's gross sales for the month and for the year-to-date. Are the sales figures stored on-line or off-line?

Lesson 22-4

4. Sally Daggett, a sales clerk, enters the data of a customer's purchase into the store terminal. The computer produces the sales receipt shown. What will each record for the parts purchased look like after the computer updates the inventory file?

```
            THE PIANO WORKS
QUAN  ITEM    STOCK #  PRICE   TOTAL
 7    WH KEY   5816    $0.89   $6.23
 4    BL KEY   5817    $0.89   $3.56
                    SUBTOTAL $ 9.79
                    TAX (3%)    0.29
              >>>>>>>>>>TOTAL $10.08
  THANK YOU - PLEASE CALL AGAIN
```

Lesson 22-5

5. Clara Miller, employee number 413, earns $7.55 an hour. What were her gross earnings for January 9?

Lesson 22-6

6. Paul Monet received a report from the Data Processing Department. The portion of the report shown shows a summary of the company's operating costs for the month of December. What percent of the total operating cost are employee wages? What percent of the total operating cost was their advertising budget for the month?

```
OPERATING COSTS
Summary for December

RENT           $ 1500.00
UTILITIES      $ 3882.46
INSURANCE      $  205.80
TAXES          $ 5595.63
WAGES          $12168.16
ADVERTISING    $  800.00
PARTS          $ 3809.19
MAINTENANCE    $ 1052.76
```

Cumulative Review (Skills/Applications) Units 17–22

Skills

Round to the nearest dollar.

(Skill 2)
1. \$4288.15 2. \$9.67265 3. \$62.473 4. \$36.458

Solve. Round to the nearest cent.

(Skill 3)
5. 32 + 87 6. 329 + 57 7. \$428 + \$375
8. \$1593 + \$3125 9. \$79 + \$144 10. \$1596 + \$15,895 + \$90

(Skill 4)
11. \$85 − \$29 12. \$337 − \$89 13. \$400 − \$196
14. \$1345 − \$1031 15. \$8740 − \$927 16. \$1946 − \$1221

(Skill 5)
17. \$14.95 + \$3.49 18. \$1.19 + \$0.27 19. \$25.25 + \$59.45
20. \$1500 + \$18.74 + \$3.52 21. \$4329.05 + \$751.94 + \$100.53

(Skill 6)
22. \$30.39 − \$14.20 23. \$500.99 − \$32.43
24. \$1543.60 − \$7.43 25. \$416.60 − \$18.79 26. \$1327.81 − \$942.79

(Skill 7)
27. \$43 × 5 28. \$51 × 9 29. \$1500 × 7
30. \$579 × 84 31. \$585 × 67 32. \$3591 × 53

(Skill 8)
33. \$140 × 0.05 34. \$2.85 × 0.13 35. \$26.24 × 0.94
36. \$793.02 × 3.40 37. \$42.24 × 66 38. \$1.59 × 43

(Skill 10)
39. \$75 ÷ 3 40. \$427 ÷ 12 41. \$5512 ÷ 26
42. \$329 ÷ 329 43. \$5275 ÷ 25 44. \$43,615 ÷ 305

(Skill 11)
45. \$30.65 ÷ 5 46. \$479.30 ÷ 18 47. \$5191.20 ÷ 6.30
48. \$8921.70 ÷ 3.45 49. \$12,291.60 ÷ 14.5

(Skill 20)
50. $\$85 \times \frac{3}{4}$ 51. $\$96 \times \frac{7}{8}$ 52. $\$32 \times \frac{5}{8}$
53. $\$40{,}817 \times \frac{7}{10}$ 54. $\frac{6}{15} \times \$439$ 55. $\$2274 \times \frac{1}{3}$

Write as a decimal.

(Skill 28)
56. 4.05% 57. 12% 58. 19.75% 59. 54.25%

Solve. Round to the nearest cent.

(Skill 30)
60. 10% of \$250 61. 58% of \$985 62. 40% of \$9453
63. $5\frac{1}{4}\%$ of \$441 64. $9\frac{1}{2}\%$ of \$12,720 65. $3\frac{1}{2}\%$ of \$243,784

Solve. Round to the nearest tenth of a percent.

(Skill 31)
66. \$35 is what percent of \$76? 67. \$150 is what percent of \$245?
68. \$15,000 is what percent of \$64,000? 69. \$43,470 is what percent of \$120,000?

Applications

Find the tax bracket.

(Application C) **70.** Taxable income is $45,690.

71. Taxable income is $79,500.

72. Taxable income is $63,750.

73. Taxable income is $19,900.

Taxable Income		Your TAX is:	
Over—	But not over—		Of the amount over—
0	$25,000	17%	0
$25,000	50,000	$4,250 + 20%	$25,000
50,000	75,000	9,250 + 30%	50,000
75,000	100,000	16,750 + 40%	75,000
100,000	—	26,750 + 46%	100,000

Find the charge per 100 pounds.

74. Weight: 5946 pounds
Distance: 623 miles

75. Weight: 2420 pounds
Distance: 1300 miles

Basic Rates Per 100 Pounds		
Weight Group (in pounds)	Distance (in miles)	
	500–1000	1001–1500
2001–5000	$40.25	$66.20
5001–10,000	$38.00	$53.00

Find the average cost to the nearest cent.

(Application Q) **76.** 2 units at $378 each and 3 units at $45.30 each.

77. 30 units at $2.35 each and 21 units at $1.77 each.

78. 75 units at $3.12 each and 14 units at $47.50 each.

79. 423 units at $0.56 each and 180 units at $1.20 each.

Find the area.

(Application X) **80.** 29 m by 8 m **81.** 15 ft by 9 ft **82.** 580 ft by 33 ft

83. 727 in. by 83 in. **84.** 7.5 ft by 9.5 ft **85.** 110 cm by 12 cm

(Application Z) Find the volume.

86. 7 ft by 3 ft by 12 ft **87.** 2.7 m by 8 m by 9.5 m

88. 9 in. by $13\frac{1}{2}$ in. by 4 in. **89.** $1\frac{3}{4}$ ft by 3 ft by $5\frac{1}{2}$ ft

Terms

Write your own definition for each term.

90. Inventory **91.** U.S. Treasury Bill **92.** Income statement

93. Posting **94.** Apportion **95.** Data analysis

96. Owner's equity **97.** Book value **98.** Net proceeds

99. Maturity value **100.** Balance sheet **101.** Commercial paper

 Refer to your reference files at the back of the book if you need help.

Cumulative Review Tests Units 17–22

Lesson 17-1

1. Office Equipment Company manufactures some 2-drawer file cabinets. Each cabinet is stored in a box measuring 2.5 feet high, 1.5 feet wide, and 2.5 feet long. How many cubic feet of space does Office Equipment need to store 1200 file cabinets?

Lesson 17-2

2. How many StraightAway models does Clarion Motor Coach have on its lot on October 1?

Item: StraightAway
Stock Number: JR 2201 F

Month of:	Opening Balance	Receipts	Issues	Inventory at End of Month
August	75	12	8	?
September	?	15	40	?
October	?			

Lesson 17-6

3. Hyland Company is shipping 984 pounds of photographic equipment. Pierce Trucking will handle shipment. The distance is 155 miles. What is the total shipping cost?

THERE-NOW TRUCKING Class 100 Items
Basic Rates Per 100 Pounds

Weight Group (in pounds)	Distance (in miles)		
	1–100	101–200	201–300
0–500	$4.23	$5.20	$6.18
501–1000	3.65	4.43	5.10

Lesson 18-2

4. Motel Seven hired 4 high school students to clean its swimming pool. Each student worked for 2.0 hours at an hourly rate of $4.90. The materials charge was $28.45. What was the total charge for this service?

Lesson 18-5

5. The Huston Corporation used 15,400 kilowatt-hours of electricity with a peak load of 120 kilowatts in April. The demand charge is $6.10 per kilowatt. The energy charge per kilowatt-hour is $0.09 for the first 10,000 kilowatt-hours and $0.07 for all kilowatt-hours over 10,000. The fuel adjustment charge is $0.05 per kilowatt-hour. What is the total cost of electricity for the Huston Corporation for April?

Lesson 19-1

6. Complete the payroll register. Use the FICA tax rate of 7.15%.

Name	Gross Income	FIT	FICA	SIT	Total Deductions	Net Pay
Bisset	$300.00	$41.90	?	$4.50	?	?
Heller	435.70	62.50	?	6.54	?	?
Molnar	405.00	46.00	?	6.08	?	?
Stein	425.65	49.80	?	6.38	?	?
Total	?	?	?	?	?	?

Lesson 19-3

7. Foster Machine Company pays $90,000 per year for security. The company apportions the cost among its departments based on space. The Research Department occupies an area that measures 60 feet by 60 feet. The building contains 1,200,000 square feet. How much does the Research Department pay for security?

Lessons 19-4, 19-5

8. Central Copy purchased a new copy machine for $7350. The machine has an estimated life of 4 years. The resale value is expected to be $1350. Use the straight-line method to find the annual depreciation. Find the book value after each year of use.

Lesson 20-1

9. The Running Shop had these assets and liabilities on May 31.

Cash: $2400
Inventory: $28,000
Supplies: $356
Store furnishings: $14,460

Unpaid merchandise: $12,000
Taxes owed: $572
Bank loan: $8500

What are the total assets? What are the total liabilities? What is the owner's equity?

Lesson 20-3

10. Shellar Manufacturing, Inc., prepares an annual income statement. This past year, Stellar had net sales of $800,000. The cost of goods sold totaled $360,750. Operating expenses included: wages and salaries totaling $165,970; depreciation and amortization totaling $19,590; interest paid totaling $15,540; product recall totaling $3596; and miscellaneous operating expenses of $6570. Prepare an income statement for Shellar Manufacturing.

Lesson 21-2

11. Solar Energy, Inc., is issuing 500,000 shares of stock. Each share will be sold at $35. The underwriting commission is 3% of the value of the stock. The other expenses are expected to be 0.6% of the value of the stock. If all the shares of stock are sold, what net proceeds will Solar Energy receive?

Lesson 21-3

12. Hereford Ranch borrowed $48,500 for 120 days at the prime rate of interest. When Hereford Haven borrowed the money, the prime rate was 18.5%. The bank charged ordinary interest at exact time. What was the maturity value of the loan?

Lesson 21-4

13. Carlton Manufacturing had $80,000 in surplus cash. The financial manager decided to invest in a United States Treasury Bill with a face value of $80,000. The bill matured in 126 days. The interest rate was 14.9% ordinary interest at exact time. The bank service fee was $25. What was the cost of the Treasury Bill?

Use the report on page 473 to answer the following.

Lesson 22-1

14. Judy Nelson ordered 6 bushings, part number 2310. What does the current record for this part look like in the inventory file?

Copyfax Center: A Simulation

You are the manager of Copyfax Center, a small business that does copying, printing, and word processing. Copyfax Center has 3 employees. Here is a chart of the hours they work each day, Monday through Friday.

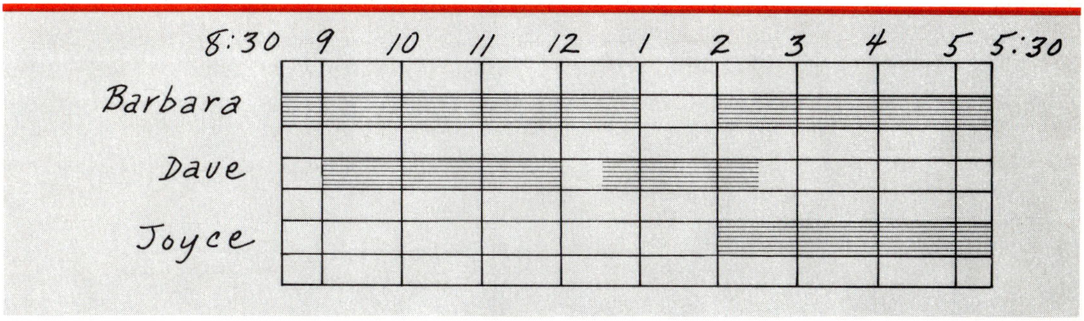

Each week you make up the weekly payroll. Each employee must pay 7.15% for FICA and 5% for state tax. You pay employees time and a half for overtime.

Use these worksheets to calculate this week's net pay for each employee. Use the tax tables on pages 566–567 to determine the federal income tax (FIT). This week Barbara worked 3 hours overtime. Dave and Joyce did not work overtime.

Employee	Hourly Pay	Hours Worked		Gross Pay
		Regular	Overtime	
Barbara Luna	$7.30			
Dave Elwood	$6.35			
Joyce Tozzi	$3.50			

Employee	Tax Status	FIT	FICA	State Tax	Health Insur.	Total Ded.	Net Pay
Barbara Luna	Married 3 allow.				$4.50		
Dave Elwood	Single 2 allow.				$3.50		
Joyce Tozzi	Single 1 allow.	0			—		

Pricing

The price that Copyfax Center charges for copies depends on the number of copies made. This chart shows the current prices.

Number of Copies	Price per Page*
1–24	$0.05
25–49	$0.035
50–99	$0.03
100–199	$0.026
200–499	$0.022
over 500	$0.019

*plus 5% sales tax

Regular customers get discounts. A few receive a trade discount of 30%. Others get a cash discount of 2% if they pay within 10 days. Both discounts are deducted before the sales tax is calculated.

Use these invoices to calculate the prices charged these customers.

COPYFAX CENTER Invoice no. 2257
Date *June 15, 19—*

Customer	*Christine's Hardware*
Number of copies	*600*
Price per page	$
Total price	$
Discount: ___% ☒ trade ☐ cash	$
Net price	$
Sales tax	$
Invoice price	$

COPYFAX CENTER Invoice no. 2289
Date *June 17, 19—*

Customer	*Mike's Travel Service*
Number of copies	*137*
Price per page	$
Total price	$
Discount: ___% ☐ trade ☒ cash	$
Net price	$
Sales tax	$
Invoice price	$

Depreciation

Depreciation of Copyfax Center's equipment is considered a business expense.

Use this worksheet to calculate the depreciation of these items, using the straight-line method.

	Copier	Word Processor	Typewriter	Delivery Van
Year purchased	1986	1987	1985	1986
Original cost	$8500	$7500	$975	$11,248
Salvage value	$1000	$750	$225	$1600
Total depreciation				
Estimated life	5 yr	5 yr	6 yr	6 yr
Annual depreciation				

As part of your long-range planning, you decide to calculate how much each item will be worth each year.

Use this worksheet to calculate the value of each item in each of the years listed. (If Copyfax Center does not own an item one year, write "N/A" for Not Applicable.) Then calculate the total value of Copyfax's equipment for each year.

Year	Copier	Word Processor	Typewriter	Delivery Van	Total Value
1986					
1987					
1988					
1989					
1990					
1991					

Balance Sheet

Each month you prepare a balance sheet. The balance sheet lists Copyfax Center's assets (what it owns), its liabilities (what it owes), and the owner's equity (assets minus liabilities).

You start by listing your assets and liabilities in a running account like this:

Copyfax Center June 30, 19—
Cash on hand $1400
Accounts receivable $5000
Accounts payable $6000
Equipment (less accumulated depreciation) $9260
Supplies (paper, etc.) $1275
Prepaid insurance (3-yr policy) $1800
Miscellaneous assets $520
Taxes owed $50

Use this form to complete Copyfax Center's balance sheet. Enter the assets and liabilities from your running account.

COPYFAX CENTER
Balance Sheet June 30, 19--

ASSETS		LIABILITIES	
	$		$
		Total liabilities	
		Owner's equity	
Total assets		Total liabilities and equity	

Income Statement

Each month you also prepare Copyfax Center's income statement. Here are the sales and expenses you have recorded during the month.

Copyfax Center	June 19–
Wages $2125	Insurance $50
advertising $135	Taxes $18
Delivery $72	Depreciation $333
Postage $38	Supplies $936
Rent $350	Gross sales $5300
Utilities $70	Sales discounts $672

Use this form to prepare an income statement.

COPYFAX CENTER

Income Statement for Month Ended June 30, 19--

Income:	Gross sales	
	Less sales discounts	
	Net sales	
Expenses:		
	Total operating expenses	
	NET INCOME	

Annual Report and Comparative Analysis

At the end of each year you prepare an annual income statement. You can then compare Copyfax Center's finances for two years, in order to plan for next year.

This chart shows Copyfax Center's income and expenses for last year and this year. Fill in the missing numbers. Then, for each amount, calculate the percent increase from last year to this year.

COPYFAX CENTER Annual Income Statement		Last Year	This Year	Percent Increase
Income:	Gross sales	$63,000	$79,600	
	Less sales discounts	8,000	10,100	
	Net Sales			
Expenses:	Wages	24,746	33,548	
	Advertising	1,575	3,700	
	Delivery	825	1,040	
	Postage	463	565	
	Rent	4,200	4,500	
	Utilities	945	1,130	
	Insurance	600	600	
	Taxes	214	250	
	Depreciation	4,000	4,000	
	Supplies	11,239	13,850	
	Total operating expenses			
	NET INCOME			

Complete the following statements comparing last year's and this year's figures. The expense with the largest percent increase was __________ . The next largest was ________ . Net sales increased by _______ %, total operating expenses by _______ %, and net income by _______ %.

The Bottom Line

What would you do differently next year to improve Copyfax Center's financial picture?

Your Reference Files

Skills Thirty-two computational skills, from whole number and decimal operations through rates and percents, provide the basics you will need as a consumer and future business person.

Applications Twenty-six mathematical applications, covering money, time, measurement, graphs, and more, provide the tools you will need for consumer and business use.

Terms Key terms with their definitions from all lessons provide the background you will need in consumer and business situations.

Skill 1 Numbers

Place Value

In 4532.869, give the place and value of each digit.

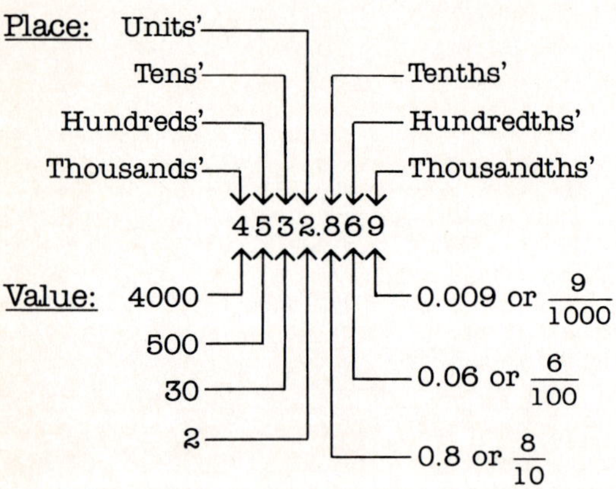

Give the place and value of the underlined digit.

1. $6\underline{5}$ **2.** $9\underline{6}$ **3.** $4\underline{7}2$

4. $\underline{2}3$ **5.** $10\underline{8}$ **6.** $\underline{5}36$

7. $15\underline{0}6$ **8.** $2\underline{8}21$ **9.** $7\underline{7}84$

10. $\underline{1}492$ **11.** $100\underline{9}$ **12.** $1\underline{9}26$

13. $0.\underline{3}7$ **14.** $1.6\underline{1}$ **15.** $2.73\underline{9}$

16. $\underline{6}.3$ **17.** $9.4\underline{2}$ **18.** $4.3\underline{7}$

19. $2\underline{4}.04$ **20.** $37.\underline{3}29$ **21.** $1.8\underline{2}4$

22. $\underline{4}93.89$ **23.** $90.25\underline{7}$ **24.** $23.\underline{5}72$

25. $12.76\underline{3}$ **26.** $0.\underline{0}78$ **27.** $5.46\underline{1}$

28. $63\underline{9}.8$ **29.** $1\underline{0}00$ **30.** $\underline{9}19$

31. $\underline{7}513$ **32.** $49.5\underline{8}3$ **33.** $0.21\underline{2}$

34. $21.2\underline{4}7$ **35.** $9.\underline{2}66$ **36.** $\underline{1}02.5$

Numbers as Words

Write each number in words.

Number		Words
105	⟶	One hundred five
26	⟶	Twenty-six
17	⟶	Seventeen
\$98.09	⟶	Ninety-eight and $\frac{9}{100}$ dollars
\$33.13	⟶	Thirty-three and $\frac{13}{100}$ dollars

Write in words.

37. 18 **38.** 34 **39.** 159

40. 78 **41.** 103 **42.** 842

43. 207 **44.** 5012 **45.** 6005

46. 119 **47.** 72 **48.** 1240

49. 5102 **50.** 194 **51.** 6590

52. \$25.00 **53.** \$6.24 **54.** \$17.09

55. \$112.35 **56.** \$120.17 **57.** \$4.37

58. \$7749 **59.** \$65.90 **60.** \$0.50

61. \$26.42 **62.** \$7.50 **63.** \$500.42

64. \$100 **65.** \$1.05 **66.** \$37.49

67. \$247.61 **68.** \$1806 **69.** \$7.01

Compare Whole Numbers

Which number is greater: 5428 or 5431?

$$5428 = 5000 + 400 + 20 + 8$$

$$5431 = 5000 + 400 + 30 + 1$$

(5000: Same; 400: Same; 30 is greater than 20.)

So 5431 is greater than 5428.

Which number is greater?

70. 23 or 32
71. 54 or 45
72. 459 or 462
73. 741 or 835
74. 810 or 735
75. 125 or 211
76. 3450 or 6450
77. 5763 or 925
78. 1000 or 999
79. 444 or 4444
80. 3002 or 4000
81. 1236 or 820
82. 150 or 149
83. 493 or 650
84. $2000 or $1997
85. $101 or $99
86. $482 or $600
87. $39 or $93
88. $1686 or $1668
89. $568 or $742
90. $86,432 or $101,000
91. $791,000 or $768,000

Compare Decimals

Which number is greater: 24.93 or 24.86?

$$24.93 = 20 + 4 + 0.9 + 0.03$$

$$24.86 = 20 + 4 + 0.8 + 0.06$$

(20: Same; 4: Same; 0.9 is greater than 0.8.)

So 24.93 is greater than 24.86.

Which number is greater?

92. 3.1 or 1.3
93. 1.2 or 2.0
94. 4.50 or 4.05
95. 25.1 or 20.8
96. 18.43 or 17.88
97. 56.84 or 58
98. 0.4 or 0.6
99. 0.01 or 0.1
100. 0.82 or 0.28
101. 0.5 or 0.06
102. 8.739 or 10
103. 0.002 or 0.020
104. $5.99 or $5.00
105. $10 or $8.50
106. $23.85 or $19.84
107. $11.19 or $19
108. $6.98 or $7.50
109. $83.59 or $600
110. $4327.75 or $6297.86
111. $8391.34 or $9521.39
112. $4640.66 or $4646.40
113. $2000.00 or $1997.98

Skill 2 Round Numbers

Whole Numbers

Round 7863 to the nearest hundred.

↓ Hundreds' place
7863
↑ 5 or more? Yes.

↓ Add 1 to hundreds' place.
7900
↑ Change to zeros.

Round.

To the nearest ten: **1.** 26

2. 37 **3.** 68 **4.** 195

5. 217 **6.** 302 **7.** 8099

To the nearest hundred: **8.** 119

9. 649 **10.** 2175 **11.** 6042

To the nearest thousand: **12.** 7423

13. 15,602 **14.** 22,094 **15.** 750

Decimals

Round 0.6843 to the nearest thousandth.

↓ Thousandths' place
0.6843
↑ 5 or more? No.

↓ Do not change.
0.684
↑ Drop the final digit.

Round.

To the nearest tenth: **16.** 0.63

17. 0.091 **18.** 0.407 **19.** 0.452

To the nearest hundredth: **20.** 0.652

21. 0.474 **22.** 0.168 **23.** 0.355

To the nearest thousandth: **24.** 0.4291

25. 0.6007 **26.** 0.0097 **27.** 0.2126

28. 6.3942 **29.** 137.4920 **30.** 9.9999

Mixed Practice

Round to the place value shown.

Nearest thousand:	**31.** 37,874	**32.** 19,266	**33.** 48,092
Nearest hundred:	**34.** 751	**35.** 919	**36.** 6771
Nearest ten:	**37.** 26	**38.** 6533	**39.** 575
Nearest unit:	**40.** 6.2	**41.** 35.73	**42.** 17.392
Nearest tenth:	**43.** 189.673	**44.** 10.009	**45.** 0.07
Nearest hundredth:	**46.** 0.392	**47.** 152.430	**48.** 0.6974
Nearest thousandth:	**49.** 0.1791	**50.** 16.0005	**51.** 108.437

Skill 3 Add Whole Numbers

Without Carrying

Add.

$$\begin{array}{r} 723 \\ 154 \\ +212 \\ \hline \end{array} \rightarrow \begin{array}{r} 723 \\ 154 \\ +212 \\ \hline 1089 \end{array}$$

1. 65 + 41
2. 76 + 32
3. 97 + 40
4. 32 + 25
5. 352 + 837
6. 361 + 834
7. 448 + 351
8. 125 + 604
9. 864 + 33
10. 721 + 77
11. 423 + 65
12. 108 + 91
13. 9037 + 1841
14. 9520 + 1379
15. 3924 + 5063
16. 2840 + 1152

With Carrying

Add.

$$\begin{array}{r} 8679 \\ +9748 \\ \hline \end{array} \rightarrow \begin{array}{r} \scriptstyle 1\ 11 \\ 8\,679 \\ +9\,748 \\ \hline 18{,}427 \end{array}$$

17. 32 + 39
18. 54 + 48
19. 187 + 23
20. 49 + 86
21. 728 + 169
22. 527 + 284
23. 845 + 697
24. 697 + 546
25. 3046 + 1592
26. 7801 + 3564
27. 5246 + 6978
28. 8347 + 1528
29. 8448 + 3753
30. 108 + 7665
31. 9179 + 3608
32. 982 + 2165

Mixed Practice

33. 1481 + 2317
34. 8495 + 1417
35. 5783 + 6535
36. 3950 + 1615
37. 6259 + 1893
38. 8347 + 1528
39. 6845 + 2639
40. 5692 + 1204
41. 2642 + 4135
42. 7921 + 2639
43. 7884 + 7069
44. 46,234 + 11,325
45. 17,694 + 15,893
46. 37,491 + 21,308
47. 59,641 + 27,840
48. 9100 536 + 2413
49. 7749 1240 + 6010
50. 6590 2408 + 5001
51. 5783 6535 + 2132
52. 6259 503 + 1893

Skill 4 Subtract Whole Numbers

Without Borrowing

Subtract.

$$\begin{array}{r} 9876 \\ -7545 \\ \hline \end{array} \Rightarrow \begin{array}{r} 9876 \\ -7545 \\ \hline 2331 \end{array}$$

1. 784 − 453	**2.** 985 − 734	**3.** 693 − 542	**4.** 199 − 158
5. 7659 − 4217	**6.** 8436 − 6223	**7.** 5792 − 2481	**8.** 4877 − 3614
9. 6754 − 5643	**10.** 1866 − 853	**11.** 8191 − 171	**12.** 1187 − 145
13. 479 − 473	**14.** 3987 − 3085	**15.** 6358 − 127	**16.** 1721 − 720

With Borrowing

Subtract.

$$\begin{array}{r} 9672 \\ -4136 \\ \hline \end{array} \Rightarrow \begin{array}{r} 9\,6\,\overset{6}{\cancel{7}}\,\overset{12}{\cancel{2}} \\ -4\,1\,3\,6 \\ \hline 5\,5\,3\,6 \end{array}$$

$$\begin{array}{r} 8352 \\ -4584 \\ \hline \end{array} \Rightarrow \begin{array}{r} \overset{7}{\cancel{8}}\,\overset{12}{\cancel{3}}\,\overset{14}{\cancel{5}}\,\overset{12}{\cancel{2}} \\ -4\,5\,8\,4 \\ \hline 3\,7\,6\,8 \end{array}$$

17. 100 − 36	**18.** 512 − 43	**19.** 602 − 503	**20.** 250 − 162
21. 6932 − 4674	**22.** 8724 − 2932	**23.** 4329 − 3163	**24.** 9721 − 6842
25. 6123 − 4214	**26.** 9231 − 6453	**27.** 7450 − 3783	**28.** 7734 − 5935
29. 8121 − 6846	**30.** 9000 − 7997	**31.** 9107 − 8248	**32.** 7734 − 5935

Mixed Practice

33. 6140 − 3157	**34.** 8005 − 6246	**35.** 7000 − 5432	**36.** 9297 − 9286	**37.** 9811 − 700
38. 9148 − 954	**39.** 2625 − 763	**40.** 1850 − 975	**41.** 7469 − 5231	**42.** 6342 − 5793
43. 10,743 − 7,842	**44.** 16,947 − 14,523	**45.** 22,493 − 5,967	**46.** 64,654 − 57,312	**47.** 79,850 − 42,347
48. 172,493 − 67,254	**49.** 249,657 − 123,254	**50.** 300,692 − 147,593	**51.** 647,593 − 546,972	**52.** 800,000 − 627,351

Skill 5 Add Decimals

Same Number of Places

Add.

```
 178.79         178.79
 596.24    ➧    596.24
+631.43        +631.43
               1406.46
```

1. 317.83 +161.16

2. 821.76 +178.23

3. 504.76 +296.36

4. 536.67 +197.47

5. 189.71 +601.48

6. 27.492 +31.608

7. 148.810 221.097 +173.206

8. 18.009 149.910 +251.295

9. 272.005 42.111 +502.543

Different Numbers of Places

Add.

```
 0.91           0.910
 6.647          6.647
 5.3       ➧    5.300
+16            +16.000
               28.857
```

Placeholders

10. 10 + 6.3

11. 42.9 +21.64

12. 63.21 +42.327

13. 141.21 4.136 + 32.003

14. 3.64 4.378 +0.213

15. 19.7 0.261 + 1.94

16. 18.09 49.91 +51.295

17. 0.161 18.24 + 6.059

18. 2.19 0.09 +0.7

19. 16.74 39.1 +26.492

20. 1.943 27.21 + 3.691

21. 2.5 1.692 +5.93

Mixed Practice

22. 60.148 +16.623

23. 271.195 +189.714

24. 163.2 + 37.915

25. 427.9 + 74.275

26. 693.642 +193.871

27. 973.573 81.91 + 9.20

28. 69 27.198 +178.26

29. 0.64 0.378 +0.223

30. 0.007 0.986 +0.034

31. 19.7 0.261 + 1.94

32. 17.92 23.81 42.63 +15.27

33. 103.69 27.4 83.621 + 5.9

34. 327.219 27.543 111.621 +257.619

35. 471.6 2.937 57.11 +241.005

36. 327.915 1.002 40.07 +154.6

Skill 6 Subtract Decimals

Same Number of Places

Subtract.

$$\begin{array}{r} 597.18 \\ -392.35 \\ \hline \end{array} \quad \Rightarrow \quad \begin{array}{r} \scriptstyle 6\ \ 11 \\ 59\not{7}.\not{1}8 \\ -392.35 \\ \hline 204.83 \end{array}$$

1. 65.46 − 14.31	**2.** 48.58 − 15.47	**3.** 151.02 − 16.13
4. 36.25 − 13.67	**5.** 87.56 − 82.47	**6.** 51.634 − 27.849
7. 69.37 − 43.86	**8.** 89.63 − 7.99	**9.** 109.46 − 29.78
10. 521.52 − 38.56	**11.** 321.02 − 117.18	**12.** 572.24 − 283.35

Different Numbers of Places

Subtract.

$$\begin{array}{r} 86.9 \\ -\ 3.84 \\ \hline \end{array} \quad \Rightarrow \quad \begin{array}{r} \scriptstyle 8\ 10 \\ 86.\not{9}\not{0} \\ -\ 3.84 \\ \hline 83.06 \end{array}$$

Placeholder

13. 79.6 − 8.75	**14.** 95.1 − 9.87	**15.** 100.1 − 15.78
16. 16.8 − 5.91	**17.** 36 − 16.4	**18.** 42 − 12.94
19. 17.9 − 9.83	**20.** 21 − 19.7	**21.** 67.2 − 9.76
22. 136.1 − 69.542	**23.** 771.9 − 394.27	**24.** 4578 − 878.127

Mixed Practice

25. 87.56 − 82.47	**26.** 39.27 − 18.38	**27.** 36.1 − 16.117	**28.** 4.546 − 2.558	**29.** 653.05 − 327.19
30. 198.20 − 64.897	**31.** 854.01 − 649.656	**32.** 316.07 − 118.29	**33.** 800.04 − 242.17	**34.** 985.93 − 99.794
35. 6194.9 − 978.954	**36.** 719.3 − 47.832	**37.** 5.9871 − 4.8693	**38.** 17.9328 − 6.2973	**39.** 843.002 − 64.973
40. 87.69 − 86.9975	**41.** 4.97652 − 1.37846	**42.** 3.29131 − 2.19378	**43.** 6.962 − 4.21698	**44.** 9.7 − 8.65947

Skill 7 Multiply Whole Numbers

Without Carrying

Multiply.

$$\begin{array}{r} 442 \\ \times 211 \\ \hline \end{array} \quad \Rightarrow \quad \begin{array}{rl} 442 & \\ \times 211 & \\ \hline 442 & \leftarrow 1 \times 442 \\ 4\,420 & \leftarrow 10 \times 442 \\ 88\,400 & \leftarrow 200 \times 442 \\ \hline 93{,}262 & \leftarrow 211 \times 442 \end{array}$$

1. 73 × 21
2. 42 × 22
3. 212 × 412
4. 311 × 232
5. 321 × 312
6. 223 × 323
7. 232 × 333
8. 7143 × 102
9. 8643 × 111

With Carrying

Multiply.

$$\begin{array}{r} 6524 \\ \times 273 \\ \hline \end{array} \quad \Rightarrow \quad \begin{array}{rl} 6524 & \\ \times 273 & \\ \hline 19\,572 & \leftarrow 3 \times 6524 \\ 456\,680 & \leftarrow 70 \times 6524 \\ 1\,304\,800 & \leftarrow 200 \times 6524 \\ \hline 1{,}781{,}052 & \leftarrow 273 \times 6524 \end{array}$$

10. 61 × 76
11. 78 × 36
12. 437 × 571
13. 465 × 541
14. 542 × 168
15. 8023 × 532
16. 64 × 27
17. 37 × 45
18. 68 × 71
19. 836 × 372
20. 7501 × 447
21. 5327 × 312

Mixed Practice

22. 480 × 10
23. 230 × 300
24. 641 × 237
25. 231 × 122
26. 122 × 40
27. 510 × 700
28. 8233 × 2584
29. 6010 × 6000
30. 9000 × 7011
31. 2793 × 1504
32. 19,008 × 8000
33. 8791 × 5000
34. 6743 × 27
35. 4231 × 253
36. 8427 × 19
37. 13,010 × 13
38. 14,231 × 12
39. 17,822 × 35
40. 22,300 × 15
41. 31,942 × 41
42. 27,642 × 321
43. 13,231 × 212
44. 14,402 × 121
45. 49,237 × 321
46. 64,159 × 347

Skill 8 Multiply Decimals

Decimals Greater Than One

Multiply.

$$\begin{array}{r} 17.45 \\ \times 2.7 \\ \hline \end{array} \Rightarrow \begin{array}{rl} 17.45 & \leftarrow \text{2 places} \\ \times 2.7 & \leftarrow \text{+1 place} \\ \hline 47.115 & \leftarrow \text{3 places} \end{array}$$

1. 2.5 × 1.8
2. 8.36 × 1.5
3. 10.2 × 8.61
4. 15.36 × 5.3
5. 25.14 × 7.5
6. 19.36 × 7.12
7. 27.06 × 8.53
8. 4.367 × 8.5
9. 5.564 × 7.9
10. 32.63 × 9.2
11. 31.20 × 9.21
12. 6.715 × 9.03

Decimals Less Than One

Multiply.

$$\begin{array}{r} 0.08 \\ \times 0.4 \\ \hline \end{array} \Rightarrow \begin{array}{rl} 0.08 & \leftarrow \text{2 places} \\ \times 0.4 & \leftarrow \text{+1 place} \\ \hline 0.032 & \leftarrow \text{3 places} \end{array}$$

13. 0.144 × 0.7
14. 0.86 × 0.5
15. 0.96 × 0.1
16. 0.56 × 0.07
17. 0.73 × 0.8
18. 0.05 × 0.9
19. 0.81 × 0.76
20. 0.47 × 0.84
21. 0.63 × 0.09
22. 0.57 × 0.03
23. 1.23 × 0.07
24. 0.01 × 0.05

Mixed Practice

25. 41.16 × 100
26. 0.923 × 0.49
27. 0.12 × 300
28. 67.32 × 10
29. 7.243 × 121
30. 557.4 × 100
31. 327.8 × 3.7
32. 14.923 × 0.76
33. 1.125 × 100
34. 0.009 × 1000
35. 2.014 × 40.7
36. 2.854 × 0.04
37. 6243.78 × 25.9
38. 5.9312 × 5.62
39. 0.534 × 0.293
40. 16.4591 × 51.23
41. 96.00 × 0.875
42. 0.3172 × .2008
43. 0.1543 × 0.4931
44. 0.7984 × 0.0003

Skill 9 Divide (Fractional Remainder)

Two-digit Divisor

Divide. $46\overline{)703}$

$$\begin{array}{r} 1 \\ 46\overline{)703} \\ -46 \\ \hline 24 \end{array} \quad \blacktriangleright \quad \begin{array}{r} 15 \\ 46\overline{)703} \\ -46\downarrow \\ \hline 243 \\ -230 \\ \hline 13 \end{array} \quad \blacktriangleright \quad \begin{array}{r} 15\frac{13}{46} \\ 46\overline{)703}\uparrow \\ -46\downarrow \\ \hline 243 \\ -230 \\ \hline 13 \end{array}$$

1. $27\overline{)63}$
2. $38\overline{)89}$
3. $41\overline{)84}$
4. $46\overline{)613}$
5. $53\overline{)754}$
6. $61\overline{)685}$
7. $21\overline{)1781}$
8. $55\overline{)4273}$
9. $43\overline{)2900}$
10. $73\overline{)3956}$
11. $23\overline{)1369}$
12. $34\overline{)2167}$
13. $81\overline{)5793}$
14. $93\overline{)78}$

Three-digit Divisor

Divide. $472\overline{)9463}$

$$\begin{array}{r} 2 \\ 472\overline{)9463} \\ -944 \\ \hline 2 \end{array} \quad \blacktriangleright \quad \begin{array}{r} 20\frac{23}{472} \\ 472\overline{)9463}\uparrow \\ -944\downarrow \\ \hline 23 \\ -\ 0 \\ \hline 23 \end{array}$$

15. $114\overline{)1837}$
16. $216\overline{)5417}$
17. $321\overline{)4832}$
18. $429\overline{)9038}$
19. $892\overline{)7928}$
20. $910\overline{)11,849}$
21. $409\overline{)9429}$
22. $900\overline{)7621}$
23. $625\overline{)8742}$
24. $710\overline{)15,694}$
25. $843\overline{)17,691}$
26. $937\overline{)22,474}$

Mixed Practice

27. $19\overline{)342}$
28. $46\overline{)782}$
29. $71\overline{)1136}$
30. $279\overline{)9778}$
31. $35\overline{)2520}$
32. $509\overline{)5089}$
33. $621\overline{)8716}$
34. $49\overline{)4959}$
35. $549\overline{)6937}$
36. $953\overline{)9597}$
37. $842\overline{)4975}$
38. $87\overline{)2139}$
39. $473\overline{)85,642}$
40. $192\overline{)91,845}$
41. $622\overline{)54,641}$
42. $812\overline{)55,692}$
43. $51\overline{)81,603}$
44. $23\overline{)44,364}$
45. $88\overline{)22,222}$
46. $34\overline{)27,945}$
47. $417\overline{)86,902}$
48. $71\overline{)103,205}$
49. $514\overline{)691,507}$
50. $64\overline{)237,605}$

Skill 10 Divide (Decimal Remainder)

Exact Quotient

Divide. $28\overline{)378}$

$$\begin{array}{r} 13 \\ 28\overline{)378} \\ -28 \\ \hline 98 \\ -84 \\ \hline 14 \end{array} \quad \Rightarrow \quad \begin{array}{r} 13.5 \\ 28\overline{)378.0} \\ -28 \\ \hline 98 \\ -84 \\ \hline 140 \\ -140 \\ \hline \end{array}$$

1. $8\overline{)89}$
2. $10\overline{)123}$
3. $15\overline{)318}$
4. $16\overline{)264}$
5. $20\overline{)102}$
6. $50\overline{)530}$
7. $12\overline{)237}$
8. $25\overline{)1310}$
9. $32\overline{)1104}$
10. $96\overline{)2988}$

Rounded Quotient

Divide. $39\overline{)818}$

Round to the nearest hundredth.

$$\begin{array}{r} 20 \\ 39\overline{)818} \\ -78 \\ \hline 38 \end{array} \quad \Rightarrow \quad \begin{array}{r} 20.974 \\ 39\overline{)818.000} \\ -78 \\ \hline 380 \\ -351 \\ \hline 290 \\ -273 \\ \hline 170 \\ -156 \\ \hline 14 \end{array}$$

(Thousandths' place: the 4 in 20.974)

20.974 rounded to the nearest hundredth is 20.97. ← Skill 2

Round to the place value shown.

Nearest tenth:

11. $14\overline{)319}$
12. $26\overline{)347}$
13. $23\overline{)371}$
14. $46\overline{)9415}$
15. $47\overline{)9719}$

Nearest hundredth:

16. $19\overline{)427}$
17. $83\overline{)168}$
18. $47\overline{)432}$

Nearest thousandth:

19. $37\overline{)402}$
20. $24\overline{)643}$
21. $21\overline{)452}$

Mixed Practice

Divide. Round to the nearest hundredth.

22. $28\overline{)1022}$
23. $24\overline{)209}$
24. $72\overline{)1665}$
25. $29\overline{)303}$
26. $85\overline{)1802}$
27. $24\overline{)747}$
28. $67\overline{)701}$
29. $71\overline{)400}$
30. $23\overline{)273}$
31. $36\overline{)630}$
32. $44\overline{)858}$
33. $59\overline{)852}$
34. $37\overline{)673}$
35. $41\overline{)9432}$
36. $73\overline{)1079}$
37. $24\overline{)994}$
38. $27\overline{)3365}$
39. $35\overline{)894}$
40. $42\overline{)5264}$
41. $110\overline{)4345}$

Skill 11 Divide Decimals

Divisor Greater Than One

Divide. $7.2\overline{)16.92}$

$7.2\overline{)16.92}$ ➧ $72\overline{)169.20}$

Move 1 decimal place.

$$\begin{array}{r} 2.35 \\ 72\overline{)169.20} \\ -144 \\ \hline 252 \\ -216 \\ \hline 360 \\ -360 \\ \hline \end{array}$$

Round to the nearest hundredth.

1. $2.5\overline{)11.5}$ **2.** $3.8\overline{)4.56}$

3. $3.2\overline{)18.272}$ **4.** $7.5\overline{)27.823}$

5. $3.15\overline{)53.55}$ **6.** $24.12\overline{)369.036}$

7. $4.08\overline{)26.52}$ **8.** $3.02\overline{)10.57}$

9. $4.23\overline{)181.5}$ **10.** $6.67\overline{)25.963}$

Divisor Less Than One

Divide. $0.032\overline{)14.400}$

$0.032\overline{)14.400}$ ➧ $32\overline{)14400}$

Move 3 decimal places.

$$\begin{array}{r} 450 \\ 32\overline{)14400} \\ -128 \\ \hline 160 \\ -160 \\ \hline 00 \\ -00 \\ \hline \end{array}$$

Round to the nearest hundredth.

11. $0.24\overline{)6.24}$ **12.** $0.372\overline{)6.324}$

13. $0.154\overline{)4.774}$ **14.** $0.48\overline{)2.938}$

15. $0.37\overline{)9.62}$ **16.** $0.21\overline{)1.374}$

17. $0.51\overline{)4.569}$ **18.** $0.67\overline{)2.693}$

19. $0.73\overline{)9.641}$ **20.** $0.81\overline{)11.632}$

Mixed Practice

Divide. Round to the nearest hundredth.

21. $3.12\overline{)4.386}$ **22.** $0.73\overline{)9.48}$ **23.** $0.136\overline{)33.32}$ **24.** $0.21\overline{)130.2}$

25. $6.94\overline{)8.378}$ **26.** $1.23\overline{)0.3813}$ **27.** $0.065\overline{)16.64}$ **28.** $8.34\overline{)7.416}$

29. $0.63\overline{)42.51}$ **30.** $2.91\overline{)5.932}$ **31.** $1.07\overline{)24.153}$ **32.** $1.1\overline{)29.9}$

33. $15.93\overline{)27.931}$ **34.** $12.12\overline{)36.422}$ **35.** $0.05\overline{)1.925}$ **36.** $2.03\overline{)21.249}$

37. $0.007\overline{)0.692}$ **38.** $1.05\overline{)25.421}$ **39.** $0.31\overline{)0.00354}$ **40.** $15.42\overline{)113.005}$

41. $5.02\overline{)86}$ **42.** $0.03\overline{)29}$ **43.** $5.4\overline{)0.062}$ **44.** $0.068\overline{)0.009}$

Skill 12 Equivalent Fractions

Higher Terms

Complete the equivalent fraction.

$$\frac{3}{4} = \frac{?}{20}$$

$$\frac{3}{4} \overset{\times 5}{=} \frac{?}{20} \Rightarrow \frac{3}{4} = \frac{15}{20}$$

Solve.

1. $\frac{5}{6} = \frac{?}{12}$
2. $\frac{4}{9} = \frac{?}{27}$
3. $\frac{7}{10} = \frac{35}{?}$
4. $\frac{11}{12} = \frac{22}{?}$
5. $\frac{8}{17} = \frac{?}{51}$
6. $\frac{15}{17} = \frac{90}{?}$
7. $\frac{11}{18} = \frac{?}{36}$
8. $\frac{4}{19} = \frac{12}{?}$
9. $\frac{16}{17} = \frac{48}{?}$
10. $\frac{11}{13} = \frac{?}{39}$
11. $\frac{3}{4} = \frac{?}{32}$
12. $\frac{15}{17} = \frac{60}{?}$
13. $\frac{7}{9} = \frac{?}{45}$
14. $\frac{11}{12} = \frac{?}{72}$
15. $\frac{5}{8} = \frac{45}{?}$
16. $\frac{11}{8} = \frac{33}{?}$
17. $\frac{5}{24} = \frac{?}{96}$
18. $\frac{23}{73} = \frac{?}{365}$

Lowest Terms

Reduce $\frac{12}{28}$ to lowest terms.

$$\frac{12}{28} \overset{\div 4}{=} \frac{3}{7} \Rightarrow \frac{12}{28} = \frac{3}{7}$$

Reduce to lowest terms.

19. $\frac{6}{9}$
20. $\frac{4}{16}$
21. $\frac{9}{18}$
22. $\frac{20}{22}$
23. $\frac{18}{27}$
24. $\frac{24}{32}$
25. $\frac{16}{48}$
26. $\frac{18}{42}$
27. $\frac{55}{66}$
28. $\frac{30}{50}$
29. $\frac{20}{54}$
30. $\frac{21}{27}$
31. $\frac{8}{14}$
32. $\frac{19}{57}$
33. $\frac{14}{28}$
34. $\frac{15}{25}$
35. $\frac{21}{28}$
36. $\frac{200}{365}$
37. $\frac{188}{366}$
38. $\frac{150}{365}$
39. $\frac{180}{360}$
40. $\frac{225}{365}$
41. $\frac{190}{360}$
42. $\frac{65}{75}$
43. $\frac{72}{468}$
44. $\frac{183}{366}$
45. $\frac{232}{1450}$
46. $\frac{1792}{5120}$

Skills

Skill 13 Change Mixed Numbers/ Improper Fractions

Mixed Numbers to Improper Fractions

Write $3\frac{7}{8}$ as an improper fraction.

$$3\frac{7}{8} = \frac{(3 \times 8) + 7}{8}$$

$$3\frac{7}{8} = \frac{24 + 7}{8}$$

$$3\frac{7}{8} = \frac{31}{8}$$

Write as improper fractions.

1. $6\frac{1}{4}$	**2.** $4\frac{3}{4}$	**3.** $7\frac{3}{8}$	**4.** $9\frac{1}{2}$
5. $6\frac{2}{3}$	**6.** $3\frac{5}{6}$	**7.** $7\frac{4}{5}$	**8.** $3\frac{1}{8}$
9. $4\frac{3}{10}$	**10.** $1\frac{1}{16}$	**11.** $4\frac{17}{32}$	**12.** $4\frac{1}{5}$
13. $5\frac{1}{3}$	**14.** $2\frac{1}{6}$	**15.** $2\frac{9}{10}$	**16.** $3\frac{1}{10}$
17. $3\frac{5}{16}$	**18.** $4\frac{9}{16}$	**19.** $5\frac{5}{32}$	**20.** $6\frac{3}{5}$
21. $7\frac{2}{11}$	**22.** $8\frac{5}{6}$	**23.** $11\frac{2}{5}$	**24.** $13\frac{1}{3}$

Improper Fractions to Mixed Numbers

Write $\frac{17}{3}$ as a mixed number.

$$\frac{17}{3} \Rightarrow \begin{array}{r} 5\frac{2}{3} \\ 3\overline{)17} \\ -15 \\ \hline 2 \end{array}$$

← Skill 9

$$\frac{17}{3} = 5\frac{2}{3}$$

Write as mixed numbers. Reduce any fractional parts to lowest terms.

25. $\frac{13}{2}$	**26.** $\frac{18}{4}$	**27.** $\frac{46}{8}$	**28.** $\frac{19}{4}$
29. $\frac{33}{6}$	**30.** $\frac{30}{7}$	**31.** $\frac{21}{9}$	**32.** $\frac{28}{8}$
33. $\frac{45}{18}$	**34.** $\frac{49}{21}$	**35.** $\frac{62}{21}$	**36.** $\frac{23}{3}$
37. $\frac{35}{6}$	**38.** $\frac{47}{7}$	**39.** $\frac{57}{18}$	**40.** $\frac{37}{9}$
41. $\frac{49}{8}$	**42.** $\frac{63}{2}$	**43.** $\frac{45}{7}$	**44.** $\frac{71}{4}$

Skill 14 Change Fractions/Decimals

Fraction to Decimal

Write $\frac{5}{12}$ as a decimal. Round to the nearest hundredth.

$$\frac{5}{12} \rightarrow 12\overline{)5.000} = 0.416$$ ← Skill 10

0.416 rounded to the nearest hundredth is 0.42. ← Skill 2

$$\frac{5}{12} = 0.42$$

Write as decimals. Round answers to the nearest hundredth.

1. $\frac{4}{5}$ **2.** $\frac{7}{20}$ **3.** $1\frac{1}{8}$ **4.** $\frac{3}{4}$

5. $2\frac{3}{7}$ **6.** $\frac{7}{12}$ **7.** $1\frac{11}{30}$ **8.** $\frac{2}{5}$

9. $\frac{3}{10}$ **10.** $3\frac{4}{25}$ **11.** $4\frac{3}{8}$ **12.** $\frac{7}{10}$

13. $3\frac{1}{12}$ **14.** $\frac{1}{15}$ **15.** $\frac{1}{30}$ **16.** $2\frac{9}{10}$

17. $5\frac{3}{20}$ **18.** $\frac{9}{20}$ **19.** $2\frac{3}{25}$ **20.** $\frac{1}{8}$

21. $7\frac{1}{7}$ **22.** $2\frac{5}{8}$ **23.** $\frac{4}{9}$ **24.** $1\frac{7}{8}$

Decimal to Fraction

Write 0.42 as a fraction in lowest terms.

$$0.42 = \frac{42}{100} = \frac{21}{50}$$

(Skill 1 → $\frac{42}{100}$; Skill 12 → $\frac{21}{50}$)

$$0.42 = \frac{21}{50}$$

Write as a fraction in lowest terms.

25. 0.1 **26.** 3.7 **27.** 0.30

28. 2.25 **29.** 0.03 **30.** 0.09

31. 0.53 **32.** 1.75 **33.** 0.003

34. 0.010 **35.** 0.064 **36.** 4.206

37. 4.444 **38.** 0.732 **39.** 0.469

40. 2.9 **41.** 7.5 **42.** 0.32

43. 1.08 **44.** 0.06 **45.** 2.83

46. 1.039 **47.** 0.105 **48.** 2.422

49. 0.0005 **50.** 0.2482 **51.** 1.6432

52. 0.0058 **53.** 0.0002 **54.** 6.66

Skill 15 Add Fractions, Like Denominators

Fractions

Add.

Like denominator

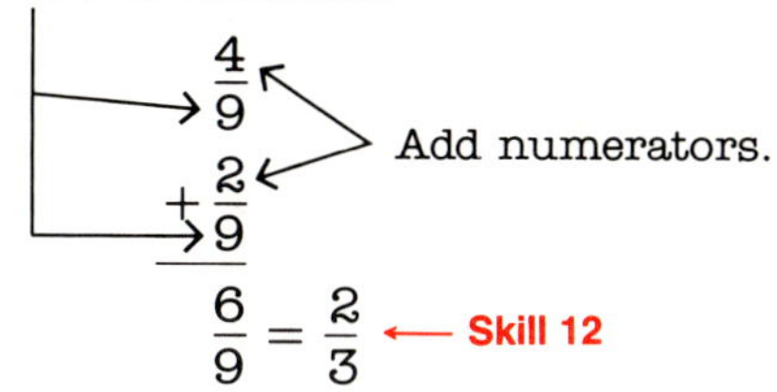

Express the sum in lowest terms.

1. $\frac{4}{5} + \frac{3}{5}$ **2.** $\frac{4}{7} + \frac{2}{7}$ **3.** $\frac{1}{8} + \frac{5}{8}$

4. $\frac{5}{9} + \frac{7}{9}$ **5.** $\frac{5}{7} + \frac{6}{7}$ **6.** $\frac{11}{12} + \frac{7}{12}$

7. $\frac{13}{25} + \frac{16}{25}$ **8.** $\frac{11}{30} + \frac{19}{30}$ **9.** $\frac{15}{32} + \frac{27}{32}$

10. $\frac{21}{40} + \frac{31}{40}$ **11.** $\frac{22}{45} + \frac{31}{45}$ **12.** $\frac{11}{50} + \frac{19}{50}$

Mixed Numbers

Add.

Like denominator

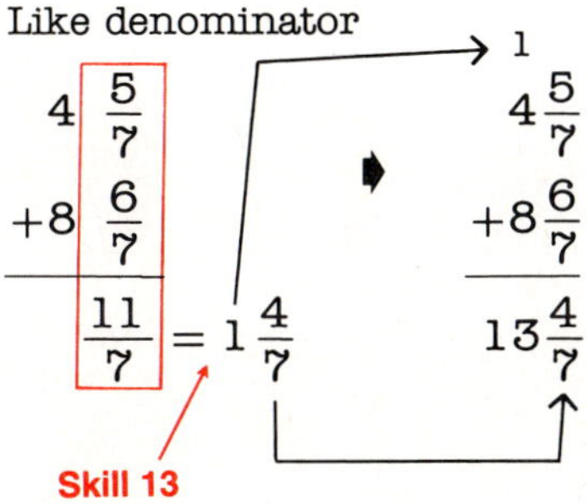

Express the sum in lowest terms.

13. $4\frac{1}{3} + 7\frac{1}{3}$ **14.** $13\frac{4}{5} + 8\frac{3}{5}$

15. $5\frac{4}{7} + 8\frac{5}{7}$ **16.** $12\frac{3}{8} + 14\frac{1}{8}$

17. $6\frac{11}{12} + 5\frac{5}{12}$ **18.** $2\frac{12}{13} + 7\frac{2}{13}$

19. $8\frac{9}{16} + 8\frac{11}{16}$ **20.** $14\frac{5}{24} + 15\frac{7}{24}$

21. $4\frac{13}{32} + 5\frac{15}{32}$ **22.** $15\frac{19}{45} + 6\frac{28}{45}$

23. $7\frac{8}{35} + 4\frac{6}{35}$ **24.** $9\frac{5}{32} + 14\frac{7}{32}$

Skill 16 Add Fractions, Unlike Denominators

Fractions

Add.

Unlike denominators → Like denominator

$$\frac{5}{8} + \frac{2}{3} \;\Rightarrow\; \frac{15}{24} + \frac{16}{24} \;\Rightarrow\; \frac{15}{24} + \frac{16}{24} = \frac{31}{24} = 1\frac{7}{24}$$

Skill 12 (rewrite with like denominators); Skill 15 (add like fractions, $\frac{31}{24}$); Skill 13 ($1\frac{7}{24}$)

Express the sum in lowest terms.

1. $\frac{1}{2} + \frac{3}{5}$ **2.** $\frac{3}{4} + \frac{1}{6}$ **3.** $\frac{2}{7} + \frac{2}{3}$

4. $\frac{3}{8} + \frac{1}{5}$ **5.** $\frac{5}{6} + \frac{1}{3}$ **6.** $\frac{7}{12} + \frac{3}{7}$

7. $\frac{9}{11} + \frac{3}{10}$ **8.** $\frac{13}{16} + \frac{9}{8}$ **9.** $\frac{7}{20} + \frac{19}{30}$

10. $\frac{5}{18} + \frac{19}{24}$ **11.** $\frac{1}{25} + \frac{13}{30}$ **12.** $\frac{2}{15} + \frac{29}{30}$

Mixed Numbers

Add.

$$12\frac{2}{5} + 9\frac{3}{4} \;\Rightarrow\; 12\frac{8}{20} + 9\frac{15}{20},\quad \frac{23}{20} = 1\frac{3}{20} \;\Rightarrow\; 12\frac{8}{20} + 9\frac{15}{20} = 22\frac{3}{20}$$

Skill 12 (rewrite $\frac{8}{20}$, $\frac{15}{20}$); Skill 15 ($22\frac{3}{20}$)

Express the sum in lowest terms.

13. $5\frac{1}{2} + 3\frac{2}{3}$ **14.** $4\frac{3}{8} + 9\frac{3}{4}$

15. $7\frac{5}{6} + 8\frac{1}{2}$ **16.** $9\frac{7}{8} + 10\frac{7}{16}$

17. $11\frac{2}{13} + 9\frac{15}{26}$ **18.** $18\frac{2}{3} + 5\frac{7}{11}$

19. $12\frac{4}{7} + 15\frac{7}{9}$ **20.** $16\frac{5}{6} + 10\frac{6}{7}$

21. $16\frac{5}{8} + 12\frac{1}{7}$ **22.** $11\frac{4}{9} + 13\frac{3}{8}$

23. $4\frac{2}{3} + 24\frac{7}{16}$ **24.** $15\frac{5}{8} + 9\frac{4}{7}$

Skill 17 Subtract Fractions, Like Denominator

Fractions

Subtract.

Like denominator

$$\frac{11}{12} - \frac{7}{12} \Rightarrow \frac{11}{12} - \frac{7}{12} = \frac{4}{12} = \frac{1}{3}$$

Subtract numerators.

$\frac{4}{12} = \frac{1}{3}$ ← Skill 12

Express answers in lowest terms.

1. $\frac{4}{9} - \frac{2}{9}$ **2.** $\frac{9}{8} - \frac{3}{8}$ **3.** $\frac{11}{12} - \frac{1}{12}$

4. $\frac{5}{6} - \frac{1}{6}$ **5.** $\frac{5}{7} - \frac{1}{7}$ **6.** $\frac{19}{27} - \frac{1}{27}$

7. $\frac{11}{16} - \frac{9}{16}$ **8.** $\frac{16}{25} - \frac{6}{25}$ **9.** $\frac{37}{40} - \frac{29}{40}$

10. $\frac{29}{36} - \frac{11}{36}$ **11.** $\frac{7}{10} - \frac{3}{10}$ **12.** $\frac{19}{36} - \frac{5}{36}$

13. $\frac{19}{24} - \frac{13}{24}$ **14.** $\frac{13}{28} - \frac{7}{28}$ **15.** $\frac{29}{60} - \frac{8}{60}$

Mixed Numbers

Subtract.

$$8\frac{11}{16} - 5\frac{9}{16} \Rightarrow 8\frac{11}{16} - 5\frac{9}{16} : \frac{2}{16} = \frac{1}{8} \Rightarrow 8\frac{11}{16} - 5\frac{9}{16} = 3\frac{1}{8}$$

Skill 12

Express answers in lowest terms.

16. $7\frac{3}{4} - 6\frac{1}{4}$ **17.** $8\frac{6}{7} - 7\frac{5}{7}$

18. $10\frac{4}{5} - 6\frac{1}{5}$ **19.** $12\frac{5}{8} - 9\frac{3}{8}$

20. $2\frac{16}{21} - 1\frac{5}{21}$ **21.** $7\frac{11}{24} - 5\frac{7}{24}$

22. $14\frac{17}{20} - 8\frac{9}{20}$ **23.** $15\frac{15}{32} - 10\frac{9}{32}$

24. $14\frac{7}{8} - 9\frac{3}{8}$ **25.** $12\frac{11}{13} - 10\frac{4}{13}$

26. $18\frac{17}{20} - 15\frac{9}{20}$ **27.** $35\frac{39}{40} - 27\frac{19}{40}$

28. $74\frac{41}{75} - 29\frac{16}{75}$ **29.** $103\frac{11}{35} - 78\frac{1}{35}$

Skill 18 Subtract Fractions, Unlike Denominators

Fractions

Subtract.

Unlike denominators

$$\begin{array}{r} \frac{5}{6} \\ -\frac{3}{10} \\ \hline \end{array} \quad \Rightarrow \quad \begin{array}{r} \frac{25}{30} \\ -\frac{9}{30} \\ \hline \end{array} \quad \Rightarrow \quad \begin{array}{r} \frac{25}{30} \\ -\frac{9}{30} \\ \hline \frac{16}{30} \end{array} = \frac{8}{15}$$

Skill 12 (rewriting $\frac{5}{6}$ and $\frac{3}{10}$ as $\frac{25}{30}$ and $\frac{9}{30}$)

Skill 17 (subtracting to get $\frac{16}{30}$)

Skill 12 (writing $\frac{8}{15}$ in lowest terms)

Express answers in lowest terms.

1. $\frac{3}{4} - \frac{3}{8}$ **2.** $\frac{1}{3} - \frac{2}{9}$ **3.** $\frac{9}{10} - \frac{3}{4}$

4. $\frac{4}{5} - \frac{2}{3}$ **5.** $\frac{5}{6} - \frac{7}{9}$ **6.** $\frac{5}{8} - \frac{5}{12}$

7. $\frac{3}{5} - \frac{8}{15}$ **8.** $\frac{11}{12} - \frac{3}{4}$ **9.** $\frac{4}{5} - \frac{13}{20}$

10. $\frac{6}{7} - \frac{10}{21}$ **11.** $\frac{1}{2} - \frac{2}{5}$ **12.** $\frac{3}{4} - \frac{11}{16}$

Mixed Numbers

Subtract.

$$\begin{array}{r} 9\frac{11}{12} \\ -7\frac{3}{8} \\ \hline \end{array} \quad \Rightarrow \quad \begin{array}{r} 9\frac{22}{24} \\ -7\frac{9}{24} \\ \hline 2\frac{13}{24} \end{array}$$

Skill 12 (rewriting as $\frac{22}{24}$ and $\frac{9}{24}$)

Skill 17 (subtracting to get $2\frac{13}{24}$)

Express answers in lowest terms.

13. $9\frac{3}{4} - 5\frac{1}{2}$ **14.** $7\frac{5}{6} - 4\frac{2}{3}$

15. $14\frac{4}{9} - 8\frac{1}{6}$ **16.** $15\frac{3}{7} - 10\frac{1}{3}$

17. $34\frac{11}{12} - 29\frac{4}{5}$ **18.** $48\frac{4}{7} - 37\frac{5}{9}$

19. $12\frac{5}{8} - 9\frac{1}{2}$ **20.** $2\frac{16}{21} - 1\frac{2}{3}$

21. $13\frac{3}{4} - 11\frac{3}{5}$ **22.** $21\frac{5}{6} - 19\frac{5}{8}$

23. $7\frac{11}{24} - 5\frac{1}{3}$ **24.** $20\frac{15}{38} - 20\frac{1}{19}$

Skill 19 Subtract Mixed Numbers, Borrowing

Whole Number and a Mixed Number

Subtract. Express answers in lowest terms.

$$\begin{array}{r} 8 \\ -4\frac{3}{5} \\ \hline \end{array} \quad \Rightarrow \quad \begin{array}{r} 7\frac{5}{5} \\ -4\frac{3}{5} \\ \hline 3\frac{2}{5} \end{array}$$

Like denominator

1. $3 - 1\frac{1}{2}$ **2.** $4 - 2\frac{3}{4}$ **3.** $10 - 7\frac{9}{10}$

4. $6 - 4\frac{3}{8}$ **5.** $12 - 9\frac{5}{11}$ **6.** $15 - 12\frac{7}{16}$

Mixed Numbers with Unlike Denominators

Subtract. Express answers in lowest terms.

Skill 12

$$\begin{array}{r} 6\frac{2}{3} \\ -2\frac{4}{5} \\ \hline \end{array} \quad \Rightarrow \quad \begin{array}{r} 6\frac{10}{15} \\ -2\frac{12}{15} \\ \hline \end{array} \quad \Rightarrow \quad \begin{array}{r} 5\frac{25}{15} \\ -2\frac{12}{15} \\ \hline 3\frac{13}{15} \end{array}$$

7. $5\frac{1}{2} - 2\frac{4}{9}$ **8.** $8\frac{1}{2} - 3\frac{5}{8}$

9. $18\frac{3}{4} - 5\frac{5}{6}$ **10.** $27\frac{2}{7} - 13\frac{3}{4}$

11. $28\frac{3}{7} - 26\frac{9}{10}$ **12.** $40\frac{19}{30} - 39\frac{3}{4}$

13. $6\frac{3}{4} - 3\frac{7}{8}$ **14.** $16\frac{1}{12} - 9\frac{3}{10}$

Mixed Practice

15. $4 - 2\frac{5}{8}$ **16.** $9 - 5\frac{11}{12}$ **17.** $12 - 11\frac{7}{8}$ **18.** $16 - 7\frac{5}{11}$

19. $8\frac{1}{3} - 4\frac{5}{9}$ **20.** $13\frac{1}{2} - 7\frac{3}{5}$ **21.** $14\frac{1}{5} - 13\frac{3}{4}$ **22.** $13 - 6\frac{5}{8}$

23. $27\frac{1}{4} - 15\frac{15}{16}$ **24.** $38\frac{3}{7} - 29\frac{9}{10}$ **25.** $18\frac{1}{12} - 3\frac{11}{36}$ **26.** $17\frac{8}{9} - 5\frac{3}{8}$

27. $19 - 5\frac{2}{7}$ **28.** $16\frac{4}{5} - 6\frac{3}{4}$ **29.** $19\frac{11}{32} - 7\frac{3}{16}$ **30.** $25 - 8\frac{7}{10}$

Skill 20 Multiply Fractions/Mixed Numbers

Fractions

Multiply.

$$\frac{5}{8} \times \frac{2}{3} = \frac{5 \times 2}{8 \times 3} = \frac{10}{24} = \frac{5}{12}$$

Skill 12

Express answers in lowest terms.

1. $\frac{1}{2} \times \frac{2}{3}$ **2.** $\frac{1}{4} \times \frac{3}{5}$ **3.** $\frac{5}{6} \times \frac{3}{4}$

4. $\frac{8}{9} \times \frac{2}{7}$ **5.** $\frac{11}{12} \times \frac{5}{11}$ **6.** $\frac{4}{13} \times \frac{8}{9}$

7. $4 \times \frac{1}{2}$ **8.** $17 \times \frac{2}{5}$ **9.** $21 \times \frac{3}{7}$

Mixed Numbers

Multiply.

$$4\frac{1}{3} \times 2\frac{1}{4} = \frac{13}{3} \times \frac{9}{4} = \frac{117}{12} = 9\frac{3}{4}$$

Skill 13

Express answers in lowest terms.

10. $1\frac{1}{2} \times 1\frac{1}{3}$ **11.** $3\frac{2}{3} \times 4\frac{2}{5}$

12. $\frac{1}{8} \times 4\frac{4}{5}$ **13.** $7 \times 8\frac{1}{3}$

14. $12\frac{1}{2} \times 1\frac{1}{2}$ **15.** $20\frac{1}{2} \times 2\frac{1}{4}$

Mixed Practice

Multiply. Express answers in lowest terms.

16. $\frac{2}{5} \times \frac{4}{7}$ **17.** $\frac{3}{8} \times \frac{4}{9}$ **18.** $\frac{2}{3} \times 1\frac{1}{8}$ **19.** $3\frac{2}{3} \times \frac{1}{4}$ **20.** $6 \times \frac{4}{5}$

21. $\frac{3}{4} \times 8$ **22.** $3\frac{1}{3} \times 4\frac{2}{5}$ **23.** $2\frac{1}{4} \times 1\frac{2}{3}$ **24.** $\frac{3}{4} \times 2\frac{5}{6}$ **25.** $6 \times 3\frac{1}{4}$

26. $18\frac{1}{2} \times 2\frac{2}{3}$ **27.** $1\frac{3}{5} \times 5\frac{2}{6}$ **28.** $5 \times 15\frac{1}{2}$ **29.** $10 \times 6\frac{2}{5}$ **30.** $2\frac{2}{7} \times 1\frac{1}{9}$

31. $3\frac{4}{9} \times 7\frac{3}{8}$ **32.** $\frac{7}{10} \times \frac{5}{9}$ **33.** $11\frac{3}{4} \times 8\frac{1}{3}$ **34.** $\frac{11}{12} \times \frac{4}{33}$ **35.** $6\frac{7}{8} \times 10\frac{1}{3}$

Skill 21 Divide Fractions/Mixed Numbers

Fractions

Divide.

$$\frac{2}{3} \div \frac{3}{8} = \frac{2}{3} \times \frac{8}{3} = 1\frac{7}{9}$$

Skill 20

Express answers in lowest terms.

1. $\frac{1}{4} \div \frac{1}{8}$ **2.** $\frac{3}{8} \div \frac{1}{4}$ **3.** $\frac{3}{4} \div \frac{2}{5}$

4. $5 \div \frac{5}{6}$ **5.** $7 \div \frac{14}{15}$ **6.** $8 \div \frac{4}{11}$

7. $\frac{3}{4} \div \frac{1}{8}$ **8.** $\frac{2}{5} \div \frac{5}{6}$ **9.** $\frac{4}{7} \div \frac{1}{28}$

Mixed Numbers

Divide.

Skill 13

$$5\frac{1}{6} \div 1\frac{1}{4} = \frac{31}{6} \div \frac{5}{4}$$

$$= \frac{31}{6} \times \frac{4}{5}$$

$$= 4\frac{2}{15}$$

Skill 20

Express answers in lowest terms.

10. $3\frac{2}{3} \div 1\frac{1}{2}$ **11.** $3\frac{5}{6} \div 1\frac{1}{5}$

12. $2\frac{1}{6} \div 3\frac{1}{3}$ **13.** $4\frac{2}{3} \div 1\frac{3}{5}$

14. $\frac{3}{4} \div 3\frac{1}{2}$ **15.** $\frac{5}{8} \div 2\frac{1}{2}$

16. $16 \div 1\frac{1}{8}$ **17.** $11\frac{1}{3} \div 2\frac{1}{5}$

18. $12 \div 2\frac{12}{17}$ **19.** $13\frac{1}{7} \div 3\frac{1}{6}$

Mixed Practice

Divide. Express answers in lowest terms.

20. $7\frac{1}{2} \div 1\frac{1}{4}$ **21.** $8\frac{1}{4} \div 5\frac{1}{2}$ **22.** $\frac{5}{11} \div 2\frac{1}{5}$ **23.** $\frac{9}{10} \div \frac{3}{5}$ **24.** $\frac{7}{12} \div \frac{1}{6}$

25. $1\frac{3}{8} \div 2\frac{1}{16}$ **26.** $13\frac{1}{3} \div \frac{1}{10}$ **27.** $8\frac{1}{3} \div \frac{5}{9}$ **28.** $\frac{19}{21} \div 3\frac{1}{2}$ **29.** $3\frac{5}{7} \div \frac{13}{21}$

30. $8 \div 1\frac{3}{5}$ **31.** $20 \div \frac{2}{3}$ **32.** $2\frac{1}{16} \div 11$ **33.** $5\frac{3}{8} \div 22$ **34.** $22 \div \frac{3}{5}$

35. $\frac{11}{15} \div 1\frac{12}{13}$ **36.** $\frac{11}{12} \div \frac{11}{12}$ **37.** $2\frac{11}{14} \div \frac{3}{7}$ **38.** $\frac{15}{16} \div \frac{5}{32}$ **39.** $19\frac{1}{2} \div 19\frac{1}{2}$

Skill 22 Write Ratios

Compare Two Numbers

What is the ratio of desks to typewriters?

OFFICE INVENTORY		
Typewriters	Calculators	Desks
6	5	15

Ratio of desks to typewriters:

15 to 6 or 15:6 or $\frac{15}{6}$

Write the ratio.

Department Sales		
Meat	Produce	Dairy
$1500	$600	$1200

1. Ratio of meat to produce
2. Ratio of meat to dairy
3. Ratio of produce to dairy
4. Ratio of dairy to produce
5. Ratio of produce to meat
6. Ratio of dairy to meat

Ratios as Fractions

Write the ratio of tables to chairs as a fraction in lowest terms.

A restaurant has 40 chairs and 16 tables.

Ratio of tables to chairs is $\frac{16}{40}$.

$$\frac{16}{40} = \frac{2}{5}$$ ← Skill 12

Write the ratio as a fraction in lowest terms.

7. Nurse-to-patient ratio in a hospital is 6 nurses to 30 patients.
8. Teacher-to-student ratio in a school is 8 teachers to 160 students.
9. Door-to-window ratio in a house is 30 doors to 45 windows.
10. Width-to-length ratio of a box is 22 cm to 30 cm.
11. Room-to-desk ratio in a school is 40 rooms to 1000 desks.
12. Land area-to-people ratio in a county is 6 km^2 to 138 people.
13. Car-to-people ratio in a town is 4000 cars to 10,000 people.
14. Hit-to-strikeout ratio of a batter is 15 hits to 25 strikeouts.

Skill 23 Proportions

Checking Proportions

Is this proportion true?

$\frac{5}{9} \stackrel{?}{=} \frac{35}{63}$

Cross multiply.

$\frac{5}{9} \stackrel{?}{=} \frac{35}{63}$ ➧ $5 \times 63 \stackrel{?}{=} 9 \times 35$

$315 = 315$

True. The products are equal.

Tell whether the proportions are true.

1. $\frac{1}{3} \stackrel{?}{=} \frac{4}{12}$ **2.** $\frac{2}{5} \stackrel{?}{=} \frac{4}{10}$ **3.** $\frac{4}{5} \stackrel{?}{=} \frac{16}{19}$

4. $\frac{5}{6} \stackrel{?}{=} \frac{25}{29}$ **5.** $\frac{3}{4} \stackrel{?}{=} \frac{6}{9}$ **6.** $\frac{1}{2} \stackrel{?}{=} \frac{11}{24}$

7. $\frac{3}{8} \stackrel{?}{=} \frac{9}{24}$ **8.** $\frac{2}{7} \stackrel{?}{=} \frac{14}{49}$ **9.** $\frac{16}{32} \stackrel{?}{=} \frac{1}{2}$

10. $\frac{13}{23} \stackrel{?}{=} \frac{1}{2}$ **11.** $\frac{10}{15} \stackrel{?}{=} \frac{5}{8}$ **12.** $\frac{22}{30} \stackrel{?}{=} \frac{11}{15}$

Solving Proportions

Solve for the number that makes the proportion true.

$\frac{7}{8} = \frac{28}{a}$

Cross multiply.

$\frac{7}{8} = \frac{28}{a}$ ➧ $7 \times a = 28 \times 8$

$7 \times a = 224$

$a = \frac{224}{7}$

$a = 32$

Solve.

13. $\frac{1}{2} = \frac{9}{h}$ **14.** $\frac{2}{3} = \frac{12}{y}$ **15.** $\frac{6}{a} = \frac{9}{11}$

16. $\frac{8}{y} = \frac{12}{15}$ **17.** $\frac{t}{6} = \frac{1}{11}$ **18.** $\frac{a}{7} = \frac{2}{19}$

19. $\frac{5}{9} = \frac{n}{10}$ **20.** $\frac{4}{11} = \frac{x}{33}$ **21.** $\frac{8}{13} = \frac{y}{20}$

22. $\frac{42}{66} = \frac{y}{11}$ **23.** $\frac{11}{33} = \frac{c}{60}$ **24.** $\frac{8}{40} = \frac{5}{n}$

25. $\frac{n}{7} = \frac{15}{21}$ **26.** $\frac{2}{3} = \frac{18}{n}$ **27.** $\frac{96}{16} = \frac{n}{4}$

28. $\frac{13}{42} = \frac{65}{c}$ **29.** $\frac{16}{3} = \frac{t}{12}$ **30.** $\frac{h}{25} = \frac{15}{125}$

31. $\frac{27}{y} = \frac{81}{45}$ **32.** $\frac{a}{21} = \frac{9}{27}$ **33.** $\frac{121}{11} = \frac{550}{h}$

Skill 24 Solve a Rate Problem

Equal Rates as a Proportion

Write the proportion.

300 words typed in 5 minutes.
How many words in 3 minutes?

$$\frac{300 \text{ words}}{5 \text{ minutes}} = \frac{n \text{ words}}{3 \text{ minutes}}$$

$$\frac{300}{5} = \frac{n}{3}$$

Write the proportions.

1. 12 oranges cost 99¢ . How much for 7 oranges?

2. A car travels 84 km on 7 L of gas. How many kilometers on 13 L of gas?

3. On a map, 5 cm represents 40 km. How many kilometers does 17 cm represent?

4. A machine uses 30 kW·h in 6 hours. How many kilowatt-hours in 45 hours?

Solve a Rate Problem

Find the number of boxes.

5000 envelopes in 10 boxes.
3000 envelopes in how many boxes?

$$\frac{5000 \text{ envelopes}}{10 \text{ boxes}} = \frac{3000 \text{ envelopes}}{n \text{ boxes}}$$

$$\frac{5000}{10} = \frac{3000}{n}$$

$n = 6$ ⟵ **Skill 23**

Solve.

5. 5 apples cost 79¢. How much for 22 apples?

6. A machine produces 121 bolts in 3 hours. How many bolts in 15 hours?

7. 36 pages typed in 4 hours. How many pages in 13 hours?

8. A car travels 19 km on 2 L of gas. How many kilometers on 76 L of gas?

Mixed Practice

Write the proportion and solve.

9. A machine produces 660 wheels in 4 hours. How many wheels are made in 9 hours?

10. Fuel costs 93¢ for 3 L. How much would it cost to fill a car with an 84 L tank?

11. The telephone rate to Europe is $6 for 3 minutes. How much for 14 minutes?

12. Sally can run 11 km in 60 minutes. How far can she run in 105 minutes?

13. Juan can read 6 pages in 5 minutes. How long will it take him to read 40 pages?

Skill 25 Compare Rates

Find the Unit Rate

Find the number of books.

4200 books in 200 boxes.
How many books in 1 box?

$$\frac{4200 \text{ books}}{200 \text{ boxes}} = \frac{n \text{ books}}{1 \text{ box}}$$

$$n = 21 \leftarrow \textbf{Skill 24}$$

The unit rate is 21 books per box.

Find the unit rate.

1. 3456 oranges in 12 crates. How many oranges in 1 crate?
2. Ida prints 4950 words in 9 minutes. How many words in 1 minute?
3. Lou drives 560 km in 7 hours. How many kilometers in 1 hour?
4. $14.25 for 15 dozen eggs. How much for 1 dozen?
5. Sam drives 384 km in 6 hours. How far does he drive in 1 hour? How far in 10 hours?
6. 2500 sheets of bond paper weigh 100 pounds. How many sheets weigh 1 pound? What will 10,000 sheets weigh?

Compare Unit Rates

Which is faster?

Car A: 280 km in 4 hours
Car B: 455 km in 7 hours

Car A	Car B
$\frac{280}{4} = 70$	$\frac{455}{7} = 65$
70 km/h	65 km/h

70 km/h is greater than 65 km/h

Car A is faster.

Compare the unit rates.

7. Which uses less fuel per kilometer?
   Car A: 52 km on 6 L
   Car B: 59 km on 7 L
8. Which costs less per kilogram?
   Carton A: $89.25 for 15 kg
   Carton B: $117.00 for 20 kg
9. Which shipment holds more per crate?
   Shipment A: 1584 apples in 11 crates
   Shipment B: 1937 apples in 13 crates
10. Which makes more cogs per hour?
    Machine A: 288 cogs in 4 hours
    Machine B: 414 cogs in 6 hours
11. Which gets more km per liter?
    Car A: 162 km on 18 L of fuel
    Car B: 220 km on 20 L of fuel

Skill 26 Write Decimals as Percents

Two or More Decimal Places

Write 0.037 as a percent.

0.037 = ?%

0.037 ➧ 0.037 (Move 2 places right.)

0.037 = 3.7%

Write as percents.

1. 0.10	**2.** 0.15	**3.** 0.25
4. 0.74	**5.** 0.82	**6.** 0.93
7. 2.13	**8.** 4.212	**9.** 5.753
10. 0.267	**11.** 0.391	**12.** 0.914
13. 12.104	**14.** 0.625	**15.** 10.82
16. 0.007	**17.** 0.106	**18.** 0.008
19. 0.04	**20.** 0.001	**21.** 0.503

Fewer Than Two Decimal Places

Write 0.5 as a percent.

0.5 = ?%

0.5 ➧ 0.50% (Use zero as placeholder. Move 2 places right.)

0.5 = 50%

Write as percents.

22. 0.4	**23.** 0.7	**24.** 0.9
25. 7.1	**26.** 9.3	**27.** 10.5
28. 11.0	**29.** 12.6	**30.** 17.1
31. 22.5	**32.** 29.0	**33.** 37.2
34. 0.1	**35.** 7.5	**36.** 1.8
37. 16.3	**38.** 0.3	**39.** 6.2

Mixed Practice

Write as percents.

40. 0.57	**41.** 3.80	**42.** 0.001	**43.** 0.67	**44.** 20.7
45. 8.8	**46.** 0.2915	**47.** 0.32	**48.** 17.4	**49.** 3.003
50. 139.25	**51.** 0.2187	**52.** 9.1	**53.** 25.00	**54.** 12.004
55. 0.14	**56.** 2.185	**57.** 0.51	**58.** 9.3	**59.** 1.868
60. 8.554	**61.** 0.003	**62.** 3.246	**63.** 0.26	**64.** 0.07
65. 0.0032	**66.** 3.642	**67.** 0.29	**68.** 0.052	**69.** 3.42
70. 0.30	**71.** 2.1	**72.** 0.0001	**73.** 1.67	**74.** 1.0
75. 2.549	**76.** 0.005	**77.** 2.077	**78.** 1.5	**79.** 0.19

Skill 27 Write Fractions/ Mixed Numbers as Percents

Fractions

Write $\frac{3}{5}$ as a percent.

$$\frac{3}{5} = ?\%$$

$$\frac{3}{5} = 0.6 = 60\%$$

Skill 14 Skill 26

Write as percents.

1. $\frac{1}{4}$
2. $\frac{2}{5}$
3. $\frac{3}{4}$
4. $\frac{7}{10}$
5. $\frac{1}{8}$
6. $\frac{3}{20}$
7. $\frac{7}{20}$
8. $\frac{12}{25}$
9. $\frac{7}{50}$
10. $\frac{7}{40}$
11. $\frac{5}{8}$
12. $\frac{11}{16}$

Mixed Numbers

Write $2\frac{3}{8}$ as a percent.

$$2\frac{3}{8} = ?\%$$

$$2\frac{3}{8} = \frac{19}{8} = 2.375 = 237.5\%$$

Skill 13 Skill 14 Skill 26

Write as percents.

13. $2\frac{3}{4}$
14. $4\frac{5}{8}$
15. $6\frac{2}{3}$
16. $7\frac{1}{6}$
17. $8\frac{5}{9}$
18. $5\frac{4}{15}$
19. $8\frac{7}{40}$
20. $14\frac{3}{16}$
21. $7\frac{11}{50}$

Mixed Practice

Write as percents.

22. $\frac{3}{5}$
23. $5\frac{3}{8}$
24. $12\frac{5}{6}$
25. $\frac{7}{12}$
26. $\frac{1}{9}$
27. $15\frac{3}{10}$
28. $\frac{9}{40}$
29. $20\frac{4}{5}$
30. $10\frac{6}{25}$
31. $18\frac{5}{8}$
32. $9\frac{3}{40}$
33. $\frac{15}{16}$
34. $8\frac{19}{50}$
35. $11\frac{5}{14}$
36. $16\frac{9}{20}$
37. $\frac{7}{9}$
38. $7\frac{3}{16}$
39. $5\frac{11}{15}$
40. $\frac{11}{30}$
41. $17\frac{7}{20}$

Skill 28 Write Percents as Decimals

Percent in Decimal Form

Write 8.75% as a decimal.

Drop % sign.

8.75% ➧ 0 08.75

Move 2 places left.

8.75% = 0.0875

Write as decimals.

1. 10.5%	**2.** 15.7%	**3.** 40%
4. 85%	**5.** 120%	**6.** 137%
7. 6.7%	**8.** 7.1%	**9.** 8.9%
10. 95%	**11.** 119%	**12.** 7.9%
13. 17.2%	**14.** 85.6%	**15.** 100%
16. 1.35%	**17.** 5.3%	**18.** 142%

Percent in Fractional Form

Write $\frac{3}{20}\%$ as a decimal.

$\frac{3}{20}\% = 0.15\%$ ➧ 0 00.15

Skill 14

$\frac{3}{20}\% = 0.0015$

Write as decimals.

19. $\frac{7}{10}\%$	**20.** $\frac{11}{20}\%$	**21.** $\frac{5}{8}\%$
22. $\frac{2}{25}\%$	**23.** $5\frac{3}{5}\%$	**24.** $7\frac{1}{10}\%$
25. $6\frac{4}{5}\%$	**26.** $22\frac{3}{4}\%$	**27.** $30\frac{2}{5}\%$
28. $89\frac{13}{20}\%$	**29.** $57\frac{21}{25}\%$	**30.** $12\frac{3}{4}\%$

Mixed Practice

Write as decimals.

31. 74.25%	**32.** $6\frac{1}{4}\%$	**33.** $18\frac{4}{5}\%$	**34.** 0.97%	**35.** $14\frac{9}{20}\%$
36. $28\frac{3}{40}\%$	**37.** 0.125%	**38.** 6.25%	**39.** $\frac{7}{50}\%$	**40.** 6.63%
41. 8.79%	**42.** 100%	**43.** $\frac{19}{20}\%$	**44.** 127%	**45.** $\frac{1}{40}\%$
46. $20\frac{1}{4}\%$	**47.** $22\frac{1}{8}\%$	**48.** 14.6%	**49.** $25\frac{1}{5}\%$	**50.** 18.9%

Skill 29 Write Percents as Fractions

Percent in Decimal Form

Write 37.5% as a fraction.

$$37.5\% = 0.375 = \frac{3}{8}$$

Skill 28 (0.375) Skill 14 ($\frac{3}{8}$)

Write as fractions in lowest terms.

1. 45%	**2.** 80%	**3.** 175%
4. 200%	**5.** 11.7%	**6.** 0.1%
7. 0.15%	**8.** 67.3%	**9.** 10.6%
10. 78.55%	**11.** 37.63%	**12.** 51.42%
13. 50%	**14.** 75%	**15.** 10.1%
16. 0.12%	**17.** 80.9%	**18.** 42.1%

Percent in Fractional Form

Write $16\frac{2}{3}\%$ as a fraction.

$$16\frac{2}{3}\% = 16\frac{2}{3} \div 100$$

$$= \frac{50}{3} \div 100 \quad \text{Skill 13}$$

$$= \frac{50}{300} \quad \text{Skill 21}$$

$$= \frac{1}{6} \quad \text{Skill 12}$$

Write as fractions in lowest terms.

19. $6\frac{1}{4}\%$	**20.** $37\frac{1}{2}\%$	**21.** $43\frac{3}{4}\%$
22. $3\frac{1}{2}\%$	**23.** $9\frac{1}{4}\%$	**24.** $10\frac{1}{3}\%$
25. $15\frac{1}{2}\%$	**26.** $20\frac{1}{2}\%$	**27.** $25\frac{1}{2}\%$
28. $15\frac{3}{4}\%$	**29.** $16\frac{2}{3}\%$	**30.** $33\frac{1}{3}\%$

Mixed Practice

Write as fractions in lowest terms.

31. 30%	**32.** 57.2%	**33.** $31\frac{1}{4}\%$	**34.** 63.5%	**35.** $12\frac{1}{2}\%$
36. $13\frac{1}{3}\%$	**37.** $26\frac{2}{3}\%$	**38.** 64.75%	**39.** $11\frac{1}{9}\%$	**40.** 78.55%
41. 50%	**42.** 75%	**43.** $1\frac{1}{4}\%$	**44.** 100%	**45.** $12\frac{3}{4}\%$
46. $4\frac{3}{4}\%$	**47.** $5\frac{3}{8}\%$	**48.** 75.91%	**49.** $6\frac{2}{3}\%$	**50.** 121%

Skill 30 Find a Percentage

Decimal Percents

Find 15.5% of 36.

$$15.5\% \text{ of } 36 = n$$

Skill 28

$$0.155 \times 36 = n$$

$$5.58 = n$$

$$15.5\% \text{ of } 36 = 5.58$$

Find the percentages.

1. 15% of 60 **2.** 30% of 72

3. 4% of 96 **4.** 9% of 122

5. 6.5% of 120 **6.** 8.3% of 150

7. 17.8% of 80 **8.** 31.2% of 140

9. 5.81% of 60 **10.** 7.32% of 45

11. 67.7% of 67 **12.** 8.92% of 35

Fractional Percents

Find $\frac{3}{4}\%$ of 1600.

$$\frac{3}{4}\% \text{ of } 1600 = n$$

Skill 29

$$\frac{3}{400} \times 1600 = n$$

$$12 = n$$

$$\frac{3}{4}\% \text{ of } 1600 = 12$$

Find the percentages.

13. $2\frac{1}{2}\%$ of 400 **14.** $4\frac{1}{2}\%$ of 200

15. $33\frac{1}{3}\%$ of 120 **16.** $3\frac{5}{6}\%$ of 600

17. $\frac{3}{4}\%$ of 800 **18.** $\frac{3}{5}\%$ of 50

19. $\frac{3}{8}\%$ of 600 **20.** $\frac{7}{8}\%$ of 80

Mixed Practice

Find the percentages.

21. 80% of 160 **22.** $8\frac{1}{3}\%$ of 72 **23.** 9.1% of 90 **24.** $15\frac{3}{5}\%$ of 90

25. $16\frac{2}{3}\%$ of 90 **26.** 45% of 72 **27.** $\frac{1}{4}\%$ of 800 **28.** 0.75% of 1000

29. 125% of 64 **30.** $12\frac{1}{4}\%$ of 65 **31.** 7.2% of 127 **32.** $6\frac{1}{4}\%$ of 1600

33. $4\frac{1}{6}\%$ of 600 **34.** 16.5% of 84 **35.** $8\frac{1}{2}\%$ of 75 **36.** 24.7% of 80

Skill 31 Find the Rate

Percentage Less Than Base

What percent of 75 is 60?

$n\%$ of 75 = 60

Write as a proportion and solve.

$$\frac{n}{100} = \frac{60}{75}$$

$75 \times n = 6000$

$n = 80$ ← Skill 23

80% of 75 = 60

Solve. Round answers to the nearest tenth of a percent.

1. $n\%$ of 40 = 20
2. $n\%$ of 60 = 15
3. $n\%$ of 90 = 18
4. $n\%$ of 60 = 9
5. $n\%$ of 70 = 25
6. $n\%$ of 90 = 65
7. $n\%$ of 65 = 58
8. $n\%$ of 84 = 75
9. $n\%$ of 113 = 79
10. $n\%$ of 410 = 295.5
11. $n\%$ of 94 = 82
12. $n\%$ of 296 = 239.76

Percentage Greater Than Base

What percent of 90 is 162?

$n\%$ of 90 = 162

Write as a proportion and solve.

$$\frac{n}{100} = \frac{162}{90}$$

$90 \times n = 16{,}200$

$n = 180$

180% of 90 = 162

Solve. Round answers to the nearest tenth of a percent.

13. $n\%$ of 20 = 30
14. $n\%$ of 36 = 45
15. $n\%$ of 50 = 60
16. $n\%$ of 70 = 91
17. $n\%$ of 60 = 80
18. $n\%$ of 72 = 96
19. $n\%$ of 24 = 31
20. $n\%$ of 56 = 77
21. $n\%$ of 37 = 148
22. $n\%$ of 60 = 90
23. $n\%$ of 81 = 415.75
24. $n\%$ of 110 = 130

Mixed Practice

Solve. Round answers to the nearest tenth of a percent.

25. $n\%$ of 125 = 100
26. $n\%$ of 100 = 115
27. $n\%$ of 130 = 60
28. $n\%$ of 56 = 96
29. $n\%$ of 89 = 49
30. $n\%$ of 84 = 108
31. $n\%$ of 115 = 92
32. $n\%$ of 42 = 52.5
33. $n\%$ of 64 = 76.8
34. $n\%$ of 64 = 24
35. $n\%$ of 204 = 51
36. $n\%$ of 42 = 6.72
37. $n\%$ of 36 = 50
38. $n\%$ of 173 = 136.25
39. $n\%$ of 18 = 36
40. $n\%$ of 21 = 63
41. $n\%$ of 120 = 80
42. $n\%$ of 15.5 = 62
43. $n\%$ of 90 = 4.5
44. $n\%$ of 125 = 18.75
45. $n\%$ of 75 = 110

Skill 32 Find the Base

Decimal Percents

42 is 37.5% of what number?

$37.5\% \text{ of } n = 42$

$0.375 \times n = 42$

Skill 28

$n = 42 \div 0.375$

$n = 112$

$37.5\% \text{ of } 112 = 42$

Find the number.

1. $12.5\% \text{ of } n = 9$
2. $62.5\% \text{ of } n = 55$
3. $8.25\% \text{ of } n = 664$
4. $7\% \text{ of } n = 3.5$
5. $5.75\% \text{ of } n = 92$
6. $9\% \text{ of } n = 8.2$
7. $11\% \text{ of } n = 37$
8. $3\% \text{ of } n = 8.7$
9. $5\% \text{ of } n = 3$
10. $12.5\% \text{ of } n = 21$
11. $37.3\% \text{ of } n = 50$
12. $81\% \text{ of } n = 14$

Fractional Percents

3 is $6\frac{1}{4}\%$ of what number?

$6\frac{1}{4}\% \text{ of } n = 3$

$\frac{1}{16} \times n = 3$

Skill 29

$n = 3 \div \frac{1}{16}$

$n = 48$

$6\frac{1}{4}\% \text{ of } 48 = 3$

Find the number.

13. $2\frac{1}{2}\% \text{ of } n = 4$
14. $3\frac{1}{8}\% \text{ of } n = 2$
15. $33\frac{1}{3}\% \text{ of } n = 20$
16. $\frac{1}{4}\% \text{ of } n = 2$
17. $37\frac{1}{2}\% \text{ of } n = 34$
18. $\frac{3}{5}\% \text{ of } n = 6$
19. $45\frac{1}{2}\% \text{ of } n = 25$
20. $\frac{3}{4}\% \text{ of } n = 7$

Mixed Practice

Find the number.

21. $80\% \text{ of } n = 60$
22. $12.5\% \text{ of } n = 39$
23. $116\% \text{ of } n = 2.9$
24. $10\% \text{ of } n = 2.9$
25. $8\% \text{ of } n = 4.2$
26. $40\% \text{ of } n = 6.8$
27. $16\frac{2}{3}\% \text{ of } n = 13$
28. $145\% \text{ of } n = 6.38$
29. $66\frac{2}{3}\% \text{ of } n = 120$
30. $25\% \text{ of } n = 36$
31. $2\frac{2}{3}\% \text{ of } n = 6$
32. $90\% \text{ of } n = 72$
33. $6\% \text{ of } n = 3.3$
34. $7.8\% \text{ of } n = 9$
35. $16\frac{3}{4}\% \text{ of } n = 33$

Application A Formulas

Substituting In a Formula

To use a formula to solve a problem, substitute known values in the formula. Then solve. Remember to perform any computations within parentheses first.

▶ An airplane travels at a rate of 960 km/h. How far can the plane travel in 7 hours?

<u>Formula:</u> Distance = Rate × Time

<u>Substitute:</u> Distance = 960 × 7

<u>Solve:</u> Distance = 6720

The plane travels 6720 km in 7 hours.

Find the distance.

	Distance = Rate × Time		
	Rate	Time	Distance
1.	100 mi/h	6 h	?
2.	55 km/h	4 h	?
3.	14 m/s	32 s	?
4.	30 mi/h	6 h	?
5.	42 m/s	5 s	?

Find the area of the triangle.

	Area of Triangle $= \frac{1}{2} \times$ Base × Height		
	Base	Height	Area
6.	1 m	0.5 m	?
7.	64 ft	38 ft	?
8.	18 in.	5 in.	?
9.	30 km	24 km	?

Find the percent of sales made.

	% of Sales $= \frac{\text{Sales Made}}{\text{Possible Sales}} \times 100$		
	Possible Sales	Sales Made	Percent
10.	$4600	$1900	?
11.	$19,100	$6000	?
12.	$23,000	$13,000	?
13.	$64,000	$15,000	?
14.	$8200	$3000	?

Find the price of the stereo system.

	Price = Amp + Turntable + Speakers			
	Amp	Turntable	Speakers	Price
15.	$199	$99.95	$139	?
16.	$395	$119.99	$259	?
17.	$685	$249.00	$495	?
18.	$449	$188.00	$350	?

Find the long distance telephone charge.

	Charge = Cost per min × Minutes		
	Minutes	Cost per min	Charge
19.	15	$.35	?
20.	9	$1.12	?
21.	31	$1.49	?
22.	22	$.54	?
23.	38	$1.32	?

Application B More Formulas

Multiplying or Dividing to Solve Formulas

To solve some formulas after substituting, you may have to MULTIPLY or DIVIDE both sides of the equation by the same number.

▶An airplane cruising at a rate of 575 km/h has traveled 2645 km. How long did it take to make this trip?

Formula: Distance = Rate × Time

Substitute: 2645 = 575 × Time

Divide: $\frac{2645}{575} = \frac{575 \times \text{Time}}{575}$

Solve: 4.6 h = Time

Find the missing amount.

	Distance = Rate × Time		
	Rate	Time	Distance
1.	88 in./min	3.5 min	?
2.	60 km/h	?	195 km
3.	675 yd/h	13 h	?
4.	?	1.75 h	78.75 km
5.	24 km/h	?	192 km
6.	?	$4\frac{1}{2}$ h	$247\frac{1}{2}$ ft
7.	3.2 m/s	?	96 m

Adding or Subtracting to Solve Formulas

To solve some formulas after substituting, you may have to ADD the same number to, or SUBTRACT it from, both sides of the equation.

▶The temperature of a solution is 212°F (degrees Fahrenheit). What is this in degrees Celsius (°C)?

Formula: °F = (1.8 × °C) + 32

Substitute: 212 = (1.8 × °C) + 32

Subtract: −32   − 32

180 = (1.8 × °C)

Divide: $\frac{180}{1.8} = \frac{1.8 \times °C}{1.8}$

Solve: 100 = °C

Find the missing amount.

	°F = (1.8 × °C) + 32	
	degrees Celsius	degrees Fahrenheit
8.	50	?
9.	?	86
10.	?	32
11.	120	?
12.	?	95
13.	?	68
14.	144	?
15.	10	?
16.	?	5

Application C Tables and Charts

Reading Tables and Charts

To read a table or chart, find the row containing one of the conditions of the information you seek. Run down the column containing the other condition until it crosses that row. Read the answer.

IMAG CORP. Sales Report			
Office	Agents	Sales	Income
Bronx	252	640	$12,000,000
Calgary	204	307	1,800,000
Columbus	92	778	1,200,000
Dallas	50	761	1,300,000
Denver	76	398	1,500,000
Lincoln	35	537	1,500,000
San Diego	128	813	14,300,000
Toronto	710	390	7,900,000

▶ How many sales were made at the Lincoln office?

The table shows that there were 537 sales made at the Lincoln office.

Find the income for each of the following offices.

1. Columbus **2.** Denver **3.** Toronto **4.** Bronx

5. Which offices made more than 500 sales? **6.** Which offices have less than 200 agents?

Using Tables and Charts

To classify an item, find the row or column that contains the given data. Then read the classification from the head of the row or column.

Parcel Post Rate Schedule				
Weight (Not to exceed)	Local	Zone 1+2	Zone 3	Zone 4
1 lb	$1.15	$1.35	$1.39	$1.58
2	1.23	1.45	1.53	1.73
3	1.29	1.50	1.65	1.82
4	1.36	1.66	1.77	1.92
5	1.42	1.71	1.84	2.01
6	1.47	1.76	1.90	2.11
7	1.51	1.80	1.97	2.20
8	1.54	1.85	2.03	2.29
9	1.57	1.89	2.10	2.39

▶ A package weighing 5 lb 5 oz costs $2.11 to deliver. What postal zone was the package delivered within?

The table shows that the package was delivered within zone 4.

Find the missing information.

7. A package weighing 4 lb costs $1.36 to send. To what postal zone was the package delivered?

8. A package costs $2.03 to deliver within postal zone 3. What is the most the package could have weighed?

9. A package costs $1.76 to deliver within postal zone 1. Between what two weights was the weight of the package?

Application D Making Change

Making Change: Building Up

To make change, build up to the next nickel, dime, quarter, and so on, until the amount presented is reached.

▶ Compute the change from $5.00 for a purchase of $3.67.

CHANGE		$3.67
C	3 pennies →	+ .03
H		3.70
A	1 nickel →	+ .05
N		3.75
G	1 quarter →	+ .25
E		4.00
	1 dollar →	+1.00
	Amount presented →	$5.00

Compute the change from $10. Use the fewest coins and bills.

1. $3.75 **2.** $7.78 **3.** $1.99
4. $3.47 **5.** $5.63 **6.** $9.19

Compute the change. Use the fewest coins and bills.

7. Amount spent: $15.05
Amount presented: $20

8. Amount spent: $3.37
Amount presented: $20

9. Amount spent: $2.29
Amount presented: $3

Making Change: Loose Coins

The amount presented may contain loose coins so that fewer coins and bills are returned in the change.

▶ Compute the change from $5.02 for a purchase of $3.67.

	Amount Presented	Purchase
a. Subtract the loose change.	$5.02	$3.67
	− .02	− .02
	$5.00	$3.65

b. Now compute the change from $5.00 for a purchase of $3.65.

CHANGE		$3.65
C	1 dime →	+ .10
H		3.75
A	1 quarter →	+ .25
N		4.00
G	1 dollar →	+1.00
E		$5.00

Compute the change using the fewest coins and bills.

10. Amount spent: $5.26
Amount presented: $6.01

11. Amount spent: $7.79
Amount presented: $10.04

12. Amount spent: $1.07
Amount presented: $5.07

13. Amount spent: $3.92
Amount presented: $10.02

14. Amount spent: $15.21
Amount presented: $20.01

15. Amount spent: $7.81
Amount presented: $10.06

16. Amount spent: $6.03
Amount presented: $21.03

17. Amount spent: $9.62
Amount presented: $21.12

Application E Rounding Time: Nearest Quarter Hour

To round to the nearest quarter hour, find the quarter-hour interval that the given time is between. Subtract the earlier quarter from the given time. Subtract the given time from the later quarter. Round to whichever is closer.

▶ Round 6:19 to the nearest quarter hour.

6:19	6:30
−6:15	−6:19
4 minutes	11 minutes

4 is less than 11; round to 6:15.

Round to the nearest quarter hour.

1. 8:39	**2.** 10:40	**3.** 6:55
4. 1:23	**5.** 11:17	**6.** 3:11
7. 12:09	**8.** 4:44	**9.** 3:32
10. 11:11	**11.** 8:51	**12.** 1:50
13. 7:35	**14.** 5:04	**15.** 9:40
16. 10:34	**17.** 12:39	**18.** 6:41
19. 6:01	**20.** 3:49	**21.** 11:36
22. 8:05	**23.** 10:10	**24.** 2:15

Application F Elapsed Time (Hours)

Finding Elapsed Time

To find the elapsed time, subtract the earlier time from the later time. If it is necessary to "borrow," rewrite the later time by subtracting 1 from the hours and adding 60 to the minutes.

▶ How much time has elapsed between 11:45 A.M. and 12:23 P.M.?

12:23 =	11:23 + :60 =	11:83
−11:45 =	−11:45 =	−11:45
		38 min

Find the elapsed time.

1. From 6:05 P.M. to 8:51 P.M.

2. From 4:18 A.M. to 6:52 A.M.

3. From 8:30 A.M. to 11:45 A.M.

4. From 6:45 P.M. to 11:59 P.M.

5. From 10:45 A.M. to 11:36 A.M.

6. From 3:45 P.M. to 5:30 P.M.

Finding Elapsed Time Spanning 1 O'Clock

To find the elapsed time when the period spans 1 o'clock, add 12 hours to the later time before subtracting.

▶ How much time has elapsed between 7:45 P.M. and 2:52 A.M.?

2:52 =	2:52 + 12:00 =	14:52
−7:45 =	−7:45	− 7:45
		7 h: 7 min

Find the elapsed time.

7. From 8:15 A.M. to 12:30 P.M.

8. From 4:55 A.M. to 1:00 P.M.

9. From 11:30 P.M. to 7:50 A.M.

10. From 9:45 P.M. to 8:00 A.M.

11. From 3:30 A.M. to 2:15 P.M.

Application G Elapsed Time (Days)

To find the number of days between two dates, or from one date to the next, use the table on page 569. Find the position of each date in the year. When the dates run from one year to the next, add 365 to the later date. Then subtract the earlier date from the later.

To find the number of days from one date through another, find the elapsed time between the dates and add 1 day.

▶ How many days have elapsed between April 4 and January 6?

Jan. 6:	$(6 + 365)$	=	371
Apr. 4:	$-$ 94	=	$-$ 94
Subtract.			277 days

Day No.	Jan.	Feb.	Mar.	Apr.	May	June	July	Aug.	Sept.
1	1	32	60	91	121	152	182	213	244
2	2	33	61	92	122	153	183	214	245
3	3	34	62	93	123	154	184	215	246
4	4	35	63	94	124	155	185	216	247
5	5	36	64	95	125	156	186	217	248
6	6	37	65	96	126	157	187	218	249

Find the elapsed time. Use the table on page 569.

1. Between August 10 and December 25
2. From March 21 through September 30
3. Between June 1 and December 30
4. From May 21 to July 2
5. From September 5 through May 20

Application H Determining Leap Years

A year is a leap year if it is exactly divisible by 4. Exceptions are years that end in '00' (1900, 2000, and so on); they must be divisible by 400.

▶ Was 1848 a leap year?

$1848 \div 4 = 462$; 1848 was a leap year.

Which are leap years?

1. 1949	2. 1980	3. 2000
4. 1912	5. 2100	6. 1954
7. 1992	8. 1951	9. 2400
10. 1882	11. 1920	12. 2004

Application I Elapsed Time in a Leap Year

When the dates span February 29 in a leap year, add 1 day to the later date.

▶ How many days elapsed between January 5, 1980 and September 3, 1980?

Sept. 3, 1980:	$(246 + 1)$	=	247
Jan. 5, 1980:	$-$ 5	=	$-$ 5
Subtract.			242 days

Find the elapsed time. Use the table on page 569.

1. January 12, 1978 to April 1, 1978
2. July 1, 1979 through March 21, 1980
3. Feb. 5, 1984 to Sept. 23, 1984
4. Jan. 9, 1988 through July 2, 1988
5. Feb. 21, 2000 to Sept. 30, 2000

Application J Fractional Parts of a Year

Changing Months into a Part of a Year

To change months into a part of a year, write as a fraction with denominator 12. Express answers in lowest terms.

▶ 6 months is what part of a year?

$$6 \text{ months} = \frac{6}{12} \text{ year} = \frac{1}{2} \text{ year}$$

Change each term to a part of a year.

1. 9 months **2.** 7 months **3.** 12 months

4. 20 months **5.** 36 months **6.** 18 months

7. 21 months **8.** 8 months **9.** 14 months

Changing Days into a Part of a Year

To change days into a part of a year, write as a fraction with denominator 365 when using the exact year or 360 when using the ordinary year. Express answers in lowest terms.

▶ 240 days is what part of an exact year?

$$240 \text{ days} = \frac{240}{365} \text{ year} = \frac{48}{73} \text{ year}$$

Change each term to a part of a year.

10. 90 days as an exact year

11. 300 days as an ordinary year

12. 120 days as an exact year

13. 175 days as an exact year

14. 146 days as an ordinary year

15. 250 days as an exact year

Application K Chronological Expressions

Common Chronological Expressions (Number of occurrences per year)		
	weekly (52)	biweekly (26)
semimonthly (24)	monthly (12)	bimonthly (6)
	quarterly (4)	
semiannually (2)	annually (1)	

▶ Monthly payments for 30 years. How many payments in all?

Monthly: 12 times a year
In 30 years: $12 \times 30 = 360$ payments

Find the number of occurrences. Use the table at the left.

1. Weekly for 1 year

2. Biweekly for 2 years

3. Quarterly for 4 years

4. Semimonthly for 1 year

5. Bimonthly for 2 years

6. Annually for 25 years

7. Monthly for $\frac{1}{2}$ year

8. Weekly for 1.5 years

9. Semimonthly for $2\frac{1}{2}$ years

Application L The Complement of a Number

To find the complement of a percent or decimal less than 1, subtract the percent from 100% or the decimal from 1.

▶ What is the complement of 65%?

The complement of 65% = 100% − 65% = 35%

▶ What is the complement of 0.25?

The complement of 0.25 = 1 − 0.25 = 0.75

Find the complement.

1. 0.15	**2.** 0.72	**3.** 0.35
4. 0.50	**5.** 0.80	**6.** 0.34
7. 91%	**8.** 85%	**9.** 47%
10. 66%	**11.** 7%	**12.** 45%
13. 3.5%	**14.** 56.6%	**15.** 75.9%

Application M Reading Bar Graphs

To read a bar graph, find the bar that represents the information you seek. Trace an imaginary line from the top of the bar to the scale on the left. Read the value represented by the bar from the scale.

▶ In 1975 how much did it cost to paint the exterior of the nine-room house?

The bar graph shows that it cost about $400 to paint the exterior in 1975.

Comparison of Building Costs (Nine-room House)

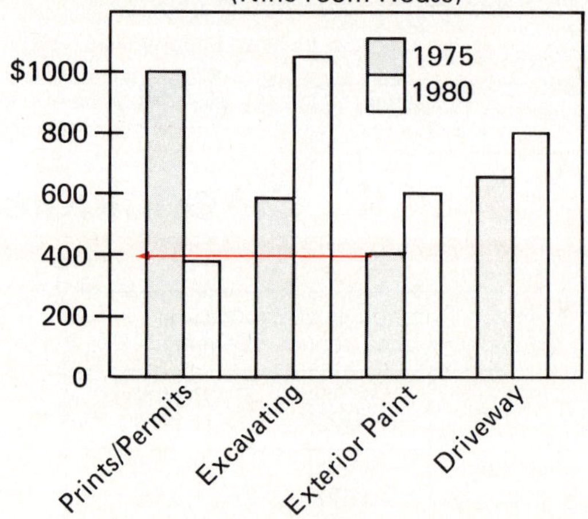

Answer the following. Use the bar graph at the right.

1. In 1975 how much did it cost to construct the driveway?

2. In 1980 how much did it cost to obtain the prints and permits?

3. In 1980 how much did it cost to excavate?

4. What type of building costs decreased rather than increased?

5. In 1980 how much more did it cost to excavate than in 1975?

6. In 1975 how much more did it cost to obtain prints and permits than in 1980?

7. In 1980 how much more did it cost to paint the exterior of the house?

Application N Reading Line Graphs

To read a line graph, find the point that represents the information you seek. Trace an imaginary line from the point to the scale on the left. Read the value represented by the point from the scale.

▶ What was the predicted and the actual amount of sales for the month of May?

The line graph shows that the predicted amount of sales was \$10,000. The actual amount was \$8000.

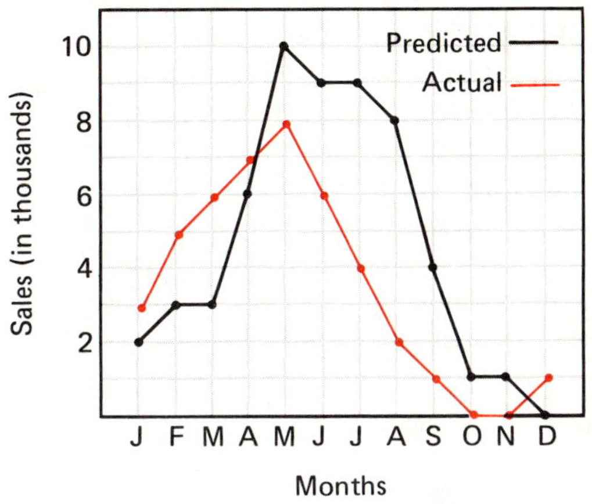

Answer the following. Use the line graph shown.

1. What was the predicted amount of sales for the month of August?
2. Which month actually had the least amount of sales?
3. What was the actual amount of sales for the month of October?
4. In which month was the actual amount of sales equal to the amount predicted?

Application O Reading Pictographs

To read a pictograph, find the scale to see how much each picture represents. Then multiply that amount times the number of pictures on each line to get the total amount.

How many cars did Motor Mart sell?

Each picture represents 100 cars sold. The Motor Mart row has 7.5 pictures.

$$7.5 \times 100 = 750 \text{ cars}$$

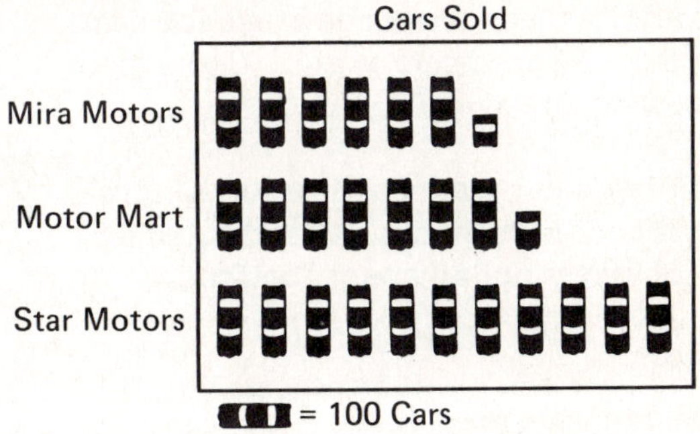

Answer the question. Use the pictograph.

1. How many cars did Mira Motors sell?
2. How many cars did Star Motors sell?
3. How many more cars did Star Motors sell than Motor Mart?

■ Applications

Application P Circle Graphs

Reading Circle Graphs

A circle graph is used to compare the parts to the whole. To find what part of the whole each section represents, multiply the amount, or percent, per section times the total amount.

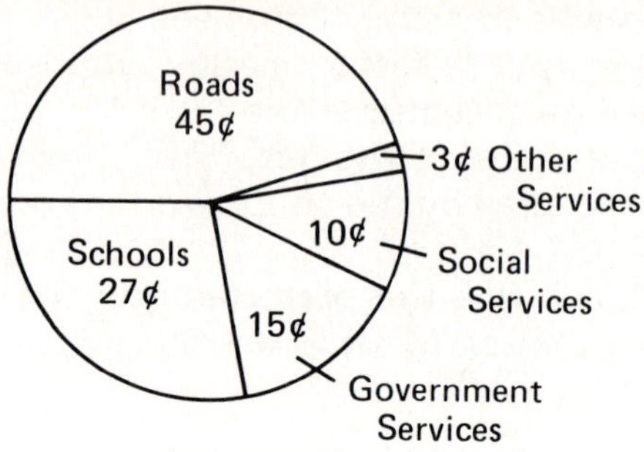

City of Westwood Tax Dollar
Annual Budget—$408,000

▶ How much of the annual budget was spent on social services?

Social Services: 10¢ of the tax dollar

$0.10 \times \$408{,}000 = \$40{,}800$

Answer the following. Use the circle graph above.

1. How much of every tax dollar was spent to operate the schools?

2. How much more of the annual budget was spent on city roads than on schools?

3. How much of the annual budget was spent to operate the schools?

4. Does the city spend more on the schools than on all other services combined?

Constructing Circle Graphs

To construct a circle graph, find the percent of the total that each section represents. Multiply each percent times 360° to find the measure of the angle for each portion of the circle graph. Use a protractor to draw each portion.

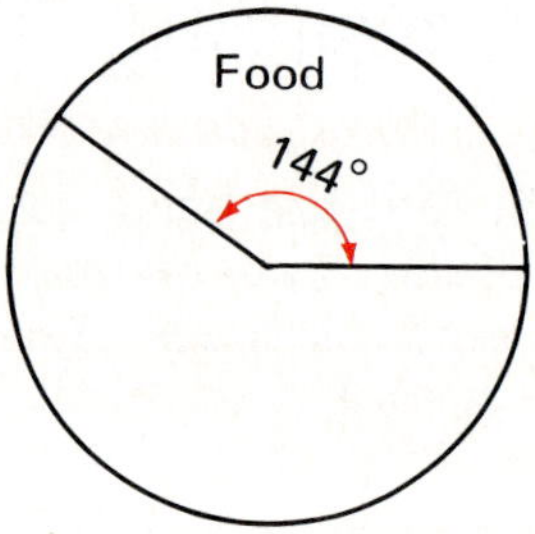

▶ Construct that portion of a circle graph that represents a spending of 40¢ out of every dollar spent for food.

$\frac{40}{100} = 40\%; \quad 0.40 \times 360° = 144°$

5. Complete the circle graph above. Use the following information.

The Bradford family budgets each dollar spent as follows: rent 30¢, books 15¢, clothing and personal items 10¢, and all other expenses 5¢.

6. Draw a circle graph. Use the following information.

A factory budgets each dollar spent as follows: salaries 35¢, raw materials 30¢, utilities 10¢, plant maintenance 15¢, and research and development 10¢.

Application Q Mean

To find the mean (average) of a group of numbers, find the sum of the group and divide it by the number of items in the group.

▶ What is the mean of the following: 454, 376, 416, 472?

$$\text{Mean} = \frac{454 + 376 + 416 + 472}{4}$$

$$= 429\frac{1}{2}$$

Find the mean for each group.

1. 43, 19, 61, 72, 81, 50
2. 116, 147, 136, 151, 123, 117, 120
3. 4615, 5918, 7437, 8937
4. 4.0, 3.5, 4.0, 3.0, 3.5, 3.0, 2.0, 0.5, 4.6, 5.0
5. $580,000; $625,000; $105,583; $733,358; $5,750,000; $255,000; $600,000

Application R Median

To find the median of a group of numbers, arrange the items in order from smallest to largest. The median is the number in the middle. If there is an even number of items, find the mean of the two middle numbers.

▶ What is the median of the following: 454, 376, 416, 472?

Arrange in order: 376, <u>416</u>, <u>454</u>, 472

$$\text{Median} = \frac{416 + 454}{2} = 435$$

Find the median for each group.

1. 141, 136, 191, 187, 149, 148
2. 17, 21, 30, 35, 27, 25, 15
3. 91, 92, 85, 98, 100, 76, 80, 75
4. 4.2, 3.7, 3.1, 4.8, 2.4, 3.0, 2.9
5. 0.07, 0.05, 0.10, 0.12, 0.09, 0.17, 0.01
6. $121,500; $49,750; $72,175; $65,449
7. 2.06, 2.04, 2.00, 2.10, 2.08, 2.24, 1.55, 2.04, 2.13, 2.08, 2.09

Application S Mode

To find the mode of a group of numbers, look for the number that appears most often. A group may have no mode or it may have more than one mode.

▶ What is the mode of the following: 92, 88, 76, 84, 92, 96, 100?

The number that appears most often, the mode, is 92.

Find the mode for each group.

1. $51, $13, $24, $62, $55, $57, $24
2. 800, 600, 800, 500, 600, 700
3. 4.1, 4.7, 4.5, 4.3, 4.2, 4.4
4. 3, 3, 5, 3, 4, 2, 3, 2, 4, 4, 4
5. 0.01, 0.1, 0.01, 1.0, 0.01, 0.11, 1.00
6. $45, $63, $27, $91, $65, $8, $43, $90

Applications

Application T Metric Length

The common units of length in the metric system are the millimeter, centimeter, meter, and the kilometer.

A millimeter(mm) is about the thickness of a U.S. dime.

A centimeter(cm) is about the thickness of a sugar cube.

A meter(m) is about the length of a baseball bat.

A kilometer(km) is 1000 m and is used to measure large distances.

Name the unit which is most commonly used to measure the object.

1. The length of a tennis court

2. The distance from Detroit to Montreal

3. The thickness of a magazine

4. The length of a standard paper clip

5. The length of your living room

6. The thickness of a soda straw

7. The length of a postage stamp

8. The width of your foot

Application U Metric Mass and Volume

The common units of mass and volume in the metric system are the gram, kilogram, metric ton, liter, and kiloliter.

A gram(g) is about the mass of a paper clip.

A kilogram(kg) is about the mass of a hammer.

A metric ton(t) is 1000 kg and is used to measure very heavy objects.

A liter(L) is about the amount of liquid in a can of motor oil.

A kiloliter(kL) is 1000 L and is used to measure large volumes.

Name the unit which is most commonly used to measure the object.

1. The mass of your best friend

2. The amount of gas in a car's tank

3. The mass of a postage stamp

4. The volume of helium in a blimp

5. The mass of an adult blue whale

6. The mass of a baseball

7. The volume of a large fuel tank

8. The mass of a dime

9. The mass of a gumdrop

10. The volume of air in a toy balloon

Application V Perimeter

The distance around a shape is its perimeter. To find the perimeter of a shape, add the lengths of all its sides.

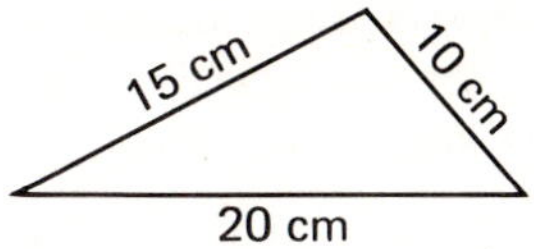

▶ What is the perimeter of the triangle at right?

Perimeter:
10 cm + 15 cm + 20 cm = 45 cm

Find the perimeter.

1. 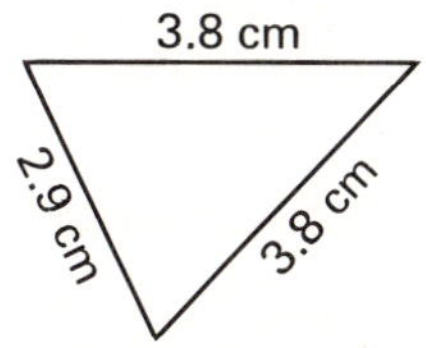

2. 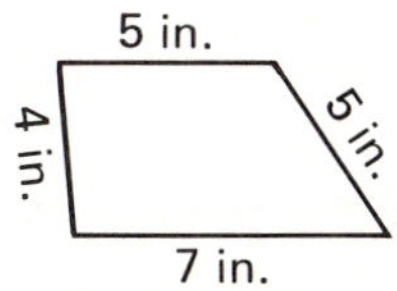

3. 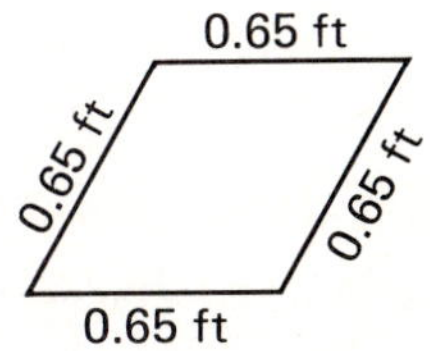

4. 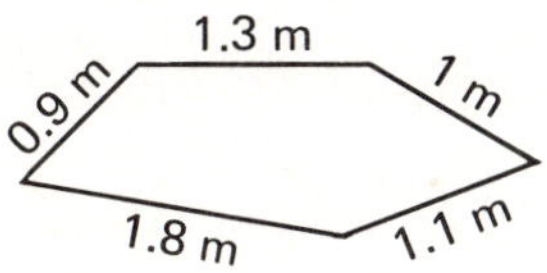

5. 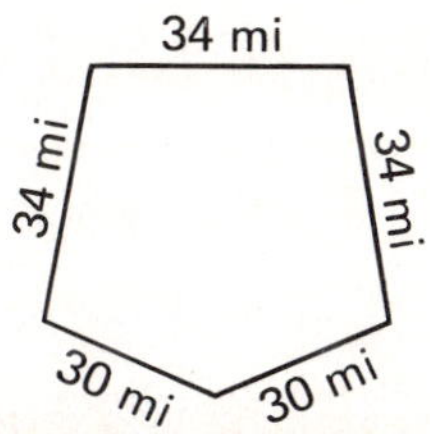

6. 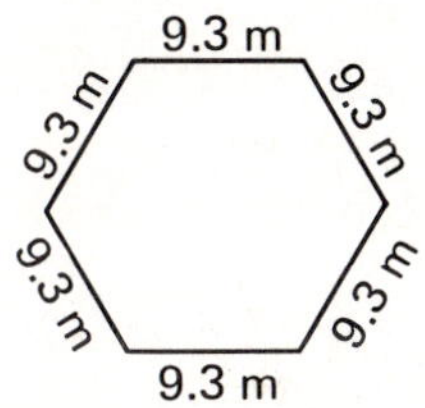

Application W Circumference

The length of a diameter of a circle is twice the length of its radius. The circumference is the distance around a circle. The circumference is estimated by multiplying 3.14 times the length of a diameter.

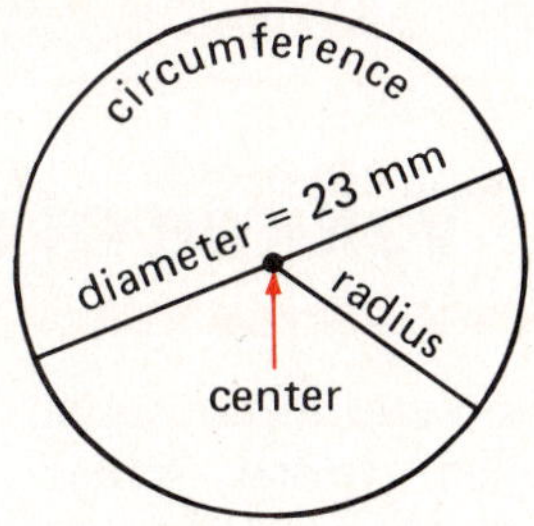

▶ The diameter of a circle is 23 mm. What is its circumference?

Circumference = 3.14 × diameter
Circumference = 3.14 × 23 mm = 72.22 mm

Find the circumference.

1. diameter: 4 mi
2. diameter: 3.5 cm
3. diameter: 0.8 m
4. radius: 35 mm
5. radius: 0.75 ft
6. radius: 13 in.
7. diameter: 18 ft
8. radius: 120 km
9. diameter: 12.5 mi

Application X Area of a Rectangle or Square

To find the area of a rectangle, multiply length times width. In a square, the length and width are equal.

▶ What is the area of a rectangular board that measures 3.5 m by 1.9 m?

Area = Length × Width
Area = 3.5 m × 1.9 m = 6.65 m^2

Find the area.

1. Rectangle: 25 in. long, 6 in. wide
2. Square: 45.4 cm per side
3. Rectangle: 0.9 m long, 0.75 m wide
4. Square: $3\frac{1}{2}$ yd per side
5. Rectangle: $4\frac{2}{3}$ ft long, $2\frac{1}{6}$ ft wide

Application Y Area of a Triangle or Circle

To find the area of a triangle or circle, use the formulas below.

Area of Triangle = $\frac{1}{2}$ × Base × Height

Area of Circle = 3.14 × Radius^2

▶ What is the area of a circle with radius 27 m?

Area of Circle = 3.14 × Radius^2
Area of Circle = 3.14 × 27^2
Area of Circle = 2289.06 m^2

Find the area.

1. Triangle: Base is 3 in., Height is 9 in.
2. Circle: Radius is 5 cm
3. Triangle: Base is 16 m, Height is 25 m
4. Triangle: Base is 132 m, Height is 9.6 m
5. Circle: Radius is $2\frac{1}{2}$ yd
6. Triangle: Base is $9\frac{1}{3}$ ft, Height is $3\frac{1}{4}$ ft
7. Circle: Radius is 86 mm

Application Z Volume of a Rectangular Solid

To find the volume of a rectangular solid, multiply length times width times height.

▶ What is the volume of a rectangular box that is 6 m long, 4 m wide, and 2 m deep?

Volume = Length × Width × Height
Volume = 6 m × 4 m × 2 m
Volume = 48 m^3

Find the volume.

1. 36 m long, 44 m wide, 328 m high
2. 1.6 m long, 0.8 m high, 0.55 m wide
3. 11 cm long, 12 cm wide, 3 cm high
4. 68 in. long, 18 in. wide, 18 in. high
5. $12\frac{1}{3}$ yd long, $4\frac{1}{4}$ yd wide, 3 yd high
6. $7\frac{1}{4}$ ft long, $3\frac{1}{2}$ ft wide, $5\frac{5}{6}$ ft high

accelerated cost recovery system (ACRS) (p. 434) A method of computing depreciation that allows depreciation according to fixed percents over the shortest period of time.
accumulated depreciation (p. 428) The total depreciation to date.
age group (p. 198) A category to which an automobile is assigned by an insurance company depending on its age. Age group affects the base premium.
air freight (p. 394) A way of shipping goods to customers.
amount (p. 119) The balance in a savings account at the end of an interest period, computed by adding the principal and interest.
amount financed (p. 178) The portion of the purchase price of an item that you still owe after making the down payment.
amount of change (p. 450) The difference between the base figure and the corresponding figure on the current income statement.
annual expenses (p. 264) Amounts for real estate taxes, insurance payments, contributions, and so on, which occur once a year.
annual interest rate (p. 117) The percent of the principal that you earn as interest based on one year.
annual percentage rate (APR) (p. 180) An index used to compare the relative cost of borrowing money.
annual premium (p. 198) The amount you pay each year for insurance coverage.
annual yield (p. 248) The total interest or return for one year expressed as a percent of the principal or amount invested.
apportion (p. 424) To distribute among several departments.
assessed value (p. 225) The value of property for tax purposes found by multiplying the market value by the rate of assessment.
assets (p. 440) The total of your cash, the items that you have purchased, and any money that your customers owe you.
average-cost method (p. 389) A way of calculating the value of an inventory based on the average cost of the goods you received.
average daily balance (pp. 164, 165) The average of the balances in a charge account at the end of each day in the billing period.

balance (p. 94) The amount of money in a bank account.
balance sheet (p. 442) A statement that shows your total assets, total liabilities, and owner's equity.
base figure (p. 450) The dollar amount on the earlier income statement when you are computing percent change.
base premium (p. 198) The portion of your annual automobile insurance premium that depends only on the amount of coverage you want.
base price (p. 192) The price of the engine, chassis, and any other piece of standard equipment for a new automobile.

basic monthly charge (p. 408) The regular monthly charge for a utility before any additions are made for extra services.
basic plan (p. 238) A health insurance program that includes only hospital and surgical-medical insurance. The basic plan does not include major medical insurance.
basic rate (p. 396) The charge, usually per 100 pounds, to ship goods by truck. It depends on the type of goods, the weight, and the distance to be shipped.
beneficiary (p. 242) The person named in a policy to receive the insurance benefits if the insured dies.
bodily injury insurance (p. 198) Insurance that protects you against financial loss if your automobile causes personal injury to someone.
bond (pp. 256, 458) A promise to pay a specified amount of money, with interest, on a certain date for lending money to a corporation or government.
book value (p. 428) The original cost of an item minus the accumulated depreciation.
break-even analysis (p. 306) An investigation of or by a manufacturing business to find out how many units of a product must be made and sold to cover production expenses.
break-even point (p. 306) The point at which income from sales equals the cost of production.
budget sheet (p. 262) An outline of your total monthly expenses that can help you plan future expenditures and savings.

canceled check (p. 97) A check that the bank has paid by deducting money from your account.
capital (p. 440) The difference between your assets and liabilities.
cash discount (p. 332) A discount from a supplier to encourage prompt payment of a bill.
cash value (p. 244) The amount of money you will receive if you cancel a whole life insurance policy.
catalogue price (p. 322) The price at which a business generally sells an item to a consumer.
certificate of deposit (p. 246) A written record of a special kind of savings account that earns higher interest than a regular account. You must leave the money on deposit for a specified time or be penalized for early withdrawal.
chain discount (p. 330) A series of trade discounts on an item.
check (p. 92) A written order directing the bank to deduct money from a checking account to make a payment.
check register (p. 94) A record of the deposits made into a checking account and of the checks written.

closing costs (p. 220) Fees paid at the time documents are signed transferring ownership of a home.
coinsurance clause (p. 240) A clause in your major medical policy stating the percent of the bill you must pay after subtracting the deductible.
collision insurance (p. 198) Insurance to pay for repairs to your automobile if it is involved in an accident.
collision waiver (p. 205) An agreement by which you can obtain complete insurance coverage on a leased automobile by paying an additional charge per day.
commercial loan (p. 460) A business loan.
commercial paper (p. 464) A 30- to 270-day promissory note issued by a company with a very high credit rating.
commission (p. 57) An amount of money paid for selling a product or service, based on the dollar value or number of sales.
commission rate (p. 57) A specified amount of money that you are paid for each sale or a percent of the total value of your sales.
complement method (p. 324) A way to find the net price when a trade discount is given.
compound interest (p. 119) Interest earned not only on the original principal but also on the interest earned during previous interest periods.
comprehensive insurance (p. 198) Insurance that protects you from loss of or damage to your automobile caused by fire, vandalism, theft, and so on.
comprehensive plan (p. 238) A health insurance plan that includes hospital, surgical-medical, and major medical insurance.
consultant (p. 412) A person who gives technical or expert advice.
consultant's fee (p. 412) The payment to a consultant for professional services.
Consumer Price Index (CPI) (p. 271) A measure of the average change in prices of a certain number of goods and services.
corporate income taxes (p. 456) Federal and state taxes on the earnings of a corporation.
corporation (p. 456) A business made up of a number of owners called stockholders.
cost (p. 340) The amount that a business pays for a product including expenses such as freight charges.
cost-of-living adjustment (p. 288) A salary raise to help an employee keep up with inflation.
coupon (p. 143) A certificate or ticket that offers the customer a special discount on an item or service.

data (p. 474) Information related to a business.
data analysis (p. 482) The study of data from a business in order to plan for and control the business.

data base (p. 480) A collection of files describing the activities of a business.

dealer's cost (p. 194) The price that the automobile dealer pays for a new car.

deductible clause (pp. 198, 240) A clause in an insurance policy stating that the insured must pay a portion of any repair bill, or a portion of the amount not covered by hospital or surgical-medical insurance.

defective (p. 308) Incorrect, broken or damaged.

demand charge (p. 410) The charge on your electric bill based on the peak load during the month.

deposit (pp. 90, 108) An amount of money put into a bank account.

depreciation (pp. 202, 426) A decrease in the value of an item because of its age or condition.

destination charge (p. 192) The cost of shipping a new automobile from the factory to the dealer.

direct labor cost (p. 304) The wages paid to the employees who manufacture an item.

direct material cost (p. 304) The cost of the goods that you use to manufacture an item.

discount **1.** (pp. 139, 354) The difference between the regular selling price of an item and its sale price. **2.** (pp. 256, 462) The difference between the face value and maturity value of a bond or U.S. Treasury Bill.

discount loan (p. 176) A loan in which the interest owed is paid in advance.

discount rate (p. 139) The discount on an item expressed as a percent of the regular selling price.

dividends (p. 252) Return on an investment in the stock of a corporation.

double-declining-balance method (p. 430) A method of computing depreciation that allows the highest possible depreciation in the first year of an item's life.

double time (p. 48) An overtime rate that is two times the regular hourly rate.

down payment (p. 178) The portion of the purchase price of an item that you pay at the time of buying when you purchase the item with an installment loan.

driver-rating factor (p. 198) A number assigned by an insurance company to each driver of an automobile to reflect the driver's age, marital status, and so on. A higher driver-rating factor results in a higher annual premium.

emergency fund (p. 267) Money set aside for unpredictable expenses, such as medical bills or repair bills.

employee benefits (p. 290) Benefits an employee receives in addition to salary, such as health insurance, pension, or paid vacation.

end-of-month dating (p. 334) Terms such as "6/15 EOM" on an invoice. In this case, a 6% cash discount is offered if a bill is paid within 15 days after the end of the month.

endowment life insurance (p. 244) Life insurance protection for a specified number of years or payment of the cash value of the policy to the insured who survives beyond that period of time.

energy charge (p. 410) The charge on an electric bill based on the total number of kilowatt-hours used during the month.

escrow account (p. 230) An account often required by the lender of a mortgage loan to guarantee payment of the home owner's real estate taxes and fire insurance premium.

estimated income taxes (p. 488) Income taxes that are paid quarterly by using the profits of the preceding quarter or quarters to estimate the taxes due for the year.

estimated life (p. 426) The length of time, usually in years, that an item is expected to last.

exact interest (p. 174) Interest calculated by basing the length of time of a loan on a 365-day year.

expenditures (p. 262) Amounts of money paid out.

face value (pp. 242, 256, 462) The amount of money printed on an insurance policy, bond, or U.S. Treasury Bill.

factor (p. 368) A company's present market share.

factor method (p. 368) A way to project sales by using a company's present market share and the total estimated sales for the market in the future.

field (p. 474) In a record, a specified area used to contain a fact about a particular aspect of a business.

FIFO(first in, first out) (p. 391) A method of valuing inventory which assumes that the first items received are the first items sold.

file (p. 476) A group of records, all containing related material.

finance charge (p. 155) Interest that is charged to a customer for delaying payment of a bill.

financial summary (p. 489) A brief report about the financial condition of a business, including earnings, operating costs, taxes, and so on.

fire protection class (p. 229) A number assigned to a home by an insurance company according to the quality of fire protection available in the area.

fixed costs **1.** (p. 202) The costs of operating and maintaining an automobile that remain about the same regardless of the number of miles driven. **2.** (p. 306) The costs of manufacturing that remain the same regardless of the number of units produced.

fixed expenses (p. 264) Amounts of money for rent or mortgage payments, installment payments, savings deposits, and so on, that do not vary from one month to the next.
format (p. 486) The arrangement of information in a report.
fuel adjustment charge (p. 410) An additional charge on your electric bill to help cover increases in the cost of fuel needed to produce electricity.

graduated commission (p. 60) A method of payment in which the commission rate changes for each of several levels of sales, increasing as sales increase.
graduated income tax (p. 71) An income tax that involves a different tax rate for each of several levels of income.
Gross National Product (GNP) (p. 357) The estimated total value of all goods and services produced by a nation during a year.
gross profit (p. 340) The difference between the selling price and the cost of a product when the selling price is greater than the cost.
group insurance (pp. 75, 238) Insurance offered to members of a group, such as the employees of a business, at a lower cost than individual insurance.
growth expenses (p. 466) Expenses from expanding a business, such as construction expenses, legal fees, or consultant's fees.

health insurance (p. 238) Insurance coverage against overwhelming medical expenses.
homeowner's insurance (p. 227) Insurance to protect yourself and your home from losses due to fire, theft, or personal liability.
hourly rate (p. 46) The amount of money earned per hour.

income statement (p. 445) A statement that shows in detail the sales, operating expenses, and net profit or loss of a business.
income tax (p. 66) A tax on the earnings, or income, of a person or business.
inflation (p. 415) An economic condition where there are price increases in the cost of goods and services.
installment loan (p. 178) A loan that is repaid in several equal payments over a specified period of time.
insurance-rating group (p. 198) A category to which an insurance company assigns an automobile depending on its size and value. Insurance-rating group affects the base premium.
interest (pp. 117, 218) The amount of money paid for the use of money.

inventory (p. 386) The items that you have in stock.
inventory card (p. 386) A record of each item in stock.
invoice (p. 332) A written record of a purchase, listing the quantities and costs of the items purchased.
issues (p. 386) The number of outgoing items that are subtracted from the previous inventory.

kilowatt (kW) (p. 410) A unit of power equal to 1000 watts.
kilowatt hour (kW•H) (p. 410) A unit to measure the consumption of electrical energy. It is equivalent to using one kilowatt of power constantly for one hour.

labor charge (p. 404) The cost of paying the people who do a job.
lease (p. 402) To agree to specific payments and other conditions in order to use property belonging to someone else for a period of time.
liabilities (p. 440) The total amount of money that you owe to creditors.
life insurance (p. 242) Financial protection that you purchase for your dependents in case of your death.
limited-payment life insurance (p. 244) Lifetime insurance protection for which you pay premiums only for a specified number of years or until you reach a certain age.
list price (p. 322) The price at which a business generally sells an item to a consumer.
living expenses (p. 264) Amounts of money spent for food, clothing, utility bills, and so on, which vary from month to month.
loan value (p. 244) The amount of money the insurance company will lend you if you have a whole life insurance policy.
loss-of-use coverage (p. 227) Insurance coverage to pay for some of the expenses of living away from home while damage to your home is being repaired.
loss (p. 254) The difference between the cost and selling price of an item when the cost is greater than the selling price.

major medical insurance (p. 240) Insurance that pays for medical costs not covered by hospital or surgical-medical insurance.
markdown (pp. 139, 354) The difference between the regular selling price and the sale price of an item.
markdown rate (pp. 139, 354) The markdown on an item expressed as a percent of the regular selling price.
market (p. 362) The total number of people who might purchase the type of product being made.

market share (p. 364) The portion of the market that purchases a specific product.
market value (p. 225) The price at which a house or other article can be bought or sold.
markup (p. 340) The difference between the selling price and the cost of a product, when the selling price is greater.
markup rate (p. 342) The markup expressed as a percent of the selling price or cost of an item.
materials charge (p. 404) The cost of the materials used to complete a job.
maturity date (p. 462) The date on which a note, bill, or bond is due and payable.
maturity value (pp. 174, 460) The total amount you repay on a loan, including both the principal borrowed and the interest owed.
merge (p. 466) To combine with another business to form a new business.
merit increase (p. 288) A raise in your salary to reward you for the quality of your work.
mill (p. 225) A unit often used to express real estate tax rates, equal to $1.00 in tax for each $1000 of assessed value.
monthly expense summary (p. 267) A totaling of actual monthly expenses in order to compare the amounts spent with the amounts budgeted.
mortgage loan (p. 216) A loan to finance the purchase of a home whereby the lender has the right to sell the property if the loan payments are not made.

net income (p. 445) The difference between the gross profit and the total operating expenses if the gross profit is greater than the operating expenses.
net interest (p. 464) The actual interest earned from investing in commercial paper after any bank service fees have been deducted.
net loss (p. 447) The difference between the markup, or gross profit, and the overhead, or operating expenses, when the overhead is greater than the markup.
net pay (p. 77) The amount of pay you have left after all tax withholdings and personal deductions have been subtracted.
net price (p. 322) The price a business actually pays for an item after discounts have been subtracted.
net-price rate (p. 330) The percent a business pays for an item found by multiplying the complements of the chain discounts.
net proceeds (p. 458) The amount that a business actually receives from the sale of stocks or bonds after subtracting all selling expenses.

net profit (pp. 344, 374, 445) The difference between the markup, or gross profit, and the overhead, or operating expenses, when the markup is greater than the overhead.
net-profit rate (p. 346) Net profit expressed as a percent of the selling price of an item.
net worth (p. 440) The difference between your assets and liabilities.
nongroup insurance (p. 238) Insurance for people not enrolled in a group.

off-line storage (p. 478) Storage of information on magnetic disks or tapes, which must be mounted onto a computer system to be used.
on-line storage (p. 478) Storage of information in the computer's internal memory for immediate access.
operating expenses (p. 344) Expenses of running a business, such as wages of employees, rent, utility charges, and taxes.
opinion-research firm (p. 360) A business that specializes in product testing and opinion surveys and the interpretation of the results.
opinion survey (p. 360) A series of questions answered by a group of volunteers. The results should approximate the opinions of a much larger group of people.
option (p. 192) An extra on a new automobile for convenience, safety, or appearance, such as a radio or air-conditioning.
ordinary dating (p. 332) Terms such as "2/10, net 30" on an invoice. In this case, a 2% cash discount is offered if the bill is paid within 10 days of date of invoice; the net price must be paid within 30 days.
ordinary interest (p. 174) Interest calculated by basing the length of time of a loan on a 360-day year.
outstanding check/deposit (p. 100) A check or deposit that appears in your check register but has not reached the bank in time to be listed on your statement.
overdraw (p. 92) To write checks in excess of the amount of money in a checking account.
overhead (p. 344) Expenses of running a business, such as wages of employees, rent, utility charges, and taxes.
overtime pay (p. 48) The amount of money you earn when you work more than your regular hours. The overtime rate may be $1\frac{1}{2}$ or 2 times the regular hourly rate.
owner's equity (p. 440) The difference between your assets and liabilities.

packaging (p. 315) Placing a product in a container for shipment.
passbook (p. 112) A book given to you in which the bank records all the deposits and withdrawals for your account and any interest earned.

payroll register (p. 418) A record of the gross income, deductions, and net income of a company's employees.

peak load (p. 410) The greatest number of kilowatts used by a consumer at one time during the month.

percent change (p. 450) The amount of change from an earlier income statement to the current one, divided by the dollar amount on the earlier statement.

periodic rate (p. 158) The monthly finance charge rate.

personal exemptions (p. 69) Subtractions made from your gross pay for yourself and every other person you support. They are used to determine your taxable wages.

personal liability and medical coverage (p. 227) Insurance coverage to protect you from financial losses if someone is injured on your property.

piecework (p. 53) Work for which you are paid according to the number of items you complete.

point (p. 221) One percent of the mortgage loan. Points are included in the closing costs.

posting (p. 474) The act of recording data manually or with the computer.

premium **1.** (p. 229) The amount paid for insurance coverage.
2. (p. 256) Bonds sold at a premium are sold for more than face value.

previous balance (p. 158) The amount owed by a credit card user on the closing date of the last statement.

prime cost (p. 304) The total of the direct material cost and the direct labor cost.

prime rate (p. 460) The lowest rate of interest available on business loans at a given time.

principal (pp. 117, 174, 222) The amount of money on which interest is paid.

proceeds (p. 176) The amount of money that you receive from a discount loan after the interest has been deducted.

product test (p. 360) An attempt to determine how well a product is going to sell by asking a group of people to try it.

profit (pp. 254, 306) The amount you receive after the costs of selling or production have been met.

profit-and-loss statement (p. 445) A statement that shows in detail your sales, operating expenses, and net profit or loss.

profit margin (p. 340) The difference between the selling price and the cost of a product.

promissory note (p. 462) A written promise to pay a certain sum of money at a specific future date.

property damage insurance (p. 198) Insurance that protects you against financial loss if your automobile damages the property of other people.

quality control (p. 308) Periodic inspection of the products you manufacture.
quality control chart (p. 308) A chart to show the percent of defective products that is allowable in a manufacturing business.
quoted price (p. 256) The actual cost of a bond, usually a percent of the face value.

rate of assessment (p. 225) The percent used to calculate the assessed value of a home, given the market value.
real estate taxes (p. 225) Fees collected from the owners of property and used to maintain government operations.
real-time processing (p. 480) Processing by a computer that keeps up with the day's events.
rebate schedule (p. 186) A table of percent refunds, used to determine the part of the finance charge refunded when an installment loan is repaid before the final due date.
receipts (p. 386) The number of incoming items that are added to the previous inventory.
reconcile (p. 100) To obtain agreement between your check register and the bank statement.
record (p. 474) A collection of related data describing one aspect of a business.
recruiting costs (p. 286) Expenses of finding a new employee, such as advertising fees, interview costs, and hiring expenses.
release time (p. 294) Time spent away from your job in training programs for which you are still paid your regular wages or salary.
rent (pp. 205, 402, 406) To pay a regular fee for the use of equipment or of space in a building.
rent escalation clause (p. 403) A statement in a lease that the rent may increase by a certain percent after a certain period of time.
replacement value (p. 227) The amount of money required to reconstruct a home if destroyed.
resale value (p. 426) The estimated trade-in or salvage value of an item at the end of its expected life.
retail price (p. 340) The price that a business expects a consumer to pay for a product.

salary (p. 55) A fixed amount of money that you earn on a regular basis.
salary scale (p. 288) A table of wages or salaries used to compare various jobs in a business.
sale price (p. 139) A purchase price which is lower than the regular selling price.

sales potential (p. 362) An estimate of the number of products that you might sell during a specified period of time.
sales projection (p. 366) An estimate of the dollar volume or unit sales expected to occur during a future time period.
sales quota (p. 61) The amount of sales expected of a sales representative for a given period of time.
sales receipt (pp. 132, 152) A sales slip or cash register tape that shows the selling price of the items purchased, any sales tax, and the total purchase price. If a credit card is used, the receipt also shows the name and signature of the purchaser and the account number.
sales tax (p. 130) A tax on the selling price of an item or service that you purchase.
salvage value (p. 426) The estimated trade-in, resale, or scrap value of an item at the end of its expected life.
sample (p. 362) A selected group of people, representative of a much greater number of people, who are asked to evaluate a new product.
selling price (p. 340) The price that a business expects a consumer to pay for a product.
service charge (p. 97) A charge by a bank for handling a checking account.
simple interest (p. 117) Interest paid only on the original principal.
single equivalent discount (SED) (p. 330) One discount that is equal to a chain discount.
single payment loan (p. 174) A loan that you repay with one payment after a specified period of time.
social security (p. 73) A federal law (F.I.C.A.) to provide hospitalization insurance for people over 65, retirement income, survivor's benefits, and disability benefits. The money for this program is provided by a tax paid by employers and employees.
statement (pp. 97, 114, 155) A record of an account sent by a bank or other business listing all transactions that have been recorded during a specified period of time.
sticker price (p. 192) The total of the base price, options, and destination charge for a new automobile.
stockbroker (p. 250) Someone who buys and sells stocks for other people.
stock certificate (p. 250) A printed form that proves part ownership of a corporation.
stocks (pp. 250, 458) Shares of ownership of the corporation issuing the stock.
storage space (p. 384) A place in which to keep products until they are needed.
straight commission (p. 57) A method of payment in which the only pay you receive is a commission on the sales you make.

straight-line method (p. 426) A method of computing depreciation which assumes that the depreciation is the same from year to year.

straight-time pay (p. 46) The total amount of money you earn for a pay period at your regular hourly rate.

sum-of-the-years'-digits method (p. 430) A method of computing depreciation that assigns a fraction of the total depreciation to each year of an item's life, with more depreciation allowed in the early years.

surgical-medical insurance (p. 240) Insurance that pays your doctor's fee for surgery, up to a certain amount.

taxable income (p. 456) The portion of the gross income of a business that remains after normal business expenses are deducted.

tax rate (p. 225) **1.** expressed in mills and used to determine real estate taxes. **2.** Expressed as a percent and used to determine sales taxes.

term (p. 174) The amount of time for which a loan is granted.

term life insurance (p. 242) A policy written for a specified amount of time, such as five years, that expires at the end of the time unless renewed.

time and a half (p. 48) An overtime rate $1\frac{1}{2}$ times the regular hourly rate.

time-study (p. 311) Determining how long a particular job should take.

trade discount (p. 322) A discount that a business receives when purchasing goods from a supplier.

trade-discount rate (p. 324) The trade discount on an item expressed as a percent of the list price.

travel expenses (p. 292) Money spent by an employee for expenses such as transportation, lodging, and meals while traveling on business.

underwriting commission (p. 458) The amount of money paid to the investment banker who helps a business distribute stocks or bonds.

unit (p. 242) One unit of insurance has a face value of $1000.

United States Treasury Bill (p. 462) A written promise by the federal government to repay the purchase price of the bill plus interest at a future date.

unit price (p. 135) The cost of an item expressed per unit of measure or count.

units-of-production method (p. 429) A method of computing depreciation based on the number of units produced.

Universal Product Code (UPC) (p. 491) A series of lines, bars, and numbers found on the packages of most products.

unpaid balance (p. 161) The portion of the balance on your previous charge account or credit card statement that you have not yet paid.

used-car guide (p. 196) A monthly publication that gives price information on used cars purchased from dealers during the previous month.

utilities costs (pp. 231, 408) Charges for public services, such as electricity or water.

variable costs **1.** (p. 202) The costs of operating and maintaining an automobile that increase as the number of miles driven increases. **2.** (p. 306) Manufacturing costs that increase as the number of units produced increases.

warehouse (p. 384) A building or part of one used to store raw materials or products until they are needed.

weekly timecard (p. 50) A record of the times you report for work and the times you leave.

whole life insurance (p. 244) Life insurance that offers financial protection for your entire life.

withdrawal (p. 110) An amount of money taken out of a savings account.

withholding allowance (p. 66) A limitation on the amount of federal income tax withheld from your pay. You may claim one allowance for yourself, one for your spouse if married, and one for every other person you support.

TABLE OF MEASURES

Time

60 seconds (s) = 1 minute (min)
60 minutes = 1 hour (h)
24 hours = 1 day
7 days = 1 week
52 weeks = 1 year
12 months = 1 year
100 years = 1 century

METRIC SYSTEM

Length

10 millimeters (mm) = 1 centimeter (cm)
100 centimeters = 1 meter (m)
1000 meters =1 kilometer (km)

Volume

1000 milliliters (mL) = 1 liter (L)
1000 cubic centimeters (cm^3) = 1 liter
10 milliliters = 1 centiliter (cL)
10 deciliters (dL) = 1 liter

Area

100 square millimeters (mm^2) = 1 square centimeter (cm^2)
10,000 square centimeters = 1 square meter (m^2)
10,000 square meters = 1 hectare (ha)

Mass

1000 milligrams (mg) = 1 gram (g)
1000 grams = 1 kilogram (kg)
1000 kilograms = 1 metric ton (t)

CUSTOMARY SYSTEM

Length

12 inches (in.) = 1 foot (ft)
3 feet = 1 yard (yd)
5280 feet = 1 mile (mi)

Volume

8 fluid ounces (oz) = 1 cup (c)
2 cups = 1 pint (pt)
2 pints = 1 quart (qt)
4 quarts = 1 gallon (gal)

Area

144 square inches ($in.^2$) = 1 square foot (ft^2)
9 square feet = 1 square yard (yd^2)
4837 square yards = 1 acre (A)

Weight

16 ounces = 1 pound (lb)
2000 pounds = 1 ton (t)

FEDERAL WITHHOLDING TAX—WEEKLY Payroll Period, Employee NOT MARRIED

And the wages are—		And the number of withholding allowances claimed is—										
At least	But less than	0	1	2	3	4	5	6	7	8	9	10 or more
		The amount of income tax to be withheld shall be—										
$130	$135	$ 14.30	$ 11.40	$ 8.50	$ 5.70	$ 3.40	$ 1.10	$ 0	$ 0	$ 0	$ 0	$ 0
135	140	15.00	12.10	9.30	6.40	4.00	1.70	0	0	0	0	0
140	145	15.80	12.90	10.00	7.10	4.60	2.30	0	0	0	0	0
145	150	16.50	13.60	10.80	7.90	5.20	2.90	.60	0	0	0	0
150	160	17.70	14.80	11.90	9.00	6.10	3.80	1.50	0	0	0	0
160	170	19.20	16.30	13.40	10.50	7.60	5.00	2.70	.40	0	0	0
170	180	20.70	17.80	14.90	12.00	9.10	6.20	3.90	1.60	0	0	0
180	190	22.20	19.30	16.40	13.50	10.60	7.70	5.10	2.80	.50	0	0
190	200	24.10	20.80	17.90	15.00	12.10	9.20	6.30	4.00	1.70	0	0
200	210	26.00	22.40	19.40	16.50	13.60	10.70	7.80	5.20	2.90	.60	0
210	220	27.90	24.30	20.90	18.00	15.10	12.20	9.30	6.50	4.10	1.80	0
220	230	29.80	26.20	22.50	19.50	16.60	13.70	10.80	8.00	5.30	3.00	.70
230	240	31.70	28.10	24.40	21.00	18.10	15.20	12.30	9.50	6.60	4.20	1.90
240	250	33.60	30.00	26.30	22.70	19.60	16.70	13.80	11.00	8.10	5.40	3.10
250	260	35.50	31.90	28.20	24.60	21.10	18.20	15.30	12.50	9.60	6.70	4.30
260	270	37.40	33.80	30.10	26.50	22.80	19.70	16.80	14.00	11.10	8.20	5.50
270	280	39.30	35.70	32.00	28.40	24.70	21.20	18.30	15.50	12.60	9.70	6.80
280	290	41.70	37.60	33.90	30.30	26.60	23.00	19.80	17.00	14.10	11.20	8.30
290	300	44.20	39.50	35.80	32.20	28.50	24.90	21.30	18.50	15.60	12.70	9.80
300	310	46.70	41.90	37.70	34.10	30.40	26.80	23.10	20.00	17.10	14.20	11.30
310	320	49.20	44.40	39.60	36.00	32.30	28.70	25.00	21.50	18.60	15.70	12.80
320	330	51.70	46.90	42.10	37.90	34.20	30.60	26.90	23.30	20.10	17.20	14.30
330	340	54.20	49.40	44.60	39.80	36.10	32.50	28.80	25.20	21.60	18.70	15.80
340	350	56.70	51.90	47.10	42.30	38.00	34.40	30.70	27.10	23.40	20.20	17.30
350	360	59.20	54.40	49.60	44.80	40.00	36.30	32.60	29.00	25.30	21.70	18.80
360	370	61.70	56.90	52.10	47.30	42.50	38.20	34.50	30.90	27.20	23.60	20.30
370	380	64.20	59.40	54.60	49.80	45.00	40.20	36.40	32.80	29.10	25.50	21.80
380	390	66.70	61.90	57.10	52.30	47.50	42.70	38.30	34.70	31.00	27.40	23.70
390	400	69.20	64.40	59.60	54.80	50.00	45.20	40.40	36.60	32.90	29.30	25.60
400	410	71.70	66.90	62.10	57.30	52.50	47.70	42.90	38.50	34.80	31.20	27.50
410	420	74.20	69.40	64.60	59.80	55.00	50.20	45.40	40.60	36.70	33.10	29.40
420	430	76.80	71.90	67.10	62.30	57.50	52.70	47.90	43.10	38.60	35.00	31.30
430	440	79.80	74.40	69.60	64.80	60.00	55.20	50.40	45.60	40.80	36.90	33.20
440	450	82.80	77.10	72.10	67.30	62.50	57.70	52.90	48.10	43.30	38.80	35.10
450	460	85.80	80.10	74.60	69.80	65.00	60.20	55.40	50.60	45.80	41.00	37.00
460	470	88.80	83.10	77.30	72.30	67.50	62.70	57.90	53.10	48.30	43.50	38.90
470	480	91.80	86.10	80.30	74.80	70.00	65.20	60.40	55.60	50.80	46.00	41.20
480	490	94.80	89.10	83.30	77.50	72.50	67.70	62.90	58.10	53.30	48.50	43.70
490	500	97.80	92.10	86.30	80.50	75.00	70.20	65.40	60.60	55.80	51.00	46.20
500	510	100.80	95.10	89.30	83.50	77.80	72.70	67.90	63.10	58.30	53.50	48.70
510	520	103.80	98.10	92.30	86.50	80.80	75.20	70.40	65.60	60.80	56.00	51.20
520	530	106.80	101.10	95.30	89.50	83.80	78.00	72.90	68.10	63.30	58.50	53.70
530	540	109.80	104.10	98.30	92.50	86.80	81.00	75.40	70.60	65.80	61.00	56.20
540	550	113.20	107.10	101.30	95.50	89.80	84.00	78.20	73.10	68.30	63.50	58.70
550	560	116.60	110.10	104.30	98.50	92.80	87.00	81.20	75.60	70.80	66.00	61.20
560	570	120.00	113.50	107.30	101.50	95.80	90.00	84.20	78.40	73.30	68.50	63.70
570	580	123.40	116.90	110.40	104.50	98.80	93.00	87.20	81.40	75.80	71.00	66.20
580	590	126.80	120.30	113.80	107.50	101.80	96.00	90.20	84.40	78.70	73.50	68.70
590	600	130.20	123.70	117.20	110.60	104.80	99.00	93.20	87.40	81.70	76.00	71.20
600	610	133.60	127.10	120.60	114.00	107.80	102.00	96.20	90.40	84.70	78.90	73.70
610	620	137.00	130.50	124.00	117.40	110.90	105.00	99.20	93.40	87.70	81.90	76.20
620	630	140.40	133.90	127.40	120.80	114.30	108.00	102.20	96.40	90.70	84.90	79.10
630	640	143.80	137.30	130.80	124.20	117.70	111.20	105.20	99.40	93.70	87.90	82.10
640	650	147.50	140.70	134.20	127.60	121.10	114.60	108.20	102.40	96.70	90.90	85.10
650	660	151.20	144.10	137.60	131.00	124.50	118.00	111.40	105.40	99.70	93.90	88.10

FEDERAL WITHHOLDING TAX—WEEKLY Payroll Period, Employee MARRIED

And the wages are—		And the number of withholding allowances claimed is—										
At least	But less than	0	1	2	3	4	5	6	7	8	9	10 or more
		The amount of income tax to be withheld shall be—										
$130	$135	$ 10.40	$ 8.10	$ 5.70	$ 3.40	$ 1.10	$ 0	$ 0	$ 0	$ 0	$ 0	$ 0
135	140	11.00	8.70	6.30	4.00	1.70	0	0	0	0	0	0
140	145	11.60	9.30	6.90	4.60	2.30	0	0	0	0	0	0
145	150	12.20	9.90	7.50	5.20	2.90	.60	0	0	0	0	0
150	160	13.10	10.80	8.40	6.10	3.80	1.50	0	0	0	0	0
160	170	14.30	12.00	9.60	7.30	5.00	2.70	.40	0	0	0	0
170	180	15.50	13.20	10.80	8.50	6.20	3.90	1.60	0	0	0	0
180	190	16.70	14.40	12.00	9.70	7.40	5.10	2.80	.50	0	0	0
190	200	18.40	15.60	13.20	10.90	8.60	6.30	4.00	1.70	0	0	0
200	210	20.10	16.80	14.40	12.10	9.80	7.50	5.20	2.90	.60	0	0
210	220	21.80	18.50	15.60	13.30	11.00	8.70	6.40	4.10	1.80	0	0
220	230	23.50	20.20	16.90	14.50	12.20	9.90	7.60	5.30	3.00	.70	0
230	240	25.20	21.90	18.60	15.70	13.40	11.10	8.80	6.50	4.20	1.90	0
240	250	26.90	23.60	20.30	17.10	14.60	12.30	10.00	7.70	5.40	3.10	.80
250	260	28.60	25.30	22.00	18.80	15.80	13.50	11.20	8.90	6.60	4.30	2.00
260	270	30.30	27.00	23.70	20.50	17.20	14.70	12.40	10.10	7.80	5.50	3.20
270	280	32.00	28.70	25.40	22.20	18.90	15.90	13.60	11.30	9.00	6.70	4.40
280	290	33.70	30.40	27.10	23.90	20.60	17.30	14.80	12.50	10.20	7.90	5.60
290	300	35.40	32.10	28.80	25.60	22.30	19.00	16.00	13.70	11.40	9.10	6.80
300	310	37.10	33.80	30.50	27.30	24.00	20.70	17.50	14.90	12.60	10.30	8.00
310	320	38.80	35.50	32.20	29.00	25.70	22.40	19.20	16.10	13.80	11.50	9.20
320	330	40.50	37.20	33.90	30.70	27.40	24.10	20.90	17.60	15.00	12.70	10.40
330	340	42.20	38.90	35.60	32.40	29.10	25.80	22.60	19.30	16.20	13.90	11.60
340	350	43.90	40.60	37.30	34.10	30.80	27.50	24.30	21.00	17.70	15.10	12.80
350	360	45.60	42.30	39.00	35.80	32.50	29.20	26.00	22.70	19.40	16.30	14.00
360	370	47.30	44.00	40.70	37.50	34.20	30.90	27.70	24.40	21.10	17.90	15.20
370	380	49.30	45.70	42.40	39.20	35.90	32.60	29.40	26.10	22.80	19.60	16.40
380	390	51.50	47.40	44.10	40.90	37.60	34.30	31.10	27.80	24.50	21.30	18.00
390	400	53.70	49.50	45.80	42.60	39.30	36.00	32.80	29.50	26.20	23.00	19.70
400	410	55.90	51.70	47.50	44.30	41.00	37.70	34.50	31.20	27.90	24.70	21.40
410	420	58.10	53.90	49.60	46.00	42.70	39.40	36.20	32.90	29.60	26.40	23.10
420	430	60.30	56.10	51.80	47.70	44.40	41.10	37.90	34.60	31.30	28.10	24.80
430	440	62.50	58.30	54.00	49.80	46.10	42.80	39.60	36.30	33.00	29.80	26.50
440	450	64.70	60.50	56.20	52.00	47.80	44.50	41.30	38.00	34.70	31.50	28.20
450	460	66.90	62.70	58.40	54.20	50.00	46.20	43.00	39.70	36.40	33.20	29.90
460	470	69.40	64.90	60.60	56.40	52.20	47.90	44.70	41.40	38.10	34.90	31.60
470	480	71.90	67.10	62.80	58.60	54.40	50.10	46.40	43.10	39.80	36.60	33.30
480	490	74.40	69.60	65.00	60.80	56.60	52.30	48.10	44.80	41.50	38.30	35.00
490	500	76.90	72.10	67.30	63.00	58.80	54.50	50.30	46.50	43.20	40.00	36.70
500	510	79.40	74.60	69.80	65.20	61.00	56.70	52.50	48.30	44.90	41.70	38.40
510	520	81.90	77.10	72.30	67.50	63.20	58.90	54.70	50.50	46.60	43.40	40.10
520	530	84.40	79.60	74.80	70.00	65.40	61.10	56.90	52.70	48.40	45.10	41.80
530	540	86.90	82.10	77.30	72.50	67.70	63.30	59.10	54.90	50.60	46.80	43.50
540	550	89.40	84.60	79.80	75.00	70.20	65.50	61.30	57.10	52.80	48.60	45.20
550	560	91.90	87.10	82.30	77.50	72.70	67.90	63.50	59.30	55.00	50.80	46.90
560	570	94.70	89.60	84.80	80.00	75.20	70.40	65.70	61.50	57.20	53.00	48.80
570	580	97.50	92.10	87.30	82.50	77.70	72.90	68.10	63.70	59.40	55.20	51.00
580	590	100.30	94.90	89.80	85.00	80.20	75.40	70.60	65.90	61.60	57.40	53.20
590	600	103.10	97.70	92.30	87.50	82.70	77.90	73.10	68.30	63.80	59.60	55.40
600	610	105.90	100.50	95.10	90.00	85.20	80.40	75.60	70.80	66.00	61.80	57.60
610	620	108.70	103.30	97.90	92.50	87.70	82.90	78.10	73.30	68.50	64.00	59.80
620	630	111.50	106.10	100.70	95.30	90.20	85.40	80.60	75.80	71.00	66.20	62.00
630	640	114.30	108.90	103.50	98.10	92.80	87.90	83.10	78.30	73.50	68.70	64.20
640	650	117.10	111.70	106.30	100.90	95.60	90.40	85.60	80.80	76.00	71.20	66.40
650	660	119.90	114.50	109.10	103.70	98.40	93.00	88.10	83.30	78.50	73.70	68.80

AMOUNT OF $1.00							
TOTAL INTEREST PERIODS	INTEREST RATE PER PERIOD						
	1.2500%	1.3750%	1.5000%	2.7500%	2.8750%	3.0000%	3.1250%
1	1.01250	1.01375	1.01500	1.02749	1.02875	1.03000	1.03125
2	1.02515	1.02768	1.03022	1.05575	1.05832	1.06090	1.06347
3	1.03797	1.04182	1.04567	1.08478	1.08875	1.09272	1.09671
4	1.05094	1.05614	1.06136	1.11462	1.12005	1.12550	1.13098
5	1.06408	1.07066	1.07728	1.14527	1.15225	1.15927	1.16632
6	1.07738	1.08538	1.09344	1.17676	1.18538	1.19405	1.20277
7	1.09085	1.10031	1.10984	1.20912	1.21946	1.22987	1.24036
8	1.10448	1.11544	1.12649	1.24237	1.25452	1.26677	1.27912
9	1.11829	1.13078	1.14339	1.27654	1.29059	1.30477	1.31909
10	1.13227	1.14632	1.16054	1.31165	1.32769	1.34391	1.36031
11	1.14642	1.16209	1.17795	1.34772	1.36586	1.38423	1.40282
12	1.16075	1.17806	1.19562	1.38478	1.40513	1.42576	1.44666
13	1.17526	1.19426	1.21355	1.42286	1.44553	1.46853	1.49187
14	1.18995	1.21068	1.23175	1.46199	1.48709	1.51258	1.53849
15	1.20482	1.22733	1.25023	1.50219	1.52984	1.55796	1.58657
16	1.21988	1.24421	1.26898	1.54350	1.57382	1.60470	1.63615
17	1.23513	1.26132	1.28802	1.58595	1.61907	1.65284	1.68728
18	1.25057	1.27866	1.30734	1.62956	1.66562	1.70243	1.74000
19	1.26620	1.29624	1.32695	1.67438	1.71351	1.75350	1.79438
20	1.28203	1.31406	1.34685	1.72042	1.76277	1.80611	1.85045
21	1.29806	1.33213	1.36706	1.76773	1.81345	1.86029	1.90828
22	1.31428	1.35045	1.38756	1.81635	1.86559	1.91610	1.96791
23	1.33071	1.36902	1.40838	1.86630	1.91922	1.97358	2.02941
24	1.34735	1.38784	1.42950	1.91762	1.97440	2.03279	2.09283

AMOUNT OF $1.00 AT 5.5%, COMPOUNDED DAILY, 365-DAY YEAR									
DAY	AMOUNT	DAY	AMOUNT	DAY	AMOUNT	DAY	AMOUNT	DAY	AMOUNT
1	1.00015	11	1.00165	21	1.00316	31	1.00468	50	1.00755
2	1.00030	12	1.00180	22	1.00331	32	1.00483	60	1.00907
3	1.00045	13	1.00196	23	1.00347	33	1.00498	70	1.01059
4	1.00060	14	1.00211	24	1.00362	34	1.00513	80	1.01212
5	1.00075	15	1.00226	25	1.00377	35	1.00528	90	1.01364
6	1.00090	16	1.00241	26	1.00392	36	1.00543	100	1.01517
7	1.00105	17	1.00256	27	1.00407	37	1.00558	110	1.01670
8	1.00120	18	1.00271	28	1.00422	38	1.00574	120	1.01823
9	1.00135	19	1.00286	29	1.00437	39	1.00589	130	1.01977
10	1.00150	20	1.00301	30	1.00452	40	1.00604	140	1.02131

THE NUMBER OF EACH DAY OF THE YEAR												
Day No.	Jan.	Feb.	Mar.	Apr.	May	June	July	Aug.	Sept.	Oct.	Nov.	Dec.
1	1	32	60	91	121	152	182	213	244	274	305	335
2	2	33	61	92	122	153	183	214	245	275	306	336
3	3	34	62	93	123	154	184	215	246	276	307	337
4	4	35	63	94	124	155	185	216	247	277	308	338
5	5	36	64	95	125	156	186	217	248	278	309	339
6	6	37	65	96	126	157	187	218	249	279	310	340
7	7	38	66	97	127	158	188	219	250	280	311	341
8	8	39	67	98	128	159	189	220	251	281	312	342
9	9	40	68	99	129	160	190	221	252	282	313	343
10	10	41	69	100	130	161	191	222	253	283	314	344
11	11	42	70	101	131	162	192	223	254	284	315	345
12	12	43	71	102	132	163	193	224	255	285	316	346
13	13	44	72	103	133	164	194	225	256	286	317	347
14	14	45	73	104	134	165	195	226	257	287	318	348
15	15	46	74	105	135	166	196	227	258	288	319	349
16	16	47	75	106	136	167	197	228	259	289	320	350
17	17	48	76	107	137	168	198	229	260	290	321	351
18	18	49	77	108	138	169	199	230	261	291	322	352
19	19	50	78	109	139	170	200	231	262	292	323	353
20	20	51	79	110	140	171	201	232	263	293	324	354
21	21	52	80	111	141	172	202	233	264	294	325	355
22	22	53	81	112	142	173	203	234	265	295	326	356
23	23	54	82	113	143	174	204	235	266	296	327	357
24	24	55	83	114	144	175	205	236	267	297	328	358
25	25	56	84	115	145	176	206	237	268	298	329	359
26	26	57	85	116	146	177	207	238	269	299	330	360
27	27	58	86	117	147	178	208	239	270	300	331	361
28	28	59	87	118	148	179	209	240	271	301	332	362
29	29	...	88	119	149	180	210	241	272	302	333	363
30	30	...	89	120	150	181	211	242	273	303	334	364
31	31	...	90	...	151	...	212	243	...	304	...	365

Monthly Payments of $1,000 Loan

Rate (Percent)	Number of Years of Loan							
	5	10	15	20	25	30	35	40
7	$19.81	$11.62	$ 8.99	$ 7.76	$ 7.07	$ 6.66	$ 6.39	$ 6.22
7½	20.04	11.88	9.28	8.06	7.39	7.00	6.75	6.59
8	20.28	12.14	9.56	8.37	7.72	7.34	7.11	6.96
8½	20.52	12.40	9.85	8.68	8.06	7.69	7.47	7.34
9	20.76	12.67	10.15	9.00	8.40	8.05	7.84	7.72
9½	21.01	12.94	10.45	9.33	8.74	8.41	8.22	8.11
10	21.25	13.22	10.75	9.66	9.09	8.78	8.60	8.50
10½	21.50	13.50	11.06	9.99	9.45	9.15	8.99	8.89
11	21.75	13.78	11.37	10.33	9.81	9.53	9.37	9.29
11½	22.00	14.06	11.69	10.67	10.17	9.91	9.77	9.69
12	22.25	14.35	12.01	11.02	10.54	10.29	10.16	10.09
12½	22.50	14.64	12.33	11.37	10.91	10.68	10.56	10.49
13	22.76	14.94	12.66	11.72	11.28	11.07	10.96	10.90
13½	23.01	15.23	12.99	12.08	11.66	11.46	11.36	11.31
14	23.27	15.53	13.32	12.44	12.04	11.85	11.76	11.72
14½	23.53	15.83	13.66	12.80	12.43	12.25	12.17	12.13
15	23.79	16.14	14.00	13.17	12.81	12.65	12.57	12.54
15½	24.06	16.45	14.34	13.54	13.20	13.05	12.98	12.95
16	24.32	16.76	14.69	13.92	13.59	13.45	13.39	13.36
17	24.86	17.38	15.40	14.67	14.38	14.26	14.21	14.19
18	25.40	18.02	16.11	15.44	15.18	15.08	15.03	15.02

ANNUAL PERCENTAGE RATES

APR	10.00%	10.25%	10.50%	10.75%	11.00%	11.25%	11.50%	11.75%	12.00%	12.25%	12.50%
TERM	FINANCE CHARGE PER $100 OF AMOUNT FINANCED										
6	2.94	3.01	3.08	3.16	3.23	3.31	3.38	3.45	3.53	3.60	3.68
12	5.50	5.64	5.78	5.92	6.06	6.20	6.34	6.48	6.62	6.76	6.90
18	8.10	8.31	8.52	8.73	8.93	9.14	9.35	9.56	9.77	9.98	10.19
24	10.75	11.02	11.30	11.58	11.86	12.14	12.42	12.70	12.98	13.26	13.54
30	13.43	13.78	14.13	14.48	14.83	15.19	15.54	15.89	16.24	16.60	16.95
36	16.16	16.58	17.01	17.43	17.86	18.29	18.71	19.14	19.57	20.00	20.43

APR	12.75%	13.00%	13.25%	13.50%	13.75%	14.00%	14.25%	14.50%	14.75%	15.00%	15.25%
6	3.75	3.83	3.90	3.97	4.05	4.12	4.20	4.27	4.35	4.42	4.49
12	7.04	7.18	7.32	7.46	7.60	7.74	7.89	8.03	8.17	8.31	8.45
18	10.40	10.61	10.82	11.03	11.24	11.45	11.66	11.87	12.08	12.29	12.50
24	13.82	14.10	14.38	14.66	14.95	15.23	15.51	15.80	16.08	16.37	16.65
30	17.31	17.66	18.02	18.38	18.74	19.10	19.45	19.81	20.17	20.54	20.90
36	20.86	21.30	21.73	22.17	22.60	23.04	23.48	23.92	24.35	24.80	25.24

APR	15.50%	15.75%	16.00%	16.25%	16.50%	16.75%	17.00%	17.25%	17.50%	17.75%	18.00%
6	4.57	4.64	4.72	4.79	4.87	4.94	5.02	5.09	5.17	5.24	5.32
12	8.59	8.74	8.88	9.02	9.16	9.30	9.45	9.59	9.73	9.87	10.02
18	12.72	12.93	13.14	13.35	13.57	13.78	13.99	14.21	14.42	14.64	14.85
24	16.94	17.22	17.51	17.80	18.09	18.37	18.66	18.95	19.24	19.53	19.82
30	21.26	21.62	21.99	22.35	22.72	23.08	23.45	23.81	24.18	24.55	24.92
36	25.68	26.12	26.57	27.01	27.46	27.90	28.35	28.80	29.25	29.70	30.15

APR	18.25%	18.50%	18.75%	19.00%	19.25%	19.50%	19.75%	20.00%	20.25%	20.50%	20.75%
6	5.39	5.46	5.54	5.61	5.69	5.76	5.84	5.91	5.99	6.06	6.14
12	10.16	10.30	10.44	10.59	10.73	10.87	11.02	11.16	11.31	11.45	11.59
18	15.06	15.28	15.49	15.71	15.93	16.14	16.36	16.57	16.79	17.01	17.22
24	20.11	20.40	20.69	20.98	21.27	21.56	21.86	22.15	22.44	22.74	23.03
30	25.29	25.66	26.03	26.40	26.77	27.14	27.52	27.89	28.26	28.64	29.01
36	30.60	31.05	31.51	31.96	32.42	32.87	33.33	33.79	34.25	34.71	35.17

APR	21.00%	21.25%	21.50%	21.75%	22.00%	22.25%	22.50%	22.75%	23.00%	23.25%	23.50%
6	6.21	6.29	6.36	6.44	6.51	6.59	6.66	6.74	6.81	6.89	6.96
12	11.74	11.88	12.02	12.17	12.31	12.46	12.60	12.75	12.89	13.04	13.18
18	17.44	17.66	17.88	18.09	18.31	18.53	18.75	18.97	19.19	19.41	19.62
24	23.33	23.62	23.92	24.21	24.51	24.80	25.10	25.40	25.70	25.99	26.29
30	29.39	29.77	30.14	30.52	30.90	31.28	31.66	32.04	32.42	32.80	33.18
36	35.63	36.09	36.56	37.02	37.49	37.95	38.42	38.89	39.36	39.82	40.29

APR	23.75%	24.00%	24.25%	24.50%	24.75%	25.00%	25.25%	25.50%	25.75%	26.00%	26.25%
6	7.04	7.12	7.19	7.27	7.34	7.42	7.49	7.57	7.64	7.72	7.79
12	13.33	13.47	13.62	13.76	13.91	14.05	14.20	14.34	14.49	14.64	14.78
18	19.84	20.06	20.28	20.50	20.72	20.95	21.17	21.39	21.61	21.83	22.05
24	26.59	26.89	27.19	27.49	27.79	28.09	28.39	28.69	29.00	29.30	29.60
30	33.57	33.95	34.33	34.72	35.10	35.49	35.88	36.26	36.65	37.04	37.43
36	40.77	41.24	41.71	42.19	42.66	43.14	43.61	44.09	44.57	45.05	45.53

Index

Selected Answers

Workshop 1 **1.** 3451 **3.** 806.025 **5.** one hundred ninety-six thousand, four hundred three **6.** eighty-nine and $\frac{12}{100}$ dollars. **7.** six hundred seventeen and thirty-five thousandths **8.** five hundred and four tenths **9.** ten and $\frac{99}{100}$ dollars **11.** 630,000 **13.** 3,770,000 **15.** 50,000 **17.** 8,618,000 **19.** 4200 **21.** 37,900 **23.** 790 **25.** 40,610 27. 487 **29.** 135 **31.** 87.4 **33.** 761 **35.** 5.48 **37.** 74.23 **39.** 70,000,000 **41.** 74,183,500 **43.** 74,200,000 **45.** $34.44 **47.** $73.85 **49.** $7.90 **51.** $100 **53.** $82 **55.** $49 **57.** $2350 **59.** $700 **61.** $19,000 **63.** $50,500 **65.** **a.** 9000, 9100, 9130; **b.** 10,000, 10,500, 10,460; **c.** 8000, 8400, 8390; **d.** 8000, 7900, 7870; **e.**7000, 6500, 6530; **f.** 10,000, 9900, 9940; **g.** 11,000, 11,000, 11,000; **h.** 9000, 8900, 8890 **67.** **a.** 290; **b.** 170; **c.** 120; **d.** 360; **e.** 430; **f.** 900; **g.** 950; **h.** 780

Workshop 2 **1.** 43 **3.** 3510 **5.** 3.87 **7.** 0.764 **9.** 9 **11.** 0.040 **13.** $8.50 **15.** $200 **17.** 5400 **19.** 5.09, 5.10, 5.13 **21.** 4.81, 4.86, 4.95 **23.** 40.004, 40.037, 40.106 **25.** 965.40, 965.51, 965.510 (same) **27.** 0.2539, 0.3239, 0.4139 **29.** 67.571, 67.6, 67.67 **31.** 8.001, 8.0135, 8.02 **34.** **a.** 873.15 and 877.142; **b.** 332.749 and 333.54 **36.** **a.** $2.50; **b.** $3.50; **c.** $3.50; **d.** $3.50; **e.** $4.45; **f.** $4.45 **g.** $4.45; **h.** $2.50; **i.** $3.00; **j.** $3.50 **38.** **a.** 3:04.258, 3:04.962, 3:05.012; **b.** 3:25.56, 3:26.04, 3:26.16; **c.** 1:01:45.0, 1:02:26.7, 1:03:13.6; **d.** 215.2; **e.** 231.2; **f.** 422.595

Workshop 3 **1.** 85.96 **3.** 1118.19 **5.** 120.919 **7.** 11.069 **9.** 193.20 **11.** 17.9862 **13.** 374.362 **15.** 680.489 **17.** 601.356 **19.** 9637.3590 **21.** $74.33 **23.** $108.27 **25.** 165.64 **27.** 622.5561 **29.** $88.15 **31.** $251.10 **33.** $139.50 **35.** $409.12 **37.** **a.** $27.31; **b.** $28.39 **39.** **a.** 50.0; **b.** 66.9; **c.** 77.7; **d.** 74.6; **e.** 43.5

Workshop 4 **1.** 24.4 **3.** 2.33 **5.** 18.80 **7.** 54.549 **9.** 15.303 **11.** 7.9553 **13.** 32.517 **15.** 610.368 **17.** 717.22 **19.** 3.641 **21.** $77.40 **23.** $775.89 **25.** $74.03 **27.** $7,679.79 **29.** 107.9 **31.** 3.51 **33.** 4.9277 **35.** $34.91 **37.** $1.37 **39.** $822.65 **41.** **a.** $0.09; **b.** $2.44; **c.** $0.13; **d.** $1.18; **e.** $0.20; **f.** $1.13; **g.** $4.27; **h.** $0.94; **i.** $0.30; **j.** $1.22 **43.** **a.** $61.45; **b.** $41.45 **45.** **a.** $126.60; **b.** $86.10 **47.** **a.** $190.05; **b.** $115.05 **49.** **a.** $249.97; **b.** $448.10; **c.** $189.85; **d.** $167.82; **e.** $29.97; **f.** $169.97; **g.** $399.88; **h.** $314.85; **i.** $0; **j.** $93.94

Workshop 5 **1.** 5.300 **3.** 17.490 **5.** 4.8140 **7.** 25.677 **9.** 0.0496 **11.** 0.0364 **13.** 0.0230 **15.** 0.001452 **17.** 0.08592 **19.** 0.71496 **21.** 0.000168 **23.** 1.614347 **25.** $44.50 **27.** $10.925 = $10.93 **29.** $19.536 = $19.54 **31.** 246 **33.** 74.35 **35.** 2842 **37.** 289.6 **39.** 53,663 **41.** 255 **43.** 98,891 **45.** 915,321 **47.** **a.** $8.85; **b.** $31.00; **c.** $4.50; **d.** $31.80; **e.** $24.95; **f.** $19.95; **g.** $39.85; **h.** $160.90 **49.** **a.** $55.90; **b.** $139.80; **c.** $124.35; **d.** $93.00; **e.** $71.40; **f.** $71.40; **g.** $749.50; **h.** $1305.35; **i.** $78.32; **j.** $1393.07

Workshop 6 **1.** 1.8 **3.** 2.7 **5.** 9.3 **7.** 7.3 **9.** 6.5 **11.** 2.8 **13.** 48.8 **15.** 31.5 **17.** 1.68 **19.** 35.91 **21.** 47.94 **23.** 0.50 **25.** $3.92 **27.** $6.92 **29.** $3.26 **31.** $17.48 **33.** 0.73 **35.** 62.681 **37.** 1.298 **39.** 34.6265 **41.** 1.8979 **43.** 0.023098 **45.** 4.345218 **47.** 0.051549 **49.** 25.16 **51.** $4.92 **53.** **a.** 16.7; **b.** 18.7; **c.** 23.6; **d.** 27.0; **e.** 13.2; **f.** 8.2

Workshop 7 (Estimates may vary) **1.** 60; 54.84 **3.** 20; 18.30 **5.** 650; 642.65 **7.** $850; $854.26 **9.** 63; 65.6196 **11.** 284; 262.0528 **13.** 880; 876.8712 **15.** 6580; 6998.288 **17.** 33; 31 **19.** 40; 38.7 **21.** 3; 2.6 **23.** 2; 2.1 **25.** 85; 83.793 **27.** 1275; 1277.75 **29.** $270; $269.85 **31.** $10,200; $10,226.51 **33.** $90; $90.96 **35.** $23; $22.66 **37.** $25; $24.70 **39.** $90; $93.20 actual **40.** **a.** 85; yes; **b.** 525; no; **c.** 230; yes; **d.** 35; no; **e.** 36,000; no; **f.** 400; no; **g.** 150; yes; **h.** 0.75; yes; **i.** 19; yes; **j.** 60; no

Workshop 8 **1.** 0.25 **3.** 0.375 **5.** 0.35 **7.** 0.15 **9.** 0.64 **11.** 0.12 **13.** 0.575 **15.** 0.048 **17.** 0.33 **19.** 0.43 **21.** 0.17 **23.** 0.42 **25.** 0.47 **27.** 0.56 **29.** 0.111 **31.** 0.667 **33.** 0.417 **35.** 0.063 **37.** 0.556 **39.** 0.063 **41.** $\frac{1}{2}$ **43.** $\frac{1}{4}$ **45.** $\frac{4}{5}$ **47.** $\frac{3}{25}$ **49.** $\frac{13}{20}$ **51.** $\frac{7}{20}$ **53.** $\frac{7}{8}$ **55.** $\frac{11}{40}$ **57.** $\frac{9}{40}$ **59.** $\frac{5}{16}$ **61.** $1\frac{3}{4}$ **63.** $2\frac{3}{25}$ **65.** $8\frac{1}{4}$ **67.** $8\frac{1}{20}$ **69.** $3\frac{5}{8}$ **71.** $9\frac{2131}{5000}$ **73.** $28\frac{2}{5}$ **75.** 0.25 **77.** $\frac{27}{40}$ **79.** $37.75; $68.38; $29.88; $30.50; $17.63; $25.13; $23.25

Workshop 9 **1.** 0.15 **3.** 0.44 **5.** 0.31 **7.** 0.12 **9.** 0.321 **11.** 0.168 **13.** 0.112 **15.** 0.755 **17.** 1.25 **19.** 3.55 **21.** 7.50 **23.** 12.44 **25.** 0.07 **27.** 0.01 **29.** 0.026 **31.** 0.014 **33.** 0.0105 **35.** 0.030125 **37.** 58% **39.** 89% **41.** 25% **43.** 75% **45.** 4% **47.** 5% **49.** 1.5% **51.** 1.25% **53.** 0.5% **55.** 0.12% **57.** 212.5% **59.** 350% **61.** 981%

63. 100.8% **65.** 0.2% **67.** 1200% **69.** 30% **71.** 34,010% **73.** 0.25 **75.** 62% **77.** 0.40 **79.** **a.** 0.1316; **b.** 0.1390; **c.** 0.1473; **d.** 0.1618; **e.** 0.1762; **f.** 0.1935; **g.** 0.1974 **81.** Atlantic Division: 77.8%; 64.2%; 62.3%; 47.3%; 46.3%; Central Division: 58.2%; 57.7%; 50.9%; 41.2%; 38.5%; 30.2%; Midwest Division: 59.3%; 52.7%; 44.6%; 43.4%; 40.0%; 38.9%; Pacific Division: 65.4%; 58.2%; 53.8%; 44.4%; 43.6%; 31.5%

Workshop 10 **1.** 13.5 **3.** 23.4 **5.** 99.9 **7.** 155.8 **9.** 377.72 **11.** 2.4 **13.** 2.52 **15.** 78.8 **17.** 3.33 **19.** 500 **21.** 1710 **23.** 3300.04 **25.** 4441.5 **27.** 39,963.76 **29.** 7.644 **31.** 5.58 **33.** 51.51 **35.** 122.4 **37.** 653.718 **39.** 238.182 **41.** $1.36 **43.** $12.60 **45.** $42.12 **47.** $0.40 **49.** $0.74 **51.** $0.43 **53.** $98 **55.** **a.**$3.00; $12.00; **b.** $3.60; $14.40; **c.** $4.40; $17.60; **d.** $5.20; $20.80 **57.** **a.** 51; **b.** 44; **c.** 33; **d.** 38; **e.** 96 **59.** **a.** $30.60; **b.** $52; **c.** $85.80; **d.** $124.20; **e.** $211.70; **f.** $229.90; **g.** $51.75; **h.** $98.36; **i.** $39.75; **j.** $135 minimum

Workshop 11 **1.** 10 **3.** 77 **5.** 407 **7.** 843 **9.** 2.3 **11.** 2.55 **13.** $13 **15.** $66.50 **17.** 4.9 **19.** 44.5 **21.** 361.9 **23.** 48.04 **25.** $53.75 = $54.00 **27.** $431.17 = $431.20 **29.** $145.76 **31.** 172 **33.** $1.60 **35.** $37.01 **37.** $10,800 **39.** Game 1: 905 total; 181 average; Game 2: 926 total; 185 average; Game 3: 870 total; 174 average; McKaig: 484 total; 161 average; Pegorsch: 606 total; 202 average; Hoskinson: 509 total; 170 average; Goldberg: 599 total; 200 average; Hallauer: 503 total; 168 average

Workshop 12 **1.** 3h:10 min **3.** 2h:13 min **5.** 4h:05 min **7.** 5h:20 min **9.** 1h:20 min **11.** 2h:50 min **13.** 1h:50 min **15.** 2h:35 min **17.** 1h:53 min **19.** 4h:27 min **21.** 5h:00 min **23.** 5h:20 min **25.** 6h:20 min **27.** 5h:18 min **29.** 5h:52 min **31.** 5h:36 min **33.** **a.** 3h:05 min; **b.** 4h:54 min; **c.** 1h:53 min; **d.** 2h:03 min; **e.** 3h:54 min; **f.** 2h:04 min; **g.** 4h:44 min

Workshop 13 **1.** $1.52 **3.** $3.31 **5.** $3.47 **7.** $2.24 **9.** $5.78 **11.** $1.34 **13.** **a.** 2 lb; **b.** 11 lb; **c.** 13 lb; **d.** 10 lb; **e.** 14 lb; **f.** 8 lb **15.** 36 or 37 **17.** 33 **19.** 29 **21.** Bangor **23.** $18.80 **25.** $8.40 **27.** $13.20 **29.** $230—$240 **31.** $230—$240 **33.** $190—$200 **35.** $12.00 **37.** $25.30 **39.** $18.40

Workshop 14 **1.** 9 ft **3.** 80 oz **5.** 240 oz **7.** 2500 m **9.** 6400 g **11.** 3 gal **13.** 2.5 gal **15.** 4 kg **17.** 2.3 L **19.** 84,600 g **21.** 0.843 kg **23.** 140 mm **25.** 76 in **27.** 13 pt **29.** 14 qt **31.** 37 pt **33.** 8 qt **35.** 16 cups **37.** 240 in **39.** 3900 **41.** 2.13 L **43.** **a.** 36 ft; **b.** 36 ft + 48 ft = 84 ft; **c.** 21 ft + 18 ft 8 in = 39 ft 8 in; **d.** 22 ft 6 in + 15 ft 4 in = 37 ft 10 in; **e.** 22 ft 8 in + 26 ft 4 in = 49 ft; **f.** 35 ft 4 in + 25 ft 6 in = 60 ft 10 in; **g.** 16 ft 10 in + 15 ft 6 in = 32 ft 4 in; total 381 ft 8 in

Unit 1 Gross Income

Pages 46–47 **1.** $280.00 **3.** $186.00 **5.** $390.63 **7.** $183.00 found by $6.10 × 30 **9.** $416.25 **11.** $129.20 found by $3.80 × 34 **13.** $290 **15.** $168.00 **17.** $206.25 **19.** 0.50 **21.** 0.10 **23.** $136.13 **25.** $471.25 **27.** 140 **29.** 100 **31.** 14,400 **33.** 1000

Pages 48–49 **1.** $440.00 found by $320.00 + $120.00 **3.** $154.80 **5.** $892.50 **7.** $91.20 found by $11.40 × 8; $372.40 found by $91.20 + **$281.20** **9.** $264.00 found by $54 + $210 **11.** $164.13 **13.** $483.84 **15.** $247.35 **17.** 1746.162 **19.** $441.60 **21.** $150 **23.** $125

Pages 51–52 **1.** $7\frac{1}{2}$ **3.** 8 **5.** $8\frac{1}{2}$ **7.** $8\frac{3}{4}$ **9.** $8\frac{3}{4}$ **11.** 9 **13.** 49 **15.** $279.65 **17.** $39\frac{1}{4}$ found by 8 + 8 + 8 + $7\frac{1}{4}$ + 8; $255.13 found by $39\frac{1}{4}$ × $6.50 **19.** $296.83 found by $182.90 + $113.925 **21.** $\frac{7}{8}$ **23.** $\frac{23}{24}$ **25.** 7:45 **27.** 3:45 **29.** 40,000 **31.** 68,000 **33.** 300

Pages 53–54 **1.** $117.00 **3.** $112.50 **5.** $132.08 **7.** $120.28 **9.** $134.40 **11.** $150.51 **13.** $84.50 **15.** $216.25 **17.** $264.06 found by 8802 × $0.03 **18.** $286.51 found by $197.01 + $40.00 + $49.50 **19.** $20.47 **21.** $251.52 **23.** $128.00 **25.** 117.00 **27.** 9.68

Pages 55–56 **1.** $600.00 **3.** $1325.00 **5.** $511.50 **7.** $425.00 **9.** $609.17 **11.** $1950.00 **13.** $1404.58 **15.** $23.00 found by $293.00 – $270.00 **17.** $1140.00 **19.** 4.84 **21.** 559.38 **23.** 27.44 **25.** 72

Pages 57–59 **1.** $83.00 **3.** $926.90 **5.** $96.75 **7.** $891.55 **8.** $2625.00 found by 7.5% of $35,000; $2625 is greater **9.** $1755.00 found by 6.5% of $27,000; $1755 is greater **11.** $1020.00 found by 8.5% of $12,000; $1020 is greater **13.** $2250 found by 7.5% of $30,000; $3500.00 is greater **15.** $237.25 **17.** $215.80 **19.** $3780.63 **20.** $654.93 found by 33.33% × $1965 **21.** $195.98 **23.** $180.25 **25.** 2.13 **27.** 6.75 **29.** 9.25 **31.** $365.50 **33.** 8.76 **35.** 109.59 **37.** 0.6 **39.** 0.8

Pages 60–61 **1.** $200; $120; $320 **3.** $105.00; $240.00; $75.00; $420.00 **5.** $667.50 **7.** $218.63

9. \$140 **11.** \$805.23 **13.** \$1034.31 **15.** 1.960 **17.** 0.1025 **19.** 0.97 **21.** \$105 **23.** \$542.50

Reviewing the Basics (Page 62)
1. \$45.67 **2.** \$469.36 **3.** \$0.05 **4.** \$53.75 **5.** \$717.33 **6.** 20.534 **7.** \$2.48 **8.** 1156.3 **9.** 0.00021 **10.** \$910 **11.** \$4660 **12.** 546.0 **13.** $19\frac{1}{2}$ **14.** 16 **15.** \$100.75 **16.** \$341 **17.** \$0.06 **18.** 0.5 **19.** 0.25 **20.** 0.75 **21.** 0.8 **22.** 0.125 **23.** 0.115 **24.** 0.0125 **25.** 0.00125 **26.** 3.85 **27.** 8 hrs. 30 min. **28.** 7 hrs. 45 min. **29.** 104 **30.** 36 **31.** 16 **32.** 48
Terms: Refer to the terms file.

Pages 78–79 **1.** \$190.63 found by \$250.50 – \$59.87 **3.** \$329.50 found by \$455 – \$125.50 **5.** \$423.076 = \$423.08 found by \$22,000 ÷ 52; \$423.08 – \$114.54 = \$308.54 **7.** \$288.88 found by \$384 – (\$31.10 + \$27.46 + \$13.44 + \$1.92 + \$16.20 + \$5) **8.** \$416 found by \$320 + \$96; \$416 – (\$42.70 + \$29.74 + \$10.68 + \$3.36 + \$25) = \$304.52 **9.** 519.439 **11.** 509.057 **13.** 0.547 **15.** 235.92 **17.** 554.25 **19.** –23.75

Reviewing the Basics (Page 80) **1.** \$239 **2.** \$200.45 **3.** \$154.81 **4.** \$25.21 **5.** \$40.10 **6.** \$1.52 **7.** \$103.31 **8.** \$28.32 **9.** \$24.01 **10.** \$8.27 **11.** \$135.43 **12.** \$74.38 **13.** \$19 **14.** \$47.80 **15.** \$4.75 **16.** \$14.14 **17.** \$4.76 **18.** \$16.009 = \$16.01 **19.** \$63.678 = \$63.68 **20.** \$28.20 **21.** 24 **22.** 52 **23.** 72

Unit 2 Net Income

Pages 66–68 **1.** \$28.80 **3.** \$33.70 **5.** \$33.70 **7.** \$22.80 **9.** \$37.70 **11.** \$94.80 **13.** \$242.10 found by \$29.60 + \$212.50 **15.** \$109.10 **17.** \$3.30 found by \$23.50 – \$20.20 **19.** ones **21.** thousands **23.** hundreds **25.** \$27 **27.** \$1.49 **29.** \$276.10 **31.** \$874.89

Pages 69–70 **1.** \$120 **3.** \$632 **5.** \$3000 **7.** \$13,400 – \$2200 = \$11,200; \$280 found by \$11,200 × 2.5% **9.** \$441 found by \$12,600 × 3.5% **11.** \$425 found by \$22,100 ÷ 52; \$3075.60 found by (\$51.80 × 52) + \$382 **13.** 65.87 **15.** 31.903 **17.** 192 **19.** 12.5

Page 72 **1.** \$3500 **3.** \$227 found by (1% × \$2000) + (3% × \$3000) + [4.5% × (\$7600 – \$5000)] **5.** \$367.50 found by \$105 + \$157.50 + \$105 **7.** \$990.50 found by \$105 + \$157.50 + \$728; \$19.048 = \$19.05 found by \$990.50 ÷ 52 **9.** 0.547 **11.** \$235.92 **13.** 16.54 **15.** 0.03 **17.** 12 **19.** 208

Pages 73–74 **1.** \$89.38 **3.** \$8.94 **5.** \$107.25 **7.** \$0 because earnings to date exceed \$42,600 **9.** \$1170 found by \$28,080 ÷ 24; \$83.655 = \$83.66 found by 0.0715 × \$1170 **11.** \$93.60 found by \$31.10 + \$62.50 **13.** October, found by \$54,000 ÷ 12 = \$4500 per month; \$42,600 ÷ \$4500 = 9.46 months **15.** 12.5 **17.** 0.15 **19.** 320 **21.** 23,980 **23.** 3800 **25.** 147,400

Pages 75–76 **1.** \$30 found by \$720 ÷ 24 **3.** \$6.25 **5.** \$1.90 **7.** \$202.50 found by \$810 × (100% – 75%); \$202.50 ÷ 24 = \$8.4375 = \$8.44; \$766.56 found by \$775 – 8.44 **9.** \$3.0615 = \$3.06 found by \$159.20 ÷ 52; \$0.7242 = \$0.72 found by \$37.66 ÷ 52; \$3.78 found by \$3.06 + \$0.72 + \$0 **11.** 13.05 **13.** 86.50 **15.** 7.68 **17.** 7.885

Unit 3 Checking Accounts

Pages 90–91 **1.** \$51.80 **3.** \$89.30 subtotal; \$69.30 total deposit **5.** \$427.95 subtotal; \$402.95 total deposit **7.** \$269.97 **9.** \$413.08 **11.** \$253.87 found by (\$265.89 + \$24.98) – \$37.00 **13.** \$1053.26 **15.** \$76.18 **17.** \$497.10 **19.** \$171.39 **21.** \$420.00 **23.** \$322.50 **25.** \$574.23

Pages 92–93 **1.** ten and $\frac{50}{100}$ **2.** three hundred seventy-five and $\frac{00}{100}$ **3.** fifty-five and $\frac{83}{100}$ **4.** nine and $\frac{72}{100}$ **5.** ninety-four and $\frac{6}{100}$ **6.** twenty-nine and $\frac{00}{100}$ **7.** four thousand one hundred ninety-six and $\frac{00}{100}$ **8.** thirteen thousand seven hundred and $\frac{00}{100}$ **9.** one hundred three and $\frac{77}{100}$ **10.** nine hundred eighty-two and $\frac{75}{100}$ **11.** nine thousand four hundred fifty-six and $\frac{86}{100}$ **12.** forty-four thousand three hundred eighty-one and $\frac{15}{100}$ **13.** sixty-eight and $\frac{92}{100}$ **15.** No, one hundred thirty-seven and $\frac{52}{100}$ **17.** No, thirty-six and $\frac{88}{100}$ **19.** \$27.15 **21.** eighteen and $\frac{35}{100}$ **22.** sixty and $\frac{54}{100}$ **23.** three hundred forty-five and $\frac{00}{100}$ **24.** seven hundred ninety-six and $\frac{72}{100}$ **25.** eight thousand one hundred seventy-eight and $\frac{52}{100}$ **26.** fourteen thousand seven hundred eighteen and $\frac{47}{100}$ **27.** one thousand six hundred ten and $\frac{25}{100}$ **28.** nineteen thousand one hundred fifty-five and $\frac{67}{100}$

Pages 94–96 **1.** \$132.63 **3.** \$18.00 **5.** \$367.11 **7.** \$239.76 **9.** check 101: \$309.71 – \$117.88 = \$191.83; check 102: \$191.83 – \$47.62 = \$144.21; check 103: \$144.21 – \$52.37 = \$91.84 **11.** check 472: \$200.42; check 473: \$100.42; plus deposit of \$345.70 = \$446.12; check 474: \$349.92; plus deposit of \$250.00 = \$599.92 **13.** check 916: \$206.47; plus deposit of \$475.00 = \$681.47; check 917: \$165.69; check 918: \$94.07; plus deposit of \$220.00 = \$314.07; plus deposit of \$165.00 = \$479.07; check 919: \$266.62; check 920:

$216.62 **15.** $845.36 **17.** $253.11 **19.** $906.45 **21.** $197.97 **23.** $47.19 **25.** $756.91

Pages 98–99 **1.** $97.77 **3.** $29.12 **5.** $8.41 **7.** $953.80 found by $183.42 + $877.65 − $105.42 − $1.85, where $215.90 + $215.90 + $445.84 = $877.65; $8.45 + $11.02 + $85.95 = $105.42 **9.** $508.61 found by $199.27 + $950.09 − $640.25 − $0.50, where $564.50 + $221.37 + $164.22 = $950.09; $147.50 + $25.00 + $417.75 + $50.00 = $640.25 **10.** present balance = $368.83 found by $247.82 + $400.00 − $276.49 − $2.50 **11.** $249.39 **13.** $794.60 **15.** $25,589.27 **17.** $520.82 **19.** $1183.51 **21.** 95.44 **23.** 1725.363

Pages 101–103 **1.** $486.95; agree **3.** new balance = $1142.44 − $1.10 = $1141.34; do not agree; adjusted balance = $1260.54 + $119.20 = $1379.74 **5.** adjusted balance = $204.20 − $55.40 + $62.40 = $211.20 where $31.20 + $24.20 = $55.40; agree **7. a.** $23.19 **b.** $250.00 **c.** $723.51 found by $496.70 − $23.19 + $250.00 **d.** $723.51 **e.** yes **9.** $411.05 **11.** $93.78 **13.** $243.77 **15.** $171.92 **17.** $46.36

Reviewing the Basics (Page 104) **1.** fifty-six and $\frac{39}{100}$ **2.** three hundred forty-seven and $\frac{33}{100}$ **3.** one thousand two hundred forty-five and $\frac{97}{100}$ **4.** thirteen thousand six hundred ninety-two and $\frac{00}{100}$ **5.** $202.67 **6.** $1457.20 **7.** $577.34 **8.** $5720.05 **9.** $441.73 **10.** $558.01 **11.** $22.12 **12.** $66.51 **13.** $244.94 **14.** $132.33 **15.** c **16.** a **17.** b **18.** f **19.** d

Unit 4 Savings Accounts

Pages 108–109 **1.** $100.40; $100.40 **3.** $101.45; $76.45 **5.** $225.17; $195.17 **7.** $89.14 **9.** $270.76 found by ($214.63 + $96.13) − $40.00 **11.** $124.45 **13.** $144.27 **15.** $247.22 **17.** $25.72 **19.** $413.34 **21.** $84.85 **23.** $162.95 **25.** $49.70

Pages 110–111 **1.** twenty and $\frac{25}{100}$ dollars **2.** eighteen and $\frac{90}{100}$ dollars **3.** sixty-five and $\frac{72}{100}$ dollars **4.** ninety-one and $\frac{5}{100}$ dollars **5.** one hundred three and $\frac{00}{100}$ dollars **6.** two hundred fifteen and $\frac{64}{100}$ dollars **7.** one thousand four hundred ninety-six and $\frac{17}{100}$ dollars **8.** two thousand seven hundred fifteen and $\frac{75}{100}$ dollars **9. a.** 17594179 **b.** $98.50 **c.** ninety-eight and $\frac{50}{100}$ dollars **11. a.** 81-0-174927 **b.** $281.92 **c.** two hundred eighty-one and $\frac{92}{100}$ dollars **13. a.** 06029175 **b.** $56.70 **c.** fifty-six and $\frac{70}{100}$ dollars **15.** seventy-four and $\frac{89}{100}$ **16.** one hundred ninety-two and $\frac{43}{100}$ **17.** two hundred sixteen and $\frac{50}{100}$ **18.** seventeen and $\frac{21}{100}$ **19.** one thousand two hundred forty-one and $\frac{80}{100}$ **21.** $562.19 **23.** $31,421.19

Pages 112–113 **1.** $431.25 **3.** $816.40 **5.** $258.15 **7.** $713.29 **9.** $325.00 **11.** $860.43 **13.** $805.00 **15.** $162.45 **17.** $1323.46

Pages 114–116 **1.** $231.00 **3.** $995.25 **5.** $1055.44 **7.** $22,479.65 found by $21,563.73 + $351.48 + ($713.47 + $663.47) − ($450.00 + $362.50) **9.** $324.46 **11.** $610.44 **13.** $302.50 **15.** $955.88 **17.** $5654.65 **19.** $4274.46 **21.** $3769.63 **23.** $10,135.54

Pages 117–118 **1.** $9.60 **3.** $55.56 **5.** $46.32 **6.** $7.8375 = $7.84 found by $570 × 5.5% × $\frac{3}{12}$ **7.** $53.89 **9.** $24.461875 = $24.46 **11.** $11.295875 = $11.30 **13.** $7.81 $32.52825 = $32.53 where March 1 to the end of November is 9 months **15.** 0.75 **17.** 0.25 **19.** 6.5 **21.** 0.055 **23.** 0.0475 **25.** 0.06625

Pages 119–120 **1.** $612.06 **3.** $54.93975 = $54.94 found by $1292.70 × 8.5% × ½; $1347.64 **5.** $2244.42 found by $24,938.00 × 9% × 1; $2446.4198 = $2446.42 found by $27,182.42 × 9% × 1 = ; $27,182.42 + $2446.42 = $29,628.84 **7.** $9.135 = $9.14 found by $609.00 × 6% × ¼; $9.2721 = $9.27 found by $618.14 × 6% × ¼; $627.41 **9.** $1573.25 **11.** $6480.12 found by $87.89 + $6392.23 **13.** $14,190.14 **15.** 0.075 **17.** 0.048 **19.** $37.50 **21.** $507.60 **23.** $1025.20 **25.** 0.2 **27.** 0.71

Pages 122–123 **1.** $92.35 **3.** $1276.83; $176.83 **5.** $3231.77; $276.17 **7.** $1134.39; $238.89 **9.** $49.09 **11.** $14.76 found by $116.64 − $101.88 **13.** $108.05 **15.** 1.250% **17.** 3.000% **19.** 3.250% **21.** 2.625% **23.** 1.875% **25.** 1.625%

Page 125 **1.** $37.70 **3.** $4231.71; $31.71 **5.** $78.28 found by $13,038.28 − $12,960 **7.** $15.89 **9.** $5.81 **10.** $30.68 where April 11 to June 30 is 80 days (181 − 101 = 80) **11.** $392.43 where August 23 to October 1 is 39 days (274 − 235 = 39) **13.** $2042.62 **15.** $605.44 **17.** $752.94 **19.** $8720.15 **21.** $75.15 **23.** $9325.63

Reviewing the Basics (Page 126) **1.** three hundred fifty-nine and $\frac{90}{100}$ dollars **2.** seventy-four and $\frac{91}{100}$ dollars **3.** four thousand twenty-four and $\frac{9}{100}$ dollars **4.** fourteen thousand six hundred forty-one and $\frac{89}{100}$ dollars **5.** $172.35 **6.** $1862.92 **7.** $398.90 **8.** $521.81 **9.** $2025 **10.** $911.27 **11.** 2.125 **12.** 1.685 **13.** 6.25 **14.** 5.5 **15.** 27.75 **16.** 12.375 **17.** 0.0675 **18.** 0.04875 **19.** 0.3995 **20.** 0.11145 **21.** 1.04567 **22.** 1.02515 **23.** 91 (151 − 60) **24.** 46 (212–166) **25.** $\frac{1}{4}$ **26.** $\frac{1}{2}$ **27.** $\frac{3}{4}$ **28.** $1\frac{1}{2}$ **29.** 8 **30.** 3 **31–36.** Refer to the terms file.

Unit 5 Cash Purchases

Pages 130–131 **1.** $0.0238 = $0.02 **3.** $0.9519 = $0.95 **5.** $41.30 **7.** $0.4016 = $0.40 found by $10.04 × 4% **9.** $0.62 **11.** $2.7531 = $2.75 **13.** $0.68 **15.** $1.75275 = $1.75 found by [$15.95 + $7.79 + (4 × $5.75)] × 3.75% **17.** 316 **19.** 0.45 **21.** 2000 **23.** 1,344,000 **25.** 45,730 **27.** 1050 **29.** $189.45 **31.** $1.38 **33.** 458.28 **35.** 432.86

Pages 132–134 **1.** $5.03 **3.** $0.5096 = $0.51;$13.25 **5.** $12.63 found by $0.49 + $12.14 **7.** $0.83 **9.** $6.60 found by $0.19 + $6.41 **11.** $23.52 **13.** $77.30 **15.** $13.92 found by (3 × $1.59) + (1½ × $1.00) + $2.99 + $1.69 + $0.75 + (3 × $0.74); $0.1122 = $0.11 found by ($2.99 + $0.75) × 3%; $14.03 found by $13.92 + $0.11 **17.** $158.01 **19.** 4.32 **21.** 0.00091 **23.** 207.71 **25.** 145.7514 **27.** 443.374 **29.** 438.05 **31.** 832.55

Pages 135–136 **1.** $0.148 = 14.8¢ **3.** $0.340 = 34¢ **5.** $0.50 = 50¢ **7.** $0.00395 = $0.004 = 0.4¢ **9.** $30.31 **11.** $0.776 = 77.6¢ **13.** $0.061875 = $0.062 = 6.2¢ **15. a.** $0.98 **b.** $0.49 **c.** $0.49 **d.** $0.196 = 19.6¢ **17.** 4.31 **19.** 0.06 **21.** 20 **23.** 1260 **25.** 141.6 **27.** 1041.7 **29.** 97.15

Pages 137–138 **1.** $0.030625 = $0.031; $0.29545 = $0.30; large size **3.** $0.521; $0.550; small size **5. a.** $0.575 **b.** $0.6225; 10-oz size **7.** 14-oz can **9.** 18-oz box **11.** 5-oz jar **13.** family size **15.** 30.12 **17.** 158.10 **19.** $1786 **21.** 9.4

Pages 140–141 **1.** $15.00 **3.** $56.84 **5.** $0.23 **7.** $85.35 **9.** $990 **10.** $11.397 = $11.40 found by $37.99 × 30% **11.** $0.4515 = $0.45 **13.** $3.55 **15.** $13.98 **17.** $64 **19.** $1.80; $1.125 = $1.13; $2.625 = $2.63; $1.275 = $1.28; $2.70 **21.** cheaper price: $5.99 @ advertisement; $8.99 @ 20% discount, $10.99 @ 20% discount, $5.99 @ advertisement, $6.99 @ advertisement **23.** 81.676 **25.** 1.764 **27.** 32.96 **29.** 312.5

Pages 142–143 **1.** $2.70; $24.30 **3.** $1.35; $7.64 **5.** $80.50; $149.50 **7.** $16.82 found by $19.79 − $2.97 **9.** $10.36 **11.** Star charges less **13.** 14.63 **15.** 15.63 **17.** 10.186 **19.** 7.08

Reviewing the Basics (Page 144) **1.** 13.4 **2.** 49.3 **3.** 16.25 **4.** $11.18 **5.** $1.64 **6.** $0.50 **7.** $4.22 **8.** $28.03 **9.** $370.50 **10.** $57.81 **11.** $68.68 **12.** $22.14 **13.** $912.60 **14.** $3.90402 = $3.90 **15.** $0.2125 **16.** $0.00625 **17.** $0.2075 **18.** $1.9425 = $1.94 **19.** $0.88 **20.** $14.5275 = $14.53 **21.** $92.4725 = $92.47 **22.** $4.9425 = $4.94 Terms **23.** g **24.** e **25.** f **26.** a **27.** d **28.** b

Unit 6 Charge Acounts and Credit Cards

Pages 153–154 **1.** $34.68 **3.** $6.985 = $6.99; $133.99 **5.** $69.12 **7.** $15.53 **9.** $17.85 **11.** $51.14 **13.** $91.00 **15.** $13.47 found by $0.52 + $12.95 **17.** $158.99 **18.** $328.95 found by $159.99 + (3 × $17.99) + (2 × $12.50) + $89.99; $13.158 = $13.16 found by $328.95 × 4%; 342.11 found by $13.16 + $328.95 **19.** $273.41 found by $15.48 + $257.93 **21.** $36.18 found by $35.30 + ($35.30 × 2.5%) **23.** 0.05 **25.** 0.055 **27.** 0.0320 **29.** $157.17 **31.** $88.17 **33.** 2.56 **35.** 2.784 **37.** $3.47

Pages 156–157 **1.** $426.00 **3.** $496.51 **5.** $206.14 **7.** $71.18 **9.** $120.19 **11.** $484.85 **13.** $151.92 **15.** $285.84 **17.** $171.69 **19.** $110.30 **21.** $97.84 **23.** $247.84

Pages 159–160 **1.** $0.60 **3.** $2.272 = $2.27 **5.** $6.7816 = $6.78 **7.** $209.85 **9.** $298.21 **11.** $799.18 **13.** $113.96 **15.** $203.75 **17.** 37.13 **19.** 8.53 **21.** 3.60 **23.** $3.85 **25.** $0.33 **27.** 9.52 **29.** 0.51584

Pages 162–163 **1.** $263.00 **3.** $243.00 found by $200 + $3.00 + $40 **5.** $711.20 **7.** $117.64 **9.** $90.96 **11.** $109.91 **13.** $248.17 **15.** $314.76 found by $197.21 + $117.55 **17.** $210.42 **19.** $241.94 **21.** $231.07 **23.** $72.51 **25.** 7.2 **27.** 19.2 **29.** 1.6 **31.** 15.54

Pages 166–169 **1.** $59 **3.** $234.838 = $234.84 found by $7280 ÷ 31 **5.** $143.46 found by $263.40 − $163.40 + $2.72 + $40.74 **7.** $67.837 = $67.84 found by $2102.96 ÷ 31; $122.26 found by $84.56 − $34.56 + $1.36 + $70.90 **9.** $226.998 = $227 found by $6809.95 ÷ 30; $236.76 found by $295.55 − $145.55 + $4.54 + $82.22 **10.** $402.694 = $402.69 found by $12,483.52 ÷ 31; $402.69 × 1.75% = $7.047 = $7.05; $448.31 found by $449.26 − $67.98 + $7.05 + $59.98 **11. a.** $108.92 found by $124.37 − $63.00 + $2.64 + $44.91; **b.** $400.357 = $400.36 found by $11,210 ÷ 28; $400.36 × 1.5% = $6.0054 = $6.01; $412.01 found by $400 − $30 + $6.01 + $36 **13.** 4050 **15.** 196.48 **17.** 260 **19.** 3562.5 **21.** 114.29 **23.** 365.20 **25.** 172 **27.** 23.402

Reviewing the Basics (Page 170) **1.** $146.14 **2.** $152.18 **3.** $36.47 **4.** $4.80 **5.** $158.98 **6.** $296.97 **7.** $96.68 **8.** $433.62 **9.** $22.64 **10.** $2240 **11.** $6480 **12.** $26,215.80 **13.** $257.12 **14.** $3180.48 **15.** $1196.79 **16.** $17.02 **17.** $65.20 **18.** $125.78 **19.** $15.25 **20.** $145.87 **21.** $654.26 **22.** 8.7 **23.** 3.452 **24.** 11.242 **25.** 26 **26.** 17 **27.** 17 **28.** 31 **29.** 46.333 **30.** 16.5 **31.** $50.60 **32.** 5.1025 **33.** c **34.** d **35.** f **36.** b **37.** a

Unit 7 Loans

Pages 174–175 **1.** $609 **3.** $5.20; $405.20 **5.** $51.5625 = $51.56; $1926.56 **7.** $5100 found by $5000 + $100 **9.** $2121.67 **11.** $9347.59 **13.** First: $120.3125 = $120.31 found by $1375 × 17.5% × $\frac{180}{360}$; $1375 + $120.31 = $1495.31; Southside: $1547.73 **15.** 0.105 **17.** 0.0925 **19.** $\frac{3}{4}$ **21.** $\frac{1}{4}$ **23.** $\frac{1}{5}$ **25.** 0.125 **27.** 0.8 **28.** 0.6

Pages 176–177 **1.** $269.50 **3.** $288; $6912 **5.** $93.50 found by $2400 × $16\frac{1}{2}$% × $\frac{85}{360}$; $2306.50 found by $2400 − $93.50 **7.** $21.75; $1178.25 **9.** $111.90; $1388.10 **11.** $1144 **13.** $210.575 = $210.58; $8189.42 proceeds; $8400 paid to bank **15.** 0.1835 **17.** 0.275 **19.** 10 **21.** 540 **23.** 295.75 **25.** 8306.24 **27.** $\frac{11}{36}$ **29.** $\frac{2}{9}$

Pages 178–179 **1.** $780 **3.** $332.50; $997.50 **5.** $872.40 found by $4362 × 20%; $3489.60 found by $4362 − $872.40 **7.** $367.20 **9.** $6417.07 **11.** $1616 **13.** $398.36 **15.** 0.2 = 20% found by $1000 ÷ $5000 **17.** 0.22 **19.** 0.15 **21.** 0.28 **23.** 67.50 **25.** 119.25 **27.** $149.58 **29.** $19.00

Pages 181–182 **1.** monthly payment: $17.58; $210.96 found by $17.58 × 12; $10.96 found by $210.96 − $200 **3.** monthly payment: $24.96; $99.04 found by $599.04 − $500 **5.** monthly payment: $90.25; $83 **7.** monthly payment: $48.48; $163.52 **9.** monthly payment: $60.05; $80.90 **11.** $168.68 found by $728.64 − $559.96 where $728.64 = $20.24 × 36 and $599.96 = $699.95 − $139.99 **13.** 15% costs less; $99.04 − $41.44 = $57.60 saved **15.** monthly payment: $161.62; $878.88 **16.** 25% costs less; $490 − $418.84 = $71.16 saved **17.** 25% costs less; $732.32 − $523.46 = $208.86 saved **19.** $585.76 found by $2585.76 − $2000 **21.** 478.98 **23.** 3439.08 **25.** 2296.80 **27.** 128.19 **29.** 400.66

Pages 184–185 **1.** 12% **3.** $14.21 found by $100 × ($426.30 ÷ $3000); 17.25% APR **5.** $19.15; 14% APR **7.** $35.25; 20.75% APR **9.** $16.95; 12.50% APR **11.** $34.805 = $34.81; 20.50% APR **13.** $23.03; 20.75% APR **15.** $16.571 = $16.57; 20% APR **16.** $74.20 found by $574.20 − $500 where $500 = ($299.99 + $249.99) − $49.98 and $574.20 = $31.90 × 18; $14.84 found by ($74.20 ÷ $500) × 100; 18% APR **17.** $474 found by $1962 − $1488 where $1962 = $54.50 × 36 and $1488 = 80% × $1860; $31.85 found by ($474 ÷ $1488) × 100; 19% APR **19.** Nathaniel: $24.312 = $24.31 found by $100 × ($1215.60 ÷ $5000); 14.75% APR; City Finance: $30.65; 18.25% APR **21.** $8.03 finance charge found by $154.98 − $146.95 where $146.95 = $245.95 − $99.00 and $154.98 = (6 × $25.83) − $146.95; $5.464 = $5.46 found by $100 × ($8.03 ÷ $146.95); 18.50% APR; $253.98 found by $154.98 + $99 **23.** $19.41 **25.** $40.30 **27.** 20.00 **29.** 56.00 **31.** 52.3 **33.** 1.25 **35.** 273 **37.** 3.16

Page 187 **1.** $4.00 **3.** 22.22%; $99.99 **5.** 14.29%; $45.728 = $45.73 **7.** 5%; $5.725 = $5.73 **9.** 12.50%; $15.50 **11.** 70.51%; $63.459 = $63.46 **13.** $24.50 found by (18 × $15.25) − $250; $1.43325 = $1.43 found by $24.50 × 5.85% **15.** 316.05 **17.** 427.90 **19.** 19.2 **21.** 83.95

Reviewing the Basics (Page 188) **1.** $242.47 **2.** $38.38 **3.** $91.11 **4.** $42.86 **5.** $117.66 **6.** $16.99 **7.** $71.24 **8.** $306 **9.** $182.64 **10.** $6836.64 **11.** 0.05185 = 0.052 **12.** 0.134375 = 0.134 **13.** 0.134826 = 0.135 **14.** $380 **15.** $154 **16.** $138.7375 = $138.74 **17.** 0.25 **18.** 0.2 **19.** 0.41666 = 0.417 **20.** 0.625 **21.** 0.2 **22.** 0.81 **23.** 0.15 **24.** 0.115 **25.** 0.195 **26.** 0.10125 **27.** 19% **28.** 25% **29.** 12.4% **30.** 26.5% **31.** 11.45% **32.** 23.5% **33.** 34.5% **34.** 138.7% **35.** 205.4% **36.** 0.51% **37.** $45.12 **38.** $261.05 **39.** $\frac{2}{3}$ **40.** $\frac{7}{12}$ **41.** $\frac{5}{12}$ **42.** $\frac{1}{4}$ **43.** $\frac{293}{365}$ **44.** $\frac{1}{3}$ **45.** $\frac{1}{8}$ **46.** $\frac{36}{73}$ **47–52.** Answers may vary.

Unit 8 Automobile Transportation

Page 193 **1.** $7500 **3.** $7845 **5.** $17,985 **7.** $6533.65 **9.** $345.37 found by $6907.45 × 5% **11.** 7048 **13.** 9174 **15.** 6206.64

Pages 194–195 **1.** $8565 **3.** $12,957 **5.** $6487 **7.** $6644.93 **9.** $15,543.40 **11.** $12,175.87 **13.** 69 **15.** 207.6 **17.** 3432 **19.** 725.42 **21.** 87.09

Pages 196–197 **1.** $2500 **3.** $2775; $2875 found by $2775 + $50 + $50 **5.** $6025 found by $6625 + $175 − $450 − $325 **7.** $9675 **9.** $9775 **11.** 10,135 **13.** 6505 **15.** 3360 **17.** 1300 **19.** 17,755 **21.** $6412

Pages 199–201 **1.** $411.60; $493.92 found by 1.20 × $411.60 **3.** $347.20; $868.00 **5.** $427.20; $427.20 **7.** $380; $703 **9.** $314; $675.10 **11.** $401.20; $1263.78 **13.** $156.80; $297.92 **15.** $123.20; $381.92 **17.** $463.60; $625.86 **19.** 1173.9 **21.** 266.3 **23.** 88.54 **25.** 111.1 **27.** 780.29 **29.** 1276.5 **31.** 122.077 **33.** 2061.36

Pages 203–204 **1.** $1800; $0.20 **3.** $3616.20; $0.246 = $0.25 found by $3616.20 ÷ 14,700 **5.** $4525; $0.362 = $0.36 **7.** $5520; $0.160 = $0.16 **9.** $0.093 = $0.09 **11.** $0.2392 = $0.24 **13.** $0.1365 = $0.14 **14.** Variable costs: $756.40 found by $513.40 + $15 + $115.71 + $112.29; fixed costs: $2279.75 found by $313 + $76.75 + ($9450 − $7560); total costs: $3036.15; $0.264 = $0.26 found by $3036.15 ÷ 11,500 **15.** $0.3827 = $0.38 found by ($1721.90 + $3559.75) ÷ 13,800 **17.** $0.0623 = $0.06 found by $1154.19 ÷ 18,500; per person: $384.73 found by $1154.19 ÷ 3 **19.** $1928.58; $0.24 **21.** 908.67 **23.** 3.81 **25.** 0.03 **27.** 1.09 **29.** 2.23 **31.** 0.12 **33.** 0.18 **35.** 0.39

Pages 206–207 **1.** $0.31 **3.** $177.44; $0.3775 = $0.38 found by $177.44 ÷ 470 **5.** $27.30; $0.091 = $0.09 **7.** $0.651 = $0.65 **9.** $0.339 = $0.34 **11.** $640.20 found by (7 × $22.50) + (2200 × $0.17) + $374 + $108.70; $0.291 = $0.29 found by $640.20 ÷ 2200 **13.** $0.548 = $0.55 **15.** $0.94 found by $131.30 ÷ 140 **17.** $0.748 = $0.75 **19.** $0.423 = $0.42 **21.** $0.7056 = $0.71 **23.** 593.63 **25.** 171.038 **27.** 15.57 **29.** 1.92 **31.** 0.77

Reviewing the Basics (Page 208) **1.** $109.84 **2.** $0.35 **3.** $364.31 **4.** $0.20 **5.** $97.65 **6.** $365.50 **7.** $55.42 **8.** $7.83 **9.** $72.08 **10.** $8.12 **11.** $291.375 = $291.38 **12.** $222.625 = $222.63 **13.** $289.6725 = $289.67 **14.** $543.865 = $543.87 **15.** $307.895 = $307.90 **16.** $276.125 = $276.13 **17.** $0.1152 = $0.12 **18.** $0.3106 = $0.31 **19.** $0.1244 = $0.12 **20.** $3600 **21.** $7435.80 **22.** $3259.33 **23.** $1700 **24.** $3350 **25.** e **26.** c **27.** a **28.** d **29.** b

Unit 9 Housing Costs

Pages 216–217 **1.** $29,600 **3.** $23,450; $70,350 **5.** $7700; $30,800 **7.** $14,705 found by $17,300 − $2595 **9.** $59,925 **11.** $69,975 **13.** $80,000 found by $12,000 ÷ 15% **15.** 6000 **17.** 34,000 **19.** 11,295 **21.** 6000 **23.** 47,250 **25.** 63,250

Pages 219 **1.** $386.40; $52,736 found by $92,736 − $40,000 **3.** $842.35; $218,246 **5.** $2234.70; $609,492 **7.** $622 found by ($50,000 ÷ $1000) × $12.44; $149,280 found by $622 × (12 × 20); $99,280 found by $149,280 − $50,000 **9.** $284,320 **11.** 11.5% has most interest; $123,060 − $98,688 = $24,372 more **13.** 492 **15.** 3988.8 **17.** 3300 **19.** 93,321.5

Pages 220–221 **1.** $1554 found by $35 + (2% × $30,000) + $60 + $100 + (0.3% × $30,000) + (2.23% × $30,000) **3.** $3728.40 **5.** $929.50 **7.** $11,034 found by $10,000 + $1034 **9.** 45 **11.** 17.64 **13.** 14.49 **15.** 216 **17.** 8.88 **19.** 700

Pages 223–224 **1.** $34,981.10 **3.** $585 found by $54,000 × 13% × $\frac{1}{12}$; $12.78 found by $597.78 − $585; $53,987.22 found by $54,000 − $12.78 **5.** $1258.67; $11.01; $94,388.99 **7.** $810 **9.** $1265.833 = $1265.83; $27.77; $97,972.23 **11.** $1000; $12; $79,988 **13.** $1069.375 = $1069.38; $30.62; $88,469.38 **15.** $1045.17; $826.875 = $826.88; $218.29; $94,281.71 **17.** $40,600; $433.608 = $433.61 found by ($40,600 ÷ $1000) × $10.68; $422.916 = $422.92 found by $40,600 × 12.5% × $\frac{1}{12}$; $40,589.31 found by $40,600 − $10.69 **19.** for payment 119: $11,595.76 found by $11,649.67 − $53.91 where $53.91 = $155.84 − $101.93; for payment 120: $11,541.38 found by $11,595.76 − $54.38 where $54.38 = $155.84 − $101.46 **21.** 91,760.45 **23.** 78,879.02 **25.** 14,807.30

Page 226 **1.** $12,600 **3.** $2587.20 found by 0.066 × $39,200 **5.** $1435 **7.** $2266.078 = $2266.08 found by 0.051385 × $44,100 **9.** $942.76 **11.** 1859.13 **13.** 58.301 **15.** 2502.498 **17.** 32,000 **19.** 5425 **21.** 14,758.80

Page 228 **1.** $40,000 **3.** $43,650; $21,825 **5.** $38,250; $3825 **7.** $41,600; $20,800 **9.** $2000 **11.** 22,500 **13.** 9825

Page 230 **1.** $171 **3.** $333 **5.** $237 **7.** 262 **8.** total costs: $13.83 + $75.22 + $463.19 = $552.24 **9.** 35,000 **11.** 84,480 **13.** 9800 **15.** 50,000 **17.** 32,800.50

Pages 231–233 **1.** $350 **3.** $553 **5.** $503.65 **7.** $531.72 found by $313.65 + $18.50 + $57 + $45.44 + $34.78 + $62.35; FHA recommended maximum: 35% × $980 = $343; exceeds maximum **9.** $1460.04; FHA recommended maximum: 35% × $4200 = $1470; within maximum **11.** $753.26; 35% × $1990 = $696.50; exceeds maximum **13.** $702.29; 35% × $2100 = $735; within maximum **15.** $427.20 found by ($40,000 ÷ $1000) × $10.68; $893.725 = $893.73 found by 0.05107 × $17,500; $689.29 found by $427.20 + $74.48 + $17.00 + $59.23 + $43.85 + $12.88 + $54.65; 35% × $2166.66 = $758.331 = $758.33; within maximum **17.** 72,236.50 **19.** 876.07 **21.** 378.4 **23.** 649.34 **25.** 1016.94 **27.** 1445.2 **29.** 1781.40 **31.** 8782.05 **33.** 67.65

Reviewing the Basics (Page 234) **1.** $1118 **2.** $1654 **3.** $30,548 **4.** $178 **5.** $54,500 **6.** $76,226 **7.** $3003.71 **8.** $1100.85 **9.** $55.93 **10.** $113.05 **11.** $483.49 **12.** $151.18 **13.** 28507.60 **14.** 634.888 = 634.89 **15.** $1160.95 **16.** $23,500 **17.** $1068.75 **18.** $1110.34 **19.** $436.40 **20.** $531.36 **21.** b **22.** c **23.** f **24.** a **25.** e **26.** d

Unit 10 Insurance and Investments

Page 239 **1.** $266 **3.** $1750; $437.50 **5.** $1726; $1035.60 **7.** $1478; $517.30; $21.55 **9.** $77.82 each **11.** 1.37 **13.** 0.05 **15.** 129.25 **17.** 313.95

Page 241 **1.** $420 **3.** $0; $0; $25 **5.** $1495 found by $1745 − $250; $549 found by $299 + $250 **7.** $240 **9.** $71 **11.** 0.65 **13.** 9483 **15.** 13 **17.** 593,560

Page 243 **1.** $22.50 **3.** 25 × $22.01 = $550.25 **5.** $138.15 **7.** $770.35 **9.** $420.75 found by ($75,000 ÷ $1000) × $5.61; $1608.75 found by (75 × $4.29) × 5; $2103.75 **11.** 259.714 **13.** 0.003 **15.** 3.400 **17.** 7.985 **19.** 4216.20 **21.** 0.12 **23.** 3.20 **25.** 689,000 **27.** 0.36

Page 245 **1.** $1698.50 **3.** $1224 **5.** $878.40 found by $408.30 + $470.10; $61.80 **7.** $45.40 annually found by ($20,000 × $1000) × $2.27; $11.58; $0.92 found by $46.32 − $45.40 **9.** 1.77 **11.** 11.95 **13.** 85.37 **15.** 555.56 **17.** 55.161 **19.** 671 **21.** 9449 **23.** 396,875

Page 247 **1.** $2786.55; $286.55 **3.** 1.593848; $1593.85; $593.85 **5.** 1.111912; $3335.736 = $3335.74; $335.74 **7.** $551.1985 = $551.20; $51.20 **9.** $11,761.52 found by $7500 × 1.568203; $11,803.72 found by $7500 × 1.573829; 11.50%; $42.20 **11.** 120 **13.** 210 **15.** 500 **17.** 0.051 **19.** 4.541 **21.** 127,246 **23.** 9610

Pages 248–249 **1.** $4581.0776 = $4581.08; $481.08; 0.117336 = 11.73% **3.** 1.099877; $54,993.85; $4993.85; 0.099877 = 9.99% **5.** $7707.86 found by $7000 × 1.101123; 0.10112 = 10.11% found by $707.86 ÷ $7000 **7.** 1.113473; 0.113474 = 11.35% **9.** 11%; $343.86; 0.11462 = 11.46% **11.** 10%; 1.103813; $103.81; 10.38% **13.** 8.91% **15.** 2550.00% **17.** 30,000 **19.** 33,960

Page 251 **1.** $733.85 **3.** $11,625; $11,743.13 **5.** $13,250; $13,382.40 **7.** $2236; $2299.75 **9.** Callon costs less: found by $3888.75 − $3882.38 = $6.37 less **11.** $12,039.67 found by $11,812.50 + (1.5% × $11,812.50) + $49.98 **13.** 4.125 **15.** 16.375 **17.** $7650 **19.** $6272

Pages 252–253 **1.** 0.05333 = 5.33% **3.** 0.12 = 12% **5.** 0.061855 = 6.19% **7.** $500; 0.05633 = 5.63% found by $1.00 ÷ $17.75 **9.** $150; 0.1111 = 11.11% **11.** 0.23225 = 23.23% found by $0.90 ÷ $3.875; 0.05 = 5% found by $0.90 ÷ $18 **13.** $623.30 found by $5623.30 − $5000; 0.12466 = 12.47% found by $623.30 ÷ $5000; 175 × $1.75 = $306.25; $1.75 ÷ $24.125 = 0.07253 = 7.25%; 12.47% is greater **15.** 1.42 **17.** 24.27 **19.** 0.7% **21.** 1342%

Pages 254–255 **1.** $3757.25; $757.25 profit **3.** $4966.40; $1041.40 profit **5.** $476 loss found by $3812.75 − $3336.75 **7.** $137.50 loss **9.** $4334 found by $4400 − $66; $571.75 profit **11.** $182.10 profit found by $2997.30 − $2815.20 **13.** 321.4 **15.** $209.90 **17.** 21.32 **19.** 0.2436 **21.** 81.9944

Page 257 **1.** 0.07446 = 7.45% **3.** $8237.50; $950; 0.11532 = 11.53% **5.** $802.50; 0.0623 = 6.23% **7.** $9037.50; $675; 0.07468 = 7.47% **9.** $2523.02 found by $2467.50 + (0.0225 × $2467.50); 3 × 0.10875 × $1000 = $326.25; 2009 **11.** $13,875 **13.** 780 **15.** 768 **17.** 460.84 **19.** 26.05 **21.** 2.70

Reviewing the Basics (Page 258) **1.** $72.13 **2.** $1.38 **3.** $7315.33 **4.** $5.00 **5.** $25,281.25 **6.** $106.68 **7.** $283,058.30 **8.** $542.75 **9.** $1937.50 **10.** $978.56 **11.** $235.73 **12.** $4.275 = $4.28 **13.** $65.21 **14.** $1303.4385 = $1303.44 **15.** $8.00 **16.** $369.2486 = $369.25 **17.** 0.1875 **18.** 0.625 **19.** 1.875 **20.** 0.21875 **21.** 1.75 **22.** 3.3 **23.** 0.9% **24.** 15% **25.** 15.3% **26.** 3.75% **27.** 103% **28.** $326.10 **29.** $982.60 **30 – 38.** Answers may vary.

Unit 11 Recordkeeping

Page 263 **1.** $3990; $798 **3.** $4861.10; $972.22 **5.** $5011.55; $1002.31 **7.** $94.3866 = $94.39 found by = $283.16 ÷ 3 **9.** $492.13 **11.** $0 **13.** $1091.875 = $1091.88 found by $6551.25 ÷ 6 **15.** $126.01 **17.** 43

Pages 265–266 **1.** $589.32 **3.** $1228.20 **5.** $1125.03 **7.** Suggested answers: fixed expenses, contributions, household expenses, credit payments **9.** $861.48 **11.** $130.29 found by $1563.48 ÷ 12 **13.** yes; $1881.83 − $1875.25 = $6.58 within net income **15.** 149.7 **17.** $19.37 **19.** 1171.175 **21.** 57 **23.** 63.1 **25.** $84.42 **27.** $125.65

Pages 267–269 **1.** less; $30.76 **3.** $0.61 more **5.** electric; telephone **7.** $12.12 less found by $150.57 − $138.45 **9.** $42.41 more **11.** no **13.** $34.13 less than amount budgeted **15.** electric; heating fuel; repairs **17.** $11.80 less **19.** $39.60 less **21.** home insurance **23.** 823.16 **25.** 147.22 **27.** 34.1 **29.** $542.38 **31.** $1277.06

Reviewing the Basics (Page 270) **1.** $41.09 **2.** $16.29 **3.** $113.54 **4.** $299.71 **5.** $482.27 **6.** $10.85 **7.** $142.78 **8.** $81.74 **9.** $113.69 **10.** $77.12 **11.** $99.51 **12.** $92.11 **13.** 18.0% **14.** 65.4%

15. 8.5% **16.** 1.3% **17.** \$70.826 = \$70.83 **18.** \$168.60 **19.** \$128.6825 = \$128.68 **20.** \$86.568 = \$86.57
Terms 21, 26; 22, 23, 25, 24

Unit 12 Personnel

Page 287 **1.** \$1773.70 **3.** \$2755.41 **5.** \$4204.78 **7.** \$3230.33 **9.** \$7061.77 found by \$1695.96 + (\$221.20 + \$104.56 + \$143.30) + (8% × \$51,000) **11.** \$4260 **13.** \$17,620 **15.** \$3362 **17.** \$3741

Page 289 **1.** \$13,776 **3.** \$16,816 **5.** \$14,193 **7.** \$13,735.98 found by \$12,510 + (5% × \$12,510) + (4.8% × \$12,510) **9.** \$13,229.06 **11.** \$13,912.50 **13.** \$18,000 **15.** \$9500 **17.** \$355.20 **19.** \$390.90 **21.** \$612.32

Pages 290–291 **1.** 25% **3.** 0.2776 = 28% **5.** 0.2950 = 30% **7.** 0.2204 = 22% found by \$3621.60 ÷ \$16,432 **9. a.** Lab technician: \$715.38; \$576.60; **\$706.80; \$1329.90; \$3328.68; Hygienist: \$480.77; \$387.50; \$475; \$893.75; \$2237.02; Receptionist: \$436; \$351.42; \$430.77; \$810.52; \$2028.71 b.** 18%; **c. \$7594.41; d. \$816.07** **11.** \$615 **13.** 0.05 **15.** 0.05 **17.** 0.07

Page 293 **1.** \$84.50 **3.** \$27; \$150.25 **5.** \$159.20; \$1013.61 **7.** \$754.52 **9.** \$486.15 **11.** \$70 **13.** \$185 **15.** \$118.80 **17.** \$88.62 **19.** \$313.14 **21.** \$1702.72 **23.** \$72 **25.** \$270

Page 295 **1.** \$718 **3.** \$18,484 **5.** \$238 **7.** \$810 **9.** \$3964 found by [(\$84 × 3) + (\$126 × 5)] × 2 + (\$250 × 8) + (\$25 × 8) **11.** \$5097.20 **13.** \$325 **15.** \$2592 **17.** \$525 **19.** \$69.20 **21.** \$936 **23.** \$4910

Reviewing the Basics (Page 296) **1.** \$1163.53 **2.** \$17,252.26 **3.** \$3588.18 **4.** \$130.15 **5.** \$1234 **6.** \$80.80 **7.** \$1899.36 **8.** \$387.60 **9.** 0.2854 = 28.5% **10.** 0.2469 = 24.7% **11.** \$23,744 **12.** \$24,710 **13.** c **14.** d **15.** b **16.** a **17.** e

Unit 13 Production

Pages 304–305 **1.** \$0.054 **3.** \$0.04; \$0.006; \$0.046 **5.** \$0.0015 = \$0.002 found by \$6.90 ÷ 4600; \$0.001; \$0.002 + \$0.001 = \$0.003 **7.** \$0.036 found by \$0.021 + \$0.015 **9.** \$0.074 **11.** \$0.145 **13.** \$0.001 **15.** 6.1 **17.** \$0.140 **19.** \$0.142 **21.** \$0.103 **23.** \$0.014 **25.** \$0.012 **27.** \$0.015

Page 307 **1.** 340,000 **3.** 450,000 **5.** 395,833 **7.** 20,000 **9.** 920,718.23 = 920,718 **11.** 5,700,000 **13.** \$0.25 **15.** \$0.35 **17.** \$2.77 **19.** 135,000 **21.** 70,000

Pages 309–310 **1.** 2%; in **3.** 0%; in **5.** 12%; out **7.** 6.8%; out **9.** 16%; out **11.** 2.5%; 0%; 7.5%; 5%; 10% **13.** 9:30: 0%; 5%; 5%; 0%; 5% 12:30: 5%; 0%; 10%; 0%; 15%; 3:30: 5%; 0%; 10%; 5%; 10% **15.** 4.3%; 2.5%; 3.5%; 4.0%; 6.7%; 2.0% **17.** 5.0% **19.** 2.7% **21.** 3.2% **23.** 3.3% **25.** 7.2% **27.** 6.7% **29.** 4% **31.** 7.2%

Page 312 **1.** 847.45 = 847 **3.** 50.22 = 50 **5.** 4.106 = 4 **7. a.** 1.7; 14.9; 1.6; 1.5; **b.** 19.7; **c.** 182.74 = 183 **9.** 21.9 **11.** 64.3 **13.** 667 **15.** 237

Pages 313–314 **1.** 31.25% **2.** 53.125% **3.** 12.5% **4.** 3.125% **5.** 43.75% **7.** 0.4267 = 42.7% **9.** 0.2333 = 23.3% **11.** 6.25%; 13.75%; 15%; 17.5%; 21.875%; 16.25%; 3.125%; 6.25% **13.** 8.0 **15.** 40 **17.** 143.4

Pages 316–317 **1.** 0.4 cm **3.** 4.4 cm **5.** 0.2 cm **7.** 5.1 cm **9.** 0.3 cm **11.** 10.7 cm **13.** Height: $27\frac{3}{8}$ in; Width: $22\frac{1}{4}$ in; Length: 38 in **15. a.** Length: 91.8 cm; Width: 31 cm; Height: 15.8 cm **b.** Length: 61.4 cm; Width: 46.2 cm; Height: 15.8 cm **16. a.** Length: $20\frac{1}{4}$ in; Width: $7\frac{1}{8}$ in; Height: 7 in *OR* Length = $10\frac{1}{2}$ in; Width: $13\frac{5}{8}$ in; Height: 7 in; **b.** Length: $20\frac{1}{4}$ in; Width: $3\frac{3}{4}$ in; Height: $\frac{1}{2}$ = 14 in **17. a.** Length: 15.6 cm; Width: 10.2 cm; Height: 12.6 cm; **b.** Answers may vary. **19.** $\frac{7}{16}$ **21.** $\frac{7}{16}$ **23.** $\frac{13}{16}$ **25.** $\frac{3}{4}$ **27.** $1\frac{3}{16}$ **29.** $\frac{3}{8}$ **31.** $\frac{3}{4}$ **33.** $\frac{3}{4}$ **35.** $3\frac{1}{2}$ **37.** 6 **39.** $3\frac{1}{4}$ **41.** $9\frac{5}{8}$

Reviewing the Basics (Page 318) **1.** \$0.05 **2.** \$1.06 **3.** \$17.65 **4.** 25.3 **5.** 54.9 **6.** \$0.37 **7.** \$5.34 **8.** \$54.75 **9.** \$0.012 **10.** 211,800 **11.** 2,540,000 **12.** $5\frac{7}{16}$ **13.** $8\frac{9}{16}$ **14.** $5\frac{7}{8}$ **15.** $\frac{3}{4}$ **16.** $1\frac{5}{16}$ **17.** 23 **18.** 35% **19.** 25% **20.** \$5500 **21.** \$5000 **22.** 6.5 **23.** 15.24 **24.** b **25.** d **26.** a **27.** e **28.** f

Unit 14 Purchasing

Pages 322–323 **1.** \$120 **3.** \$12.69; \$29.61 **5.** \$1.947 = \$1.95; \$11.03 **7.** \$17.0992 = \$17.10; \$196.64 **9.** \$114 **11. a.** \$4.998 = \$5.00; \$19.99; **b.** \$5.7475 = \$5.75; \$17.24; **c.** \$5.985 = \$5.99; \$13.96 **13.** \$902.50 found by \$3610 − \$2707.50 **15.** \$1.92 **17.** \$19.56 **19.** \$208.35 **21.** \$17.28 **23.** \$7.34

Pages 324–325 **1.** \$120.00 **3.** 85%; \$10.20 **5.** 68%; \$203.1772 = \$203.18 **7.** 73%; \$55.8669 = \$55.87 **9.** \$291.032 = \$291.03 found by 80% × \$363.79 **11. a.** \$20.4876 = \$20.49; **b.** \$2.4948 = \$2.49; **c.** \$9.4185 = \$9.42; **d.** \$30.0825 = \$30.08 **13. a.** \$95.968 = \$95.97; **b.** \$53.98 **15.** 75% **17.** 85% **19.** 65% **21.** 35% **23.** \$54.00 **25.** \$21.08 **27.** \$485.71

Pages 326–327 **1.** 25% **3.** \$0.70; 0.250896 = 25.1% **5.** \$76.46; 0.3407763 = 34.1% **7.** \$280.18; 0.330666 = 33.1% **9.** \$41.54; 0.3333 = 33.3% **11.** 25% found by \$2.20 ÷ \$8.80; 0%; 35%; 40% **13.** \$15.60 **15.** \$48.81 **17.** \$87.59 **19.** 0.2 **21.** 0.2 **23.** 0.3

Pages 328–329 **1.** \$36; \$324 **3.** \$190; \$760; \$114; \$646 **5.** \$1030; \$1545; \$309; \$1236 **7.** \$39.73 found by \$45.15 − \$5.42 where \$45.15 = \$64.50 − \$19.35 and \$5.42 = 12% × \$45.15 = \$5.418 = \$5.42 **9. a.** \$39.99 found by \$42.10 − \$2.11; **b.** \$55.01 found by \$61.12 − \$6.11; **c.** \$31.47 found by \$39.34 − \$7.87 **11. a.** \$42.95 found by \$50.53 − \$7.58; **b.** \$31.16 found by \$34.62 − \$3.46; **c.** \$16.71 found by \$17.59 − \$0.88; **d.** \$40.90 found by \$43.98 − \$3.08 **13.** \$79 **15.** \$14,146.81 **17.** 149 **19.** \$14.20 **21.** \$169.69 **23.** \$16,071.95 **25.** \$83.16 **27.** \$171.89 **29.** \$6119.71

Page 331 **1.** 28%; \$140 **3.** \$324; 32.5%; \$156 **5.** \$302.40; 49.6%; \$297.60 **7.** \$81; \$31.50 found by 28% × \$112.50 **9. a.** \$53.712 = \$53.71 found by 72% × \$74.60; **b.** \$34.56; **c.** \$8.28; SED = 28% **11. a.** 60% × 75% × 85% = 38.25%; \$42.95475 = \$42.95; **b.** 70% × 80% × 90% = 50.4%; \$49.6944 = \$49.69; **c.** 75% × 85% × 95% = 60.5625%; \$44.937375 = \$44.94; **d.** 80% × 90% × 95% = 68.4%; \$31.7034 = \$31.70 **13.** \$96.75 **15.** \$1978.49 **17.** 75% **19.** 95% **21.** 73% **23.** 17% **25.** \$70.44

Page 333 **1.** \$833 **2.** \$13.896 = \$13.90; \$449.30 **3.** \$0, no discount, $\frac{2}{19} - \frac{2}{6}$ = 13 days; \$216.75 **5.** 3% × \$1645.70 = \$49.371 = \$49.37; \$1596.33 **7.** \$36.06 **9.** \$229.61; no discount **11.** \$106.02 **13.** \$18.81 **15.** \$212.50 **17.** \$27.48 **19.** \$8.07

Page 335 **1.** \$441 **3.** \$0, no discount; \$4854.80 **5.** 2% × \$612.70 = \$12.254 = \$12.25; \$600.45 **7.** \$16.50, no discount **9.** \$321.28 **11.** \$406.41 **13.** \$14.32 **15.** \$3.74 **17.** \$5000.74

Reviewing the Basics (Page 336) **1.** \$1375.73 **2.** \$753.46 **3.** \$171.45 **4.** \$191.80 **5.** \$9.80 **6.** \$908.35 **7.** \$334.30 **8.** \$6379.68 **9.** \$19,329.22 **10.** \$300.60 **11.** \$105.89 **12.** \$48.21 **13.** \$17.24 **14.** \$276.15 **15.** \$673.90 **16.** \$300.96 **17.** 76.8% **18.** 0.414962 = 41.5% **19.** 25% **20.** 10% **21.** 85% **22.** 44% **23.** 46.45% **24.** 98% **25.** 38% **26.** f **27.** c **28.** b **29.** a **30.** d **31.** e

Unit 15 Sales

Pages 340–341 **1.** \$0.24 **3.** \$2.01 **5.** \$42.85 **7.** \$493.81 **9.** \$24.24 **11.** \$0.42 **13.** \$3.98 **15.** \$29.98 **17. a.** \$15.70; **b.** \$1.3083 = \$1.31 **19.** \$9.99 **21.** \$23.36 **23.** \$22.50 **25.** \$21.07 **27.** \$890.55

Pages 342–343 **1.** 0.5534 = 55.3% found by \$5.23 ÷ \$9.45 **3.** 0.4937 = 49.4% **5.** \$17.62; 0.5875 = 58.8% **7.** 0.4124 = 41.2% **9.** 0.2758 = 27.6% **11.** 0.3081 = 30.8% **13.** \$2.98 **15.** \$29.25 **17.** \$100.00 **19.** 20.0% **21.** 45.0% **23.** 61.1%

Pages 344–345 **1.** \$34 **3.** \$0 **5.** \$1.31 found by \$2.49 − \$1.18 **7.** \$23.88 − \$13.65 = \$10.23; \$10.0296 = \$10.03 found by 42% × \$23.88; \$10.23 − \$10.03 = \$0.20 **9.** \$17.24 **11.** \$345,050 **13.** \$4.67 **15.** \$0.31 **17.** \$9.80 **19.** \$21.89 **21.** \$17.25 **23.** \$4.67 **25.** \$1.59 **27.** \$2205.66

Pages 346–347 **1.** \$17.50; 35% **3.** \$6.15; 25.0% **5.** \$5.97; 30% **7.** 30.0% **9.** 0.38998 = 39.0% found by \$24.76 ÷ \$63.49 where \$24.76 = \$31.74 − \$6.98 **11.** 0.0767 = 7.7% found by \$1.18 ÷ \$15.38 **13.** \$0.3043 = 30.4% **15. a.** 0.2652 = 26.5%; **b.** 0.23939 = 23.9% **17.** \$25.69 **19.** \$60.00 **21.** \$255.01 **23.** 20%

Pages 348–349 **1.** \$12.50 **3.** 55%; \$20.00 **5.** 24%; \$980.958 = \$980.96 **7.** \$1.00 **9.** \$2.13 **11.** \$0.50; \$0.5909 = \$0.59; \$1.2727 = \$1.27 **13.** 68% **15.** 74% **17.** 41.3% **19.** \$40.00 **21.** \$765.00 **23.** \$2.26 **25.** \$2304.29 **27.** \$0.80

Pages 350–351 **1.** \$0.25; 20.0% **3.** \$642.66; 200.0% **4.** \$3.99; 25.0% **5.** \$179.00; 1.0948 = 109.5% **7.** 1.00719 = 100.7% **9.** 0.9229 = 92.3% **11.** 2.29167 = 229.2% **13.** 1.4068 = 140.7%; 1.2305 = 123.1%; 0.9741 = 97.4%; 1.4875 = 148.8% **15.** \$59.80 **17.** \$24.45 **19.** \$10,818.67 **21.** 66.7%

Pages 352–353 **1.** \$180.00 **3.** \$344.00; \$774.00 **5.** \$90.75; \$163.35 **7.** \$7.28 **9.** \$9.00 **11.** 51.6% × \$199.74 = \$103.06584 = \$103.07; \$302.81 **13.** \$2.07 **15. a.** \$32.52; **b.** 0.650369 = 65.0% **17.** \$105 **19.** \$561.60 **21.** \$72 **23.** \$50.60 **25.** \$0.99 **27.** \$67.33

Pages 354–355 **1.** \$3.00; 25.0% **3.** \$39.99; 50.0% **5.** 0.37617 = 37.6% **7.** 0.333 = 33.3% **9.** 0.3448 = 34.5% **11. a.** Outside: 0.1843 = 18.4%; Border: 0.2395 = 24.0%; Inside: 0.1843 = 18.4%; 24 × 24: 0.2241 = 22.4%; 16 × 16: 0.1864 = 18.6%; **b.** 24 × 24: \$0.80; **c.** Border: 24.0% **13.** \$40.00 **15.** \$4.80 **17.** 20.0% **19.** 10.0% **21.** 5.0% **23.** 35.0% **25.** 14.0% **27.** 3.4%

Reviewing the Basics (Page 356) **1.** 7.0% **2.** 141.5% **3.** 34.9% **4.** 72.0% **5.** \$337.99 **6.** \$503.28

7. \$859.45 **8.** \$205.74 **9.** \$30.21 **10.** \$35.52 **11.** \$168.68 **12.** \$66.666 = \$66.67 **13.** \$700.00 **14.** \$321.365 = \$321.37 **15.** 0.29 **16.** 0.35 **17.** 0.474 **18.** 0.487 **19.** 1.71 **20.** \$62.7076 = \$62.71 **21.** \$221.85 **22.** \$61.4298 = \$61.43 **23.** 22.0% **24.** 0.20699 = 20.7% **25.** 61% **26.** 27% **27.** 85.4% **28.** 31.9% **29.** f **30.** a **31.** c **32.** b **33.** e

Unit 16 Marketing

Page 361 **1.** 0.50 = 50% **3.** 40% **5.** 52.5% **7.** 17% **9. a.** 21.3%; 0.3517 = 35.2%; 0.2798 = 28.0%; 0.155 = 15.5%; **b.** 0.1405 = 14.1% found by 43 ÷ (43 + 75 + 97 + 91); 26.3%; 22.6%; 0.186 = 18.6%; **c.** 25 – 34 **11.** 48 **13.** 420 **15.** \$0.77

Page 363 **1.** 136,000 found by 17% × 800,000 × 1 **3.** 3%; 141,000 **5.** 12%; 46,944,000 **7.** 18,000,000 found by 40% × 15,000,000 × 3 **9.** 214,700 **11.** 1.57 **13.** \$2675

Pages 364–365 **1.** 20% **3.** 24% **5.** 9% **7.** 6.25% = 6.3% **9.** 25% **11.** 39% **13.** 45.78% = 45.8% **15.** 18.38% = 18.4% **17.** 37.5 **19.** 200% **21.** 30.7 **23.** 3.0 **25.** 13.7

Page 367 **1.** \$35,000 approx. **3.** \$41,000 approx. **5.** \$47,000 approx. **7.** \$6.5 million projected for 1989 **9.** July: 7500; August: 6000; September: 4000 **11.** \$1,240,030 **13.** 0.026 **15.** one hundred seven **17.** seven tenths **19.** fifteen thousand seven hundred forty-six

Pages 368–369 **1.** \$1,000,000 **3.** \$2,100,000 **5.** \$240,000 **7.** \$960,000 **9.** \$31,350 **11.** 22,950 **13.** 21.08 **15.** \$3720 **17.** \$1.50 **19.** 600 **21.** 12.88

Page 371 **1.** \$647.325 = \$647.33 **3.** \$119.00 **5.** \$1313.55 **7.** \$1956.15 **9.** \$598.26 **11.** 5.220 **13.** 8.000 **15.** 0.149 **17.** 42

Page 373 **1.** \$12,000 found by (\$3000 × 2) + [1 × (\$3000 × 2)] **3.** \$600,000 **5.** \$16,800 **7.** \$161,300 **9.** \$463,750 **11.** $\frac{3}{16}$ **13.** $2\frac{4}{7}$

Page 375 **1.** \$130,100 **3.** \$17.89; \$36.87; \$277,488 **5.** \$2.75 and 2,000,000 **7.** 459.18 **9.** 13.6

Reviewing the Basics (Page 376) **1.** \$3.75 **2.** \$65.13 **3.** \$1818.60 **4.** \$285.57 **5.** \$74,140 **6.** 6011.28 **7.** 429.85 **8.** 26.062 **9.** 5.55 **10.** \$435.50 **11.** \$5741.50 **12.** \$4065 **13.** 11,308.80 **14.** \$2,859,400 **15.** 24.3% **16.** 34.3% **17.** 5734 **18.** 5500 approx. **19 – 24.** Answers may vary.

Unit 17 Warehousing and Distribution

Pages 384–385 **1.** 1.105 ft^3; 773.5 ft^3 **3.** 0.049 m^3; 58.8 m^3 **5.** 1.2 m^3; 300 m^3 **7.** 87.5 ft^3 **9.** 1800 ft^3 **11.** 75,000 ft^3 **13.** 570.375 = 570.38 m^3 **15.** 27 m^3 **16.** 100 m^3 found by (1 × 1 × 0.5) × 200 **17.** 40.0 ft^3 **19.** 90.0 in^3 **21.** 52.92 cm^3 **23.** 10,000 ft^3 **25.** 77,500 in^3

Pages 386–388 **1.** 225 **3.** 300; 275 **5.** 143 **7. a.** 161; **b.** 58; **c.** 23 **9.** 115; 185; 110; 180; 93 **11.** 35; 60; 30 **13.** 95 **15.** 121; 81; 131; 96; 156 **17.** 190 **19.** 178 **21.** 568 **23.** 5522 **25.** 5948

Pages 390–391 **1.** \$2.658 = \$2.66 found by \$319.00 ÷ 120; \$74.48 found by \$2.66 × 28 **3. a.** \$180.00; \$102.00; \$282.00; **b.** \$0.13756 = \$0.14 found by \$282.00 ÷ 2050; **c.** \$140.00 **5.** \$6.00 found by \$0.25 × 24 where \$0.25 = \$118.56 ÷ 480 **7. a.** \$26.5625 = \$26.56; **b.** \$557.76 **9. a.** \$275.00; \$700.00; \$375.00; \$1350.00; **b.** \$275; \$560; \$835.00; **c.** \$140; \$375; \$515 **11.** \$458 **13.** \$325 **15.** \$309.69 **17.** \$2756.04 **19.** \$7.36 **21.** \$7.59 **23.** \$1.27 **25.** \$21.37

Pages 392–393 **1.** \$75 **3.** \$2061 **5.** \$575.376 = \$575.38 **7.** \$12,000 **9.** \$12,600 **11. a.** \$11,900; \$18,700; \$6800; \$3400; \$1700; \$850; \$850; **b.** \$170,000 ÷ 2 = \$85,000; \$22,100 found by \$44,200 – \$22,100 **13.** \$180,000 **15.** \$183,200 **17.** \$520,875 **19.** \$17,424 **21.** \$82,150 **23.** \$467.42

Page 395 **1.** \$53.17 **3.** \$605.34 **5.** \$85.07 found by \$15.00 + \$55.80 + \$11.20 + \$3.07 **7.** 470 ÷ 100 = 4.7 therefore use 5; \$103.68 found by \$16.25 + \$68.00 + \$14.50 + \$4.93 **9.** 73 ÷ 100 = 0.73 therefore use 1; \$24.00 **11.** \$6.90 **13.** \$6.11 **15.** \$182.88 **17.** \$135.00

Page 397 **1.** \$36.81 found by (450 ÷ 100) × \$8.18 **3.** \$172.31 **5.** \$325.85 **7.** \$135.72 **9.** \$59.99 **11.** \$49.2773 = \$49.28 **13.** \$33.0261 = \$33.03 **15.** \$37.57 **17.** \$166.57 **19.** \$36.90 **21.** \$165.25 **23.** \$444.05 **25.** \$167.05 **27.** \$178.35 **29.** \$50.65 **31.** \$184.15

Reviewing the Basics (Page 398) **1.** 500 **2.** 2200 **3.** 1900 **4.** 3700 **5.** 110 **6.** 814 **7.** 872 **8.** 23 **9.** 366 **10.** 78 **11.** \$149.10 **12.** \$8699.13 **13.** \$11,189.96 **14.** \$8.54 **15.** \$55.69 **16.** \$82.30 **17.** \$21.18 **18.** \$76,000 **19.** \$16.70 **20.** \$1707.44 **21.** \$2883.19 **22.** \$2.4166 = \$2.42 **23.** \$0.38 **24.** 168 ft^3 **25.** 9108 m^3 **26.** b **27.** a **28.** d **29.** c **30.** e

Unit 18 Services

Pages 402–403 **1.** 400 sq. ft; \$200 **3.** 13,000 sq. ft; \$2492 **5.** \$1187.50 = \$1188 found by (1500 ×

$9.50) ÷ 12 **7.** $1958.33 = $1958 **9.** $870.833 = $871 found by [(20 × 40 × $2.75) + (5% × $165,000)] ÷ 12 **11.** $1225.35 = $1225 **13.** $1500 **15.** $313 **17.** $26,789

Pages 404–405 **1.** $252; $622 **3.** $132.30; $957.30 **5.** $246.40 **7.** $2421.45 found by $973.95 + $1447.50 **9.** $585.08 **10.** $1663.49 estimated total; $1471.26 employee total; have the employees do the job **11.** $144.21 found by $84.83 + (70% × $84.83) where $84.83 = $65.25 + $19.58 **13.** 396.50 **15.** 174 **17.** 132.30

Page 407 **1.** $10.35; $42.64 **3.** $105.30; $895.05 **5.** $1035.65 **7.** $406 **9.** $1870.81 found by ($162.75 × 11) + $80.56 **10.** Direct Office Rental **11.** $23.20 **13.** $10.47

Page 409 **1.** $17.61 **3.** $11; $1.60; $54.95 **5.** $22.45 found by $19.30 + $2.50 + $0.65 **7.** $36.07 **9.** $51.40 found by $46.90 + $3.00 + $1.50 **11.** $64.83 **13.** $836.39

Page 411 **1.** $2090 found by $550 + $1100 + $440 **3.** $1355 **5.** $1450.09 **7.** $842.60 found by $21.25 + $51.75 + $100 + $78 + $119 + $180.60 + $190 + $102 **9.** 530 **11.** 135

Pages 412–413 **1.** $712.50 **3.** $3750 **5.** $800,000 **7.** $6900 **9.** $1,662,500 **11.** $341,391.50 found by $201,500 + $139,500 + $391.50 **13.** 1230.00 **15.** 6549.698 **17.** 720 **19.** 481.50

Reviewing the Basics (Page 414) **1.** $378 **2.** $13 **3.** $2490 **4.** $109 **5.** $926 **6.** $11,032 **7.** $66,054 **8.** 174 **9.** 57,700 **10.** 82,779 **11.** 2464.73 **12.** 1142.34 **13.** 55.93 **14.** 6201.76 **15.** 2841.72 **16.** 212.45 **17.** 7200 **18.** 1632 **19.** 5670 **20.** 24,314.04 **21.** 1800.77 **22.** 24,810.96 **23.** 215.58 **24.** 1196.75 **25.** 1888.46 **26.** 186,240 **27.** 41,216 **28.** 51,526.7 **29.** 960 sq. ft **30.** 1215 sq. ft **31.** 9718 sq. ft **32.** 84 sq. ft **33.** 806.25 sq. ft **34.** 4450.88 sq. ft **35.** d **36.** f **37.** a **38.** c **39.** b

Unit 19 Accounting

Pages 419–421 **1.** $349.81 **3.** $262.95; $567.05 **5.** $31.91; $238.09 **7.** $6891.71 **9. a.** $224 found by 32 × $7.00; FIT: $26.20; $16.016 = $16.02 found by $224 × 7.15%; $44.26 found by $26.20 + $16.02 + $2.04; $179.74 found by $224 − $44.26; **b.** $252; $35.50; $18.018 = $18.02; $55.82; $196.18; **c.** $231; $28.10; $16.516 = $16.52; $46.72; $184.28; **d.** $238; $31.70; $17.017 = $17.02; $50.89; $187.11; **e.** $224; $26.20; $16.016 = $16.02; $44.26; $179.74; **f.** $245; $30; $17.5175 = $17.52; $49.75; $195.25; **g.** $266; $37.40; $19.019 = $19.02; $58.84; $207.16; **h.** $1680; $215.10; $120.14; $350.54; $1329.46 **11. a.** $228.80 found by ($3.00 × 40) + (17% × $640); FIT: $26.20; $16.36; $47.56 found by $26.20 + $16.36 + $5.00; $181.24 found by $228.80 − $47.56; **b.** $219.45; $27.90; $15.69; $48.59; $170.86; **c.** ($241.95; $26.30; $17.30; $48.60; $193.35; **d.** $214.18; $24.30; $15.31; $44.61; $169.57; **e.** $212.10; $27.90; $15.17; $48.07; $164.03; **f.** $236; $21; $16.87; $42.87; $193.13; **g.** $246.55; $30; $17.63; $52.63; $193.92; **h.** $4659; $1599.03; $183.60; $114.33; $35; $332.93; $1266.10 **13.** $204.16 **15.** 1.078 **17.** $94.34

Page 423 **1.** 12.5% **3.** 7.2% **5.** 58.5% **7.** $238,000; 75.1% found by $178,800 ÷ $238,000 **9.** $270,000 **11.** 145.1441 **13.** 62.5% **15.** 50.0% **17.** 9.0%

Pages 424–425 **1.** $3000 **3.** $700 **5.** 0.011 found by (35 × 45) ÷ 150,000; $13.97 found by 0.011 × $1270 **7. a.** 0.020; $14,490; **b.** 0.007; $5071.50 **9. a.** $1300.80 found by 0.271 × $4800; $5691 found by 0.271 × $21,000; $910.56 **b.** $1598.40; $6993; $1118.88; **c.** $600; $2625; $420 **11.** 21.3 **13.** 10.5 **15.** 0.1665 **17.** 937.3

Pages 426–427 **1.** $10,000 **3.** $13,650 **5.** $4000 found by ($25,000 − $5000) ÷ 5 **7.** $11,300 **9.** $12,000 **11.** $12,450 found by $4150 × 3 **12.** $590 found by $170 + $420 **13.** $1291 **15.** $390 **17.** 10.8 **19.** 4.8 **21.** $18.20 **23.** $2.61

Pages 428–429 **1.** $10,500; $10,500; $5500 **3.** $2760; $2760; $2240 **5.** $4000 found by $5800 − (3 × $600) **7.** $645.85 found by $2375 − $1729.15 where $1729.15 = $345.83 × 5 **9.** $4375; $4375; $18,125; $4375; $8750; $13,750; $4375; $13,125; $9375; $4375; $17,500; $5000 **11.** 9200 **13.** 1674 **15.** 12,960 **17.** 6,562,400 **19.** 187,450

Page 431 **1.** $1600 **3.** $\frac{2}{3}$; $12,000; $\frac{1}{3}$ × ($24,000 − $6000) = $6000 **5.** $2400 found by $\frac{2}{3}$ × ($4500 − $900) **7.** $1800 found by $\frac{4}{10}$ × $4500; $1350 found by $\frac{3}{10}$ × $4500; $900; $450 **9.** $19,000; $19,000; $31,000; $14,250; $33,250; $16,750; $9500; $42,750; $7250; $4750; $47,500; $2500 **11.** $\frac{9}{16}$ **13.** $3\frac{23}{32}$ **15.** $52\frac{5}{16}$ **17.** $\frac{1}{18}$

Page 433 **1.** $2080 **3.** $29997 found by 33.33% × $90,000 *or* $\frac{1}{3}$ × $90,000 = $30,000 **5.** 33.33% × $9600 = $3199.68 *or* $\frac{1}{3}$ × $9600 = $3200; $6400 found by $9600 − $3200 **7.** $30,000 found by 50% × $60,000; $30,000 found by $60,000 − $30,000; $45,000; $15,000; $7500; $52,500; $7500; $3750; $56,250; $3750 **8.** $2080; $3120; $1248; $3328; $1872; $748.8 = $749; $4077; $1123; $449.2 = $449; $4526; $674;

$269.6 (but disregard—depreciation must be $324, so that the accumulated depreciation is $4850 and book value is $350) **9.** $945 **11.** 0.36 **13.** 19.35 **15.** 0.01

Page 435 **1.** $13,320 **3.** $45,000; $68,400; $66,600 **5.** $7250; $11,020; $10,730 **7.** $9625; $14,630; $14,245 **9.** $7185; $10,538; $10,059; $10,059; $10,059 **11.** $26,000; $299,000; $45,500; $253,500; $39,000; $214,500; $32,500; $182,000 **13.** $465 **15.** $0.45

Reviewing the Basics (Page 436) **1.** $8473 **2.** $10,440 **3.** $21,965 **4.** $13,874 **5.** $122,520 **6.** $108 **7.** $46.78 **8.** $1845.87 **9.** $5.20 **10.** $2350.87 **11.** $4711.11 **12.** $1778 **13.** $61,200 **14.** $5280 **15.** $644.67 **16.** 30 **17.** $198 **18.** $5536.80 **19.** $315 **20.** $6285 **21.** 0.095 **22.** 0.024 **23.** 0.226 **24.** $13,750.00 **25.** $272.50 **26.** $39.98 **27.** $17,568.75 **28.** $25.58 **29.** $7.92 **30.** 0.01597 = 1.6% **31.** 0.13557 = 13.6% **32.** 0.01643 = 1.6% **33.** 0.03111 = 3.1% **34.** 4440 sq. ft **35.** 20,825 sq. ft **36–42.** Answers will vary.

Unit 20 Accounting

Page 441 **1.** $41,869.26 **3.** $46,707.77 **5.** $43,773.40 found by $14,640 + $27,671.50 + $1461.90; $16,750 + $671.85 = $17,421.85; $26,351.55 found by $43,773.40 − $17,421.85 **7.** $56,428.37 found by $130,418.87 − $73,990.50 **9.** 63.522 **11.** 122.397 **13.** 90.754 **15.** 176.149

Pages 443–444 **1.** $25,500; $9400; $16,100 **3.** $116,000; $63,000; $53,000 **5.** $30,000 found by $52,000 − $22,000 **7.** $77,514.92; $56,554.07; $20,960.85 found by $77,514.92 − $56,554.07; $77,514.92 **9.** $196,270,000; $155,900,000; $40,370,000; $196,270,000 **11.** 19,369 **13.** 607,116 **15.** 115,686

Pages 446–447 **1.** $500 **3.** $88,229; $42,155; $33,982 **5.** $700,000; $300,000; $20,000 **7.** $410,000 net sales; $110,000 gross profit; $110,000 − $190,000 = ($80,000) net loss **9.** $2775; $2154; $621 **11.** $32.2; $27.5; $4.7 **13.** $578.11 **15.** $11,570.93 **17.** $32,291 **19.** -$400

Pages 448–449 **1.** 60% **3.** 15.4% **5.** **a.** $9717; **b.** $6157; **c.** 33.6%; **d.** 66.4%; **e.** 24.3%; **f.** 42.1% **7.** 50 **9.** 300.0 **11.** 25.0

Pages 450–451 **1.** 18.6% **3.** $11,595; 24.5% **5.** $500; 50.0% **7.** -$29,274 found by $186,375 − $215,649; − 13.6% **9.** **a.** $2,388,454; 0.6%; **b.** $27,832,832; 10.7%; **c.** − $25,444,378; − 18.2%; **d.** − $14,863,948; − 11.5%; **e.** − $10,580,430; − 99.9% **11.** 70.0 **13.** 77.8 **15.** 16.7

Reviewing the Basics (Page 452) **1.** $1,950,914 **2.** $92,182,763 **3.** $10,887 **4.** $10,764 **5.** $11,514 **6.** $8739 **7.** $4421.58 **8.** $41,963.67 **9.** $7821.42 **10.** $25,279.79 **11.** $61,222.15 **12.** $33,717.49 **13.** 19.5% **14.** 29.4% **15.** 30% **16.** 423.8% **17.** $8068 **18.** $24,806 **19.** $50,000 **20.** $110 **21.** c **22.** e **23.** b **24.** d

Unit 21 Financial

Page 457 **1.** $2291.60 **3.** $92,910; $23,914 found by $16,750 + [($92,910 − $75,000) × 40%)] **5.** $48,275; $8905 **7.** $69,288.50 found by $26,750 + [($192,475 − $100,000) × 46%] **9.** $124,178.95 found by $187,200 − $63,021.05; $37,872.32 found by $26,750 + [($124,178.95 − $100,000) × 46%] **11.** 5.14 **13.** 650.00 **15.** 19.40 **17.** 31.68 **19.** 123,998.22

Page 459 **1.** $667,000; $7,333,000 **3.** $910,000; $1,022,500; $12,977,500 **5.** $454,080 found by ($4,730,000 × 7.7%) + $89,870; $4,275,920 found by $4,730,000 − $454,080 **7.** $353,900 found by $500,000 − $146,100 **9.** 150,000 × $22.70 = $3,405,000; $3,405,000 × 4.5% = $153,225; 3,405,000 × 1.4% = $47,670; $153,225 + $47,670 = $200,895; $3,405,000 − $200,895 = $3,204,105 **11.** $3060.00 **13.** 69.12

Page 461 **1.** $800; $20,800 **3.** $1402.50; $26,902.50 **5.** $2385; $28,885 **7.** $537.15 found by $6500 × 17% × $\frac{175}{360}$; $7037.15 **9.** **a.** $33,587.50 found by $15,075 + $11,981.25 + $2156.25 + $4375; **b.** $293,587.50 found by $120,000 + $90,000 + $50,000 + $33,587.50 **11.** 2.00 **13.** 0.003

Page 463 **1.** $1444.44; $38,585.56 **3.** $4025; $66,009 **5.** $1100; $28,930 **7.** $2000; $58,020 **9.** $19,376.39 **10.** $202,913.33 found by $96,250 + $106,663.33 **11.** 136.98 **13.** 0.195 **15.** 10.75 **17.** 14.12

Page 465 **1.** $383.33; $363.33 **3.** $22,750; $22,750 **5.** $1019.44 found by $1044.44 − $25 **7.** $9596.83 found by $4837.50 + $4400 + $383.33 − $24.44 **9.** 411,600 **11.** 21.42

Page 467 **1.** $132.50 **3.** $9364 found by ($964 × 3) + ($11,360 × 20%) + $4200 **5.** 1800 sq ft ÷ 9 = 200 yd 2; $9989.25 found by ($14.50 × 200) + (25% × $3617) + $80 + $945 + $2420 + $2740 **7.** $1,901,053.58 **9.** 502.204 **11.** 2.808

Reviewing the Basics (Page 468) **1.** $73,975.86 **2.** $150,471.05 **3.** $303,750 **4.** $571,200 **5.** $201,600 **6.** $48,600,000 **7.** $972.22 **8.** $1888.89 **9.** $2666.67 **10.** 0.123 **11.** 0.0725 **12.** 0.1075

13. 34.00 **14.** $3995.00 **15.** $5717.80 **16.** 0.56 **17.** 156.0 **18.** 140.0 **19.** 0.125 **20.** $4250 + 20% of amount over $25,000 **21.** $16,750 + 40% of amount over $75,000 **22.–27.** Answers will vary.

Unit 22 Information Management

Page 475 **1.** $0.90; $1.20 **3.** Blocks; 3; $78; June 23, 1984 **5.** 01/12/86 172R9 15 2420 1; 01/12/86 172R9 15 2010 3; 01/12/86 172R9 15 6023 1; 01/12/86 172R9 15 8642 1; 01/12/86 172R9 15 8643 1 **7.** 2683 3; 3102 2; 2070 2; 201 2; 6023 1; 2080 3; 2315 2; 2310 1; 845 4; 8641 1; 413 8 **9.** $0.59; $0.69; $0.29; $0.29; $1.60; $1.00; $1.81 Maintaining Your Skills: 2670 2671 2683 2693 2709 2714 2768 2783

Page 477 **1.** Status of inventory **3.** Production costs **5.** 2008 T-Bolt 11 3.18 3.75 011586 2; Between 1850 TUNFORK and 2009 BOLTS **7.** Status of inventory; 2009 Bolts 20 **8.** Yes; 2315 AGRAFFES 39 0.90 1.20 06/23/84 20 **9.** $1.05 **11.** $1.11 **13.** $8.50 **15.** $148.64

Pages 478–479 **1.** 50 hours **3.** 48 hours **5.** #5816 White Key; #5817 Black Key; #5818 Keytops; #5819 Keytops; #5907 Wedges; #2070 Shims; #2503 Strings **7.** 1004, 1428, 2010, 2080, 3101, 3102, 3153, 6023, 6274 **9.** 156 **11.** 195

Page 481 **1.** TOGGLES **3.** HAMMER; $5.49; $16.47 **5.** 2419 WIRE 21 5.00 6.00 08/12/84 15; 5893 DAMP FELT 35 0.43 0.98 12/20/84 20; 1373 POLISH 11 2.90 3.50 08/12/84 6 **7.** 1428 PIVOT 30 7.60 8.65 01/15/86 12; 3101 GROMMETS 78 1.84 2.10 01/15/86 20; 3102 CLIPS 65 1.40 1.65 01/15/86 20; 6274 SPRINGS 40 2.50 3.29 01/15/86 20 **9.** $104.70 **11.** $414.72 **13.** $1837.80 **15.** $116,320

Pages 483–485 **1. a.** 16; **b.** $125.60 **3.** 3; 48; 160; 62; 62; 63; 36; 36; 64 **5.** Wells: 4–8 years, 3 weeks; Hoi: 1–3 years, 2 weeks; Villa: 1–3 years, 2 weeks **7.** 0.90 **9.** 0.119 **11.** 0.075 **13.** 9.52 **15.** 9.858

Pages 487–489 **1.** $180.80 **3.** 15.9% **5.** 9 **7.** Model H22-C **9.** 38.8% **11. a.** $187,340; **b.** $66,926.40; **c.** $55,150.84; **d.** $11,775.56 **13.** 5.25 **15.** 18.7741 **17.** 18,507 **19.** 61,551

Reviewing the Basics (Page 490) **1.** $482,900 **2.** $561,500 **3.** 16.3% **4.** $7.50 (1984); $12.50 (1985) **5.** b **6.** f **7.** i **8.** g **9.** c **10.** h **11.** a **12.** e **13.** d

Skills File

Pages 504–505 **1.** ones; 5 **2.** ones; 6 **3.** tens; 70 **4.** tens; 20 **5.** ones; 8 **6.** hundreds; 500 **7.** tens; 0 **8.** hundreds; 800 **9.** hundreds; 700 **10.** thousands; 1000 **11.** ones; 9 **12.** hundreds; 900 **13.** tenths; 0.3 **14.** hundredths; 0.01 **15.** thousandths; 0.009 **16.** ones; 6 **17.** hundredths; 0.02 **18.** hundredths; 0.07 **19.** ones; 4 **20.** tenths; 0.3 **21.** hundredths; 0.02 **22.** hundreds; 400 **23.** thousandths; 0.007 **24.** tenths; 0.5 **25.** thousandths; 0.003 **26.** tenths; 0.0 **27.** thousandths; 0.001 **28.** ones; 9 **29.** hundreds; 0 **30.** hundreds; 900 **31.** thousands; 7000 **32.** hundredths; 0.08 **33.** thousandths; 0.002 **34.** hundredths; 0.04 **35.** tenths; 0.2 **36.** hundreds; 100 **37.** Eighteen **38.** Thirty-four **39.** One hundred fifty-nine **40.** Seventy-eight **41.** One hundred three **42.** Eight hundred forty-two **43.** Two hundred seven **44.** Five thousand twelve **45.** Six thousand five **46.** One hundred nineteen **47.** Seventy-two **48.** One thousand two hundred forty **49.** Five thousand one hundred two **50.** One hundred ninety-four **51.** Six thousand five hundred ninety **52.** Twenty-five dollars **53.** Six and $\frac{24}{100}$ dollars **54.** Seventeen and $\frac{9}{100}$ dollars **55.** One hundred twelve and $\frac{35}{100}$ dollars **56.** One hundred twenty and $\frac{17}{100}$ dollars **57.** Four and $\frac{37}{100}$ dollars **58.** Seven thousand seven hundred forty-nine dollars **59.** Sixty-five and $\frac{90}{100}$ dollars **60.** Fifty cents or $\frac{50}{100}$ dollars **61.** Twenty-six and $\frac{42}{100}$ dollars **62.** Seven and $\frac{50}{100}$ dollars **63.** Five hundred and $\frac{42}{100}$ dollars **64.** One hundred dollars **65.** One and $\frac{5}{100}$ dollars **66.** Thirty-seven and $\frac{49}{100}$ dollars **67.** Two hundred forty-seven and $\frac{61}{100}$ dollars **68.** One thousand eight hundred six dollars **69.** Seven and $\frac{1}{100}$ dollars **70.** 32 **71.** 54 **72.** 462 **73.** 835 **74.** 810 **75.** 211 **76.** 6450 **77.** 5763 **78.** 1000 **79.** 4444 **80.** 4000 **81.** 1236 **82.** 150 **83.** 650 **84.** $2000 **85.** $101 **86.** $600 **87.** $93 **88.** $1686 **89.** $742 **90.** $101,000 **91.** $791,000 **92.** 3.1 **93.** 2.0 **94.** 4.50 **95.** 25.1 **96.** 18.43 **97.** 58 **98.** 0.6 **99.** 0.1 **100.** 0.82 **101.** 0.5 **102.** 10 **103.** 0.020 **104.** $5.99 **105.** $10 **106.** $23.85 **107.** $19 **108.** $7.50 **109.** $600 **110.** $6297.86 **111.** $9521.39 **112.** $4646.40 **113.** $2000.00

Page 506 **1.** 30 **2.** 40 **3.** 70 **4.** 200 **5.** 220 **6.** 300 **7.** 8100 **8.** 100 **9.** 600 **10.** 2200 **11.** 6000 **12.** 7000 **13.** 16,000 **14.** 22,000 **15.** 1000 **16.** 0.6 **17.** 0.1 **18.** 0.4 **19.** 0.5 **20.** 0.65 **21.** 0.47 **22.** 0.17 **23.** 0.36 **24.** 0.429 **25.** 0.601 **26.** 0.010 **27.** 0.213 **28.** 6.394 **29.** 137.492 **30.** 10.000 **31.** 38,000 **32.** 19,000 **33.** 48,000 **34.** 800 **35.** 900 **36.** 6800 **37.** 30 **38.** 6530 **39.** 580 **40.** 6 **41.** 36 **42.** 17 **43.** 189.7 **44.** 10.0 **45.** 0.1 **46.** 0.39 **47.** 152.43 **48.** 0.70 **49.** 0.179 **50.** 16.001 **51.** 108.437

Page 507 **1.** 106 **2.** 108 **3.** 137 **4.** 57 **5.** 1189 **6.** 1195 **7.** 799 **8.** 729 **9.** 897 **10.** 798 **11.** 488 **12.** 199 **13.** 10,878 **14.** 10,899 **15.** 8987 **16.** 3992 **17.** 71 **18.** 102 **19.** 210 **20.** 135 **21.** 897 **22.** 811 **23.** 1542 **24.** 1243 **25.** 4638 **26.** 11,365 **27.** 12,224 **28.** 9875 **29.** 12,201 **30.** 7773 **31.** 12,787 **32.** 3147 **33.** 3798 **34.** 9912 **35.** 12,318 **36.** 5565 **37.** 8152 **38.** 9875 **39.** 9484 **40.** 6896 **41.** 6777 **42.** 10,560 **43.** 14,953 **44,** 57,559 **45.** 33,587 **46.** 58,799 **47.** 87,481 **48.** 12,049 **49.** 14,999 **50.** 13,999 **51.** 14,450 **52.** 8655

Page 508 **1.** 331 **2.** 251 **3.** 151 **4.** 41 **5.** 3442 **6.** 2213 **7.** 3311 **8.** 1263 **9. 1111** **10.** 1013 **11.** 8020 **12.** 1042 **13.** 6 **14.** 902 **15.** 6231 **16.** 1001 **17.** 64 **18.** 469 **19.** 99 **20.** 88 **21.** 2258 **22.** 5792 **23.** 1166 **24.** 2879 **25.** 1909 **26.** 2778 **27.** 3667 **28.** 1799 **29.** 1275 **30.** 1003 **31.** 859 **32.** 1799 **33.** 2983 **34.** 1759 **35.** 1568 **36.** 11 **37.** 9111 **38.** 8194 **39.** 1862 **40.** 875 **41.** 2238 **42.** 549 **43.** 2901 **44.** 2424 **45.** 16,526 **46.** 7342 **47.** 37,503 **48.** 105,239 **49.** 126,403 **50.** 153,099 **51.** 100,621 **52.** 172,649

Page 509 **1.** 478.99 **2.** 999.99 **3.** 801.12 **4.** 734.14 **5.** 791.19 **6.** 59.100 **7.** 543.113 **8.** 419.214 **9.** 816.659 **10.** 16.3 **11.** 64.54 **12.** 105.537 **13.** 177.349 **14.** 8.231 **15.** 21.901 **16.** 119.295 **17.** 24.460 **18.** 2.98 **19.** 82.332 **20.** 32.844 **21.** 10.122 **22.** 76.771 **23.** 460.909 **24.** 201.115 **25.** 502.175 **26.** 887.513 **27.** 1064.683 **28.** 274.458 **29.** 1.241 **30.** 1.027 **31.** 21.901 **32.** 99.63 **33.** 220.611 **34.** 724.002 **35.** 772.652 **36.** 523.587

Page 510 **1.** 51.15 **2.** 33.11 **3.** 134.89 **4.** 22.58 **5.** 5.09 **6.** 23.785 **7.** 25.51 **8.** 81.64 **9.** 79.68 **10.** 482.96 **11.** 203.84 **12.** 288.89 **13.** 70.85 **14.** 85.23 **15.** 84.32 **16.** 10.89 **17.** 19.6 **18.** 29.06 **19.** 8.07 **20.** 1.3 **21.** 57.44 **22.** 66.558 **23.** 377.63 **24.** 3699.873 **25.** 5.09 **26.** 20.89 **27.** 19.983 **28.** 1.988 **29.** 325.86 **30.** 133.303 **31.** 204.354 **32.** 197.78 **33.** 557.87 **34.** 886.136 **35.** 5215.946 **36.** 671.468 **37.** 1.1178 **38.** 11.6355 **39.** 778.029 **40.** 0.6925 **41.** 3.59806 **42.** 1.09753 **43.** 2.74502 **44.** 1.04053

Page 511 **1.** 1533 **2.** 924 **3.** 87,344 **4.** 72,152 **5.** 100,152 **6.** 72,029 **7.** 77,256 **8.** 728,586 **9.** 959,373 **10.** 4636 **11.** 2808 **12.** 249,527 **13.** 251,565 **14.** 91,056 **15.** 4,268,236 **16.** 1728 **17.** 1665 **18.** 4828 **19.** 310,992 **20.** 3,352,947 **21.** 1,662,024 **22.** 4800 **23.** 69,000 **24.** 151,917 **25.** 28,182 **26.** 4880 **27.** 357,000 **28.** 21,274,072 **29.** 36,060,000 **30.** 63,099,000 **31.** 4,200,672 **32.** 152,064,000 **33.** 43,955,000 **34.** 182,061 **35.** 1,070,443 **36.** 160,113 **37.** 169,130 **38.** 170,772 **39.** 623,770 **40.** 334,500 **41.** 1,309,622 **42.** 8,873,082 **43.** 2,804,972 **44.** 1,742,642 **45.** 15,805,077 **46.** 22,263,173

Page 512 **1.** 4.5 **2.** 12.54 **3.** 87.822 **4.** 81.408 **5.** 188.55 **6.** 137.8432 **7.** 230.8218 **8.** 37.1195 **9.** 43.9556 **10.** 300.196 **11.** 287.352 **12.** 60.63645 **13.** 0.1008 **14.** 0.43 **15.** 0.096 **16.** 0.0392 **17.** 0.584 **18.** 0.045 **19.** 0.6156 **20.** 0.3948 **21.** 0.0567 **22.** 0.0171 **23.** 0.0861 **24.** 0.0005 **25.** 4116 **26.** 0.45227 **27.** 36 **28.** 673.2 **29.** 876.403 **30.** 55,740 **31.** 1212.86 **32.** 11.34148 **33.** 112.5 **34.** 9 **35.** 81.9698 **36.** 0.11416 **37.** 161,713.902 **38.** 33.333344 **39.** 0.156462 **40.** 843.199693 **41.** 84 **42.** 0.06369376 **43.** 0.07608533 **44.** 0.00023952

Page 513 **1.** $2\frac{9}{27}$ **2.** $2\frac{13}{38}$ **3.** $2\frac{2}{41}$ **4.** $13\frac{15}{46}$ **5.** $14\frac{12}{53}$ **6.** $11\frac{14}{61}$ **7.** $84\frac{17}{21}$ **8.** $77\frac{38}{55}$ **9.** $67\frac{19}{43}$ **10.** $54\frac{14}{73}$ **11.** $59\frac{12}{23}$ **12.** $63\frac{25}{34}$ **13.** $71\frac{42}{81}$ **14.** $\frac{78}{93}$ **15.** $16\frac{13}{114}$ **16.** $25\frac{17}{216}$ **17.** $15\frac{17}{321}$ **18.** $21\frac{29}{429}$ **19.** $8\frac{792}{892}$ **20.** $13\frac{19}{910}$ **21.** $23\frac{22}{409}$ **22.** $8\frac{421}{900}$ **23.** $13\frac{617}{625}$ **24.** $22\frac{74}{710}$ **25.** $20\frac{831}{843}$ **26.** $23\frac{923}{937}$ **27.** 18 **28.** 17 **29.** 16 **30.** $36\frac{13}{279}$

31. 72 **32.** $9\frac{508}{509}$ **33.** $14\frac{22}{621}$ **34.** $101\frac{10}{49}$ **35.** $12\frac{349}{549}$ **36.** $10\frac{67}{953}$ **37.** $5\frac{765}{842}$ **38.** $24\frac{51}{87}$ **39.** $181\frac{29}{473}$ **40.** $478\frac{69}{192}$ **41.** $87\frac{527}{622}$ **42.** $68\frac{476}{812}$ **43.** $1600\frac{3}{51}$ **44.** $1928\frac{20}{23}$ **45.** $252\frac{46}{88}$ **46.** $821\frac{31}{34}$ **47.** $208\frac{166}{417}$ **48.** $1453\frac{42}{71}$ **49.** $1345\frac{177}{514}$ **50.** $3712\frac{37}{64}$

Page 514 **1.** 11.125 **2.** 12.3 **3.** 21.2 **4.** 16.5 **5.** 5.1 **6.** 10.6 **7.** 19.75 **8.** 52.4 **9.** 34.5 **10.** 31.125 **11.** 22.8 **12.** 13.3 **13.** 16.1 **14.** 204.7 **15.** 206.8 **16.** 22.47 **17.** 2.02 **18.** 9.19 **19.** 10.865 **20.** 26.792 **21.** 21.524 **22.** 36.5 **23.** 8.71 **24.** 23.13 **25.** 10.45 **26.** 21.2 **27.** 31.13 **28.** 10.46 **29.** 5.63 **30.** 11.87 **31.** 17.5 **32.** 19.5 **33.** 14.44 **34.** 18.19 **35.** 230.05 **36.** 14.78 **37.** 41.42 **38.** 124.63 **39.** 25.54 **40.** 125.33 **41.** 39.5

Page 515 **1.** 4.6 **2.** 1.2 **3.** 5.71 **4.** 3.71 **5.** 17 **6.** 15.3 **7.** 6.5 **8.** 3.5 **9.** 42.91 **10.** 3.89 **11.** 26 **12.** 17 **13.** 31 **14.** 6.12 **15.** 26 **16.** 6.54 **17.** 8.96 **18.** 4.02 **19.** 13.21 **20.** 14.36 **21.** 1.41 **22.** 12.99 **23.** 245 **24.** 620 **25.** 1.21 **26.** 0.31 **27.** 256 **28.** 0.89 **29.** 67.48 **30.** 2.04 **31.** 22.57 **32.** 27.18 **33.** 1.75 **34.** 3.01 **35.** 38.5 **36.** 10.47 **37.** 98.86 **38.** 24.21 **39.** 0.01 **40.** 7.33 **41.** 17.13 **42.** 966.67 **43.** 0.01 **44.** 0.13

Page 516 **1.** 10 **2.** 12 **3.** 50 **4.** 24 **5.** 24 **6.** 102 **7.** 22 **8.** 57 **9.** 51 **10.** 33 **11.** 24 **12.** 68 **13.** 35 **14.** 66 **15.** 72 **16.** 24 **17.** 20 **18.** 115 **19.** $\frac{2}{3}$ **20.** $\frac{1}{4}$ **21.** $\frac{1}{2}$ **22.** $\frac{10}{11}$ **23.** $\frac{2}{3}$ **24.** $\frac{3}{4}$ **25.** $\frac{1}{3}$ **26.** $\frac{3}{7}$ **27.** $\frac{5}{6}$ **28.** $\frac{3}{5}$ **29.** $\frac{10}{27}$ **30.** $\frac{7}{9}$ **31.** $\frac{4}{7}$ **32.** $\frac{1}{3}$ **33.** $\frac{1}{2}$ **34.** $\frac{3}{5}$ **35.** $\frac{3}{4}$ **36.** $\frac{40}{73}$ **37.** $\frac{94}{183}$ **38.** $\frac{30}{73}$ **39.** $\frac{1}{2}$ **40.** $\frac{45}{73}$ **41.** $\frac{19}{36}$ **42.** $\frac{13}{15}$ **43.** $\frac{2}{13}$ **44.** $\frac{1}{2}$ **45.** $\frac{4}{25}$ **46.** $\frac{7}{20}$

Page 517 **1.** $\frac{25}{4}$ **2.** $\frac{19}{4}$ **3.** $\frac{59}{8}$ **4.** $\frac{19}{2}$ **5.** $\frac{20}{3}$ **6.** $\frac{23}{6}$ **7.** $\frac{39}{5}$ **8.** $\frac{25}{8}$ **9.** $\frac{43}{10}$ **10.** $\frac{17}{16}$ **11.** $\frac{145}{32}$ **12.** $\frac{21}{5}$ **13.** $\frac{16}{3}$ **14.** $\frac{13}{6}$ **15.** $\frac{29}{10}$ **16.** $\frac{31}{10}$ **17.** $\frac{53}{16}$ **18.** $\frac{73}{16}$ **19.** $\frac{165}{32}$ **20.** $\frac{33}{5}$ **21.** $\frac{79}{11}$ **22.** $\frac{53}{6}$ **23.** $\frac{57}{7}$ **24.** $\frac{40}{3}$ **25.** $6\frac{1}{2}$ **26.** $4\frac{1}{2}$ **27.** $5\frac{3}{4}$ **28.** $4\frac{3}{4}$ **29.** $5\frac{1}{2}$ **30.** $4\frac{2}{7}$ **31.** $2\frac{1}{3}$ **32.** $3\frac{1}{2}$ **33.** $2\frac{1}{2}$ **34.** $2\frac{1}{3}$ **35.** $2\frac{20}{21}$ **36.** $7\frac{2}{3}$ **37.** $5\frac{5}{6}$ **38.** $6\frac{5}{7}$ **39.** $3\frac{1}{6}$ **40.** $4\frac{1}{9}$ **41.** $6\frac{1}{8}$ **42.** $31\frac{1}{2}$ **43.** $6\frac{3}{7}$ **44.** $17\frac{3}{4}$

Page 518 **1.** 0.8 **2.** 0.35 **3.** 1.13 **4.** 0.75 **5.** 2.43 **6.** 0.58 **7.** 1.37 **8.** 0.4 **9.** 0.3 **10.** 3.16 **11.** 4.38 **12.** 0.7 **13.** 3.08 **14.** 0.07 **15.** 0.03 **16.** 2.9 **17.** 5.15 **18.** 0.45 **19.** 2.12 **20.** 0.13 **21.** 7.14 **22.** 2.63 **23.** 0.44 **24.** 1.88 **25.** $\frac{1}{10}$ **26.** $3\frac{7}{10}$ **27.** $\frac{3}{10}$ **28.** $2\frac{1}{4}$ **29.** $\frac{3}{100}$ **30.** $\frac{9}{100}$ **31.** $\frac{53}{100}$ **32.** $1\frac{3}{4}$ **33.** $\frac{3}{1000}$ **34.** $\frac{1}{100}$ **35.** $\frac{8}{125}$ **36.** $4\frac{103}{500}$ **37.** $4\frac{111}{250}$ **38.** $\frac{183}{250}$ **39.** $\frac{469}{1000}$ **40.** $2\frac{9}{10}$ **41.** $7\frac{1}{2}$ **42.** $\frac{8}{25}$ **43.** $1\frac{2}{25}$ **44.** $\frac{3}{50}$ **45.** $2\frac{83}{100}$ **46.** $1\frac{39}{1000}$ **47.** $\frac{21}{200}$ **48.** $2\frac{211}{500}$ **49.** $\frac{1}{2000}$ **50.** $\frac{1241}{5000}$ **51.** $1\frac{402}{625}$ **52.** $\frac{29}{5000}$ **53.** $\frac{1}{5000}$ **54.** $6\frac{33}{50}$

Page 519 **1.** $1\frac{2}{5}$ **2.** $\frac{6}{7}$ **3.** $\frac{3}{4}$ **4.** $1\frac{1}{3}$ **5.** $1\frac{4}{7}$ **6.** $1\frac{1}{2}$ **7.** $1\frac{4}{25}$ **8.** 1 **9.** $1\frac{5}{16}$ **10.** $1\frac{3}{10}$ **11.** $1\frac{8}{45}$ **12.** $\frac{3}{5}$ **13.** $11\frac{2}{3}$ **14.** $22\frac{2}{5}$ **15.** $14\frac{2}{7}$ **16.** $26\frac{1}{2}$ **17.** $12\frac{1}{3}$ **18.** $10\frac{1}{13}$ **19.** $17\frac{1}{4}$ **20.** $29\frac{1}{2}$ **21.** $9\frac{7}{8}$ **22.** $22\frac{2}{45}$ **23.** $11\frac{2}{5}$ **24.** $23\frac{3}{8}$

Page 520 **1.** $1\frac{1}{10}$ **2.** $\frac{11}{12}$ **3.** $\frac{20}{21}$ **4.** $\frac{23}{40}$ **5.** $1\frac{1}{6}$ **6.** $1\frac{1}{84}$ **7.** $1\frac{13}{110}$ **8.** $1\frac{15}{16}$ **9.** $\frac{59}{60}$ **10.** $1\frac{5}{72}$ **11.** $\frac{71}{150}$ **12.** $1\frac{1}{10}$ **13.** $9\frac{1}{6}$ **14.** $14\frac{1}{8}$ **15.** $16\frac{1}{3}$ **16.** $20\frac{5}{16}$ **17.** $20\frac{19}{26}$ **18.** $24\frac{10}{33}$ **19.** $28\frac{22}{63}$ **20.** $27\frac{29}{42}$ **21.** $28\frac{43}{56}$ **22.** $24\frac{59}{72}$ **23.** $29\frac{5}{48}$ **24.** $25\frac{11}{56}$

Page 521 **1.** $\frac{2}{9}$ **2.** $\frac{3}{4}$ **3.** $\frac{5}{6}$ **4.** $\frac{2}{3}$ **5.** $\frac{4}{7}$ **6.** $\frac{2}{3}$ **7.** $\frac{1}{8}$ **8.** $\frac{2}{5}$ **9.** $\frac{1}{5}$ **10.** $\frac{1}{2}$ **11.** $\frac{2}{5}$ **12.** $\frac{7}{18}$ **13.** $\frac{1}{4}$ **14.** $\frac{3}{14}$ **15.** $\frac{7}{20}$ **16.** $1\frac{1}{2}$ **17.** $1\frac{1}{7}$ **18.** $4\frac{3}{5}$ **19.** $3\frac{1}{4}$ **20.** $1\frac{11}{21}$ **21.** $2\frac{1}{6}$ **22.** $6\frac{2}{5}$ **23.** $5\frac{3}{16}$ **24.** $5\frac{1}{2}$ **25.** $2\frac{7}{13}$ **26.** $3\frac{2}{5}$ **27.** $8\frac{1}{2}$ **28.** $45\frac{1}{3}$ **29.** $25\frac{2}{7}$

Page 522 **1.** $\frac{3}{8}$ **2.** $\frac{1}{9}$ **3.** $\frac{3}{20}$ **4.** $\frac{2}{15}$ **5.** $\frac{1}{18}$ **6.** $\frac{5}{24}$ **7.** $\frac{1}{15}$ **8.** $\frac{1}{6}$ **9.** $\frac{3}{20}$ **10.** $\frac{8}{21}$ **11.** $\frac{1}{10}$ **12.** $\frac{1}{16}$ **13.** $4\frac{1}{4}$ **14.** $3\frac{1}{6}$
15. $6\frac{5}{18}$ **16.** $5\frac{2}{21}$ **17.** $5\frac{7}{60}$ **18.** $11\frac{1}{63}$ **19.** $3\frac{1}{8}$ **20.** $1\frac{2}{21}$ **21.** $2\frac{3}{20}$ **22.** $2\frac{5}{24}$ **23.** $2\frac{1}{8}$ **24.** $\frac{13}{38}$

Page 523 **1.** $1\frac{1}{2}$ **2.** $1\frac{1}{4}$ **3.** $2\frac{1}{10}$ **4.** $1\frac{5}{8}$ **5.** $2\frac{6}{11}$ **6.** $2\frac{9}{16}$ **7.** $3\frac{1}{18}$ **8.** $4\frac{7}{8}$ **9.** $12\frac{11}{12}$ **10.** $13\frac{15}{28}$ **11.** $1\frac{37}{70}$ **12.** $\frac{53}{60}$
13. $2\frac{7}{8}$ **14.** $6\frac{47}{60}$ **15.** $1\frac{3}{8}$ **16.** $3\frac{1}{12}$ **17.** $\frac{1}{8}$ **18.** $8\frac{6}{11}$ **19.** $3\frac{7}{9}$ **20.** $5\frac{9}{10}$ **21.** $\frac{9}{20}$ **22.** $6\frac{3}{8}$ **23.** $11\frac{5}{16}$ **24.** $8\frac{37}{70}$
25. $14\frac{7}{9}$ **26.** $12\frac{37}{72}$ **27.** $13\frac{5}{7}$ **28.** $10\frac{1}{20}$ **29.** $12\frac{5}{32}$ **30.** $16\frac{3}{10}$

Page 524 **1.** $\frac{1}{3}$ **2.** $\frac{3}{20}$ **3.** $\frac{5}{8}$ **4.** $\frac{16}{63}$ **5.** $\frac{5}{12}$ **6.** $\frac{32}{117}$ **7.** 2 **8.** $6\frac{4}{5}$ **9.** 9 **10.** 2 **11.** $16\frac{2}{15}$ **12.** $\frac{3}{5}$ **13.** $58\frac{1}{3}$
14. $18\frac{3}{4}$ **15.** $46\frac{1}{8}$ **16.** $\frac{8}{35}$ **17.** $\frac{1}{6}$ **18.** $\frac{3}{4}$ **19.** $\frac{11}{12}$ **20.** $4\frac{4}{5}$ **21.** 6 **22.** $14\frac{2}{3}$ **23.** $3\frac{3}{4}$ **24.** $2\frac{1}{8}$ **25.** $19\frac{1}{2}$
26. $49\frac{1}{3}$ **27.** $8\frac{8}{15}$ **28.** $77\frac{1}{2}$ **29.** 64 **30.** $2\frac{34}{63}$ **31.** $25\frac{29}{72}$ **32.** $\frac{7}{18}$ **33.** $97\frac{11}{12}$ **34.** $\frac{1}{9}$ **35.** $71\frac{1}{24}$

Page 525 **1.** 2 **2.** $1\frac{1}{2}$ **3.** $1\frac{7}{8}$ **4.** 6 **5.** $7\frac{1}{2}$ **6.** 22 **7.** 6 **8.** $\frac{12}{25}$ **9.** 16 **10.** $2\frac{4}{9}$ **11.** $3\frac{7}{36}$ **12.** $\frac{13}{20}$ **13.** $2\frac{11}{12}$
14. $\frac{3}{14}$ **15.** $\frac{1}{4}$ **16.** $14\frac{2}{9}$ **17.** $5\frac{5}{33}$ **18.** $4\frac{10}{23}$ **19.** $4\frac{20}{133}$ **20.** 6 **21.** $1\frac{1}{2}$ **22.** $\frac{25}{121}$ **23.** $1\frac{1}{2}$ **24.** $3\frac{1}{2}$ **25.** $\frac{2}{3}$
26. $133\frac{1}{3}$ **27.** 15 **28.** $\frac{38}{147}$ **29.** 6 **30.** 5 **31.** 30 **32.** $\frac{3}{16}$ **33.** $\frac{43}{176}$ **34.** $36\frac{2}{3}$ **35.** $\frac{143}{375}$ **36.** 1 **37.** $6\frac{1}{2}$ **38.** 6 **39.** 1

Page 526 **1.** 1500:600 **2.** 1500:1200 **3.** 600:1200 **4.** 1200:600 **5.** 600:1500 **6.** 1200:1500 **7.** $\frac{1}{5}$ **8.** $\frac{1}{20}$
9. $\frac{2}{3}$ **10.** $\frac{11}{15}$ **11.** $\frac{1}{25}$ **12.** $\frac{1}{23}$ **13.** $\frac{2}{5}$ **14.** $\frac{3}{5}$

Page 527 **1.** T **2.** T **3.** F **4.** F **5.** F **6.** F **7.** T **8.** T **9.** T **10.** F **11.** F **12.** T **13.** 18 **14.** 18
15. $7\frac{1}{3}$ **16.** 10 **17.** $\frac{6}{11}$ **18.** $\frac{14}{19}$ **19.** $5\frac{5}{9}$ **20.** 12 **21.** $12\frac{4}{13}$ **22.** 7 **23.** 20 **24.** 25 **25.** 5 **26.** 27
27. 24 **28.** 210 **29.** 64 **30.** 3 **31.** 15 **32.** 7 **33.** 50

Page 528 **1.** $\frac{99}{12} = \frac{x}{7}$ **2.** $\frac{84}{7} = \frac{x}{13}$ **3.** $\frac{40}{5} = \frac{x}{17}$ **4.** $\frac{30}{6} = \frac{x}{45}$ **5.** $3.48 **6.** 605 bolts **7.** 117 pages
8. 722 km **9.** 1485 wheels **10.** $26.04 **11.** $28 **12.** 19.25 km **13.** $33\frac{1}{3}$ minutes

Page 529 **1.** 288 oranges **2.** 550 words **3.** 80 km **4.** $.95 **5.** 64 km; 640 km **6.** 25 sheets, 400 lbs
7. A **8.** B **9.** B **10.** A **11.** B

Page 530 **1.** 10% **2.** 15% **3.** 25% **4.** 74% **5.** 82% **6.** 93% **7.** 213% **8.** 421.2% **9.** 575.3%
10. 26.7% **11.** 39.1% **12.** 91.4% **13.** 1210.4% **14.** 62.5% **15.** 1082% **16.** 0.7% **17.** 10.6% **18.** 0.8%
19. 4% **20.** 0.1% **21.** 50.3% **22.** 40% **23.** 70% **24.** 90% **25.** 710% **26.** 930% **27.** 1050% **28.** 1100%
29. 1260% **30.** 1710% **31.** 2250% **32.** 2900% **33.** 3720% **34.** 10% **35.** 750% **36.** 180% **37.** 1630%
38. 30% **39.** 620% **40.** 57% **41.** 380% **42.** 0.1% **43.** 67% **44.** 2070% **45.** 880% **46.** 29.15%
47. 32% **48.** 1740% **49.** 300.3% **50.** 13,925% **51.** 21.87% **52.** 910% **53.** 2500% **54.** 1200.4%
55. 14% **56.** 218.5% **57.** 51% **58.** 930% **59.** 186.8% **60.** 855.4% **61.** 0.3% **62.** 324.6% **63.** 26%
64. 7% **65.** 0.32% **66.** 364.2% **67.** 29% **68.** 5.2% **69.** 342% **70.** 30% **71.** 210% **72.** 0.01%
73. 167% **74.** 100% **75.** 254.9% **76.** 0.5% **77.** 207.7% **78.** 150% **79.** 19%

Page 531 **1.** 25% **2.** 40% **3.** 75% **4.** 70% **5.** 12.5% **6.** 15% **7.** 35% **8.** 48% **9.** 14% **10.** 17.5%
11. 62.5% **12.** 68.8% **13.** 275% **14.** 462.5% **15.** 666.7% **16.** 716.7% **17.** 855.6% **18.** 526.7%
19. 817.5% **20.** 1418.8% **21.** 722% **22.** 60% **23.** 537.5% **24.** 1283.3% **25.** 58.3% **26.** 11.1%
27. 1530% **28.** 22.5% **29.** 2080% **30.** 1024% **31.** 1862.5% **32.** 907.5% **33.** 93.8% **34.** 838%
35. 1135.7% **36.** 1645% **37.** 77.8% **38.** 718.8% **39.** 573.3% **40.** 36.7% **41.** 1735%

Page 532 **1.** 0.105 **2.** 0.157 **3.** 0.40 **4.** 0.85 **5.** 1.20 **6.** 1.37 **7.** 0.067 **8.** 0.071 **9.** 0.089 **10.** 0.95
11. 1.19 **12.** 0.079 **13.** 0.172 **14.** 0.856 **15.** 1.00 **16.** 0.0135 **17.** 0.053 **18.** 1.42 **19.** 0.007
20. 0.0055 **21.** 0.00625 **22.** 0.0008 **23.** 0.056 **24.** 0.071 **25.** 0.068 **26.** 0.2275 **27.** 0.304 **28.** 0.8965
29. 0.5784 **30.** 0.1275 **31.** 0.7425 **32.** 0.0625 **33.** 0.1880 **34.** 0.0097 **35.** 0.1445 **36.** 0.28075

37. 0.00125 **38.** 0.0625 **39.** 0.0014 **40.** 0.0663 **41.** 0.0879 **42.** 1.00 **43.** 0.0095 **44.** 1.27 **45.** 0.00025 **46.** 0.2025 **47.** 0.22125 **48.** 0.146 **49.** 0.252 **50.** 0.189

Page 533 **1.** $\frac{9}{20}$ **2.** $\frac{4}{5}$ **3.** $1\frac{3}{4}$ **4.** 2 **5.** $\frac{117}{1000}$ **6.** $\frac{1}{1000}$ **7.** $\frac{3}{2000}$ **8.** $\frac{673}{1000}$ **9.** $\frac{53}{500}$ **10.** $\frac{1571}{2000}$ **11.** $\frac{3763}{10,000}$ **12.** $\frac{2571}{5000}$ **13.** $\frac{1}{2}$ **14.** $\frac{3}{4}$ **15.** $\frac{101}{1000}$ **16.** $\frac{3}{2500}$ **17.** $\frac{809}{1000}$ **18.** $\frac{421}{1000}$ **19.** $\frac{1}{16}$ **20.** $\frac{3}{8}$ **21.** $\frac{7}{16}$ **22.** $\frac{7}{200}$ **23.** $\frac{37}{400}$ **24.** $\frac{31}{300}$ **25.** $\frac{31}{200}$ **26.** $\frac{41}{200}$ **27.** $\frac{51}{200}$ **28.** $\frac{63}{400}$ **29.** $\frac{1}{6}$ **30.** $\frac{1}{3}$ **31.** $\frac{3}{10}$ **32.** $\frac{143}{250}$ **33.** $\frac{5}{16}$ **34.** $\frac{127}{200}$ **35.** $\frac{1}{8}$ **36.** $\frac{2}{15}$ **37.** $\frac{4}{15}$ **38.** $\frac{259}{400}$ **39.** $\frac{1}{9}$ **40.** $\frac{1571}{2000}$ **41.** $\frac{1}{2}$ **42.** $\frac{3}{4}$ **43.** $\frac{1}{80}$ **44.** 1 **45.** $\frac{51}{400}$ **46.** $\frac{19}{400}$ **47.** $\frac{43}{800}$ **48.** $\frac{7591}{10,000}$ **49.** $\frac{1}{15}$ **50.** $1\frac{21}{100}$

Page 534 **1.** 9 **2.** 21.6 **3.** 3.84 **4.** 10.98 **5.** 7.8 **6.** 12.45 **7.** 14.24 **8.** 43.68 **9.** 3.486 **10.** 3.294 **11.** 45.359 **12.** 3.122 **13.** 10 **14.** 9 **15.** 40 **16.** 23 **17.** 6 **18.** $\frac{3}{10}$ **19.** $2\frac{1}{4}$ **20.** $\frac{7}{10}$ **21.** 128 **22.** 6 **23.** 8.19 **24.** $14\frac{1}{25}$ **25.** 15 **26.** 32.4 **27.** 2 **28.** 7.5 **29.** 80 **30.** $7\frac{77}{80}$ **31.** 9.144 **32.** 100 **33.** 25 **34.** 13.86 **35.** $6\frac{3}{8}$ **36.** 19.76

Page 535 **1.** 50% **2.** 25% **3.** 20% **4.** 15% **5.** 35.7% **6.** 72.2% **7.** 89.2% **8.** 89.3% **9.** 69.9% **10.** 72.1% **11.** 87.2% **12.** 81% **13.** 150% **14.** 125% **15.** 120% **16.** 130% **17.** 133.3% **18.** 133.3% **19.** 129.2% **20.** 137.5% **21.** 400% **22.** 150% **23.** 513.3% **24.** 118.2% **25.** 80% **26.** 115% **27.** 46.2% **28.** 171.4% **29.** 55.1% **30.** 128.6% **31.** 80% **32.** 125% **33.** 120% **34.** 37.5% **35.** 25% **36.** 16% **37.** 138.9% **38.** 78.8% **39.** 200% **40.** 300% **41.** 66.7% **42.** 400% **43.** 5% **44.** 15% **45.** 146.7%

Page 536 **1.** 72 **2.** 88 **3.** 8048.5 **4.** 50 **5.** 1600 **6.** 91.1 **7.** 336.4 **8.** 290 **9.** 60 **10.** 168 **11.** 134 **12.** 17.3 **13.** 160 **14.** 64 **15.** 60 **16.** 800 **17.** $90\frac{2}{3}$ **18.** 1000 **19.** $54\frac{86}{91}$ **20.** $933\frac{1}{3}$ **21.** 75 **22.** 312 **23.** 2.5 **24.** 29 **25.** 52.5 **26.** 17 **27.** 78 **28.** 4.4 **29.** 180 **30.** 144 **31.** 225 **32.** 80 **33.** 55 **34.** 115.4 **35.** $197\frac{1}{67}$

Applications File

Page 537 **APPLICATION A** **1.** 600 mL **2.** 220 km **3.** 448 m **4.** 180 mi **5.** 210 m **6.** 0.25 m **7.** 1216 ft^2 **8.** 45 $in.^2$ **9.** 360 km^2 **10.** 41.3% **11.** 31.4% **12.** 56.5% **13.** 23.4% **14.** 36.6% **15.** $437.95 **16.** $773.99 **17.** $1429 **18.** $987 **19.** $5.25 **20.** $10.08 **21.** $46.19 **22.** $11.88 **23.** $50.16

Page 538 **APPLICATION B** **1.** 308 in. **2.** 3.25 h **3.** 8775 yd **4.** 45 km/h **5.** 8 h **6.** 55 ft/h **7.** 30 s **8.** 122°F **9.** 30°C **10.** 0°C **11.** 248°F **12.** 35°C **13.** 20°C **14.** 291.2°F **15.** 50°F **16.** −15°C

Page 539 **APPLICATION C** **1.** $1,200,000 **2.** $1,500,000 **3.** $7,900,000 **4.** $12,000,000 **5.** Bronx, Columbus, Dallas, Lincoln, San Diego **6.** Columbus, Dallas, Denver, Lincoln, San Diego **7.** Local **8.** 8 lb **9.** 5 and 6 lb

Page 540 **APPLICATION D** **1.** 1 quarter, 1 dollar bill, 1 five-dollar bill **2.** 2 pennies, 2 dimes, 2 dollar bills **3.** 1 penny, 3 dollar bills, 1 five-dollar bill **4.** 3 pennies, 2 quarters, 1 dollar bill, 1 five-dollar bill **5.** 2 pennies, 1 dime, 1 quarter, 4 dollar bills **6.** 1 penny, 1 nickel, 3 quarters **7.** 2 dimes, 3 quarters, 4 dollar bills **8.** 3 pennies, 1 dime, 2 quarters, 1 dollar bill, 1 five-dollar bill, 1 ten-dollar bill **9.** 1 penny, 2 dimes, 2 quarters **10.** 3 quarters **11.** 1 quarter, 2 dollar bills **12.** 4 dollar bills **13.** 1 dime, 1 dollar bill, 1 five-dollar bill **14.** 1 nickel, 3 quarters, 4 dollar bills **15.** 1 quarter, 2 dollar bills **16.** 1 five-dollar bill, 1 ten-dollar bill **17.** 2 quarters, 1 dollar bill, 1 ten-dollar bill

Page 541 **APPLICATION E** **1.** 8:45 **2.** 10:45 **3.** 7:00 **4.** 1:30 **5.** 11:15 **6.** 3:15 **7.** 12:15 **8.** 4:45 **9.** 3:30 **10.** 11:15 **11.** 8:45 **12.** 1:45 **13.** 7:30 **14.** 5:00 **15.** 9:45 **16.** 10:30 **17.** 12:45 **18.** 6:45 **19.** 6:00 **20.** 3:45 **21.** 11:30 **22.** 8:00 **23.** 10:15 **24.** 2:15 **APPLICATION F** **1.** 2 h 46 min **2.** 2 h 34 min **3.** 3 h 15 min **4.** 5 h 14 min **5.** 51 min **6.** 1 h 45 min **7.** 4 h 15 min **8.** 8 h 5 min **9.** 8 h 20 min **10.** 10 h 15 min **11.** 10 h 45 min

Page 542 **APPLICATION G** **1.** 137 days **2.** 194 days **3.** 212 days **4.** 42 days **5.** 258 days **APPLICATION H** **1.** No **2.** Yes **3.** Yes **4.** Yes **5.** No **6.** No **7.** Yes **8.** No **9.** Yes **10.** No **11.** Yes **12.** Yes **APPLICATION I** **1.** 79 d **2.** 265 d **3.** 231 d **4.** 176 d **5.** 222 d

Page 543 **APPLICATION J** **1.** $\frac{3}{4}$ yr. **2.** $\frac{7}{12}$ yr. **3.** 1 yr. **4.** $1\frac{2}{3}$ yr. **5.** 3 yr. **6.** $1\frac{1}{2}$ yr. **7.** $1\frac{3}{4}$ yr. **8.** $\frac{2}{3}$ yr. **9.** $1\frac{1}{6}$ yr. **10.** $\frac{18}{73}$ yr. **11.** $\frac{5}{6}$ yr. **12.** $\frac{24}{73}$ yr. **13.** $\frac{35}{73}$ yr. **14.** $\frac{73}{180}$ yr. **15.** $\frac{50}{73}$ yr. **APPLICATION K** **1.** 52 **2.** 52 **3.** 16 **4.** 24 **5.** 12 **6.** 25 **7.** 6 **8.** 78 **9.** 60

Page 544 **APPLICATION L** **1.** 0.85 **2.** 0.28 **3.** 0.65 **4.** 0.50 **5.** 0.20 **6.** 0.66 **7.** 9% **8.** 15% **9.** 53% **10.** 34% **11.** 93% **12.** 55% **13.** 96.5% **14.** 43.4% **15.** 24.1% **APPLICATION M** **1.** $600 **2.** $375 **3.** $1050 **4.** Prints/Permits **5.** $450 **6.** $625 **7.** $200

Page 545 **APPLICATION N** **1.** $8000 **2.** Oct., Nov. **3.** $0 **4.** None **APPLICATION O** **1.** 625 cars **2.** 1100 cars **3.** 350 cars

Page 546 **APPLICATION P** **1.** $.27 **2.** $73,440 **3.** $110,160 **4.** No **5.** rent 108°, books 54°, clothing and personal items 36°, all other expenses 18° **6.** salaries 126°, raw materials 108°, utilities 36°, plant maintenance 54°, research and development 36°

Page 547 **APPLICATION Q** **1.** $54\frac{1}{3}$ **2.** 130 **3.** $6726\frac{3}{4}$ **4.** 3.31 **5.** $1,235,563 **APPLICATION R** **1.** $148\frac{1}{2}$ **2.** 25 **3.** 88 **4.** 3.1 **5.** 0.09 **6.** $68,812 **7.** 2.08 **APPLICATION S** **1.** $24 **2.** 600,800 **3.** None **4.** 3, 4 **5.** 0.01 **6.** None

Page 548 **APPLICATION T** **1.** m **2.** km **3.** mm **4.** cm **5.** m **6.** mm **7.** mm **8.** cm **APPLICATION U** **1.** kg **2.** L **3.** g **4.** kL **5.** t **6.** kg **7.** kL **8.** g **9.** g **10.** L

Page 549 **APPLICATION V** **1.** 10.5 mm **2.** 21 in. **3.** 2.6 ft **4.** 6.1 m **5.** 162 mi **6.** 55.8 m **APPLICATION W** (answers are approximate) **1.** 12.56 mi **2.** 10.99 cm **3.** 2.512 m **4.** 219.8 mm **5.** 4.71 ft **6.** 81.64 in. **7.** 56.52 ft **8.** 753.6 km **9.** 39.25 mi

Page 550 **APPLICATION X** **1.** 150 $in.^2$ **2.** 2061.16 cm^2 **3.** 0.675 m^2 **4.** $12\frac{1}{4}$ yd^2 **5.** $10\frac{1}{9}$ ft^2 **APPLICATION Y** **1.** 13.5 $in.^2$ **2.** approx. 78.50 cm^2 **3.** 200 m^2 **4.** 633.6 m^2 **5.** approx. $19\frac{5}{8}$ yd^2 **6.** $15\frac{1}{6}$ ft^2 **7.** 23,223.44 mm^2 **APPLICATION Z** **1.** 519,552 m^3 **2.** 0.704 m^3 **3.** 396 cm^3 **4.** 22,032 $in.^3$ **5.** $157\frac{1}{4}$ yd^3 **6.** $148\frac{1}{48}$ ft^3

Answers to A Microcomputer Application: Payroll Deductions

1.

EMPLOYEE	GROSS	FEDERAL	CITY	FICA	NET
1001	144.5	13.18	2.17	10.38	118.77
1002	148.75	16.7	2.23	10.68	119.14
1003	114.75	8.72	1.72	8.24	96.07
1004	136	14.79	2.04	9.770001	109.4
1005	127.5	10.63	1.91	9.16	105.8
1006	170	17.01	2.55	12.21	138.23
1007	165.75	19.25	2.49	11.91	132.1
0					

2.

EMPLOYEE	GROSS	FEDERAL	CITY	FICA	NET
1001	153	14.46	2.29	10.99	125.26
1002	165.75	19.25	2.49	11.91	132.1
1003	106.25	7.44	1.59	7.63	89.58999
1004	144.5	16.07	2.17	10.38	115.88
1005	93.5	5.67	1.4	6.71	79.72
1006	110.5	8.08	1.66	7.94	92.82
1007	153	17.34	2.29	10.99	122.38
0					

3.

EMPLOYEE	GROSS	FEDERAL	CITY	FICA	NET
1001	161.5	15.73	2.42	11.6	131.75
1002	170	19.89	2.55	12.21	135.35
1003	157.25	15.09	2.36	11.29	128.51
1004	153	17.34	2.29	10.99	122.38
1005	114.75	8.72	1.72	8.24	96.07
1006	106.25	7.44	1.59	7.63	89.58999
1007	165.75	19.25	2.49	11.91	132.1
0					

CREDITS

Mechanical art by ANCO and Textbook Art Associates.
Cover concept and photo by Andrew Brilliant.

Photographs:

45: U.S. Department of Labor
65: Jean-Claude Lejeune/STOCK BOSTON
89: Charles Gupton/STOCK BOSTON
107: Peter Menzel/STOCK BOSTON
129: Ray Ellis/PHOTO RESEARCHERS INC.
151: Janice Fullman/THE PICTURE CUBE
173: Read D. Brugger/THE PICTURE CUBE
191: © Joel Gordon 1984
215: Fredrik D. Bodin
237: Michael Weisbrot & Family/STOCK BOSTON
261: Tim Courlas
285: Robert Schoen Photography
303: Peter Menzel/STOCK BOSTON
321: © Joel Gordon 1981
339: Peter Vandermark/STOCK BOSTON
359: Robert Schoen Photography
383: Gary Wolinsky/STOCK BOSTON
401: Donald Dietz/STOCK BOSTON
417: Guy Gillette/PHOTO RESEARCHERS INC.
439: Ann Chwatsky/LEO DE WYS INC.
455: Sharon A. Bazarian/THE PICTURE CUBE
471: © Joel Gordon 1979